THE UNIVERSE AND LIFE

ORIGINS AND EVOLUTION

Joy and amazement at the beauty and grandeur of this world of which man can just form a faint notion.

Albert Einstein
(1879–1955)

THE UNIVERSE AND LIFE

ORIGINS AND EVOLUTION

G. Siegfried Kutter

The Evergreen State College

Jones and Bartlett Publishers, Inc.

Boston/Portola Valley

Editorial offices:
  Jones and Bartlett Publishers, Inc.
  30 Granada Court
  Portola Valley, CA 94025

Sales and customer service offices:
  Jones and Bartlett Publishers, Inc.
  20 Park Plaza
  Boston, MA 02116

Printed in the United States of America
10 9 8 7 6 5 4 3 2 1

Library of Congress Cataloging-in-Publication Data
Kutter, G. Siegfried.
    The universe and life.

    Bibliography: p.
    Includes index.
      1. Cosmology.   2. Life — Origin.   3. Evolution.
I. Title.
QB981.K93   1987   523.1   85–23156

ISBN: 0-86720-033-2

Manuscript editor: Elizabeth Sorenson
Proofreader: Margaret E. Hill
Production editor: Elizabeth W. Thomson
Illustrations:
    Drawings: Meredith Edgcomb, John G. Hamwey, Len Shalansky, Matthew Zimet
    Maps: Cynthia C. Harris
Photo research: Traci A. Sobocinski
Text and cover design: Rafael Millán
Typesetting: Publication Services, Inc.

Cover photograph: The Eta Carinae Nebula ("Keyhole Nebula").
This enormous cloud of luminous interstellar gas and dark dust lanes, located some 9,000 light-years from Earth, is one of many regions in the Milky Way in which stars have recently formed. Cerro Tololo Inter-American Observatory, Curtis Schmidt telescope photograph (Courtesy National Optical Astronomy Observatories).

To the memory of my parents

Foreword

PROFESSOR KUTTER has actually accomplished what most scientist-scholars experience only in their wildest dreams: a readable treatise on cosmic evolution. Are we the only intelligent (interstellar-communicating) life in the universe? The calculation of the probability of the existence of the book you hold in your hands follows a logic similar to that Kutter leads us through in his fascinating Epilogue on the probability of other intelligent life in the universe. I base my rough estimate that Kutter's *The Universe and Life: Origins and Evolution* could ever exist at all on the following probabilities.

1. The probability of the rise of modern astronomy which employs the properties of light (this includes of course the construction and use of telescopes, spectroscopes, satellites, and the rest.)

2. The probability of the entire geological (geophysical and oceanographic) revolution. I refer to the radical shift of our thinking about the surface of the Earth that followed from the plate tectonics explanation of Wegener's concept of continental drift.

3. The probability of emergence of our awareness of the "biosphere-Gaia" phenomenon. What I refer to here is the recognition of life as a planetary surface phenomenon, one that modulates its own environment. Specifically I am thinking of V. I. Vernadsky's understanding of life as a geological force coupled with J. E. Lovelock's Gaia hypothesis, according to which life produces and regulates the chemistry, the salt concentration, the temperature and other aspects of its immediate environment. These local phenomena have led to a global system making the Earth different from its planetary neighbors, Mars and Venus. Life has lead to Earth surface regulation over immense periods of time.

4. In figuring out the probability of the existence of this book we must factor in the chances that the Darwinian revolution, including its most recent manifestations, would ever have occurred. I would include here the new recognition of the antiquity of Archean and Proterozoic life, the glimpse of understanding of symbiosis, karyotypic fissioning, and other kinds of "macromutation" as important mechanisms in the Darwinian evolutionary process. Here too is the recent success in the paleontological reconstruction of primate history and the "DNA revolution." We now understand far more clearly the cell and molecular basis of evolution through the techniques of biochemistry, molecular genetics, electron microscopy, and the like. Even

techniques developed by the petroleum geochemical industry have been integrated into our investigations of the evolution of early life.

5. We must include in our analysis of the chance of the existence of this book the probability that curious scientists, each limited to a tiny aspect of these grand themes, would be supported in their investigations to the extent that their work reaches a stage of publication. No one, after all, even the most dedicated philosopher is paid to contemplate cosmic evolution on the scale of the Kutter treatise. And Kutter, in all his wisdom, could never have woven together these scientific threads without the lost weekends and all-night labors in every corner of the world of thousands of obsessive and compulsively communicative scientists.

6. Given that the scientific information required to produce this long essay on evolution of the universe even exists anywhere, we must include in our calculation the probability that a G. Siegfried Kutter also can exist. That is, we must first estimate the chances that an academic institution encourages a research astronomer to read the disparate, contradictory, sometimes boring and inaccurate literature of organic chemistry, biology, paleontology and anthropology. Then we must ask what the probability is that such an institution — The Evergreen State College in Kutter's case — fosters intercommunication among scientists and students in entirely different specialties and departments. Remembering that science is paid for by those who want to see the development of new drugs, weapons, communication systems and other marketable commodities, we must decrease the estimate of our probability of the existence of this book. Finally, we must ask what are the chances that such a single G. S. Kutter, who can assimilate, evaluate, integrate and formulate the slippery information into a scientific story line can make his tale compelling enough that it is readable with interest by the rest of us. This is not all. In our calculation of probabilities we must inquire: even if our able, articulate, hard-working scientist-scholar-writer exists, what are the chances that his personal situation will permit the time and effort needed to actually create a book? Even if we assume that such a G. S. Kutter can produce a scientifically accurate yet fascinating manuscript, we have to ask another probability question. What are the odds that some publisher (knowing that few other scientist-scholars are educated well enough even to *teach* from such a manuscript) will undertake the financial risk in bringing out a large textbook for a course that does not exist?

Like Kutter, although I believe that quantification is essential to science, I want to limit the amount of arithmetic forced upon the reader. In any case it is clear that the probability that the book you now hold in your hands can exist at all is vanishingly small. A conservative scientist would tend to dismiss the entire problem: estimating the probability to be so close to zero he would assume that no such book could ever have been published. Yet here it is.

Will Kutter's text be too broad, too intellectual and too difficult for the undergraduate and other readers for which it was written? Will its erudition and sense-of-the-whole frighten teachers of undergraduate general science courses who are unaccustomed to instruct in a mode more familiar to literature and humanities experts? We hope not. Indeed we expect an alternative scenario. Broadly-based university and college courses on science that, from lack of an adequate text, could never be taught in the past, fertilized by Kutter, should spring up like mushrooms in September after rain. Science, for the first time since 19th century academic programs in Natural Theology, will be integrated. Science will begin to be recognized as the humanity it is. On this basis, not of preconceived religious views, but of hard-earned answers to questions asked directly to nature, science and scientific teaching will be ushered through a renaissance. Science will become a reliable way of knowing about the whole. Science is not just a method for the calculation of the targets of missiles, or a means for developing flexible plastics and cures for infectious diseases. Science can become the way of knowing about the eternal questions of the philosophers.

I cannot help but notice that Kutter as a biologist still has some learning to do. I am sure that any anthropologist, vertebrate zoologist, geophysicist, organic chemist, enzymologist or sedimentologist will see small errors and differences in interpretation. My response to this inevitability is that all of you specialists, students and teachers using this tome celebrate the magnitude of Kutter's effort by bringing to his attention your criticisms. The first edition of the book is a unique event in scientific publishing; a later edition, responding to comments and corrections by readers, might provide a product of concern to everyone in the academic world. From bureaucrat to parent, dean and faculty member . . . everyone lamenting our scientific illiteracy and ignorance will have cause to celebrate. The accomplishment is within view and one expects many imitators. The definitive, scientifically accurate, integrated one-year course of study at the college level revealing (in T. H. Huxley's terms) "man's place in nature" can now be brought into the curriculum with an appropriate course number, to be taught at an appropriate level. Why now? Because such a course, formerly a product only of a scientist's wildest dream, now has a textbook.

Lynn Margulis
University Professor
Boston University

March 1987

Preface

LONG BEFORE THE FORMATION of the Solar System, a series of cosmic events took place that set the stage for the eventual origin of life on Earth and its evolution toward the enormous diversity we witness today. These events began some 13 billion years ago when space and time as well as all the matter and energy contained in the Universe came into being in one gigantic explosion, the *Big Bang*. Chaos reigned everywhere. Energy was largely in the form of radiation, while matter consisted almost entirely of gaseous hydrogen and helium. The heavier elements such as carbon, nitrogen, oxygen, silicon, phosphorus, sulfur, and the metals — so crucial for life and for a solid Earth — had not yet appeared.

In the course of subsequent billions of years, the gas coalesced into galaxies and the first generations of stars. In the cores of the stars, nuclear reactions fused hydrogen into the heavier elements. As the most massive stars reached the end of their lives, they exploded and threw the newly synthesized elements back into space. There the elements intermingled with primordial hydrogen and helium that remained in the form of giant clouds. From this mixture new generations of stars were born, some of which also exploded and further contributed to the gradual buildup of heavy elements in the Universe.

Among the billions of galaxies populating the Universe, one was the Milky Way. In one of its clouds, enriched with newly synthesized heavy elements, the Sun and its planets formed approximately 4.5 billion years ago. The third planet from the Sun was Earth. During its formation, Earth lost most of its hydrogen and helium. What remained was a solid globe surrounded by a protective atmosphere and a crust that contained plenty of water and other molecules rich in carbon, nitrogen, oxygen, and other elements necessary for the eventual origin of life. From the Sun, Earth received a steady supply of radiation, which provided warmth and energy.

Conditions on the surface of the early Earth were well suited for the occurrence of countless kinds of chemical reactions. In time, the molecules created by these reactions became larger and ever more complex. Proteins, nucleic acids, sugars, and other organic molecules accumulated in the ocean and gave rise to solutions of interacting carbon-nitrogen compounds, the so-called primordial soup. The interactions between the molecules increased in diversity and eventually some sort of order emerged. Within thousands to, perhaps, hundreds of millions of years, the initial relatively simple chemical and physical interactions led to the origin of self-maintaining systems — the first organisms on Earth.

These earliest organisms lived in the juvenile ocean and resembled today's bacteria. They were capable of neither photosynthesis nor respiration, but — in the absence of oxygen — obtained their nourishment by fermentation of organic molecules accidentally synthesized in the soup. Furthermore, they did not reproduce sexually by the mating of males and females, but by direct asexual cellular division. However, natural selection — the maintenance of genetic diversity and the preferential survival of those individuals best adapted to the prevailing conditions in the environment — exerted itself right from the beginning. The early organisms became more complex and, in the course of some 4 billion years, evolved into all of the diversity of life inhabiting Earth today — modern bacteria, protists, fungi, plants, and animals, including humans.

This, in a nutshell, is the story of the origin and evolution of life on Earth as it is currently understood. The story reveals that we — mankind and the rest of earth-bound life — are subject to the same laws as are the stars and galaxies. It further reveals that our existence and evolution are a natural outgrowth of the Universe and its evolution: the story of our origin and evolution is part of the larger story of the Universe's evolution. Therefore, to gain a comprehensive perspective of life on Earth, including ourselves, we need to study life in the context of the evolution of the Universe as a whole. To do so is the aim of this book.

The book is divided into two parts: *Physical Evolution* and *Biological Evolution*. *Part One* (*Physical Evolution*) begins with the birth of the Universe and traces the evolution of inanimate matter to the present. I emphasize those physical events that were crucial for the eventual genesis of life on Earth: the Big Bang origin of the Universe, the brief initial era of nucleosynthesis that determined the primordial composition of matter, the condensation of matter into galaxies and stars, the synthesis of the heavy elements, and the formation of the Sun and its planets.

Part Two (*Biological Evolution*) begins with the chemical reactions in the ocean and atmosphere of the young Earth that led to the emergence of the earliest organisms. I then follow the major stages of life's evolution through the ages to the present: the increase in metabolic and reproductive complexity of early life, the development of the first large and complex multicelled organisms, the relatively rapid diversification of animal and plant life into numerous phyla, the evolution of the vertebrates, and, finally, the appearance of human beings.

Throughout Part Two, I emphasize that biological evolution, unlike physical evolution earlier, never followed an independent course. Right from the start life interacted strongly with its physical environment. The early organisms were responsible for our oxygen-bearing atmosphere, they contributed to the shaping of Earth's geological surface features, and they affected the ocean and the climate. At the same time, the physical environment played an important role in the origin of life, the evolution of metabolic pathways, the rise and fall of species, the spreading of life from water to land to air, and the origin of the primates and the *Homo* lineage.

This book uses a chronological approach, which presents two challenges — one philosophical, the other pedagogical. The first challenge arises from the order of presentation of the topics. This order might mislead the reader into thinking that there is purpose to evolution, whose ultimate aim is the making of mankind. Nothing could be further from the truth, at least as far as is known from science. The laws of nature came into being along with space and time, matter and energy during the Big Bang, and they have governed the evolution of the Universe and all it contains ever since. The appearance of life on Earth, including mankind, is a consequence of these laws, just as the formation of galaxies, stars, and planets is, no more and no less. Science does not accord any special status to life or to mankind. This conclusion had been reached as early as 1677 by Benedict de Spinoza (1632–1677), when he wrote in his famous metaphysical treatise, *Ethica*, "Nature has no goal in view, and final causes are only human imaginings" (*Ethica*, Part I, Appendix).

Of course, we — human beings with feelings and emotions — do accord special status to our existence. Most of us are more interested in knowing about ourselves than, say, about stars or spiders. That is why the final chapters of this book focus on vertebrate, primate, and hominid evolution and not, for example, on mollusks or arthropods. However, to repeat, this is not to be interpreted as meaning that we, *Homo sapiens*, are the crowning achievement of evolution. We have certain unique attributes not found among other animals, such as the ability to communicate through language, think logically, develop science, read and write books, and so forth. But we are still just one species among many, each with its own special attributes. And all species are the result of the same evolutionary processes, processes that, no doubt, will continue to lead to extinction and give rise to new species in the future.

The pedagogical challenge arises from the fact that this book is written on the introductory level, yet, because of its chronological approach, starts with complex and difficult topics. For example, chapter 1 deals with the Big Bang, during which conditions were far from simple and as different from our experiences as can be imagined. The same is true of the origin and evolution of galaxies and stars, the topics of chapters 2 and 3. Only in chapters 4 and 5 (the last two of Part One) are the Earth and the rest of the Solar System introduced. Chapter 6 begins with a discussion of conditions on Earth during its final formative stages and chemical reactions that are believed to have taken place then. The evolution of early life and the origin of the major animal phyla are presented in chapters 7 and 8. Again, these topics are probably unfamiliar, at least in part, to most readers. Finally in chapters 9 and 10 (the last two of Part Two) the animals most closely related in an evolutionary sense to us appear — fishes, amphibians, reptiles, and mammals. Clearly, a chronological approach does not allow me to start with the familiar or simple and then develop the subject matter in a systematic manner, as is common in traditional texts.

In order to minimize the difficulties the chronological approach may present to the reader, introductions are provided for Parts One and Two that provide essential background information. Furthermore, each chapter contains numerous comments and short essays that are boxed to set them off from the main text. They offer additional information that, if included in the text, would interrupt the flow of the presentation. To keep the book a manageable length, I have carefully selected those topics most crucial to our history. Finally, I have tried to give roughly equal emphasis to each of the two parts of the book.

Every effort has been made to ensure that the presentation is clear, interesting, and enjoyable for the reader, and to meet the particular challenges that writing a book such as this one presents. Above all, I hope that upon finishing the book the reader will be left with a sense of satisfaction in having taken a comprehensive view at our evolutionary heritage.

This book is written on the introductory level for the scientifically curious laymen and for college and university students. It does not require a strong science or mathematics background. With the exception of exponential notation, the use of mathematics is avoided. However, chemical notation and a few nuclear and chemical equations are used to elucidate and shorten the description of nuclear and chemical processes.

As a text, the book is intended for a number of different students: liberal arts majors seeking a broad exposure to the sciences; freshmen and sophomores interested in the natural sciences who, before specializing, want to survey the broad spectrum of exciting topics confronting present-day astronomy, physics, geology, biology, and anthropology; and science students who already have selected their specialized fields, but who want to acquaint themselves with some of the scientific problems and issues beyond their immediate areas of concern.

The book is written for courses of various lengths. In a one-term course, the instructor may wish to emphasize some chapters more than others. This is possible because, although the chapters are arranged in a chronological order, most of them do not depend strongly on earlier material as a prerequisite for understanding. For example, chapters 1, 2, 6, and 10 could serve as the basis for a one-term interdisciplinary course on the *Origins of the Universe, Life, and Humankind*; chapters 4, 5, 6, 7, and 8 could be used for a one-term course on the *Origin of Life on Earth*; or the Epilogue, along with selected topics from Parts One and Two, could form the content of a course on the *Search for Extraterrestrial Intelligence*. In a two-term course the entire book can be covered comfortably, with the two parts of the book providing a natural division between the first and second terms. In courses lasting more than two terms, the instructor may wish to supplement the text with extra reading assignments, selected from the up-to-date bibliographies at the ends of the chapters. Furthermore, because of their discussions of evolutionary developments, selected chapters may be useful as supplementary reading in traditional courses in astronomy, geology, biology, and anthropology.

The breadth of the material covered in this book presented another challenge. I am by training an astrophysicist, with research experience in stellar evolution, so my background extended initially only over the content of the first part of the book. Since 1972 I have been a Member of the Faculty at The Evergreen State College, an innovative institution that stresses interdisciplinary teaching. Through joint teaching assignments with colleagues in other disciplines I began to concern myself seriously with fields beyond physics, astronomy, and mathematics. Once again I became a student, listening to friends and colleagues, learning through teaching new subjects and through seminars with my students, and searching for the threads that tie diverse disciplines together. Without these experiences this book could not have been written.

Burton S. Guttman and David H. Milne are the colleagues with whom I was most closely and frequently associated. They patiently taught me organic chemistry and biology, and helped me again and again when I had difficulties with some of the concepts in their fields. They also read and critiqued large parts of the manuscript. I am deeply indebted to them and express my most sincere appreciation.

Many other Evergreen colleagues supported me by answering questions, discussing various topics, reading and critiquing chapters and boxes, providing references, and giving encouragement. They are Clyde H. Barlow, Michael W. Beug, Richard A. Cellarius, Robert S. Cole, Donald G. Humphrey, Willard C. Humphreys, Jeffrey J. Kelly, Elizabeth M. Kutter, Mark L. Papworth, Jacob B. Romero, James M. Stroh, Charles B. Teske, and Byron L. Youtz. Robert H. Knapp and Barbara L. Smith rearranged the teaching schedule for 1983–84, allowing me to go on a sabbatical leave to work on the manuscript. And Joan M. Allard, Kelley E. Emmons, Mary M. Huston, Ernestine G. Kimbro, Patricia L. Matheny-White, Frank C. Motley, M. Deborah Robinson, and Malcolm H. Stilson, the college's reference librarians, were always willing to look up facts, search for references, and support me in many other ways in my library research. All of these colleagues contributed immeasurably to my writing efforts, and I thank them heartily.

From 1978 to 1980—and again in 1986 to 1987—I was a National Research Council Research Associate at the NASA Goddard Space Flight Center. These appointments allowed me to stay up-to-date in the very fast-moving fields of cosmology, galactic structure, stellar evolution, and planetary physics, which strongly influenced my writing. I would like to thank especially Robert D. Chapman, John A. O'Keefe, and Allen V. Sweigart of NASA, Ken'ichi Nomoto (University of Tokyo), Malcolm P. Savedoff (University of Rochester), and Warren M. Sparks (Los Alamos Scientific Laboratory) for the many exciting and far-ranging discussions we had on topics in physics, astronomy, the Solar System, and scientific issues in general. Furthermore, I would like to thank Drs. Chapman, O'Keefe, E. Howard Scott, Sparks, and Sweigart for reading and critiquing chapters from the first part of the book.

Since the summer of 1981, I have received steady encouragement and support from my editor, Arthur C. Bartlett, and his product manager, Anne C. Gingras. I am greatly indebted to both of them and wish to express my most sincere thanks.

In particular, I am indebted to Mr. Bartlett for tirelessly soliciting chapter reviews from research scientists across the nation and abroad: John Billingham, Patrick M. Cassen, and Harold P. Klein (all at the NASA Ames Research Center), Cedric I. Davern (University of Utah), Gordon J. Edlin (University of California, Davis), Clair Edwin Folsome (University of Hawaii), Lynn Margulis (Boston University), Jayant V. Narlikar (Tata Institute of Fundamental Research, Bombay, India), Everett C. Olson (University of California, Los Angeles), Dorion S. Sagan, Robert W. Sussman (Washington University), and David H. White (University of Santa Clara). Their thorough, professional critiques, helped eliminate many serious flaws from the original manuscript and bring it up-to-date, and I am sincerely grateful to them all.

The editorial production stages of the book have been particularly satisfying because of the competent, dedicated, and cheerful support given by the editorial, art, and production staffs of Jones and Bartlett: Maureen A. Cunningham-Neumann (Director of Production), Meredith Edgcomb, John G. Hamwey, Cynthia C. Harris, Margaret E. Hill, Rafael Millán, Len Shalansky, Traci A. Sobocinski, Elizabeth Sorenson, Elizabeth W. Thomson, and Matthew Zimet. I wish to thank them all.

Despite the considerable support I have received, errors and inaccuracies undoubtedly remain and topics that should have been included are not. For these shortcomings I take full responsibility.

Finally, I would like to thank my family for faithfully standing by me during the years of writing, encouraging me again and again through their love and through their interest in the project, and never complaining about the many hours I was kept away from them.

Olympia, Washington
January 1987

G. Siegfried Kutter

Table of Contents

Expanded Table of Contents

Chapter 2: Galaxies   81

Part Two: Biological Evolution

Introduction to Biological Evolution   253

Photograph Credits

350   Fig. 7.6—Stanley M. Awramik, Department of Geological Sciences, University of California, Santa Barbara, CA 93106
351   Fig. 7.7—Stanley M. Awramik, Department of Geological Sciences, University of California, Santa Barbara, CA 93106
351   Fig. 7.8—Smithsonian Institution, Photo No. 86-13573
362   Fig. 7.11—David Chase, Sepulveda Veterans Administration Hospital, Sepulveda, CA 91343. Print #3138
363   Fig. 7.12—M. M. Allen, Wellesley College, Wellesley, MA 02181
371   Fig. 7.17—Dr. Antoinette Ryter, Institute Pasteur
375   Fig. 7.20—Jones and Bartlett Publishers, Inc., from Shih and Kessel: *Living Images* (Science Books International). 1982, pp. 152–155
381   Fig. 7.25—Dr. Bernard John
382   Fig. 7.26—Jones and Bartlett Publishers, Inc., from Shih and Kessel: *Living Images* (Science Books International). 1982, p. 150
383   Fig. 7.27—T. F. Anderson, Institute for Cancer Research, Philadelphia, PA and E. L. Wollmann and F. Jacob, Pasteur Institute, Paris, France

Chapter 8: Origin and Evolution of the Major Animal Phyla
Page 389   Fig. 8.1— British Museum (Natural History)
390   Fig. 8.2—*Discover the Invisible* by Eric V. Gravé © 1984 by Prentice-Hall, Inc., Englewood Cliffs, NJ 07632. Reprinted by permission of the publisher, Prentice-Hall, Inc.
397   Fig. 8.9—Animals/Animals: Oxford Scientific Films
405   Fig. 8.18—Professor M. F. Glaessner, University of Adelaide, from "The Ediacarian Period and System: Metazo Inherit the Earth," Cloud, P. and Glaessner, M. F., *SCIENCE* Vol. 217 (Fig. 2a, p. 785), 27 August 1982. Copyright 1982 AAAS
406   Fig. 8.19—Professor M. F. Glaessner, University of Adelaide, from "The Ediacarian Period and System: Metazo Inherit the Earth," Cloud, P. and Glaessner, M. F., *SCIENCE* Vol. 217, 27 August 1982. Copyright 1982 American Association for the Advancement of Science
419   Fig. 8.24—Carolina Biological Supply
420   Fig. 8.26—Marjorie L. Reaka
426   Fig. 8.31—Geological Enterprises, Inc., Ardmore, OK
427   Fig. 8.32—Smithsonian Institution. Photo No.: Raymond Rye

Chapter 9: The Evolution of the Vertebrates
Page 435   Fig. 9.1—British Museum (Natural History)
438   Fig. 9.4—Animals/Animals: Oxford Scientific Films
447   Fig. 9.11—Animals/Animals: Zig Leszczynski
450   Fig. 9.14—Diane Edwards
450   Fig. 9.15—Robert Carr: Bruce Coleman, Inc.
451   Fig. 9.16—Field Museum of Natural History, Chicago
452   Fig. 9.18—Smithsonian Institution. Photo No. 74-4513
454   Fig. 9.20—Siber & Siber, Ltd., 8607 Aathal, Switzerland
463   Fig. 9.22—Field Museum of Natural History, Chicago
467   Fig. 9.25—Black Hills Institute of Geological Research, Hill City, SD
475   Fig. 9.30—Bildarchiv Preussischer Kulturbesitz, 1000 Berlin 61
483   Fig. 9.35 and Fig. 9.36—Australian Information Service

Chapter 10: Origin and Evolution of the Primates
Page 506   Fig. 10.6—Animals/Animals: Photo by George Roos
523   Fig. 10.15—Bob Campbell, © 1979, National Geographic Society
541   Fig. 10.29—Musée d'Aquitaine, Bordeaux, France. Photo courtesy of *The Washington Post*
543   Fig. 10.30—Casa Editrice Scode, S.p.A.
544   Fig. 10.31—Casa Editrice Scode, S.p.A.
545   Fig. 10.32—The University Museum, University of Pennsylvania
546   Fig. 10.33—The University Museum, University of Pennsylvania

PART ONE

PHYSICAL EVOLUTION

Introduction to Physical Evolution

In the beginning, this world was nothing at all,
heaven was not, nor earth, nor space.
Because it was not, it bethought itself:
I will be. It emitted heat.
  Ancient Egyptian text

There is nothing permanent except change.
  Heraclitus of Ephesus (ca. 540–480 B.C.)

If simple perfect laws uniquely rule the universe, should not pure thought be capable of uncovering this perfect set of laws without having to lean on the crutches of tediously assembled observations? True, the laws to be discovered may be perfect, but the human brain is not. Left on its own, it is prone to stray, as many past examples sadly prove. In fact, we have missed few chances to err until new data freshly gleaned from nature set us right again for the next steps. Thus pillars rather than crutches are the observations on which we base our theories . . .
  Martin Schwarzschild (Professor Emeritus, Princeton University)

Earth as viewed from the *Apollo 13* spacecraft on its return from the Moon (April 1970). Views such as this one prompted Apollo astronaut James Lovell to declare: "We do not realize what we have on Earth until we leave it."

I

Introduction
to Physical Evolution

WHEN THE APOLLO astronauts stood on the Moon more than a decade ago, the most beautiful object they saw anywhere in the sky was Earth. Satellite photographs of our planet confirm this for all of us (figures I.1–I.4): its spherical symmetry, its cover of clouds and oceans, the outlines of the continents, the meandering of the rivers, the green of the forests, and the yellow and purple of the deserts. Earth is our home—our refuge—in the vastness of space. Its solid crust provides a stable platform upon which we and the rest of life have come into existence and are evolving, its atmosphere and magnetic field offer protection from the hostile environment of outer space, and its distance from the Sun is such that it absorbs just enough radiant energy for warmth and comfort. How could such a planet, with conditions so well suited for life, have come to be? Or, reversing the perspective, how could life that is so well suited for this planet have come to be? Answering the first of these two questions will be the main goal of the first part of this book. Answering the second question will be the main goal of the second part.

The Earth, the Sun, and the rest of the Solar System are not isolated in space and they were not created in isolation. They originated as byproducts of the general physical evolution of the Universe. Thus, in order to understand our origin, we need to look beyond our immediate surroundings and learn about the origin and evolution of the Universe and its major components—the stars, galaxies, and clusters of galaxies. In our thinking we must switch from the familiar scales of inches, feet, and miles to scales that are measured in light-years. At the same time, we must also consider the microscopic world of molecules, atoms, and nuclei of atoms, for many of the large-scale features of the Universe are rooted in processes occurring on the smallest scale. For instance, all the matter in the Universe today is the result of countless microscopic interactions that occurred under conditions of immense temperatures and pressures during the first few seconds of the Universe's existence more than 10 billion years ago. And nearly all the light emitted by stars and galaxies originates

Figure I.1. Earth as seen from the *Apollo 8* Command Module, 110 km above the Moon's surface (December 1968). *Apollo 8* was the first manned spacecraft to orbit the Moon.

Figure I.2. The eastern Mediterranean, showing the Nile Delta, Gulf of Suez (bottom right), Sinai Peninsula, Israel, Dead Sea, Lebanon, Syria, and Cyprus (top middle). The photograph was taken from *Gemini 7* (December 1965), at an altitude of 300 km.

Figure I.3. The "top of the world," as seen from an altitude of 250 km by the *Apollo 7* astronauts. The Himalayan Mountains bisect this photograph diagonally, with the Ganges River of India to the left and Tibet to the right. The circle indicates Mt. Everest, Earth's highest peak (8848 m).

Figure I.4. Hurricane Gladys off the west coast of Florida, as photographed from *Apollo 7* on October 17, 1968. The clouds are aligned along an enormous spiral pattern, extending over hundreds of square kilometers and focused on the eye of the hurricane. The eye itself is hidden under a layer of clouds that is pushed by a vigorous updraft against the cold, stable air of the tropopause (at 16,500 m).

from the energy liberated by the conversion of atomic nuclei of one kind into those of another kind deep in the interior of stars.

Despite the psychological obstacles we may face in thinking simultaneously of the very large and the very small, our task is made easier by the fact that the fundamental laws of nature are few in number and they apply on all scales, large and small. Furthermore, there are only five stable elementary particles that make up the known content of the Universe: *electrons, protons, neutrons, neutrinos,* and *photons* (but see box I.1). The first three—electrons, protons, and neutrons—are the basic building blocks of the material content of the Universe, from galaxies and stars down to molecules and atoms. Neutrinos are very elusive particles created in certain nuclear reactions, and photons are quanta of light. Both neutrinos and photons play important roles in nature, as I shall explain in the following chapter. Mostly, however, this book deals with just electrons, protons, and neutrons, and the various structures created from them. This introduction describes some of the properties of these five elementary particles and the laws by which they interact. The second part of this book— Biological Evolution—shows that living organisms are also made of electrons, protons, and neutrons and hence are subject to the same natural laws as inanimate matter. Thus, the laws of nature and the existence of a few elementary particles allow us to view the origin and evolution of the Universe—including life on Earth—in a unified way.

BOX I.1.
Quarks and Gravitons

SINCE the 1930s physicists have been discovering a host of "elementary" particles other than the five mentioned above. These particles were first detected among the cosmic rays and then produced with particle accelerators. They are all extremely short-lived and decay into stable elementary particles fractions of seconds after they are produced.

Recent theoretical work suggests that protons and neutrons are perhaps not really elementary, but consist of still more fundamental particles—the *quarks*. The name *quark* was given to these particles by the American physicist Murray Gell-Mann, after the phrase "three quarks for Muster Mark" in James Joyce's *Finnegan's Wake*. Six quarks are believed to exist—*up, down, charm, strange, top*, and *bottom* quarks. A triad of up and down quarks makes up protons and neutrons. Much effort is currently devoted to the detection of quarks in laboratories around the world.

If quarks do exist, are they the ultimate building blocks of matter? This is one of the most profound questions confronting science today and there is at present no answer.

There may exist a sixth stable elementary particle—the *graviton*. Gravitons have never been detected. They have merely been predicted by theory. However, the theoretical arguments for their existence are very compelling. They would be the gravitational analogue of electromagnetic radiation (photons), be massless, and travel with the speed of light. They are believed to be created whenever mass is violently accelerated, as during the gravitational collapse of a star, the falling of matter into a black hole, or the Big Bang origin of the Universe.

Because the gravitational force is the weakest of all known forces in nature, gravitons—if they exist—would interact only exceedingly weakly with the other particles. That may be the reason why they have not yet been detected, despite concerted efforts to do so.

THE FUNDAMENTAL BUILDING BLOCKS OF MATTER

To begin the discussion of elementary particles, consider a familiar object, a snowflake (see figure I.5). It measures about one centimeter across (see box I.2 for comparison of the metric and English systems of units). A look through a high-power microscope reveals its very beautiful and intricate structure, one that a master architect might have put together. Myriads of particles, called *water molecules*, are joined together in very precise ways so as to give the snowflake its characteristic macroscopic appearance. Each water molecule is about 100 million times smaller than the snowflake. The water molecules are composed of still smaller particles called *atoms* (from the Greek, meaning "not cut" [a-tom] or "indivisible"). Two atoms of hydrogen and one of oxygen, bonded together as shown in figure I.5, make up one water molecule. Finally, the atoms are made of electrons, protons, and neutrons. Protons and neutrons are 100,000 times smaller than the atoms, and electrons are smaller still. These particles are so small that even the most powerful electron microscopes cannot reveal their existence. Exotic instruments such as X-ray diffraction machines or particle accelerators are required to "see" them. In figure I.6 the sizes of the elementary particles, atoms, and molecules are compared with those of some of the larger objects far and near in the Universe.

If, instead of a snowflake, we examine the internal architecture of a water droplet, a lunar rock, a piece of gold (see figure I.7), tissues from our own bodies, or any other kind of matter, we find that each of them, too, is constructed of molecules and atoms. These, in turn, are always made of the same fundamental building blocks—electrons, protons, and neutrons. Thus, all matter is made of the same stuff. The only difference between one kind of matter and another lies in the ways in which the building blocks are assembled.

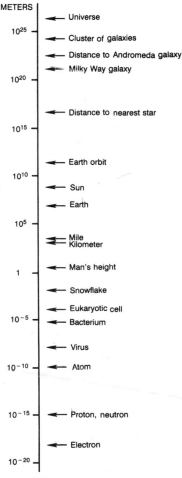

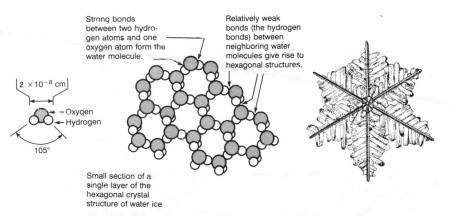

Figure I.5. Snowflakes, liquid water, and steam are composed of the same building blocks—water molecules. Each water molecule consists of one oxygen atom and two hydrogen atoms, firmly bonded together by the electric force. In addition, weak attractive forces (also electric) exist between the oxygen atom of one water molecule and the hydrogen atoms of neighboring water molecules. At temperatures below freezing, these weak attractions allow water molecules to arrange themselves spontaneously into hexagonal crystal structures. Snowflakes and ice crystals are evidence on the macroscopic scale of the precision of structure existing on the molecular level.

Figure I.6. Range of sizes of some of the objects dealt with in this book.

BOX I.2.
Units of Measurement Used in This Book

THE natural sciences are experimental and observational in nature. That means scientific theories are not purely intellectual constructs, but are based on experiments and observations. In the natural sciences, nature is the ultimate judge.

Scientific experimentation and observations require the carrying out of measurements: measuring distance or a time interval, reading the temperature or pressure on an instrument dial, determining the mass or electric charge of a particle, comparing the intensity of a light signal against some standard signal, counting objects or events, and so forth. In order to carry out such measurements, the scales or units to be used must be defined carefully. Are the units for distance in inches, meters, or light-years? Is temperature expressed in degrees Kelvin, Celsius, or Fahrenheit? Is the pressure scale in units of atmospheres, centimeters of mercury, or dynes per square centimeters? It must be clear from the outset what units are being used in order to avoid confusion.

In this text I shall follow the practice of the scientific community. Lengths, masses, and time intervals will be expressed in units of centimeters, grams, and seconds or in convenient multiples of these units. Temperature will be referred to in degrees Kelvin or Celsius. Energy and pressure will be described by making comparisons with familiar standards, such as the energy output of the Sun and the terrestrial atmospheric pressure at sea level.

The unit of length—the *centimeter* (cm)—is described by the markings on the following line:

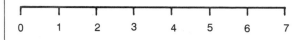

The multiples of the centimeter used in the text are:

$$\begin{array}{lcl}
1/10{,}000 \text{ cm} & = & 1 \text{ micrometer} \\
 & & \text{(abbreviated as } \mu m) \\
1/10 \text{ cm} & = & 1 \text{ millimeter (mm)} \\
100 \text{ cm} & = & 1 \text{ meter (m)} \\
100{,}000 \text{ cm} & = & 1 \text{ kilometer (km)} \\
6.38 \times 10^{8} \text{ cm} & = & 1 \text{ earth radius} \\
 & & \text{(equatorial)} \\
7.0 \times 10^{10} \text{ cm} & = & 1 \text{ solar radius} (R_{\odot}) \\
1.5 \times 10^{13} \text{ cm} & = & 1 \text{ astronomical unit (AU),} \\
 & & \text{the average Sun–Earth} \\
 & & \text{distance} \\
9.5 \times 10^{17} \text{ cm} & = & 1 \text{ light-year (LY)}
\end{array}$$

The unit of mass—the *gram* (g)—is the mass (or weight, if the weighing is done at sea level on Earth) of one cubic centimeter of water. The multiples of the gram used are:

$$\begin{array}{lcl}
1000 \text{ g} & = & 1 \text{ kilogram (kg)} \\
6.0 \times 10^{27} \text{ g} & = & 1 \text{ earth mass} \\
2.0 \times 10^{33} \text{ g} & = & 1 \text{ solar mass } (M_{\odot})
\end{array}$$

The metric system of length and mass compares with the English system as follows:

$$\begin{array}{llll}
1 \text{ cm} = 0.3937 & \text{inch} & 1 \text{ inch} & = \quad 2.540 \text{ cm} \\
1 \text{ m} = 3.281 & \text{feet} & 1 \text{ foot} & = \quad 30.48 \text{ cm} \\
1 \text{ km} = 0.6214 & \text{mile} & 1 \text{ mile} & = \quad 1.609 \text{ km} \\
\\
1 \text{ g} = 0.03527 & \text{ounce} & 1 \text{ ounce} & = \quad 28.35 \text{ g} \\
1 \text{ kg} = 2.205 & \text{pound} & 1 \text{ pound} & = 453.6 \quad \text{g}
\end{array}$$

The unit of time—the *second* (sec)—is the same in the metric and English systems of units. One *year* contains 3.16×10^{7} seconds.

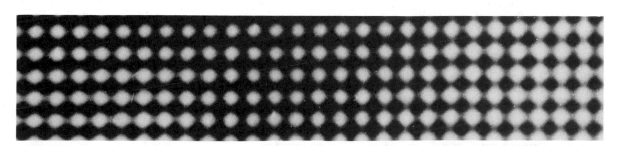

Figure I.7. A thin layer of gold, as seen through an electron microscope, magnified 40 million times. The regularly spaced bright dots are individual gold atoms. The distance between adjacent gold atoms is about 0.0002 μm.

The universality of electrons, protons, and neutrons raises the question: How can just three kinds of particles produce the staggering variety of objects found in the Universe? The answer is that each of the three elementary particles possesses unique properties that determine how the particles interact with each other and how they combine into larger objects. The number of properties of the particles and the number of kinds of interactions are small, but the number of ways in which the properties and interactions allow the particles to combine is virtually infinite. That is why the material objects all around us have such a great variety of shapes, forms, and sizes.

There are four known ways in which the elementary particles interact with each other. They are the gravitational, electromagnetic, nuclear, and weak interactions or forces.* We are all familiar with the first two of them because we are continuously exposed to them. The nuclear and weak interactions are somewhat more remote because generally they remain hidden deep within the atoms. They do affect us, however, for they are responsible for such phenomena as the energy production of the Sun, the explosive power of hydrogen bombs, and the generation of energy in nuclear power plants. The main features of the four interactions are described in the next section.

THE FOUR FORCES OF NATURE

Gravitational Force

Gravity is the force that holds us to the surface of the Earth. Without it we, the atmosphere, and everything else on our planet would fly apart and off into space. Gravity also holds the Sun, the stars, and the galaxies together. It keeps the Moon in orbit around the Earth and the Earth in orbit around the Sun. In short, *gravity is an attractive force that acts on all matter in the Universe*. It arises because all matter has mass. (Gravity also acts on photons, even though they carry no mass, as explained in chapter 1.)

The mass of an object is the property that we refer to colloquially as weight and that we measure in such units as pounds or grams (see box I.2). The greater the masses of particles or objects, the greater is the gravitational attraction between them. The greater their separation, the weaker is the gravitational force. More precisely, the gravitational force between two particles or objects (such as two elementary particles, the Earth and the Moon, the Sun and the Galaxy) is directly proportional to the product of the masses (M_1 and M_2) of the particles or objects and inversely proportional to the square of their separation (R):

$$F_{gravity} \propto \frac{M_1 M_2}{R^2}.$$

However, the gravitational force never becomes zero, no matter how great the separation between the two objects. For instance, the Solar System is approximately 30,000 light-years from the center of the Milky Way galaxy. Yet the

---

*The words *interaction* and *force* have the same meaning. They refer to such actions as pulling, pushing, pressing, and lifting. For instance, a person pulling a cart or lifting a box does so by exerting a force.

gravitational force exerted by the concentration of stars at the Galaxy's center prevents the Sun and the planets from flying off into intergalactic space despite their distance from the center. Similarly, galaxies are separated by millions and billions of light-years, but they still influence each other through gravity. The gravitational force, even though it weakens with distance, has an infinite range.

Compared to the other forces, gravity is extremely weak. We feel it strongly on Earth only because the mass of the Earth is large and we stand right on the planet's surface. Similarly, the gravitational force exerted by the center of our Galaxy is able to keep the Solar System in orbit because the concentration of mass at the Galaxy's center is so great. And galaxies affect each other gravitationally over the immense distance separating them because each contains a mass of millions to many billions of suns. However, the gravitational attraction between two elementary particles is virtually negligible, even at distances comparable to their sizes, for their masses are so small.

Electromagnetic Force

The second force that we are continuously exposed to is the *electromagnetic force*. It is responsible for the properties of all the objects in our immediate, macroscopic environment. It makes rocks hard, water liquid, and air gaseous. It participates in such processes as the dissolving of a sugar cube in a cup of tea, the burning of wood, and the digestion of food. Quite generally, the *electromagnetic force is responsible for all chemistry, including the chemistry of life*. In addition, it makes electrical appliances work, from TVs to electric motors and pocket calculators. Finally, it is responsible for light and radio waves. It is a very versatile kind of force.

As the name implies, the electromagnetic force consists of two parts—the electric force and the magnetic force. The electric force exists because two of the three elementary particles have electric charges and, hence, are surrounded by electric fields: *The proton is positively charged and the electron is negatively charged*. Because their charges are opposite, protons and electrons attract each other. Electrons repel each other and so do protons, because their charges are alike. The electric force does not affect the neutron because it has no electric charge. It is electrically neutral, as its name implies.

The mathematical expression for the strength of the electric force between two charged particles is similar to that of the gravitational force, except that the electric charge takes the place of mass. It is directly proportional to the product of the charges of the particles (Q_1, Q_2) and inversely proportional to the square of their separation (R):

$$F_{electric} \propto \frac{Q_1 Q_2}{R^2}.$$

And, like the gravitational force, the electric force has infinite range even though it weakens with distance.

The electric force comes into play in a flashlight (see figure I.8). Pressure on the button closes the flashlight's circuit and an electric current flows from the negative terminal of the battery through a wire or the metal casing to the bulb and on to the positive terminal. This current is just a flow of electrons. They flow because the first terminal of the battery is negatively charged and repels the electrons, while the second terminal is positively charged and attracts them. In the process the electrons are forced through the thin, high-resistance filament of the bulb and heat it to incandescence by friction.

Another, more spectacular demonstration of the electric force is lightning (see figure I.9). When the high winds of a thunderstorm sweep past each other, the resulting friction gives the lower regions of the rain clouds a negative charge (excess of electrons) and their upper regions, as well as the ground right below the clouds, a positive charge (excess of protons). As the charges build up, the attractive electric force between the negatively and positively charged regions increases until suddenly a surge of electrons neutralizes them, which we observe as lightning. This discharge of electricity between the upper and lower regions of the clouds, or between the clouds and the ground, is accompanied by the release of a great deal of energy in the form of light, thunder, and localized heat.

Whenever charged particles are in motion they create a magnetic field, and objects possessing magnetic fields interact by the magnetic force. This force is evident when we play a radio near a power line. The electric current in the power line creates a magnetic field that interferes with the radio reception and

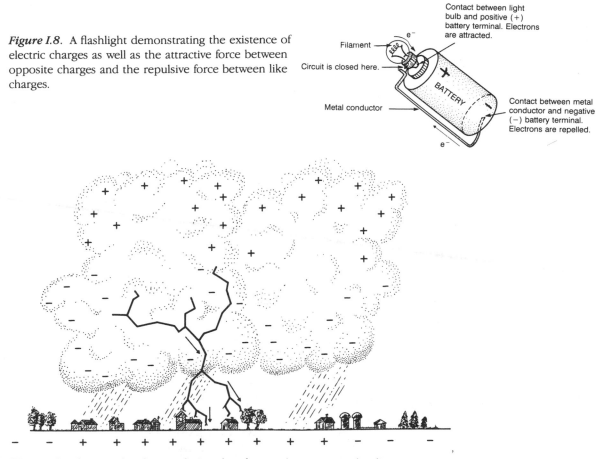

Figure I.8. A flashlight demonstrating the existence of electric charges as well as the attractive force between opposite charges and the repulsive force between like charges.

Figure I.9. Electron distribution during thunderstorms. Arrows in the diagram indicate the flow of electrons during a cloud–ground discharge.

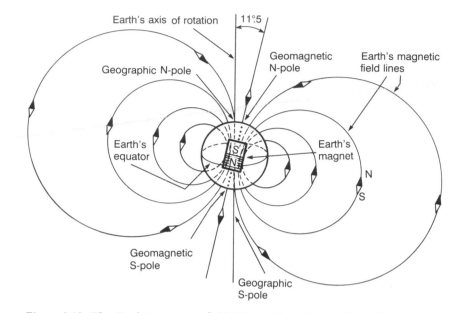

Figure I.10. The Earth's magnetic field. The configuration of the field is as if a short, strong bar magnet were buried at its center. The S-pole of the magnet points north and its N-pole points south. The magnet's axis is inclined by about 11.5° to the Earth's axis of rotation. In reality, the magnet is the result of motions of electrically conducting matter deep in the Earth's interior. Compass needles align themselves along the Earth's magnetic field lines such that their N-poles point toward the geomagnetic N-pole and their S-poles point toward the geomagnetic S-pole.

causes static. Two bar magnets provide a more direct experience of the magnetic force. When the magnets are close together, their north poles (and their south poles) repel each other, but the N-pole of one bar magnet attracts the S-pole of the other. A third example of the action of the magnetic force is the alignment of a compass needle along the magnetic field of the Earth (see figure I.10). The Earth's magnetic field is caused by the motion of electrically conducting matter deep in its interior. The compass needle also possesses a magnetic field, caused by microscopic motions of electrons in the atoms of the needle. The two magnetic fields interact via the magnetic force such that the needle's N-pole points toward the geomagnetic N-pole and its S-pole points toward the geomagnetic S-pole. Other examples of the presence of the magnetic force are the auroras known as the northern and southern lights. Their shimmering patterns occur when electrons and protons arriving from the Sun (the solar wind) are funneled by the Earth's magnetic field into the upper atmosphere near the polar regions and interact with nitrogen and oxygen atoms.

Moving electric charges give rise to magnetic fields and magnetic forces. Similarly, moving magnetic poles give rise to electric fields and electric forces. Thus, the electric and magnetic forces are inextricably linked and that is why collectively they are called the electromagnetic force. Later in this introduction I shall discuss one more important application of this linkage, namely light and radio waves.

The electromagnetic force between an electron and a proton is far stronger than the gravitational force between them. Yet, large aggregates of matter, like rocks or our bodies, interact very little with each other by the electromagnetic force. The reason is that they contain roughly equal numbers of electrons and protons and hence their electrical charges and magnetic poles cancel each other. The magnetic field of the Earth and electrically charged clouds in thunderstorms are exceptions.

No such canceling occurs with mass, for mass always has a positive sign. No negative mass exists. That is why the masses of the electrons, protons, and neutrons of our bodies, the Earth, the Sun, and the galaxies add up to large total masses. They create large gravitational effects, even though the contribution of a single elementary particle is very small.

Nuclear Force

Two forces remain to be discussed—the *nuclear* and the *weak forces*. Both participate in nuclear reactions, which are reactions between the nuclei of atoms (more about the structure of atomic nuclei in the following section; for a discussion of nuclear reactions see chapters 1 and 3). The roles the two forces play in these reactions are quite different.

The nuclear force (also referred to as the *strong* or *strong nuclear* force) is responsible for the large amounts of energy released in nuclear reactions, such as the energy generated by nuclear power stations, the tremendous explosive power of atomic and hydrogen bombs (figure I.11), and the more gentle,

Figure I.11. The first hydrogen bomb explosion, at Bikini Atoll in the Pacific Ocean (November 1, 1952). The explosion was so powerful that it obliterated a small island and left a crater more than one mile in diameter. The energy of hydrogen bomb explosions results from the conversion of hydrogen (in particular, the isotopes ^{2}H and ^{3}H) into helium, similar to the energy generation in the core of the Sun (where the nuclear fuel is ^{1}H).

though still potentially harmful, release of energy by radioactive decay (see box 5.1). Most important, it is responsible for the nuclear reactions that take place at the center of the Sun and that are the source of the radiant energy the Sun pours forth so prodigiously year after year. Without this steady supply of energy, life on Earth would quickly come to an end.

Processes involving the nuclear force are very energetic because this force is by far the strongest of the four forces. It is an attractive force that acts between protons and neutrons and holds them together to form the nuclei of atoms. The nuclear force does not affect electrons, neutrinos, or photons.

Protons and neutrons interact by the nuclear force only over distances comparable to their own sizes. At greater distances the nuclear force drops quickly to zero. Thus, unlike the gravitational and electromagnetic forces, the range of the nuclear force is very short. That is most fortunate, for if there were no limit on its range, this force would overpower all other forces and compress matter together into tightly bound lumps. Extended objects like living organisms, planets, stars, and galaxies could not exist.

Weak Force

In contrast to the nuclear force, the weak force (also referred to as the *weak nuclear* force) does not lead to the release of large amounts of energy in nuclear reactions. As its name implies, this is a very weak force with a strength that falls between that of the electromagnetic force and the gravitational force. Its range is even shorter than the range of the nuclear force.

The weak force comes into play only in the presence of a neutrino, the fourth elementary particle introduced above. The name *neutrino* means "little neutron" and was given to the particle by the great Italian physicist Enrico Fermi (1904–1954). It is a most appropriate name because the neutrino is electrically neutral like the neutron, but has a much smaller mass. In fact, up to the mid-1970s it was assumed that the neutrino resembled the photon (see below) by having no mass at all. This is now questioned because of new theoretical results, according to which the neutrino has a tiny mass, corresponding to about 1/100,000 of an electron's mass. This conjecture has not yet been confirmed by measurement.

Neutrinos are restless and elusive particles, and they are not part of ordinary matter. They come spontaneously into existence during the conversion of protons into neutrons or neutrons into protons (more about this in later chapters). As soon as they are created, neutrinos fly off into space with nearly the speed of light. When they come upon a star, a planet, or any other chunk of matter, they pass right through with only a negligible chance of interacting. In fact, neutrinos pass through our bodies all the time, yet we never feel them. It is very difficult to detect these particles.

Neutrinos are so elusive because they interact by the weak force, which has a very short range and is exceedingly weak. They do not interact by the electromagnetic or nuclear forces. Like all particles, they do interact by the gravitational force. However, the gravitational effect of the neutrinos passing through our bodies or the Earth is entirely negligible. Still, as discussed in chapter 1, the total number of neutrinos in the Universe may be so great that

Table I.1. Properties of the Stable Elementary Particles

Particle	Symbol	Electric Charge	Mass	Size
proton	p^+	+	1.7×10^{-27} kg	10^{-15} m
neutron	n	0	1.7×10^{-27} kg	10^{-15} m
electron	e^-	−	9.1×10^{-31} kg	10^{-18} m
neutrino	ν*	0	approx. 10^{-35} kg	less than 10^{-18} m
photon	γ*	0	0	no fixed size, depends on wavelengths

*ν and γ are the Greek letters *nu* and *gamma*, corresponding to Roman *n* and *g*.

their combined gravitational effect could have a decisive influence on the eventual fate of the Universe.

Summary

Table I.1 summarizes the most important properties of the elementary particles discussed here and in the previous section—the proton, neutron, electron, and neutrino. The table also includes the fifth elementary particle—the photon—which will be discussed later in this introduction. The symbols used as abbreviations for the proton, neutron, and electron are self-explanatory. The symbols for the neutrino and photon, ν and γ, are the Greek letters *nu* and *gamma* and correspond to the Roman letters *n* and *g*. The Greek letter γ is used because it was chosen when gamma-radiation was named (see box 5.1).

Three of the elementary particles in the table have zero electric charge—the neutron, neutrino, and photon. The charges of the proton and electron are *exactly* equal, but opposite in sign. These charges are very small and it takes a great number of them to produce a noticeable, macroscopic effect. For instance, when we turn on a 150-watt light, the electric current carries 10^{19} electrons* through the filament of the bulb every second.

Of the five elementary particles, one is without mass—the photon—and the others have nearly vanishingly small masses, as indicated by the large negative exponents in the fourth column of table I.1. It may help you comprehend just how small these masses are if you know that with every breath you inhale approximately ¾ of a gram of air and that each ¾ gram consists of more than 10^{23} protons, neutrons, and electrons. Thus the mass of any one of these particles is truly vanishingly small. Also note that the masses of the proton and neutron are nearly equal, and that the mass of the electron is almost 2000 times smaller. The estimated mass of the neutrino is smaller still by roughly another factor of 100,000.

*The expression 10^{19} is a shorthand notation for the number 1 followed by 19 zeros. This notation is used in the sciences whenever very large or very small numbers are encountered. It is explained in box I.3.

Table I.2. The Four Known Fundamental Forces of Nature

Force	Relative Strength At 10^{-15} m	Range	Particles Affected	Applications
Gravitational	10^{-38}	infinite	all particles	holding together cosmic structures like the Earth, Sun, Solar System, Galaxy, Universe
Weak	10^{-13}	less than 10^{-18} m	ν	Beta decay: conversion of protons into neutrons or neutrons into protons
Electromagnetic	10^{-2}	infinite in principle, but limited by the canceling between positive and negative electric charges and between north and south magnetic poles	e^-, p^+	atoms and molecules, chemical reactions, electric appliances, magnetic field of Earth, electromagnetic radiation (e.g., light, radiowaves)
Nuclear	1	10^{-15} m	p^+, n	nuclear reactions (e.g., stellar energy sources, atomic energy, radioactive decay)

Table I.2 summarizes the essential features of the four forces. The gravitational force is the weakest force, followed by the weak, the electromagnetic, and the nuclear forces. Their range varies from very short-range to infinite. Only gravity acts on all particles; the weak force plays a role only when a neutrino is present; the electromagnetic force acts only on matter with electric charge or magnetic fields; and the nuclear force acts only between protons and neutrons. Because of the differences in strength and range, each of the four forces has very specific applications in nature, from holding stars and galaxies together to making chemistry work, producing light and radio waves, and releasing enormous amounts of energy in nuclear reactions.

And now a word of caution. Though at present only four fundamental forces are well established experimentally and theoretically, others may exist — perhaps forces that are very weak or of short range, and thus have escaped detection. In fact, the existence of an additional fundamental force has been suggested repeatedly for the past two decades, most recently by Ephraim Fischbach, Daniel Sudarsky, Aaron Szafer, and Carrick Talmadge of Purdue University and by S. H. Aronson of Brookhaven National Laboratory.

These researchers re-analyzed the data of a famous experiment carried out earlier in this century by Baron von Eötvös (1848–1919), a Hungarian physicist. Eötvös and his coworkers carefully measured the gravitational acceleration — that is, the rate at which bodies fall near the surface of the Earth — of bodies made of different materials and possessing different masses, among them metals, water, wood, tallow, and several kinds of minerals. Their aim was to determine whether gravitational acceleration depends on the composition and mass of falling bodies or not. They reported no such dependences. This result confirmed the traditional view of physicists that a feather and coin dropped from the same height in a vacuum fall at the same rate. This view dates back to

Galileo Galilei (1564–1642), who supposedly experimented with different sized cannon balls by dropping them from the Leaning Tower of Pisa.

When Fischbach and his colleagues re-examined the original data of the Eötvös experiment, they found small unexplained discrepancies. The Hungarian scientists had also been aware of these discrepancies, but thought they were unavoidable experimental errors and ignored them. Upon very careful analysis, the Fischbach team interpreted the discrepancies as having arisen from a fifth fundamental force, which acted in addition to the force of gravity. They suggest that this force is much weaker than gravity, that it is repulsive, and that it has a range of about 200 meters. Furthermore, they suggest that this force depends on a property of matter called *hypercharge*, which is related to the number of protons and neutrons present in atomic nuclei. This latter property means that the hypercharge force is composition-dependent. For example, in a vacuum a feather should fall more rapidly than a coin.

At present it is not known whether the suggestion of a fifth fundamental force is correct. The Eötvös experiment needs to be repeated with greater sensitivity and with a variety of materials of precisely known composition to find out. If it turns out to be real, "the force is not likely to change calculations on the scale of planetary systems. [However,] when you see something as fundamental as a new force, it's likely to change many things. We will have to rethink many views of particle physics and cosmology." (Fischbach, quoted in *The New York Times*, January 8, 1986, 1:2.) But even if the hypercharge force is not confirmed by new experiments, the fact that a team of competent scientists suggested its possible existence indicates that current scientific theories are not necessarily complete and final. Scientists have barely begun to investigate nature and, no doubt, many currently held views of physics and the other sciences will have to be rethought in the future.

THE ARCHITECTURE OF MATTER

Now that we are acquainted with the fundamental building blocks of matter and the forces by which they interact, let us examine the architecture of matter in more detail. In particular, let us look at the structure of atomic nuclei, atoms, and molecules. As noted earlier, these particles are far too small to be seen with the unaided eye or felt by our sense of touch. That is why we have no direct experience with them. In order to acquire some understanding of them, let us for the moment give our imagination free rein and do a *Gedanken* or "thought" experiment. We invent a machine that can magnify objects indefinitely and show all their details. (In this regard, the machine is superior to any electron microscope, X-ray diffraction machine, or particle accelerator.) Then let us magnify a water droplet and "see" what it is made of.

As the droplet becomes larger, there is at first no apparent change in its appearance. Even when it has grown to the size of a football field, it still seems to contain clear, liquid water, except for some giant bacteria, protists, and other impurities, which do not interest us at the moment. However, when the droplet becomes greater than one kilometer across, details suddenly emerge. We no longer see water, but an immense swarm of particles that are in continuous disorganized motion. Let us stop the magnification at 100 million, when the

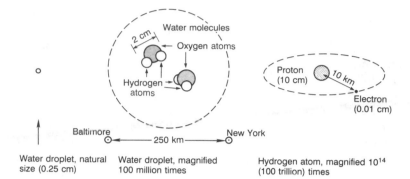

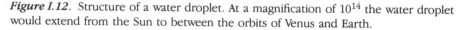

Figure I.12. Structure of a water droplet. At a magnification of 10^{14} the water droplet would extend from the Sun to between the orbits of Venus and Earth.

droplet has grown to a size of 250 km, roughly the distance from New York City to Baltimore (see figure I.12). Each of the particles is now about two centimeters across and they are all alike. If we had sufficient patience to count them, we would find approximately 10^{20} of them in that one water droplet.

Each of the particles making up the water droplet consists of three blurry-looking spheres, which are bonded together by the electric force and are not easily broken up. These particles are water molecules and the spheres are atoms. The central and slightly larger spheres are oxygen atoms and the two peripheral ones are hydrogen atoms. For brevity, hydrogen and oxygen atoms are abbreviated by the capital letters H and O, and water molecules by H_2O.*

At this magnification we cannot recognize any structural details in the atoms. For that we need to enlarge the water droplet further until it is 250,000 km in size and the total magnification is 100 billion. The water droplet is now nearly one-fifth the size of the Sun, the molecules are roughly 20 m across, and the atoms are about 10 m. The atoms turn out to consist mostly of empty space. At each of their centers is a tiny dense particle, just barely visible despite the enormous magnification. They are about 1/100 cm in size and are the nuclei of the atoms. The nuclei of the hydrogen atoms are single protons, while those of the oxygen atoms consist of clusters of eight protons and eight neutrons, held tightly together by the nuclear force.

Circling these nuclei are electrons at distances of up to approximately 10 m, corresponding to the sizes of the atoms. The hydrogen atom has only one electron, but the oxygen atom has eight. The electrons are smaller still than the nuclei, and it would require an additional magnification of one thousand to make them visible.

The electrons are held in their orbits by the electric force, which pulls them (since they are negatively charged) toward the positively charged nuclei. However, the electrons do not fall into the nuclei because of the forward component of their orbital motions, which would carry them off into space if the pull toward the central nucleus were suddenly to disappear. This is quite analogous to the

*The subscript 2 in H_2O means that there are two hydrogen atoms in every water molecule. The absence of a subscript on the O means that only one oxygen atom is present.

BOX I.3.
Exponential Notation

MANY concepts in mathematics and in the sciences are more clearly expressed by numbers or symbols than by words. For instance, "one thousand nine hundred eighty-six" requires considerably more effort to write as well as to read than the numerical notation "1986."

When we deal with very large or very small numbers, even Arabic numerals become cumbersome. For example, "150,000,000,000" and "0.000,000,000,000,000,000,000,000,000,91" require that we count zeros before we know the values of these numbers. To ease the task of writing and reading very large and very small numbers, mathematicians have invented *exponential notation*. This notation has the form a^m and is read "*a* to the power *m*," where *a* and *m* stand for numbers and where *a* is called the *base* and *m* the *exponent*. In this book, *a* will always be the number 10 and *m* will always be a positive or negative integer or zero. (In general, however, *a* and *m* may be any number, including fractions and irrational numbers.)

Thus, the exponential notation used in this book is of the form 10^9, 10^8, 10^{-6}, and so forth. This notation is to be interpreted as illustrated by the following examples:

$$
\begin{array}{lll}
\vdots & \quad\quad\quad \vdots & \quad\quad\quad\quad \vdots \\
10^9 & = 1,000,000,000 \quad \text{one billion} & \text{(a one followed by } \textit{nine} \text{ zeros.)} \\
10^8 & = 100,000,000 \quad \text{one hundred million} & (- " - \textit{eight} ") \\
\vdots & \quad\quad\quad \vdots & \quad\quad\quad\quad \vdots \\
10^3 & = 1,000 \quad \text{one thousand} & (- " - \textit{three} ") \\
10^2 & = 100 \quad \text{one hundred} & (- " - \textit{two} ") \\
10^1 & = 10 \quad \text{ten} & (- " - \textit{one} ") \\
10^0 & = 1 \quad \text{one} & (- " - \textit{zero} ") \\
10^{-1} & = 0.1 \text{one-tenth} & \text{(a one with the decimal point shifted} \\
& & \textit{one} \text{ place to the left.)} \\
10^{-2} & = 0.01 \text{one-hundredth} & (\textit{two} \text{ places} - " -) \\
\vdots & \quad\quad\quad \vdots & \quad\quad\quad\quad \vdots \\
10^{-6} & = 0.000,001 \text{one-millionth} & (\textit{six} \text{ places} - " -) \\
\vdots & \quad\quad\quad \vdots & \quad\quad\quad\quad \vdots
\end{array}
$$

In exponential notation, the two numbers in the second paragraph above can now be written clearly and compactly as 1.5×10^{11} and 9.1×10^{-31}. These two numbers describe the Sun–Earth distance in meters and the mass of the electron in kilograms.

motions of the planets around the Sun or of satellites around the Earth (see box I.4).

It is interesting to carry a bit further the comparison between electrons orbiting around their nuclei and planets circling around the Sun. For instance, electron orbits are approximately 50,000 times larger than the sizes of the atomic nuclei. In contrast, the Earth's orbit is only about 200 times larger than

BOX I.4.
Centripetal and Centrifugal Forces

THE orbital motions of electrons around the central nucleus or of planets around the Sun are sometimes explained as being due to a balance between two forces: the *centripetal (center-seeking) force* that pulls the electrons or planets toward the centers of their orbits and the *centrifugal (center-fleeing) force* that pulls them away from the centers of their orbits. Strictly speaking, this is not a correct way of describing orbital motions because only the centripetal forces (that is, the electric and gravitational forces in these two examples) are real. The centrifugal forces are not. There are no forces that push the electrons or planets away from the centers of their orbits. The centrifugal force merely constitutes a particle's resistance to having the direction of its motion changed. It is not a force on par with the four fundamental forces I introduced in the text. Should the centripetal force suddenly disappear, then the centrifugal force would disappear as well and the particle would fly off in a straight line. (This can be tested easily by whirling a rock in a circle at the end of a string and suddenly cutting the string, or, more unpleasantly, by driving a car around a curve and suddenly skidding on an oil slick or patch of ice.)

Despite its fictitious nature, the concept of centrifugal force is a useful one, particularly when analyzing the behavior of rotating gases and fluids. However, the use of the concept of centrifugal force is logically consistent with the rest of physics only in a reference-frame that rotates along with the gases or orbiting objects. A simple example will explain this. Let us assume we step onto the platform of a merry-go-round and sit on one of the horses. As the merry-go-round starts turning, we feel a force that tends to push us off the horses in the direction away from the merry-go-round's center. That force is the centrifugal force. To avoid falling off we need to exert an equally strong force toward the center of the merry-go-round, the centripetal force. We do this by holding on to the horses or by leaning inward. As long as the two forces, the centrifugal and centripetal forces, are in balance, we will not fall off. Of course, the faster the merry-go-round turns or the farther from its center we sit, the stronger we have to hold on or lean to maintain this balance. The reason is that the centrifugal force increases with rotation speed and distance from the rotation axis.

Note that only we who are on the rotating merry-go-round feel the centrifugal force. Spectators on the ground are not aware of it. All they see is that we are holding on to the horse or leaning inward. Therefore, as far as they are concerned, only the centripetal force is real, the centrifugal force is not. Nevertheless, in the context just described (that is, from the point of view of a rotating reference frame), the concept of centrifugal force is a perfectly good one and many research scientists make use of it. This text will too in discussions of the physics of rotating galaxies, interstellar gas clouds, and the origin of the Solar System.

the Sun. Thus, atoms are, in relative terms, more extended by far than the Solar System. Expressed in units of volume, atoms contain roughly 10 million times more empty space than the Solar System.

There are many different kinds of atoms in nature—hydrogen, carbon, nitrogen, oxygen, iron, gold, and uranium atoms, to name just a few. Quite often, scientists are not so much interested in individual atoms, but in collections of them, like the carbon atoms in our bodies or the oxygen atoms in the air. They then speak of the *element* carbon, the *element* oxygen, or more briefly, just of carbon or oxygen. The total number of elements occurring in nature is 92 and about a dozen more (the transuranic elements) have been created in nuclear accelerator experiments. Of the naturally occurring elements, the simplest is hydrogen and the most complex is uranium. Only a handful of them are of real interest to us in this book: hydrogen, helium, carbon, nitrogen, oxygen, phosphorus, and sulfur, for example.

Every element is characterized by unique physical and chemical properties, such as color, degree of hardness, density, ability to conduct heat or electricity, tendency to combine with other elements and so forth. For instance, the oxygen in the present Earth atmosphere exists in molecular form, O_2, and it is gaseous; hydrogen reacts readily with oxygen to form water, H_2O; iron is a solid up to approximately 1800 K,[*] but at higher temperatures it melts and becomes a liquid; and gold is a very dense, malleable, noncorroding and yellow metal, with high heat and electrical conductivity.

Why do the elements have unique properties that distinguish them from each other? The answer is that the atoms of every element have their own unique structures: each has a unique nucleus that (in the electrically neutral state) is surrounded by a unique number of electrons. For instance, the nuclei of hydrogen, helium, carbon, nitrogen, and oxygen contain, respectively, one, two, six, seven, and eight protons (see figure I.13). In the electrically neutral state, the protons (because of their positive charge) keep an equal number of electrons in orbit; thus, the nucleus of hydrogen is surrounded by one electron, that of helium by two, that of carbon by six, and so forth.

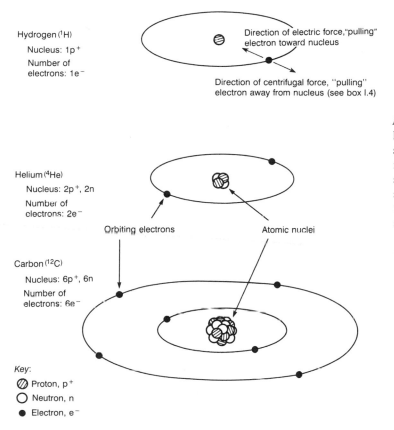

Hydrogen (¹H)
Nucleus: 1p⁺
Number of electrons: 1e⁻

Direction of electric force,"pulling" electron toward nucleus

Direction of centrifugal force, "pulling" electron away from nucleus (see box I.4)

Helium (⁴He)
Nucleus: 2p⁺, 2n
Number of electrons: 2e⁻

Orbiting electrons   Atomic nuclei

Carbon (¹²C)
Nucleus: 6p⁺, 6n
Number of electrons: 6e⁻

Key:
⊘ Proton, p⁺
○ Neutron, n
● Electron, e⁻

Figure I.13. Diagrams of hydrogen, helium, and carbon atoms (not to scale). The protons and neutrons are roughly the same size. The electrons are approximately 1000 times smaller and their orbits are approximately 100,000 times larger than the protons and neutrons.

*This book uses degrees *Kelvin* or *Celsius* to measure temperature, the temperature scales used in science. Box 1.2 discusses the meaning of temperature and compares degrees Kelvin and Celsius to the Fahrenheit scale used in the United States.

Electrically neutral atoms are the only ones under consideration here. In later chapters I describe how atoms may lose one or more electrons or they may gain an electron. They then acquire a positive or negative charge and are called *ions*. The process by which atoms lose or gain electrons is called *ionization*. It is important to recognize that ionization does not change an atom from one kind to another (such as a nitrogen atom into a carbon or oxygen atom). That can be accomplished only by altering the number of protons in the nucleus.

Remember, *the number of protons present in the nucleus determines the nature of an atom*. Thus, the above description of hydrogen, helium, and carbon and so on may be turned around and stated like this: An atom whose nucleus contains one proton is a hydrogen atom; an atom whose nucleus contains two protons is a helium atom; an atom whose nucleus contains six protons is a carbon atom; and likewise for all the other atoms.

In addition to protons, atomic nuclei usually contain neutrons. The neutrons are necessary to make the nuclei stable, but because they are electrically neutral, neutrons have little effect on the chemical properties of the elements and they are not considered in the naming of the elements.

Most elements come in several versions that differ from each other merely in the number of neutrons present in their nuclei. Such different versions of the same element are called *isotopes*. For instance, there are three isotopes of hydrogen: Their nuclei consist of *one* proton (the most common of the hydrogen isotopes), *one* proton and one neutron, and *one* proton and two neutrons. There are also three isotopes of carbon: Their nuclei consist of *six* protons and six neutrons (the most common of the carbon isotopes), *six* protons and seven neutrons, and *six* protons and eight neutrons. The third carbon isotope is not stable, but decays into nitrogen. (Radioactive decay of unstable isotopes is discussed further in box 5.1.)

There is a shorthand notation for the elements and their isotopes. It consists of one- or two-letter symbols, together with superscripts and subscripts, as shown in table I.3, the Periodic Table of the Elements. The letters are abbreviations of the elements' names; the superscripts refer to the total number of protons *plus* neutrons present in the nuclei; and the subscripts are reminders of the number of protons present in the nuclei. For instance, the notation for the three isotopes of hydrogen is ^1_1H, ^2_1H, ^3_1H, and that for the three isotopes of carbon is $^{12}_6C$, $^{13}_6C$, and $^{14}_6C$.* Only the most common isotopes of the elements are listed in the table.

Generally it is not necessary to distinguish between different isotopes, and in this book most discussions will use the symbols H, C, and so forth without superscripts. Only in discussing nuclear reactions or decays will it be necessary to refer to specific isotopes and to use the superscripts. The subscripts will

---

*Usually the letter abbreviations are taken from the elements' Latin names. When the English and the Latin names of an element do not coincide, strange abbreviations result. For instance, Na is the abbreviation for sodium and is derived from this element's Latin name *natrium*; Ag is the abbreviation for silver (*argentum*); and Au is the abbreviation for gold (*aurum*).

The isotope ^2_1H is called hydrogen-2 or deuterium and the isotope ^3_1H is called hydrogen-3 or tritium. The isotope $^{12}_6C$ of carbon is called carbon-12, $^{13}_6C$ is called carbon-13, and $^{14}_6C$ is carbon-14. The isotopes of all the other elements are referred to similarly: helium-4 for ^4_2He, oxygen-16 for $^{16}_8O$, and so forth.

Table I.3. Periodic Table of the Elements. Only the most common isotopes of the elements are listed.

1 (H)	2											13	14	15	16	17	18
1 **H** 1 Hydrogen																	4 **He** 2 Helium
7 **Li** 3 Lithium	9 **Be** 4 Beryllium											11 **B** 5 Boron	12 **C** 6 Carbon	14 **N** 7 Nitrogen	16 **O** 8 Oxygen	19 **F** 9 Fluorine	20 **Ne** 10 Neon
23 **Na** 11 Sodium	24 **Mg** 12 Magnesium											27 **Al** 13 Aluminum	28 **Si** 14 Silicon	31 **P** 15 Phosphorus	32 **S** 16 Sulfur	35 **Cl** 17 Chlorine	40 **Ar** 18 Argon
39 **K** 19 Potassium	40 **Ca** 20 Calcium	45 **Sc** 21 Scandium	48 **Ti** 22 Titanium	51 **V** 23 Vanadium	52 **Cr** 24 Chromium	55 **Mn** 25 Manganese	56 **Fe** 26 Iron	59 **Co** 27 Cobalt	58 **Ni** 28 Nickel	63 **Cu** 29 Copper	64 **Zn** 30 Zinc	69 **Ga** 31 Gallium	74 **Ge** 32 Germanium	75 **As** 33 Arsenic	80 **Se** 34 Selenium	79 **Br** 35 Bromine	84 **Kr** 36 Krypton
85 **Rb** 37 Rubidium	88 **Sr** 38 Strontium	89 **Y** 39 Yttrium	90 **Zr** 40 Zirconium	93 **Nb** 41 Niobium	98 **Mo** 42 Molybdenum	97 **Tc** 43 Technetium	102 **Ru** 44 Ruthenium	103 **Rh** 45 Rhodium	106 **Pd** 46 Palladium	107 **Ag** 47 Silver	114 **Cd** 48 Cadmium	115 **In** 49 Indium	120 **Sn** 50 Tin	121 **Sb** 51 Antimony	130 **Te** 52 Tellurium	127 **I** 53 Iodine	132 **Xe** 54 Xenon
133 **Cs** 55 Cesium	138 **Ba** 56 Barium	139 **La** 57 Lanthanum	180 **Hf** 72 Hafnium	181 **Ta** 73 Tantalum	184 **W** 74 Tungsten	187 **Re** 75 Rhenium	192 **Os** 76 Osmium	193 **Ir** 77 Iridium	195 **Pt** 78 Platinum	197 **Au** 79 Gold	202 **Hg** 80 Mercury	205 **Tl** 81 Thallium	208 **Pb** 82 Lead	209 **Bi** 83 Bismuth	209 **Po** 84 Polonium	210 **At** 85 Astatine	222 **Rn** 86 Radon
223 **Fr** 87 Francium	226 **Ra** 88 Radium	227 **Ac** 89 Actinium															

Lanthanides:

140 **Ce** 58 Cerium	141 **Pr** 59 Praseodymium	142 **Nd** 60 Neodymium	145 **Pm** 61 Promethium	152 **Sm** 62 Samarium	153 **Eu** 63 Europium	158 **Gd** 64 Gadolinium	159 **Tb** 65 Terbium	164 **Dy** 66 Dysprosium	165 **Ho** 67 Holmium	166 **Er** 68 Erbium	169 **Tm** 69 Thulium	174 **Yb** 70 Ytterbium	175 **Lu** 71 Lutetium

Actinides:

232 **Th** 90 Thorium	231 **Pa** 91 Protactinium	238 **U** 92 Uranium

always be left off because they merely spell out information about protons that is implied by the letter symbols and given explicitly, for reference purposes, in the table.

ELECTROMAGNETIC RADIATION

The light from the Sun and the stars, radio waves emitted by broadcasting stations, radar used in air traffic control and by the police, and X rays used in medicine are all examples of the same phenomenon—electromagnetic radiation. Clearly, the various kinds of electromagnetic radiation are not all alike. What is the difference?

Electromagnetic radiation consists of waves of oscillating electric and magnetic fields traveling with the speed of light. In order to understand this statement and to answer the above question, let us first consider more familiar waves, namely water waves on a lake or ocean.

When water waves travel across the surface of a lake or the ocean, water does not flow horizontally along with the wave. Instead, the water's motion is a rhythmic up and down oscillation. We can observe this by watching a piece of cork bobbing up and down as a wave passes it or, better yet, by going into the water and enjoying the sensation of being lifted way up and then brought way down by a big wave. The up and down oscillations, which constitute a water wave, vary according to a regular pattern. One swimmer may just reach the crest of the wave, while a little farther away another swimmer is still on the way up and a third is already being carried down. With time, the rhythmic up and down motions of the swimmers repeat as wave after wave passes by (see figure I.14).

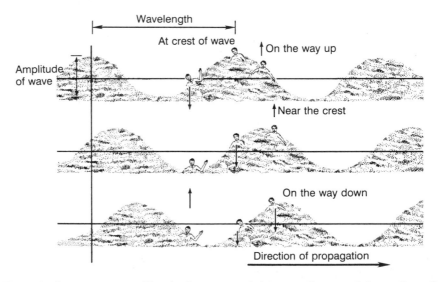

Figure I.14. A water wave. The rhythmic up and down oscillations of the surface of the water are illustrated by three swimmers. (Top) The swimmer on the right is being carried up by the wave; (center) the right-hand swimmer has almost reached the crest of the wave; and (bottom) the crest of the wave has passed and the swimmer is on the way down. Between the top and bottom diagrams, the wave travels one-third of a wavelength to the right.

Water waves are not all alike. Not only do they have different amplitudes, they also differ in wavelength, oscillation frequency, and the speed with which they progress. The wavelength is the distance from crest to crest of the wave and the oscillation frequency is the rate at which the water moves up and down at a given location (expressed in cycles per second). The speed of the wave is equal to the product of wavelength and oscillation frequency. Large ocean waves have wavelengths exceeding 100 meters; their oscillation frequencies are roughly one cycle every 10 seconds; and their speeds are about 30 to 50 kilometers per hour. On a lake, the wavelength may be as short as a fraction of a meter, the oscillation frequency of the order of one cycle per second, and the speed a few kilometers per hour.

The various kinds of electromagnetic radiation resemble water waves in that they, too, consist of rhythmic back and forth motions, and they are characterized by wavelength, oscillation frequency, and the speed of propagation. The main difference is that in the case of electromagnetic radiation the back and forth motions do not involve water or any other material substance. They are oscillations of electric and magnetic fields. The two fields oscillate in unison and they are aligned at right angles to each other and are perpendicular to the direction of wave propagation, as shown in figure I.15. As in the case of water

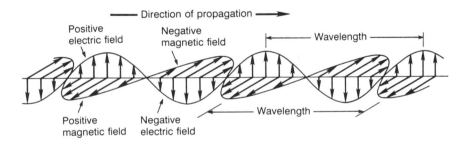

Figure I.15. Schematic diagram of electromagnetic radiation.

waves, the product of wavelength and oscillation frequency of electromagnetic radiation gives the speed of propagation—*300,000 km/sec*. However, unlike the speed of water waves, this speed is the same for all electromagnetic waves. It is the *speed of light*. More precisely, it is the speed of light in vacuum. When electromagnetic radiation passes through a material medium, such as air or glass, the speed of propagation is slightly reduced.

The only difference between the various kinds of electromagnetic radiation is in wavelength and oscillation frequency, as shown in figure I.16. The wavelengths range from 10^{-12} m for gamma rays to several kilometers for the longest radio waves. The rest—namely X rays, ultraviolet (UV) rays, visible light, infrared (IR) radiation, and the various kinds of radio waves (microwaves, radar, and FM, TV, and AM broadcasting waves)—lie between these two extremes.

Visible light has wavelengths between 4×10^{-7} and 7×10^{-7} m, which is a relatively narrow range compared with the full range of electromagnetic radiation (see figure I.17). Yet this range contains all the color components to which human eyes are sensitive. These components become apparent when white light, like that from the Sun, is sent through a prism, which splits it into the familiar spectrum of rainbow colors from violet to blue, green, yellow, orange, and red. The same splitting occurs in a rainbow, where raindrops act as

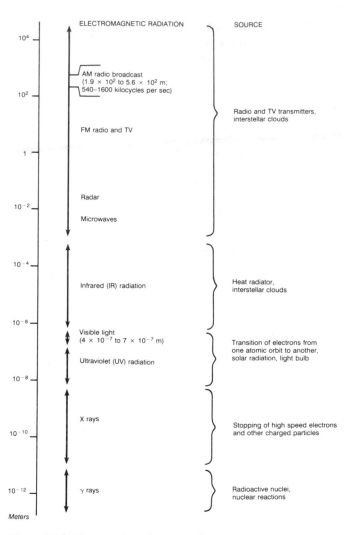

Figure I.16. The wavelength range of electromagnetic radiation.

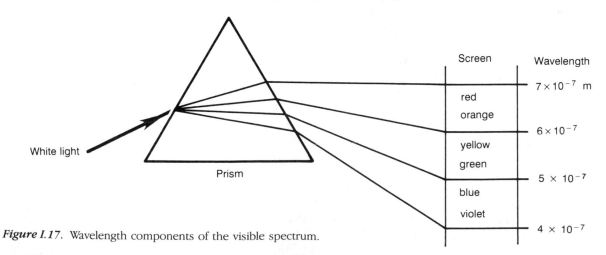

Figure I.17. Wavelength components of the visible spectrum.

prisms. Each of the color components of visible light has its own wavelength, with violet light having the shortest and red light the longest.

In an earlier discussion I referred to electromagnetic radiation as being an elementary particle—the photon. Yet it is also a wave, consisting of oscillating electric and magnetic fields, propagating with the speed of light. How can the same entity be both a particle and a wave?

To answer this question it is important to realize that when scientists speak of elementary particles, of atoms and molecules, and of other natural phenomena, they are constructing models that they hope will correspond to nature. The models are human creations, not reality. Any model that is consistent with observations or experiments is a good model. Photons and electromagnetic waves are two different models—a particle model and a wave model—of the same phenomenon. Both models are equally consistent with what is observed in nature. Hence, both are good models. It turns out that the particle model is more useful when considering the emission or absorption of electromagnetic radiation, while the wave model is more useful when considering the propagation of electromagnetic radiation. Thus scientists need both models and later chapters will refer to both. This is because the emission and absorption as well as the propagation of electromagnetic radiation are important processes in nature and affect the structure and evolution of matter.

ELECTROMAGNETIC SPECTRA

All material bodies—solids, liquids, and gases—emit electromagnetic radiation. The emission is usually the result of very complex processes, involving the random motions of a body's constituent atoms, molecules, and other particles. Generally, if a body is opaque it will emit radiation in all wavelengths. Furthermore, if it is very hot, such as the Sun, burning wood, or a stove-top burner turned on high, most of the wavelengths tend to be short (which is equivalent to saying that individual photons carry a lot of energy) and can be seen by the human eye or by UV and X-ray detectors. If a body is cool, such as our bodies, the ground, or ice, most of the wavelengths tend to be long (that is, the individual photons carry little energy) and can be detected only by IR sensors or radio antennae.

The intensity of the radiation emitted by an opaque body is not the same in all wavelengths but shows a distinct wavelength dependence. A plot representing this dependence is said to be a body's *continuum spectrum*. In figure I.18, continuum spectra are shown for several bodies, each at a different temperature. The curves indicate that the radiation from a hot body is more intense than that from a cool body.

Two features of the curves deserve special notice. First, the peaks of the curves shift to shorter wavelengths with increasing temperature. We all are familiar with this: A piece of iron withdrawn from a very hot fire at first emits whitish-blue radiation (short wavelengths), but as it cools its radiation turns orange and then deep red (long wavelengths). Second, the areas under the curves increase with temperature, meaning that the total energy emitted per unit area every second by the bodies increases as their temperatures increase. We are familiar with this as well: The radiation from the Sun is more intense

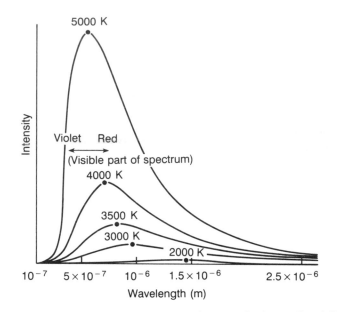

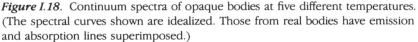

Figure I.18. Continuum spectra of opaque bodies at five different temperatures. (The spectral curves shown are idealized. Those from real bodies have emission and absorption lines superimposed.)

than that from cooler bodies such as burning coals or the ground. (*Intensity* is defined as the energy in a beam of electromagnetic radiation passing through a unit normal area per unit time.)

Exactly how is electromagnetic radiation emitted or absorbed by atoms? First we need to recognize that unlike planets or satellites, which may assume orbits of any size, electrons circling a nucleus can occupy only distinct, or quantized, orbits. If an electron is in the smallest allowed orbit, it is said to be in the *ground state*. If it is in one of the higher orbits, it is said to be in an *excited state*. Normally, the electrons are in the ground state. However, collisions resulting from the random motions of a body's constituent atoms and molecules tend to kick the outermost electrons of atoms into higher orbits (this may also happen by the absorption of photons, see below). When an electron has been raised into a higher orbit, it is not long before it jumps back to lower orbits. Sometimes it jumps back to the ground state in one step. At other times it returns to the ground state in several steps by first jumping to intermediate orbits. As it does so, electromagnetic radiation of specific wavelengths is emitted, corresponding to the energy differences between the higher and lower orbits (see figure I.19).

When the radiation from a large collection of identical atoms, whose electrons are in excited states, is sent through a prism and projected onto a screen or photographic plate, a distinct wavelength pattern, called an *emission spectrum*, can be seen. Emission spectra are unique for each element (because the allowed electron orbits of each element are unique) and can be used to deduce which elements are present in a radiating body, their relative abundances, the body's temperature, and much other information.

When the radiation from a hot, opaque body (that is, radiation that is fairly intense and has a continuum spectrum) passes through a thin layer of gas, it

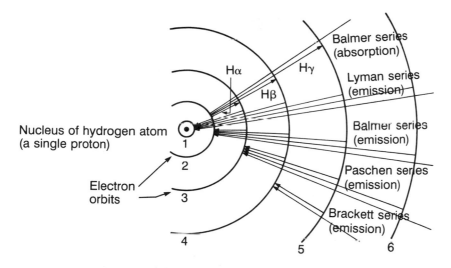

Figure I.19. Emission and absorption of electromagnetic radiation by the hydrogen atom. Six electron orbits are shown. Orbit 1 represents the ground state; the others are excited states. The inward-pointing arrows indicate possible jumps by the electron from higher to lower orbits, during which photons of distinct wavelengths are emitted. The outward-pointing arrows indicate possible jumps by the electron from the first excited state to higher orbits, during which photons of distinct wavelengths are absorbed. The names *Lyman, Balmer, Paschen*, and *Bracket* are labels for the series of emission and absorption lines (that is, emission and absorption spectra) resulting from the transitions indicated in the diagram. The arrows labelled Hα, Hβ, and Hγ refer to the Balmer series of hydrogen and correspond to prominent absorption lines seen in the solar spectrum, as shown in figure I.20. (*Note*: In reality, electron orbits are not quite as simple as indicated in this diagram.)

tends to kick electrons of the gas's constituent atoms into excited states. This process requires energy of specific, quantized values, corresponding to the energy differences between the electron's lower and higher orbits. Thus, only photons having wavelengths that correspond to these energy differences can kick the electrons into excited states. Consequently, these photons are absorbed (that is, they disappear due to the kicking), which creates a distinct pattern of valleys in the original continuum spectrum. This pattern is called an *absorption spectrum*. Figure I.20 shows that absorption spectra, too, are unique for each element. They are commonly observed in the electromagnetic radiation from the Sun. They are also observed in the radiations we receive from distant stars, from galaxies, and from gas clouds lying between these distant objects and us.

If you are somewhat uneasy about this rather brief discussion of continuum, emission, and absorption spectra, do not worry. This is a difficult subject and would require several chapters and advanced physical theory to be dealt with thoroughly. The main point is that by analyzing the spectra of the electromagnetic radiation coming from stars, interstellar gas clouds, galaxies, and other distant objects, astronomers are able to deduce a great deal of information. In the next five chapters I shall discuss some of this information.

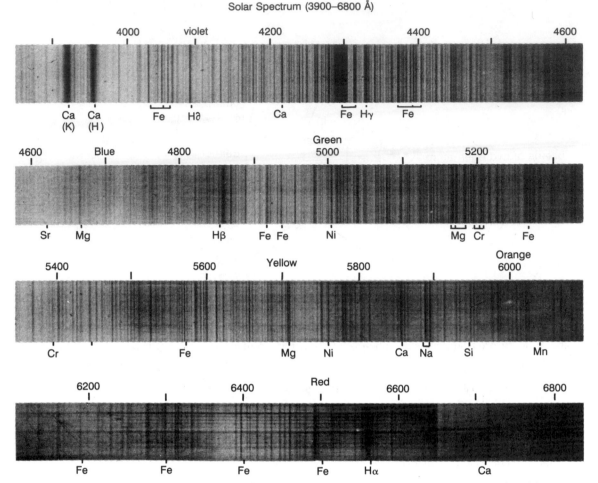

Figure I.20. The solar absorption spectrum. This spectrum is produced by continuous radiation from the Sun's interior as it passes through the solar atmosphere. Elements in the solar atmosphere selectively absorb radiation of specific wavelengths and create the pattern illustrated here. The numbers shown above the spectrum refer to wavelength in Ångstroms (Å; one Å equals 10^{-8} cm or 10^{-4} μm). The symbols shown below the spectrum refer to the elements responsible for the absorption lines. (Note the absorption lines of hydrogen [labeled Hα, Hβ, Hγ, and Hδ] and the H and K lines of calcium. They will be referred to in later chapters.)

SCIENTIFIC THEORY

In this Introduction to Part One I presented the five kinds of stable elementary particles that make up the matter and energy content of the Universe — electrons, protons, neutrons, neutrinos, and photons. I then introduced the four forces by which these particles interact with each other — the gravitational, electromagnetic, nuclear, and weak forces — and then described how protons and neutrons combine to form atomic nuclei and how these nuclei surround themselves with electrons to form atoms. Finally, I described electromagnetic radiation, which is really just another way (the wave model) of representing photons (the particle model).

Knowledge of the stable elementary particles and their interactions provides sufficient background to begin the study of the origin and evolution of the Universe. However, before we do so, let us concern ourselves with the question, "How closely do the theories discussed in this book correspond to the evolution that really happened?"

To answer this question, we need to recognize the basis of the theories being discussed. In brief, their basis is scientific knowledge that has been acquired over several centuries by many people. This knowledge is rooted in careful observations of natural phenomena as well as in countless experiments designed to help us understand specific aspects of nature. Scientists use the results of observations and experiments to construct theories that they hope will explain natural phenomena, relate different phenomena to each other, and in general contribute to an understanding of nature.

When scientific theories are first put forth, they are always tentative. They are then tested by additional observations and experiments, often carried out by many different scientists or teams of scientists and, whenever possible, under a great range of conditions. Depending on the results of the new observations and experiments, the original theories are refined, altered, or discarded.

This process of observation, experimentation, development of theories, communication of the theories to other scientists, testing, and critiquing lies at the heart of doing science. It is a very social process, one that involves many people with differing degrees of talent, expertise, background, temperament, style, and motivation. And, like all human endeavors, it progresses in very nonlinear ways. Quite often a given path turns out to be wrong, even though at first it appeared to be promising, and it needs to be either redirected or abandoned. Observational and experimental data are sometimes misinterpreted. Some individuals acquire high status and their opinions are accepted as established facts. However, misinterpretations and mere opinions usually do not survive very long in science. They do not last because scientific theories are continually reexamined, questioned in the light of new discoveries in other areas of science, compared with alternative theories, and tested through the use of new and often more sensitive observational and experimental tools. Ultimately, neither dogma nor opinion makes scientific theories acceptable. It is how well these theories correspond to what is observed in nature that determines their acceptance by the scientific community. That is why Martin Schwarzschild (1958) referred to observations as the " . . . pillars . . . on which we base our theories. . . . "

The correspondence of a given theory to nature is never perfect. At best the correspondence is limited by the unavoidable uncertainties present in all observational and experimental results. Additional uncertainties usually arise from incomplete observational and experimental data. They may also arise from our intellectual limitations in interpreting correctly the data we do have and from the particular models or biases we start with when carrying out our observations or setting up our experiments. Werner K. Heisenberg (1901–1976), one of the great physicists of our century, referred to these last sources of uncertainties when he said, "What we observe is not nature itself, but nature exposed to our method of questioning" (1958). Thus, in general, scientific theories span the whole spectrum from being very good, when they correspond extremely closely to all relevant observational and experimental data, to being rather tentative and speculative.

Many of the theories presented in this book are good ones and are generally accepted by the scientific community. But because many of the topics deal with events that took place long ago, at great distances, or under uncommon circumstances, a good part of the book's discussion is of a tentative nature. Thus, the answer to the question "How closely do this book's theories correspond to the evolution that really happened?" is that some (maybe many) of the discussions of evolution probably correspond quite closely to what really happened. In the case of others it is impossible to be so certain. No doubt, some of the theories will have to be significantly revised or even discarded as continued scientific inquiry provides new information.

In most instances in which a theory is clearly speculative, I point this out in the text. Even in cases that are not so qualified, I would like to caution the reader not to think that the theories represent the final word on the subject. Science is a multifaceted and rapidly developing human enterprise and no one can predict in which directions our understanding of nature will progress.

> If a man will begin with certainties, he shall end in doubts; but if he will be content to begin with doubts he shall end in certainties.
> **Francis Bacon** (1561–1626)
> *The Advancement of Learning, Book I*

Exercises

1 Describe how the macroscopic structure of matter, such as that of ice crystals, snowflakes, and diamonds, is determined by the microscopic properties of matter.

2 (a) Confirm through your own measurements that the density of liquid water is 1 g/cm^3.
(b) Why is this value so much less than the density of a proton or a neutron, which is roughly 10^{14} g/cm^3?
(c) Determine, also through your own measurements, the densities of a piece of wood, a rock, and a steel bolt.

3 Make a short list of applications of the gravitational and electromagnetic forces, in addition to the applications found in table I.2.

4 (a) Explain the difference between chemical and nuclear reactions.
(b) Why are nuclear reactions far more energetic, in general, than chemical reactions?

5 Explain the meanings of CO, CO_2, N_2, O_2, CH_4.

6 (a) Explain the meanings of the symbols $^3_2$He, $^4_2$He, $^{12}_6$C, $^{13}_6$C, $^{14}_6$C, $^{14}_7$N, $^{16}_8$O, $^{24}_{12}$Mg, $^{28}_{14}$Si, $^{31}_{15}$P, $^{32}_{16}$S, $^{56}_{26}$Fe, $^{197}_{79}$Au, $^{235}_{92}$U, $^{238}_{92}$U.

(b) Which of these symbols stand for isotopes of the same element, and which stand for different elements?

7 (a) Induce a wave in a taut rope. Measure the wavelength, amplitude, period (in seconds), frequency (in cycles per second), and speed of the wave.
(b) Repeat for a wave induced on a lake.

8 (a) What is meant by *electromagnetic spectrum*?
(b) Distinguish between X rays, UV radiation, IR radiation, microwaves, and radar waves, and give examples of sources of these kinds of radiation.

9 Look up the Greek alphabet in a dictionary and familiarize yourself with both its lower and upper case letters.

10 Look up the meanings of *micro-*, *milli-*, *kilo-*, and the other prefixes below in a dictionary and explain the following terms: micrometer, millimeter, centimeter, decimeter, decameter, hectometer, kilometer, megalight-year, milliliter, milligram, kilogram, nanosecond, microsecond, millisecond.

11 Use the metric units indicated in parentheses to measure: the thicknesses and diameters of United States coins (mm, cm); the width and length of a

dollar bill (cm); your height (cm, m); the heights of a few familiar mountains (m); the length of the Olympic marathon (km); the distances between several important United States cities (km); and the circumference of the Earth (km).

12 (*a*) What makes a scientific theory a "good" theory?
(*b*) How is such a theory developed?

Suggestions for Further Reading

Abell, G. O. 1982. *Exploration of the Universe.* 4th ed. New York: CBS College Publishing.* College-level astronomy text.

Berman, L., and **J. C. Evans.** 1983. *Exploring the Cosmos.* 4th ed. Boston: Little, Brown and Co.* College-level astronomy text.

Boslough, J. 1985. "Worlds Within the Atom." *National Geographic*, May, 634–63. Current research into the ultimate structure of matter with giant machines and elegant mathematics.

Chaisson, E. 1981. *Cosmic Dawn: The Origins of Matter and Life.* Boston: Little, Brown and Co.* An enjoyable book for the general reader, covering much of the same material as this one.

Cloud, P. 1978. *Cosmos, Earth, and Man: A Short History of the Universe.* New Haven and London: Yale University Press.* An excitingly written book by an eminent geologist and paleontologist about what he has "learned from a lifetime of research, reading, and reflection about our planet, its history and cosmic connections, the development of life on it, and its capability to sustain our descendants."

Feinberg, G. 1967. "Ordinary Matter." *Scientific American*, May, 126–34. Ordinary matter consists of only three particles: electrons, protons, and neutrons. In addition to explaining the meaning of this statement, the article follows the history of human thought that has led to this contemporary model of matter.

Georgi, H. 1981. "A Unified Theory of Elementary Particles and Forces." *Scientific American*, April, 48–63. The author explains that at a range of 10^{-29} cm the world may be much simpler than the one we experience, with just one kind of elementary particle and one important force.

Gribbin, J. 1981. *Genesis: The Origins of Man and the Universe.* New York: Delacorte Press. A book covering roughly the same topics as this one, in a style suitable for casual reading.

Hallett, J. 1984. "How Snow Crystals Grow." *American Scientist*, November–December, 582–89.

Huens, J.-L., and **T. Y. Canby.** 1974. "Pioneers in Man's Search for the Universe." *National Geographic*, May, 626–33.* Illustrated biographies of Copernicus, Galileo, Kepler, Newton, Herschel, Einstein, and Hubble.

Learned, J. G., and **D. Eichler.** 1981. "A Deep-Sea Neutrino Telescope." *Scientific American*, February, 138–52.

McCusker, B. 1983. *The Quest for Quarks.* Cambridge University Press. A short, popularly written account of elementary particle structure, in particular about the question, "Are quarks the fundamental building blocks of protons and neutrons?"

Murdin, P., and **D. Allen.** 1979. *Catalogue of the Universe.* New York: Crown Publishers.* An atlas of exquisite photographs of the Solar System, the Milky Way, and galaxies, including up-to-date and concise descriptions.

Pauling, L., and **R. Hayward.** 1964. *The Architecture of Molecules.* San Francisco and London: W. H. Freeman and Co. A beautifully illustrated book about atoms, the periodic table, and many of the molecules and larger structures into which atoms can be arranged. The book is easy to read, topically arranged, and particularly suitable for the chemistry novice.

Quigg, C. 1985. "Elementary Particles and Forces." *Scientific American*, April, 84–95.

Shapiro, S. L., R. F. Stark, and **S. A. Teukolsky.** 1985. "The Search for Gravitational Waves." *American Scientist*, May–June, 248–57.

Struve, O., and **V. Zebergs.** 1962. *Astronomy of the 20th Century.* New York: The MacMillan Co.* A comprehensive account of astronomical discoveries between 1900 and 1960. The book is written in a lively style and reflects the touch of Otto Struve (1897–1963), one of the foremost astronomers of the

*The asterisk denotes books and articles that are general references for Part One.

twentieth century. A must for anyone interested in the history of astronomy.

Zeilik, M. 1985. *Astronomy: The Evolving Universe*. 4th ed. New York: Harper & Row, Publishers.* College-level astronomy text.

Additional References

Heisenberg, K. 1958. *Physics and Philosophy*. New York: Harper & Row.

Schwarzschild, M. 1958. *Structure and Evolution of the Stars*. Princeton: Princeton University Press.

Wald, G. 1954. "The Origin of Life." *Scientific American*, August, 44–53.

Figure 1.1. The central part of the Coma cluster of galaxies, at a distance of about 450 million LY.

1

Cosmology

H OW DID THE UNIVERSE begin? How will it end? Did it, in fact, have a beginning and will it have an end? What does it contain? What is our place—both in space and in time—in the Universe? These are questions humans have been asking in one way or another for millenia. They have asked them because the answers offer a perspective on how we—humans and the rest of Earth-bound life—fit into the larger scheme of things. They give meaning and content to our existence.

The answers found over the ages are expressed in art, myths, and religions. They have become part of our cultural heritage. In this and the chapters to come, we, too, shall seek answers to the above questions. We shall do this by turning to modern science, which is the product of our own culture. In the present chapter, we shall turn to *cosmology*, the *study of the large-scale structure and evolution of the Universe. Large-scale structure* means that this chapter is concerned with the cosmic distribution of matter, in which clusters of galaxies measuring many millions of light-years across are the basic units. Details on a finer scale, such as individual galaxies, stars, planets, and life are left for other branches of science and will be discussed in later chapters. *Evolution of the Universe* means that this chapter will try to explain how the distribution of the matter changes with time. In particular, we shall seek answers in this chapter to questions about the origin of the Universe and about its eventual fate.

In our study we shall find that our galaxy does not occupy any special place in the Universe—it is just one of billions of galaxies. The natural laws that operate in our part of the Universe and that have made life on Earth possible are, to the best of our knowledge, the same everywhere, and they probably have been so almost since the beginning. Our bodies are made of the same kind of matter and are subject to the same forces as the stars, the galaxies, and the Universe at large. In short, we believe that the laws of nature are universal and that they apply equally to the very near and very far, the small and large, animate and inanimate.

THE STRUCTURE OF THE UNIVERSE

To begin the study of cosmology, this section describes the large-scale distribution of matter in the Universe as it is seen today. How the matter originated and evolved will be discussed in later sections. Let us start by taking an imaginary journey from Earth to the distant galaxies and see what we find along the way. We will pretend that we travel in a spaceship that moves at the speed of light. Traveling at that speed allows us to tell how far we have come just by looking at a clock. Every second of time we will cover a distance of one light-second, which equals 300,000 km or seven and a half trips around the Earth. Every year we will cover a distance of one light-year (abbreviated LY), which equals 9.5×10^{12} km or 32,000 round trips from the Earth to the Sun. This imaginary journey will last billions of years of time and it will take us across billions of light-years of space.

The Milky Way Galaxy

Let us begin the journey. Within four minutes of lift-off from Earth we are already crossing the orbit of Mars and ten minutes later we are passing through the

Figure 1.2. The open double-star cluster h and Chi Persei, at a distance of about 7000 LY from Earth. Each of the two clusters contains close to 300 stars.

Asteroid Belt, between Mars and Jupiter. Soon the giant outer planets — Jupiter, Saturn, Uranus, and Neptune — briefly fill our viewing window. And in just a little over five hours we have left all of the Sun's nine planets behind. During the next few weeks and months we sweep through the outer fringes of the Solar System, coming upon an occasional comet and other debris left over from the Sun's birth. All along, the Sun has been growing fainter and is now just one of many bright stars in the sky.

It will be four years before we encounter the first stars beyond the Sun. They are a triple star system — Proxima Centauri and Alpha Centauri A and B. Alpha Centauri A resembles the Sun in size and temperature, while the other two are somewhat smaller and cooler. Every few years we pass another star. They brighten up as we approach them and then fade in the distance. After just 50 years we lose sight of our Sun. It has become too faint to be seen with the unaided eye.

The centuries and millenia of our journey tick away. We are passing thousands upon thousands of stars. Many are alone, others are double and triple star systems, and still others are grouped in clusters (figures 1.2 and 1.3). Some of the stars are very young and are still surrounded by the clouds of gas and dust (figure 1.4) from which they were born. Others are old and nearing the end of their active lives. Occasionally we pass a star that pulsates or one that explodes, spewing gas in all directions and momentarily blinding us with its brilliance.

Figure 1.3. The globular star cluster Messier 13, in the constellation Hercules. This beautiful cluster lies at a distance of about 25,000 LY and is the largest globular cluster in the northern sky. It contains perhaps half a million stars.

Figure 1.4. The Horsehead nebula. This dark and cold interstellar cloud of dust, in the shape of a horse's head and neck, protrudes into the bright emission nebula known as IC 434. It is about 1000 LY distant and lies in the constellation Orion.

Eventually the concentration of stars lessens and we break out of the obscuring layer of gas and dust associated with them. Looking back we see an enormous number of stars, star clusters, and clouds, arranged in a huge but very flat disk-like distribution. At its center the disk thickens into a very prominent and bright spherical bulge, like the hub of a giant wheel. From the bulge a distinct spiral pattern of luminous, dark matter winds outward to the edge of the disk. The luminous matter consists of clouds of gas brightly lit by massive stars, while the interspersed dark patches are cold, light-absorbing interstellar dust.

What we are seeing is the Milky Way galaxy—our home in the vastness of intergalactic space. It is a *spiral galaxy* like many others that we shall encounter on our journey. It measures about 100,000 light-years across and contains close to 150 billion stars. Apart from its central bulge or nucleus, its most distinctive feature is the pattern of spiral arms. In fact, our journey began from an inconspicuous spot at the inner edge of one of them, roughly 30,000 light-years from the center of the Galaxy.

The Local Group

As we leave our galaxy, we still pass an occasional star or star cluster, whose motions carry them far above the Galaxy's disk. But most of the Milky Way lies behind us. We are entering intergalactic space.

Ahead of us lies the Universe filled with many billions of galaxies. Each is like an island universe, separated from the others by many light-years of emptiness. Many of the galaxies are spirals. Others have spherical or elliptical structures and are called *elliptical galaxies*. Still others look highly asymmetric and distorted; they are known as *irregular galaxies*. The galaxies are not distributed randomly through space, but are grouped together in small and large clusters. The small clusters contain up to a few dozen galaxies, while the large clusters have thousands of them.

The Milky Way belongs to a small cluster of galaxies, known as the Local Group, that contains some thirty members. The two nearest members are the Large and Small Magellanic Clouds (figure 1.5), and it is in their direction that we are now steering our ship. Both clouds are composed of young blue stars, bright gaseous nebulae, and dark dust lanes, the same matter that dominates the spiral arms in our galaxy. However, that matter is not distributed in well-defined spiral arms. Powerful forces seem to distort it into rather disorganized shapes. In fact, the Small Magellanic Cloud appears to have been torn into two misshapen components, perhaps by a collision with the Large Magellanic Cloud. Furthermore, an extended stream of hydrogen gas trails behind the clouds, as if it had been pulled out by the forces of such a collision and by tidal interactions with the Milky Way. Because of the clouds' disorganized shapes, astronomers have classified the Magellanic Clouds as irregular galaxies. From Earth they are visible only from the Southern Hemisphere and Europeans did not become aware of them until they began their worldwide explorations in the sixteenth century. The two clouds were named in honor of the Portuguese sea captain Ferdinand Magellan (1480–1521), who led the first circumnavigation of the Earth.

Beyond the Magellanic Clouds and after roughly 750 thousand years we pass two dwarf elliptical galaxies—Leo I and Leo II (figure 1.6). They impress us with their near perfect symmetry and uniformity. Gas and dust seem to be absent in them, and their light comes entirely from faint, old stars. As our journey continues, we will find that by far the largest number of galaxies in the Universe are dwarf ellipticals.

We are now leaving the Milky Way's immediate neighborhood (see figure 1.7) and are changing course toward the most spectacular galaxy in the Local Group, the great spiral galaxy in Andromeda (figure 1.8). We pass this galaxy in about 2 million years. From Earth it appears as a small and faint haze of light when viewed with the unaided eye. However, from close up we are able to see many details that remind us of the Milky Way: an almost spherical and very bright nucleus, a great disk of stars, gas, and dust, and an impressive pattern of spiral arms. Like the Milky Way, it also has two nearby neighbors, the small elliptical galaxies M 32 and NGC 205.*

Figure 1.5. The Large (left) and Small Magellanic Clouds. These two irregular galaxies, at distances of roughly 160,000 LY and 200,000 LY, respectively, are our closest galactic neighbors.

*M 32 and NGC 205 are labels astronomers use to refer to galaxies and other celestial objects. M 32 is the 32nd object listed in the *Messier Catalogue*, compiled by the eighteenth-century French astronomer Charles Messier (1730–1817). NGC 205 is the 205th object listed in the *New General Catalogue of Nebulae and Clusters of Stars*, compiled at the end of the nineteenth century by the Danish astronomer Johan L. E. Dreyer (1852–1926).

Figure 1.6. Galaxy Leo II. This dwarf elliptical galaxy (of type E1; see also figure 2.10) at a distance of about 750,000 LY is highly resolved into stars and except for its greater size resembles a globular star cluster.

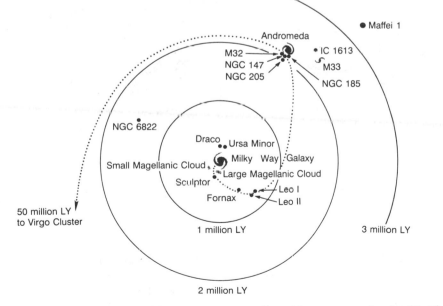
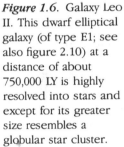

Figure 1.7. Flight path of the cosmic journey (dotted line; see text for details). The diagram illustrates the layout of the Local Group of galaxies in two dimensions. The third dimension—into and out of the page—is not shown.

Figure 1.8. The Andromeda galaxy. This galaxy (of type Sb) is tilted so that its outlines appear elliptical. Near its upper edge lies M 32, and slightly below it lies NGC 205. M 32 and NGC 205 are two dwarf ellipticals that are close companions of the Andromeda galaxy.

The Virgo Cluster

After we have inspected at close range the Andromeda galaxy, M 32, NGC 205, and several other galaxies in the same vicinity, we change course once again and leave the Local Group. Our next destination is the nearest large cluster of galaxies, the Virgo cluster. This cluster is still some 50 million light-years distant, and along the way we come upon several other smaller clusters similar to the Local Group. Each of them contains one or two large galaxies about which other smaller ones are gathered.

One of the first galaxies we pass is NGC 3115 (figure 1.9), a highly flattened elliptical. It closely resembles a spiral galaxy, except that gas and dust and, therefore, arms are absent. After roughly 16 million years we fly past a very unusual galaxy, Centaurus A. It has the spherical shape of an elliptical galaxy, but it also contains a disk of gas and dust. From Earth we see the disk edge-on, making the dark, light-absorbing dust contrast sharply against the bright stellar component (figure 1.10). A number of violent explosions have occurred in the recent past in the nucleus of Centaurus A, ejecting jets of gas at up to 5% of the speed of light. A pair of these jets, tossed out in opposite directions roughly 30 million years ago, are powerful sources of radio emission. Centaurus A's nucleus is also a strong and variable emitter of X rays, indicating that violent events are still taking place today. Astronomers suspect that the source of the radiation and

Figure 1.9. Galaxy NGC 3115. This highly flattened galaxy (of type E7 or type S0), at a distance of about 13 million LY, has a mass distribution intermediate between that of ellipticals and spirals.

Figure 1.10. Centaurus A (NGC 5128). This unusual galaxy lies at a distance of about 16 million LY, and is the nearest of the giant radio galaxies.

of the explosions is a gigantic *black hole** at the very center of the galaxy, which periodically tears up passing stars and draws their matter into its bottomless gravitational well.

After about 45 million years the frequency of galaxies increases as we approach the Virgo cluster (figure 1.11). This cluster contains thousands of members. They are of all sizes and types — spirals, ellipticals, irregulars — and they are rather randomly scattered about. Near the cluster's center lie several giant elliptical galaxies. They form the gravitational hub of the Virgo cluster, about which the other galaxies trace their orbits.

The largest of the elliptical galaxies near the center of the Virgo cluster is M 87 (figure 1.12). It is an impressive and remarkably symmetric galaxy, containing close to one hundred times as many stars as the Milky Way. This makes it the most massive galaxy known. The radiation emitted from the main body of M 87 is so intense that few details can be discerned. However, toward the outer fringes of the galaxy, the light gradually fades revealing a great number of star clusters. Each contains hundreds of thousands to millions of stars and resembles the globular star clusters in our galaxy (more about them in the following chapter). Short exposure photographs of the nucleus of M 87 show a jet-like structure, with a faint indication of a second one protruding in the opposite direction (see figure 2.19). The jets must have resulted from an explosion of enormous magnitude. Is the source of this violence a black hole, like the one

---

*A black hole is a star that collapsed under its own gravity. Its surface gravity is so strong that nothing, including light, can escape. Matter can, however, fall into it, releasing enormous amounts of energy. Black holes will be discussed in more detail in the next two chapters.

Figure 1.11. A part of the Virgo cluster of galaxies. With thousands of members, this cluster is the nearest rich cluster of galaxies and is believed to be the gravitational hub of the local supercluster of galaxies.

Figure 1.12. Galaxy M 87. This giant elliptical galaxy (of type E0), with a nearly perfectly spherical structure, is the brightest of the galaxies in the Virgo cluster and, like Centaurus A, a powerful emitter of radio waves.

believed to form the center of Centaurus A? Many astronomers are leaning toward this interpretation. They base their opinion on evidence that within less than 300 LY of M 87's very center exists a mass equal to at least 5 billion suns. Only a black hole can contain so much mass in such a small volume and also provide the energy to power the observed events.

It takes us well over 10 million years to thread our way through the heart of the Virgo cluster. Besides the kinds of galaxies already described, we pass some that are colliding or have recently (that means during the past billion or so years) undergone collisions. Many of these galaxies are grotesquely distorted by the gravitational forces they exert on each other. Often they are connected by long, thin bridges of gas and dust that they have torn off each other. Others are surrounded by gaseous and dusty rings. Some spiral galaxies are stripped entirely of their gas and dust by these interactions and have become what are known as *SO galaxies* (see figure 2.10). The Virgo cluster is a world of great fascination and violence.

As we travel on, the concentration of galaxies eventually begins to thin again and, as earlier, we encounter only now and then a small cluster of galaxies. About 100 million years after leaving the Milky Way their frequency diminishes still

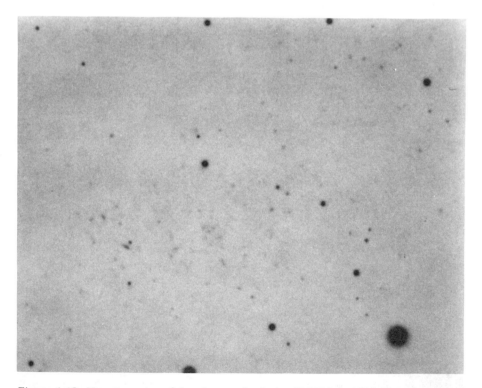

Figure 1.13. Negative print of the cluster of galaxies CI 0024 + 1654. This cluster is one of the most remote ones known, at a distance of about 7 billion LY. The light received on Earth today from this cluster was emitted roughly 2.5 billion years before the birth of the Solar System. Due to their great distance from Earth, the individual galaxies are barely distinguishable from the images of stars within our galaxy.

further and we find ourselves in mostly empty space. We have left our local cosmic neighborhood.

Superclusters of Galaxies

Looking back we see that we have come from a huge *supercluster of galaxies*, measuring more that 100 million light-years across. Its center is dominated by the Virgo cluster with its rich concentration of galaxies. Surrounding it like a halo is a sparse distribution of smaller clusters, one of which is the Local Group. These small clusters are weakly tied to the Virgo cluster by gravity and circle in giant orbits around it. The whole system contains millions of galaxies of all sizes and types.

The Virgo supercluster is not alone in the Universe. As our journey continues, we encounter such superclusters of galaxies again and again. The Coma cluster (figure 1.1) and Cluster CI 0024 + 1654 (figure 1.13) are just two examples. No matter how far we travel or in which direction, the pattern remains similar. On the average, we cross another supercluster every 100 to 200 million years, although occasionally we see stretched-out aggregates of clusters of galaxies that are two or three times that size. The superclusters of galaxies form the basic structural units of the Universe, and between them exist great voids nearly as vast as the superclusters themselves.

Let us end our journey at this point and summarize what we have discovered: The matter of the Universe is concentrated in large superclusters of galaxies, each measuring, on the average, 100 to 200 million light-years across and containing millions of galaxies. There are differences in the exact number of galaxies, in the individual appearance of these galaxies, and in their spatial arrangement. However, if we disregard such fine-scale details, we find that the superclusters are very much alike. Furthermore, they are rather evenly distributed through space, as the map of nearly a million galaxies seen in the northern sky (figure 1.14) illustrates. We therefore conclude that on the largest scale the Universe is homogeneous and isotropic (*homogeneous* means that the Universe has the same structure and composition everywhere; *isotropic* means that it looks the same in all directions).

The Fourth Dimension

Our imaginary journey to the distant galaxies has left unanswered some compelling questions, such as "Where is the center of the Universe?", "How far is its edge?" and "Where exactly are we located?" We did not find answers to such questions because there aren't any. Such questions are perfectly valid when asked in the context of the three-dimensional Earth environment. However, our local experiences do not necessarily apply to the large-scale structure of the Universe, and frequently they lead us completely astray. In the study of the Universe as a whole, a fourth dimension — time — must be included along with the three spatial dimensions. The above questions then lose their meaning, for it turns out that in four-dimensional space–time the Universe has neither a center nor an edge. It only has a beginning in time and, possibly, an end. Some of these bewildering features of the Universe are discussed later in this chapter.

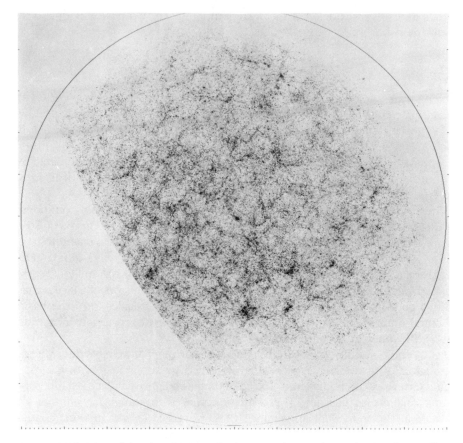

Figure 1.14. Map of the distribution of galaxies in the northern sky. Nearly a million galaxies are included. The map illustrates that on the large scale the distribution is uniform, though it shows clumpiness where galaxies are concentrated in clusters.

THE EXPANSION OF THE UNIVERSE

Up to 500 years ago, most humans believed in a geocentric Universe, with Earth at the center and the Sun, the planets, and the stars circling around it. The Polish astronomer Nicolaus Copernicus (1473–1543) revolutionized that point of view by placing the Sun at the center of the Universe and making Earth one of the planets. In the following centuries, astronomers gradually realized that the Sun is just one star among many millions in the Milky Way and that it is far from the center of the Galaxy. The prevailing view was that the Milky Way galaxy comprised the entire Universe. However, by the eighteenth century a few scientists and philosophers, such as William Herschel (1738–1822) in England and Immanuel Kant (1724–1804) in Germany, dared to consider that the Universe may extend beyond these limits. They suggested that the many faint nebulae visible through telescopes might be separate galactic systems.

Only in the early part of the twentieth century, with the development of modern reflecting telescopes with large light-gathering mirrors, were astronomers able to estimate accurately the distances to those nebulae. The new telescopes allowed detailed study of individual stars in many nebulae. From the faintness of their light, astronomers calculated that they were at distances of hundreds of thousands to many millions of light-years, which places them far beyond the Milky Way. This proved beyond the shadow of a doubt that these nebulae were separate galaxies* and that the Universe was much bigger than anyone had imagined. Despite some initial dispute among astronomers concerning this conclusion, by the mid–1920s it was generally accepted and modern cosmology was born.

The Hubble Constant

Among the pioneers who participated in these discoveries was the great American astronomer Edwin P. Hubble (1889–1953). After the true nature of the galaxies had been established, he turned his attention to their motions. He was motivated to do this by a puzzling report by Vesto M. Slipher (1875–1969) of the Lowell Observatory in Arizona that many of the faint nebulae are moving away from us with velocities of hundreds and thousands of kilometers per second (a velocity of 1000 km/sec equals 3.6 million km/hr or 2.2 million mi/hr). That seemed peculiar, for the stars in the Milky Way have velocities much smaller than that and some move away while others move towards us. In contrast, the velocities of nearly all galaxies, with the exception of a few nearby ones, are always directed away from us. Slipher had made his observations during the 1910s, before anyone was certain that these nebulae were really galaxies far beyond the Milky Way. Hence, he did not know what to make of their unusual velocities. However, by the mid–1920s Hubble did know that the galaxies were separate galactic systems and he began a systematic study of the relation between their velocities and their distances.

Working with the new 2.5-meter (100-inch) telescope at the Mount Wilson Observatory in southern California (figure 1.15), Hubble and his colleague

Figure 1.15. The 2.5-meter (100-inch) Hooker telescope at the Mount Wilson Observatory, California. This telescope, completed in 1917, was the first of the large reflectors that contributed to the modern perspective of the structure and evolution of the Universe.

*Upon the discovery that many of the faint nebulae were separate stellar systems beyond the Milky Way, astronomers began calling them *galaxies*. The word *nebula* is now used solely to refer to clouds of gas and dust that, along with stars and star clusters, are the major components of the spiral and irregular galaxies.

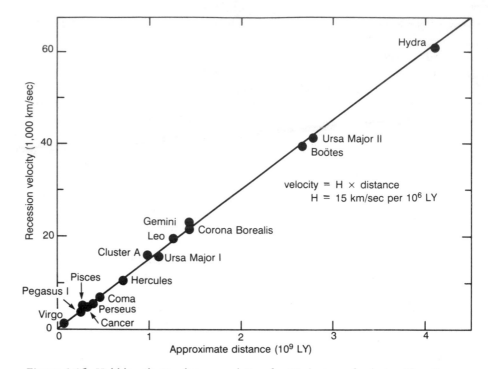

Figure 1.16. Hubble velocity–distance relation for 15 clusters of galaxies. The alignment of the data along a nearly straight line implies that recession velocity varies regularly with distance. For instance, galaxies at 1 billion LY have a recession velocity of 15,000 km/sec, those at 2 billion LY have a recession velocity of 30,000 km/sec, and so forth.

Milton L. Humason (1891–1972) let the faint light of distant galaxies pass through a spectrograph, an instrument that splits electromagnetic radiation into its spectral components (some spectrographs use prisms to accomplish this, see figure I.16). In almost every case they found that the light was shifted toward the color red. This meant that the galaxies are receding from us and confirmed Slipher's earlier result. (How recession velocities follow from redshifts is explained in box 1.1.) Furthermore, Hubble and Humason noticed that these recession velocities vary regularly with distances: Galaxies that have two, three, or ten times greater recession velocities than nearby ones are approximately two, three, or ten times farther away. A plot of recession velocity versus distance (see figure 1.17) clearly shows that, on the average, for every increase in distance by 1 million light-years there exists a corresponding increase in velocity by 15 km/sec. Today astronomers refer to the rate of increase of recession velocity with distance as the *Hubble constant* or, more briefly, as *H*.*

---

*Hubble and Humason's estimate for *H* was 170 km/sec per million LY. They obtained such a large value because in the 1920s and 1930s the distances to the galaxies were underestimated by roughly a factor of ten. The distances to the galaxies are still uncertain today and the modern estimate of *H* reflects this. It lies in the range from 12 to 30 km/sec per million LY. In this book, I have adopted the value of 15 km/sec per million LY.

BOX 1.1.
The Doppler Effect: An Explanation of the Cosmological Redshift

THE cosmological redshift refers to the displacement of light from distant galaxies toward longer wavelengths, as discovered by Vesto M. Slipher in the 1910s. Edwin P. Hubble interpreted this displacement as being due to recession velocities of the galaxies, and virtually all astronomers today agree. The larger the redshift, the faster the galaxies are receding.

The same phenomenon is observed in the case of stars in our Galaxy, except with them the shift in color is smaller because their motions relative to us are smaller. Furthermore, in the case of stars, both redshifts and blueshifts (that is, displacements toward shorter wavelengths) are observed. Stars whose light is redshifted move away from us and stars whose light is blueshifted move toward us. A striking example of this can be observed in the case of binary star systems, in which two stars orbit around a common center of mass and alternately approach and recede along our line of sight. As a result, the light from these stars is alternately blue- and redshifted, as was first discovered by the Austrian physicist Christian Johann Doppler (1803–1853). In his honor, red- and blueshifts of light, due to the motion of the source, are now known as the *Doppler effect*.

The Doppler effect is not limited to light, but occurs with all wave phenomena in which the source and observer are in relative motion along the line joining them. One example of the Doppler effect that we have all experienced is the shift in pitch of the whistle of a train passing at high speed. During the approach, the whistle has a higher pitch than normal. After the train has passed and is moving away, the pitch is lower than normal.

What is happening is this: Sound is a perturbation of the air and consists of pressure waves that spread out in all directions from a source. The spreading of a sound wave may be compared with the spreading of the circular rings of water waves from some source of perturbation on a lake, such as a rock thrown into the lake. The major difference between water and sound waves is the following: Water waves consist of up and down motions of the water surface. These motions are *transverse* to the direction in which the waves propagate and, hence, the water waves are called *transverse waves*. Electromagnetic waves are another example of transverse waves. Sound waves consist of back and forth motions of air molecules *along* the direction of wave propagation. Hence, sound waves are called *longitudinal waves* (see figure A).

Figure A. Comparison of transverse and longitudinal waves.

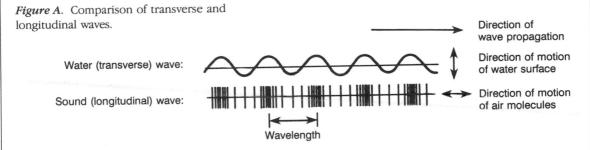

Water (transverse) wave:

Sound (longitudinal) wave:

Direction of wave propagation

Direction of motion of water surface

Direction of motion of air molecules

Wavelength

The pitch of sound is related to the rate at which the high pressure crests of the waves pass us. If they pass at a high rate, we hear a high pitch; if they pass at a low rate, we hear a low pitch. If the source of sound is moving toward us, successive pressure crests have less far to travel to reach us. Hence, the wavelengths are shortened and successive pressure crests pass us at a higher rate than those from a source at rest relative to us. That is why we hear the whistle of an approaching train at a higher pitch than normal. If

the source of sound is moving away from us, successive pressure crests have further to travel to reach us. Hence, the wavelengths are stretched and successive pressure crests pass us at a lower rate than those from a source at rest relative to us. That is why the whistle of a receding train has a lower pitch than normal (see figure B).

The red- and blueshifts of light are subject to the same physical explanation as the one just given for sound. If a light source approaches, the wavelengths

BOX 1.1   continued

Figure B. The Doppler effect in the case of sound. At top, the train and the observers are standing still; both observers hear a normal pitch. At bottom, the observers stand still but the train moves. The observer whom the train is leaving hears a lower pitch; the observer whom the train approaches hears a higher pitch.

are shortened and we perceive a blueshift. If a light source recedes, the wavelengths are stretched and we perceive a redshift. This explanation of red- and blueshifts of light was first given by the French physicist Armand H. L. Fizeau (1819–1896).

The cosmological redshift is illustrated in figure C, a composite photograph of the spectra of light from five galaxy clusters. For instance the shift in wavelength of light from the Hydra cluster corresponds to a recession velocity of 61,000 km/sec and, assuming $H = 15$ km/sec per million LY, to a distance of 4 billion light-years. Light from the most distant quasar discovered to date has a wavelength shift corresponding to a recession velocity of 92% the speed of light and to a distance of approximately 12 billion light-years (see chapter 2). The most extreme example of the cosmological redshift is that of the cosmic background radiation. Today we observe this radiation at wavelengths of roughly 0.1 cm. However, when that radiation decoupled from matter (see text), its wavelengths were approximately 1000 times shorter. They were Doppler-shifted by the cosmic expansion of the Universe.

Cluster nebula in	Distance in Million LY	Redshifts
Virgo	50	1,200 km/sec
Ursa Major	1,000	15,000 km/sec
Corona Borealis	1,400	22,000 km/sec
Boötes	2,500	39,000 km/sec
Hydra	3,960	61,000 km/sec

Figure C. The relationship of recession velocities and distances for selected galaxies in clusters. The recession velocities are deduced from the redshifts of the H and K lines of calcium in the spectra of the galaxies. Distances are deduced from the faintness of the galaxies in each cluster.

The Big Bang

The relation between recession velocities of galaxies and their distances has powerful implications for the origin and evolution of the Universe. It suggests that the entire Universe is in a state of expansion. This in turn implies that in the past the Universe was much more compact than it is today and that, possibly, at some long-ago time it came into existence in one gigantic explosion. The explosion sent matter flying in all directions. Matter that initially acquired large velocities relative to us is consequently very distant from us today, and matter that acquired a small velocity is comparatively still quite close. The situation may be compared with the explosion of a grenade. The energy of the explosion breaks the grenade into many fragments, some of which acquire high velocities and others low velocities. As the fragments fly apart, the ones with high velocities fly farther than those with smaller velocities. This analogy should, however, not be taken too literally, for there exists a definite center in the case of the explosion of the grenade. There is no such center in the case of the Universe (more about this apparent paradox later).

Observers in other galaxies (if, indeed, there are any) see the expansion of the Universe as we do, except from their vantage point it is our galaxy that is moving away from them. And the more distant we are from them, the faster they see the Milky Way receding. Hence, they share our impression that the Universe was born in a gigantic explosion — the *Big Bang*.

Age of the Universe: First Estimate

The relation between the velocities and distances of galaxies allows us to estimate the age of the Universe by the following reasoning. From the Hubble constant and also from the graph of figure 1.16 we know that, on the average, galaxies at a distance of 100 million light-years move away from us with a velocity of 1500 km/sec, those at 1 billion light-years with 15,000 km/sec, and so forth. If, for example, we assume that galaxies at 100 million light-years have been receding from us with a velocity of 1500 km/sec since the Big Bang, we can compute how long it took them to reach that distance:

$$\text{time} = \frac{\text{distance}}{\text{velocity}} = \frac{100 \text{ million LY}}{1500 \text{ km/sec}} = 20 \text{ billion years.}^*$$

Thus, we may conclude that the Universe was born approximately 20 billion years ago.

A later section will show that this simple arithmetic leads to overestimating the age of the Universe. Gravity is continuously slowing down the expansion of the Universe; hence, the value of the Hubble constant — that is, the rate of the Universe's expansion — must have been greater in the past than it is today. Consequently, the Universe probably reached its present size in less than 20 billion years.)

---

*This calculation is the same as that used to find how long it takes to drive 180 miles at 60 mi/hr:

$$\text{time} = \frac{\text{distance}}{\text{velocity}} = \frac{180 \text{ mi}}{60 \text{ mi/hr}} = 3 \text{ hr.}$$

The only added complication in the astronomical calculation is that light-years need to be converted to kilometers and seconds to years (see box I.2 for a list of conversion factors).

BOX 1.2.
The Meaning of Temperature

TEMPERATURE is a physical parameter that expresses how hot or cold a substance is. It also gives an indication of how much heat energy a substance possesses. The higher the temperature, the greater is the heat energy.

What is meant by heat energy? If the substance is a solid, the atoms and molecules are tightly bonded to each other. However, the atoms and molecules are not at rest, but oscillate back and forth relative to each other. The heat energy is the sum of the oscillation energies of all of the atoms and molecules that make up the solid. If the substance is a liquid or a gas, in which the atoms and molecules are free to move about, the heat energy is the sum of the energies that the atoms and molecules possess due to their various kinds of motions—translational motions (that is, motions from one place to another) and, in the case of molecules, rotations and internal oscillations. (Atoms are too small to have measurable motions of rotation and internal oscillation; but molecules may rotate and oscillate like weights connected by springs.)

To illustrate the correspondence between heat energy, temperature, and the structure of matter, consider what happens when we heat or cool a familiar object, namely an ice cube. At temperatures below the freezing point of water, the water molecules of the ice cube, which are tightly bonded together into a solid configuration, oscillate back and forth. The closer the temperature is to the melting point, the faster are the oscillations and the greater is the heat content of the ice cube. At the melting point, the oscillations are so strong that the bonds between neighboring molecules break. If we continue to add heat, molecule after molecule breaks free and the water changes from the solid to the liquid state. We say "The ice melts."

The water molecules of liquid water are not bonded to each other as they are in ice. They are free to move about, collide with each other, bounce off each other, speed up and slow down, rotate, and oscillate. They do this in a random, helter-skelter manner, which gives liquid water its fluid properties. If we heat the liquid water and raise its temperature, the various kinds of motions of the water molecules become faster. Eventually they become so fast that the water molecules form gaseous bubbles that rise to the liquid's surface (because of the bubbles' lower densi-

ties), and from there they fly off into the surrounding atmosphere. The water changes from the liquid to the gaseous state (that is, into steam), a process we call boiling. In the gaseous state, the water molecules are separated much farther from each other and move about even more freely than in the liquid state.

If we heat the steam, the molecular motions become still faster and the temperature climbs. Eventually, the collisions between the water molecules become so violent that they break up into their constituent hydrogen and oxygen atoms—the gaseous water changes into hydrogen gas and oxygen gas. If we continued to add heat, we would eventually reach the point beyond which even the atoms cannot survive. The collisions would knock the electrons off the atomic nuclei and produce a gas of free electrons and ions. This kind of ionization, due to thermal motions, occurs at temperatures corresponding to stellar atmospheres and interiors.

The process just described—from the melting of the ice cube to the boiling of water, the breakup of the water molecules into atoms, and the ionization of the atoms—requires the input of heat energy and occurs at progressively higher temperatures. If we go in the opposite direction by cooling the ice cube (that is, by taking away heat energy), the oscillations of the water molecules become slower and the temperature drops. Eventually, the molecules reach the state at which they oscillate at their lowest possible frequency (*note*: the oscillations can never come entirely to rest). According to well-tested laws of quantum mechanics, the oscillations cannot be slowed down further and no more energy can be taken away from the ice. The temperature has now reached its lowest possible value, called the *absolute zero point*.

Temperature is usually measured in degrees according to scales whose zero points and gradations are defined arbitrarily. The most common temperature scales are the Celsius, Kelvin, and Fahrenheit scales.* The Celsius scale (formerly called the centigrade scale) is used in most parts of the world and in

---

*The three temperature scales are named after Anders Celsius (1701–1744, a Swedish astronomer), Lord Kelvin (1824–1907, a British physicist), and Gabriel D. Fahrenheit (1686–1736, a German physicist).

BOX 1.2 continued

many scientific applications; the Kelvin scale is preferred by physical scientists; and the Fahrenheit scale is used mainly in the United States (see figure A).

On the Celsius scale the zero point (written as 0° C) corresponds to the melting point of ice at sea level and 100° C corresponds to the boiling point of water at sea level. Absolute zero occurs at −273° C.

The Kelvin scale has the same size gradations as the Celsius scale, but its zero point is different. It coincides with absolute zero (0 K), and that is why this scale is preferred by physical scientists. It not only simplifies calculations involving temperature, but in this scale zero degrees refers to the lowest possible temperature to which material objects can be cooled. Degrees Kelvin (T_K) and degrees Celsius (T_C) are related according to the formula

$$T_K = T_C + 273.$$

At sea level the melting point of ice and boiling point of water occur at 273 K and 373 K, respectively.

On the Fahrenheit scale, the melting and boiling points of water fall at 32° F and 212° F, respectively. The difference between these two points is 180 degrees, which means that one degree Fahrenheit equals 5/9 degree Celsius or Kelvin (5/9 = 100/180). As a consequence, degrees Fahrenheit (T_F) and degrees Celsius (T_C) are related by the formula

$$T_C = \tfrac{5}{9}(T_F - 32).$$

Using this formula, we find that the zero point of the Fahrenheit scale (the lowest temperature Gabriel Fahrenheit could reach with a mixture of ice and common salt) equals −17.8° C, absolute zero occurs at −459° F, the normal human body temperature of 98.6° F equals 37° C, and −40° F is the same as −40° C.

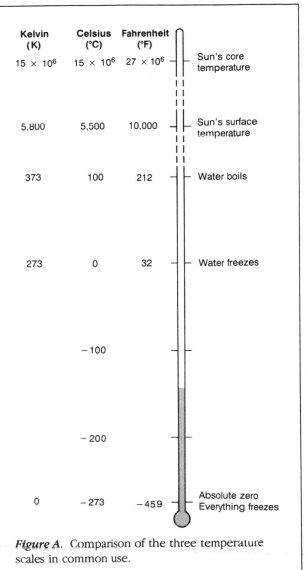

Kelvin (K)	Celsius (°C)	Fahrenheit (°F)	
15×10^6	15×10^6	27×10^6	Sun's core temperature
5,800	5,500	10,000	Sun's surface temperature
373	100	212	Water boils
273	0	32	Water freezes
	−100		
	−200		
0	−273	−459	Absolute zero Everything freezes

Figure A. Comparison of the three temperature scales in common use.

COSMOLOGICAL PRINCIPLE

Remember that during our imaginary journey through space we discovered that on the average *the Universe is homogeneous and isotropic*. This observation (which in reality, of course, was discovered by observations through Earth-based telescopes) plays such a fundamental role in the construction of theories about the origin and evolution of the Universe that it is referred to as the *Cosmological Principle*.

Does the expansion of the Universe not invalidate the Cosmological Principle? Will there not be changes with time that will destroy the homogeneity and isotropy? There will be changes with time, but because the Universe expands uniformly in all directions, those changes will be the same everywhere. A galaxy that today is twice as far away as another will still be that much farther away in a million or billion years because its recession velocity is also twice as great. Therefore, even though the Universe expands and ages, it remains homogeneous and isotropic.

To observe the Universe's homogeneity and isotropy we must compare different parts of it at the same time. When we look at galaxies through telescopes we don't do that because we see them not as they are today but as they were when the light that we pick up with our telescopes left them. The most distant galaxies visible through present-day telescopes are approximately 12 billion light-years away. This means that the light we receive from them was emitted 12 billion years ago and has been in transit every since. It gives us information about those galaxies at that distant time. According to the Cosmological Principle, however, the conditions in our local cosmic neighborhood 12 billion years ago were, on the average, the same as those that we observe by looking at galaxies 12 billion light-years away.

The delay in transmitting information across intergalactic space is due to the finite speed of light. A similar delay in the transmission of electromagnetic radiation occurs here on Earth, except that it is negligible because the distances are so small. Light or telephone signals take less than 2/100 of a second to travel from the Atlantic to the Pacific coast of the United States. This interval is too short to affect communication. However, when the astronauts landed on the Moon, noticeable pauses existed between the transmission of radio messages from Earth and the reception of answers. They were due to the 2.5-second travel time of radio waves from Earth to the Moon and back.

A message from the nearest star would take approximately four years to reach Earth and from the Andromeda galaxy it would take 2 million years. We do not expect that things would greatly change on Alpha or Proxima Centauri in just four years or in the Andromeda galaxy in 2 million years. However, with the distant galaxies this is not the case. In the course of several billion years and longer, significant evolutionary changes can take place in the Universe. Thus, observing distant objects gives astronomers an important tool for probing the Universe's past.

COSMIC BACKGROUND RADIATION

If we extrapolate the present structure and expansion of the Universe back in time, we find that the galaxies were closer together. Eight billion years ago they were approximately twice as close as they are today and 10 billion years ago they were three times as close. At the time of the Big Bang—well before the galaxies were formed—the matter must have been compressed to exceedingly high densities and heated to enormous temperatures. If the Universe started from such extreme conditions, some remnant of the early, intense heat might still be around today in the form of electromagnetic radiation and perhaps it can be detected.

This was the kind of reasoning that in the 1940s lead George Gamow (1904–1968), a Russian physicist who had emigrated to the United States, to calculate what the cosmic radiation left over from the Big Bang should be like today. He assumed that the radiation, if it existed at all, should be as much part of the Universe as is matter. Hence, it should participate in the expansion of the Universe and all its wavelengths should have become greatly stretched, just as the distances between galaxies became stretched. From the expansion rate of the Universe and its present density of matter, Gamow estimated that the present average wavelength of the cosmic radiation should be about 0.06 centimeters. This places it into the microwave range and gives it a temperature of about 5 degrees Kelvin (for a discussion of temperature scales see box 1.2). The radiation should be very faint, fill the entire Universe, and come to us equally from all directions (very much like white light comes equally from all directions during a whiteout in a snowstorm or heavy fog). Unfortunately, at the time no one took Gamow's bold predictions seriously, and no efforts were made to search for cosmic radiation.

The story continues in 1964 with two scientists of the Bell Telephone Laboratories, Arno Penzias and Robert Wilson. They were observing radio emission from our galaxy using a highly sensitive 20-foot radio antenna built for communication with the *Echo* and *Telstar* satellites. To their dismay, their instrument picked up low intensity microwave noise that resembled the static heard on a radio with poor reception. The noise was very weak and isotropic. At first they suspected that it was produced by faulty equipment, but careful checking showed that this was not the case.

When Penzias and Wilson discussed their problem with other scientists, they learned that Robert Dicke and James Peebles, two theoreticians at Princeton University, might have an explanation for their radio noise. Dicke and Peebles had taken up Gamow's old idea of cosmic background radiation and repeated his calculations. The two teams got together and upon comparing the observations with the theoretical predictions they realized that the microwave noise could only be the relic radiation from the Big Bang. Since then the *cosmic background radiation*, as it is now known, has been observed by radio astronomers around the world at wavelengths from fractions of a centimeter to 50 centimeters. It peaks near one-tenth of a centimeter, which is in the microwave range and corresponds to a temperature of about 3 degrees Kelvin, just a bit lower than Gamow's ingenious prediction 20 years earlier.

In measuring the cosmic background radiation, astronomers have paid much attention to its isotropy. Does it come equally from all directions as the Cosmological Principle predicts? The answer is no. In 1977 several researchers reported that radiation coming approximately from the direction of the Virgo Cluster is slightly blueshifted and radiation coming from the opposite direction is slightly redshifted. Because the two directions are exactly opposite each other and the blue- and redshifts are nearly equal in magnitude, only one conclusion can be drawn: The cosmic background radiation is isotropic, but Earth is traveling relative to the radiation. The speed of travel is about 390 km/sec, which is the combined result of the motion of the Sun within the Milky Way and the motion of the Milky Way through the Universe. If we subtract from this value the Sun's orbital velocity around the center of the Milky Way, we find that our galaxy is falling at approximately 600 km/sec towards the Virgo cluster, the gravitational center of our local supercluster of galaxies.

This result, and the otherwise near-isotropic nature of the cosmic background radiation, suggests that this radiation defines an absolute reference frame for the Universe, relative to which the motions of the galaxies and all other material objects can be measured. The frame is an expanding one, because the radiation is expanding right along with the expansion of the Universe. Any motion relative to the frame is a motion relative to the expansion of the Universe.

It is commonly believed that the superclusters of galaxies have no appreciable velocity relative to the frame defined by the cosmic background radiation.* However, individual galaxies move relative to the frame, because each of them has an appreciable velocity within the supercluster to which it belongs.

The discovery of the cosmic background radiation must be ranked among the great achievements of twentieth century science. In 1978 the Swedish Academy recognized this and awarded the Nobel Prize to Penzias and Wilson, commenting that "their discovery made it possible to obtain information about cosmic processes that took place a very long time ago, at the time of the creation of the universe."

PRIMORDIAL ABUNDANCE OF THE ELEMENTS

Since the discovery that the Universe is expanding, the Big Bang theory has had a number of rivals, such as the steady-state theory.† However, none of the other theories is able to explain both the expansion of the Universe and the existence of the cosmic background radiation. Nor are they able to explain an additional discovery astronomers made during the past two decades: the discovery of the primordial abundance of the elements.

The primordial abundance of the elements refers to the composition matter acquired when it was created by the Big Bang. Astronomers have estimated this abundance by examining the spectrum of light emitted by the oldest stars in our and other galaxies. They found that out of every 100 atoms, approximately 93 are ^{1}H and seven are ^{4}He. By mass, this amounts to approximately 76% ^{1}H and 24% ^{4}He. (^{4}He has four times the mass of ^{1}H, hence the difference in the percentage values by number and by mass.) Elements heavier than helium are present in only trace amounts. As explained in the final section of this chapter, this ratio of helium to hydrogen in the oldest stars is consistent with the Big

---

*The traditional view that superclusters of galaxies are distributed homogeneously and that they have no appreciable velocities relative to the frame defined by the cosmic background radiation has recently been questioned by astronomers in the United States and England. Their measurements indicate far greater clumping and larger drift velocities than previously suspected (*Scientific American*, June 1986, pp. 66, 67; *Science News*, July 5, 1986, p. 7). No convincing theoretical explanations of these results have been offered so far.

†The steady-state theory of the Universe was first proposed in 1920 by Sir James Jeans (1877–1946) and then again in the late 1940s by Hermann Bondi, Thomas Gold, and Fred Hoyle. It proposes that the Universe is expanding forever, yet it never changes. Matter is created continuously and new galaxies are formed at just the rate at which old galaxies recede due to the expansion of the Universe. A steady-state Universe has, therefore, neither a beginning nor an end.

Bang model of creation. The Universe started out explosively from a very hot and dense state and quickly cooled as it expanded. The hot and dense conditions lasted long enough for some hydrogen to fuse into helium, but not so long as to allow the production of significant amounts of the heavier elements. They were made much later in the interiors of massive stars.

COSMOLOGICAL THEORY

So far we have been largely concerned with observational information about the Universe: the distribution of galaxies, the expansion of the Universe, the cosmic background radiation, and the primordial abundance of the elements. These observations led to the conclusion that the Universe was created some 20 billion years ago by a gigantic explosion — the Big Bang — and has been expanding ever since. We now face the obvious question: Will the Universe go on expanding forever or will it eventually fall back onto itself? To answer this question, observational information must be merged with theoretical predictions.

The expansion of the Universe may be compared to the flight path of a rocket launched from the surface of a planet. Whether or not the rocket is able to escape the gravitational pull of the planet depends on the planet's mass and radius and on the rocket's velocity at take-off. In the case of the Earth, the rocket's velocity must exceed 11 km/sec (about 40,000 km/hr or 25,000 mi/hr) to escape. If it is less than that, the Earth's gravity is sufficient to reverse the rocket's motion. The rocket will rise to a maximum height and then fall back to Earth and crash.

Likewise, the eventual fate of the Universe depends on how much matter it contains, on its present size, and on how fast the matter is expanding. The more matter there is, the greater are the gravitational forces holding the Universe together, and the more likely it is that the Universe is finite and will eventually collapse back on itself. The greater the present size of the Universe, the further are the galaxies and the rest of the matter separated from each other and the weaker is the gravitational attraction. This increases the likelihood that the Universe is infinite and will expand forever. Finally, the greater the velocity of expansion, the harder it is for gravity to slow the expansion to zero and to reverse it. Again, this contributes toward making the Universe infinite and ever-expanding.

The task of this section is to determine the relative contribution of each of these three factors — cosmic mass, size, and expansion velocity — to the evolution of the Universe and then to use them to decide on theoretical grounds what the destiny of the Universe will be. Before going on, it is important to understand two differences that exist between the launching of a rocket and the expansion of the Universe.

First, in the case of the rocket, there are two separate entities — rocket and Earth. The rocket is flying away from Earth. In the case of the Universe, there is just one entity — the Universe. It is not individual galaxies that fly away from the Universe. It is the entire Universe and all it contains that fly apart.

Second, when calculating the flight path of a rocket it is safe to assume that space is described by the three-dimensional geometry of Euclid (third century B.C.), with time as a separate parameter. Furthermore, the physics of the flight path may be assumed to be the classical one developed by Isaac Newton in the seventeenth century.* Neither of these assumptions is appropriate when dealing with the Universe at large. Here space and time must be linked into the four-dimensional geometry of space–time and the physics must be based on the General Theory of Relativity. This theory, developed early in the present century by the German-American physicist Albert Einstein (1879–1955), is considered one of the great intellectual accomplishments of our time.

Einstein's General Theory of Relativity

In Einstein's theory, massive objects warp the geometry of space–time and this warping gives rise to gravitational forces. The greater the warping, the greater are the gravitational forces. The Earth and Sun warp space–time only slightly because they are not very massive and they are rather extended. In their vicinity classical physics is a good approximation. However, in the case of the large-scale structure of the Universe or near highly compact objects like neutron stars and black holes, the curvature of space–time is appreciable. Classical physics breaks down and general relativity applies.

General relativity predicts a number of effects that classical physics cannot explain (see figure 1.17). It predicts that light travels through the Universe and past stars not in straight lines, but along curved paths; that time does not tick away at equal rates everywhere, but ticks more slowly where gravity is strong; and that light emitted from a star or galaxy loses energy and its wavelengths are stretched to longer values (which means that the wavelengths are shifted toward the red part of the electromagnetic spectrum, an effect that is called the *gravitational redshift*). The greater the warping of space–time, the greater are these effects. Since Einstein first published this theory in 1915, each of these predictions has been observed and verified as occurring in nature.

Perhaps the most dramatic confirmation of the predictions of general relativity has been the discovery in 1979 of the gravitational lens effect (figure 1.18). The light from two distant quasars (objects at great distances, characterized by luminous, point-like sources of light—see chapter 2), QSO 0957 + 561 A and B, was found to originate from a single quasar and to be bent by the mass of an intervening galaxy so as to produce two separate images 6.15 seconds of arc apart (figure 1.19). That the two images (A and B) are, in fact, of the same source is demonstrated by their identical redshifts (velocity of recession = 200,000 km/sec, corresponding to a distance of about 9 billion LY) and identical spectral features. Five other examples of gravitational lensing of distant quasars have been discovered since 1979, and many more are suspected to exist.

---

*Sir Isaac Newton (1642–1727) is considered one of the fathers of modern science. Among many other contributions, he invented the reflecting telescope, discovered the splitting of white light into its color components by prisms, suggested the corpuscle (photon) theory of light, was one of the co-developers of calculus, derived the classical theory of gravitation, and laid the foundation of mechanics with his three celebrated laws of motion (see box 4.2). In 1687 Newton published *Philosophiae naturalis principia mathematica* (Mathematical Principles of Natural Philosophy), a book considered by many to have had the single most important influence in the development of modern science.

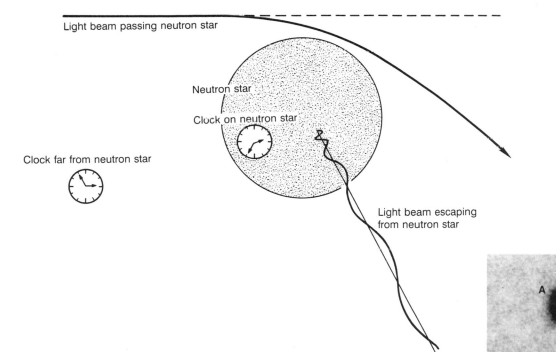

Light beam passing neutron star

Neutron star

Clock on neutron star

Clock far from neutron star

Light beam escaping from neutron star

Figure 1.17. Illustration of three of the predictions of the General Theory of Relativity: 1. Light travels past a star not in a straight line, but along a curved path. 2. Clocks tick more slowly on a star than in free space. 3. The wavelengths of light emitted from a star become stretched (called the *gravitational redshift*). According to the General Theory of Relativity, all three effects are caused by the warping of space-time in the vicinity of the star.

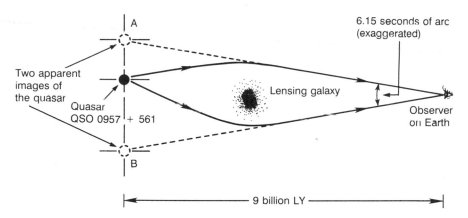

Figure 1.18. One of the most dramatic confirmations of Einstein's General Theory of Relativity, the gravitational lens effect. The light from a quasar (QSO 0957 + 561) was found to be bent by the gravitational field of a massive intervening galaxy so as to produce two images.

Figure 1.19. Quasar QSO 0957 + 561 A and B, an example of the gravitational lens effect. Three images, labeled A, B, and G, are shown. Images A and B are of the same object, namely the quasar. Image G is that of an intervening galaxy whose gravitational field produces the splitting of the quasar image.

Friedmann's Three Models of the Universe

After developing his General Theory of Relativity, Einstein and others applied it to the study of the large-scale structure and evolution of the Universe. The first major success in these efforts came in the early 1920s when the Russian mathematician Alexander A. Friedmann (1888–1925) obtained three different solutions to Einstein's equations. Each solution predicts that the Universe began with an explosion, but they differ on the subsequent evolution. In the first solution, the Universe expands for some time, reaches a maximum size, and then contracts. This model predicts a Universe that is *finite and closed*. In the other two solutions, the Universe expands forever, making it infinite. In one case, the Universe is still expanding even after all the galaxies and the rest of its content have become infinitely dispersed. This model universe is said to be *open*. In the other case, the Universe is just barely able to disperse to infinity, its expansion velocity slowing to zero in the process. This model universe lies at the border between the other two and is called the *flat universe*.

Einstein developed his General Theory of Relativity in order to describe in a consistent manner physical phenomena occurring near stars and other gravitating bodies. It is remarkable that this theory, derived from local phenomena, allowed Friedmann to make predictions about the evolution of the entire Universe, including the prediction of the cosmic expansion several years before Hubble confirmed it by observation. This result attests to the broad applicability of the theory, as well as to the genius of its inventor.

The theoretical predictions of Friedmann are most clearly illustrated by a graph of the scale* of the Universe versus time (see figure 1.20). All three models start from a very compact state—the Big Bang origin of the Universe. For some time thereafter they evolve nearly identically, and their graphs show this. However, as the rate of expansion is slowed by the decelerating effect of gravity, differences appear. The graph of the closed universe reaches a maximum height and then descends. Eventually, the graph dips all the way down as this universe collapses into the compact state of its birth. There may occur a bounce from which the closed universe re-emerges for yet another cycle in an ongoing evolution, though this is not a prediction that follows from general relativity. The other two models of the Universe expand to infinity, as shown by their open-ended graphs. The graph of the open universe remains steeper and lies above those of the other two. This indicates that its expansion velocity is the largest of the three models and that it is still expanding even after it has become infinitely dispersed. The graph of the flat universe becomes less and less steep until, after an infinite time, it becomes horizontal while simultaneously reaching infinite height. This means that the flat universe slows more and more until the expansion rate approaches zero while simultaneously dispersing to infinity.

Now we must decide which of Friedmann's three models corresponds to the actual Universe. As stated earlier, this decision needs to be based on observa-

---

*In this context it might at first appear that "size of the Universe" should be plotted versus time. However, "size" loses its meaning if the Universe is infinite. Therefore, the new concept "scale of the Universe" is introduced. It refers to the distance between *any* two superclusters of galaxies, as, for instance, the distance between the Virgo and the Coma clusters. As the Universe expands and the distance between the two superclusters increases, the scale increases proportionately. It does not matter whether the Universe is finite or infinite.

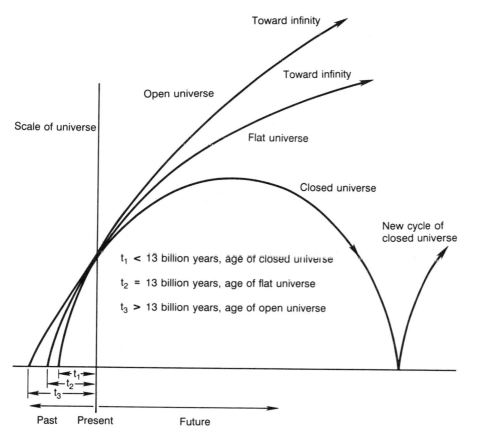

Figure 1.20. Evolution of Friedmann's three models of the Universe. The distances between clusters of galaxies (labeled *Scale of Universe*) remain finite in the closed universe, but they become infinite in the flat and open universes. The ages indicated are based on $H = 15$ km/sec per million light-years.

tional data and on theory. The observational data pertinent to the problem are the present average density of matter in the Universe and the present rate of cosmic expansion. The theory is Einstein's General Theory of Relativity. It will tell us whether or not the cosmic expansion rate is sufficient to overcome the gravitational pull of the matter.

The present rate of expansion of the Universe is given by the Hubble constant — 15 km/sec per million light-years. For this rate of expansion Einstein's theory predicts that the Universe is finite and closed if the average density of matter is greater than 5×10^{-30} g/cm^3; it is infinite and open if the average density is less than that value; and it is infinite and flat if the average density is equal to it.

The value of 5×10^{-30} g/cm^3, which corresponds to about three hydrogen atoms per cubic meter, is called the *critical density* of the Universe.* To decide

---

*The critical density is derived from theory and based on the value adopted for the Hubble constant. For a value of H twice as great as the one used above, namely $H = 30$ km/sec per million light-years, the critical density is 2×10^{-29} g/cm^3. In other words, if the Universe expands twice as fast as we have assumed, the density required to close it is four times greater.

in what kind of universe we live—open, flat, or closed—we need to compare the average density of the actual Universe with the critical density. The average density of matter is the amount of mass contained in a volume large enough to embrace a representative sample of the Universe. Such a volume must extend over several superclusters of galaxies. By adding up the masses of the galaxies of the superclusters in our vicinity of the Universe and dividing by the corresponding volume, it turns out that the average density of matter is approximately 7.5×10^{-32} g/cm^3, or four to five hydrogen atoms per 100 cubic meters of space. This is roughly 70 times less than the critical density and suggests that the matter in the Universe is insufficient for reversing the cosmic expansion. The Universe is infinite and open.

Could it be that our estimate of the average density of matter in the Universe is too low? Might there not exist somewhere in the vastness of space matter not yet discovered that is sufficient to close the Universe? For example, there are many more ways for matter to be invisible, or at least hard to see, than there are for it to be visible. Visible matter is either in the form of luminous stars or gas clouds illuminated by stars, but dark or hard-to-see matter may be in such different forms as rocks, planet-like objects, faint burnt-out stars, black holes, dark gaseous halos surrounding galaxies, intergalactic clouds of neutral hydrogen that never condensed into stars or galaxies, or even neutrinos (if they possess mass).

Although dark matter cannot be seen directly, its presence can be deduced from its gravitational effect on visible matter. This is done by measuring the velocities of stars and gas clouds within galaxies and of galaxies within galaxy clusters. The velocities tell how much mass (luminous and dark) is present to keep the fast moving objects bound by gravity and prevent them from flying off.

Measurements of stellar motions within galaxies indicate that, depending on the galaxy type, their true masses are anywhere from a factor of 3 to 80 times greater than the masses of the visible matter, with a factor of 6 to 7 being most common. If we include this invisible matter of the galaxies in our estimate, we obtain very approximately a value of 5×10^{-31} g/cm^3 for the average density of the Universe. This value falls still short by a factor of 10 of the critical density and again suggests that the Universe is infinite and open.

Measurements of the velocities of double galaxies around each other and of galaxies within clusters of galaxies indicate that additional matter exists in the space between the galaxies, not accounted for by the galaxies themselves. Without that matter, the clusters would have become dispersed long ago. This is the *missing mass*, whose existence was first suspected nearly 50 years ago by Fritz Zwicky (1898–1974) of the California Institute of Technology. Of course, the mass is not really missing, it is just not visible. As suggested above, this dark mass might be in the form of black holes or intergalactic gas clouds. Some of it could be in the form of small faint galaxies or globular star clusters that have escaped detection over intergalactic distances. Another contribution to the missing mass could arise from the hypercharge force that has recently been suggested by Ephraim Fischbach and his colleagues as a fifth fundamental force of nature (see the Introduction to Part One). The missing mass might also exist in the form of neutrinos, those elusive particles that, according to current theories, have a mass of about 1/100,000 the mass of an electron. There may, in fact, be so many neutrinos that, despite their tiny mass, they add up to ten times the ordinary mass found in galaxies. If this is correct, the average density of matter in the Universe would be very close to the critical density. That is a very interesting result. It means that the Universe is very nearly flat. However, at present the estimate of the Universe's average density is too uncertain to allow us to be more specific. We do not know whether the Universe is exactly flat, slightly open, or slightly closed.

Age of the Universe: Revised Estimate

The models derived by Friedmann allow us to estimate the age of the Universe more accurately than we did earlier. The estimate of 20 billion years was based on the assumption that the Universe has always been expanding at its current rate of 15 km/sec per million light-years. The steepness of the graphs in figure 1.20 shows, however, that this rate was much greater in the past. Hence, it took the Universe less time to reach its present scale than we estimated. If the Universe is flat, its age turns out (assuming $H = 15$ km/sec per million light-years) to be two-thirds of our earlier result—13 billion years. If it is open, its age is somewhat greater. And if it is closed, its age is a little less. Because it seems that the Universe is very nearly flat, 13 billion years is the estimate for the age of the Universe assumed in this text.

GEOMETRICAL INTERPRETATION OF THE MODEL UNIVERSES

Much can be learned by considering the geometrical characteristics of Friedmann's three models. To do this we must draw diagrams of their space–time structures. However, it is not possible to draw diagrams of four-dimensional structures. What we can do is "cut through"* one of the space dimensions and then draw the remaining three—one time and two spatial dimensions. Presumably, physical events happening along the space dimension deleted by cutting are, on the average, like the events happening along the two space dimensions that remain. Hence, what we learn from three-dimensional space–time diagrams should be correct in principle, if not in detail.

Flat Universe

The simplest geometry is that of the flat universe: it is the Euclidean geometry with which we are all familiar from daily life. Cutting through one space dimension results in three-dimensional space–time. The two remaining spatial dimensions are a flat surface that extends to infinity in all directions, and time, which runs perpendicularly to that surface.

---

*By "cutting" we reduce the number of dimensions of a real object or a geometrical figure. For instance, by cutting through a three-dimensional object, we obtain a two-dimensional surface; by cutting a two-dimensional surface, we obtain a one-dimensional line; and by cutting a line, we obtain a zero-dimensional point.

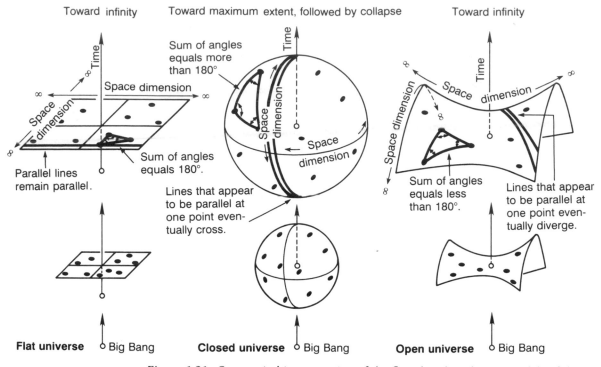

Figure 1.21. Geometrical interpretation of the flat, closed, and open models of the Universe. Two space and the time dimensions are drawn. The third space dimension has been deleted. All three models begin from a highly compact state — the Big Bang (shown by open circles). As time progresses, all three models expand and clusters of galaxies (shown by dots) recede from each other.

Let us focus on a finite square that lies on that flat surface obtained by cutting and see how it and its contents evolve (see figure 1.21, flat universe). Dots on the square represent clusters of galaxies. As time goes on, the square moves along the time axis and expands; the dots move farther and farther apart; and after an infinite length of time, all the dots become dispersed to infinity. These changes in the size and internal structure of the square simulate the aging and expansion of the flat universe. If we go back in time, we find that the square started out from a highly compact state, in which all the dots — the galaxy clusters — were crunched together and the Universe was born.

We selected the square quite arbitrarily. There exist infinitely many other squares exactly like it, one next to the other and filling the space of the flat universe. No matter how far we go in any direction, we never come to an end or to an edge. We will also never find a center, for that would mean that one square distinguishes itself from all the others. The only edge that we ever find in the flat universe is an edge in time — namely the beginning.

In Euclidean geometry the shortest distance between two points is a straight line, parallel lines always remain parallel, and the interior angles of triangles add up to 180°. Furthermore, light travels along straight lines. However, these features are unique to the flat universe and are not shared by the other two models.

Closed Universe

The geometries of the other two model universes are non-Euclidean. Their space–time is not flat but curved. Let us consider the closed universe (figure 1.21). By cutting through one of its spatial dimensions, we obtain a finite sphere (think of a balloon). The radius of the sphere corresponds to time and its surface to curved two-dimensional space. Again, we let dots on the surface represent clusters of galaxies. As the closed universe evolves, the sphere expands until it reaches a maximum size and then it contracts again (like blowing air into the balloon and then letting it out). During the expansion phase all the dots move away from each other, and during the contraction phase they approach each other. Eventually, the sphere collapses into a very compact state and all the dots are crushed together. This represents the unavoidable, catastrophic end of the closed universe, the so-called big crunch. Going back in time, we find that the closed universe was born from a similar, equally compact state in space–time.

This universe, too, has neither a spatial edge nor a spatial center, for they do not exist on the surface of a sphere. The point at the center of the sphere refers not to space, but to space–time. It corresponds to the two unique events in the evolution of the closed universe: its creation and its end.

Our geometrical intuition does not apply to the global structure of the closed universe. As the diagram illustrates, in this universe the shortest distance between two points is a great circle,* there are no parallel lines (that is, all lines cross), and the interior angles of triangles add up to more than 180°. Furthermore, light rays will not travel along straight lines, but follow the curvature of the universe, which is represented by great circles.

Open Universe

Finally, consider the geometry of the open universe (figure 1.21). Cutting through one of the space dimensions yields a saddle-shaped surface that extends to infinity in all directions. As time marches on, the saddle expands, its curvature lessens, and the dots—representing galaxy clusters—move apart. After an infinite length of time, all the dots will be separated from each other by infinite distances. Going back in time, the section of the saddle shown in the diagram becomes smaller and the dots move closer together. During the Big Bang, they were compressed together into the same highly compact state that we encountered at the beginning of the other model universes.

As in the other universes, neither an edge nor a center exists in the space dimensions of the open universe. No matter how far we go in any direction, we will find galaxy clusters. An edge exists only in space–time, namely at the beginning.

Again, our geometrical intuition fails in the open universe. The shortest distance between two points is a curve (although it is not a great circle), there are no parallel lines (that is, all lines cross or diverge), the interior angles of triangles add up to less than 180°, and beams of light follow curved paths.

---

*A great circle is a circle formed on the surface of a sphere by the intersection of a plane that passes through the center of the sphere. For example, the equator of the Earth is a great circle. All longitudinal circles are great circles, but latitudinal circles (except the equator) are not.

In principle, we should be able to tell in which kind of universe we live by observing the geometry of our cosmic environment. Do the interior angles of triangles add up to 180°, or are they greater or smaller? Are there parallel lines, or do all lines cross or diverge? What is the shortest distance between two points? Here on Earth—by drawing lines on paper or surveying the country-side—Euclidean geometry holds as far as we can tell. However, that does not help us decide in which kind of universe we live. Earth extends over such a small portion of the Universe that even if space were curved it could not be distinguished locally from flat space. This is analogous to looking at a micro-scopic section of the surface of a large balloon or saddle. It, too, cannot be distinguished from a section of a flat surface. Likewise, the section of the Universe that we can observe with precision with present-day telescopes is too small to permit us to discover the nature of the cosmic geometry. It could be flat, spherical, or saddle-shaped. All we can say is that if space is curved, it is not curved very much. Again, as when we compared the average density of the Universe with the critical density, we have failed to discover the Universe's structure and its ultimate fate.

THE BIRTH OF THE UNIVERSE

The currently available evidence suggests that the Universe came into existence in one gigantic explosion some 13 billion years ago, as discussed in the preceding sections. During the earliest phase of this explosion, the tempera-tures and densities were so extreme that neither space and time, nor matter and energy, nor the laws of physics were as we know them today. In order to get some idea of what the conditions may have been like then, we need to start from the known conditions of today and extrapolate backward in time. The best presently available theories concerning the behavior of matter at high energies along with the general theory of relativity shall be our guides.

Such an extrapolation is daring, to say the least. It may even be grossly misleading. Physicists have learned from hard experience that extrapolating beyond solid observational and experimental evidence has generally led them astray. However, at present there are no observational or experimental data of the first instant of creation, and possibly, there never will be. Extrapolation is the only means of such an investigation. Besides, to do so is an exciting intellectual adventure.

Our extrapolation will tell us that the initial extreme conditions passed in rapid succession through a number of transitions, during which the Universe acquired the characteristics that still distinguish it today. A single *superforce*, which ruled in the beginning, split into the four forces we are accustomed to. Matter was spontaneously created and acquired the enormous expansion ve-locities that we still observe. There occurred a brief burst of primordial nu-cleosynthesis, during which hydrogen (H) was converted into helium four (^{4}He). One by one, the lightest of the elementary particles—gravitons, neutrinos, and photons—decoupled from the rest of the particles and began to evolve inde-pendently. This entire sequence of events, with the exception of photon decoupling, was over in about 5 minutes. The photons decoupled approxi-mately at age 500,000 years and subsequently evolved into today's 3 K cosmic background radiation. From then on, matter (that is, hydrogen and helium) evolved on an independent course as well and began to condense into the lumpy distributions — clusters of galaxies, galaxies, stars, and planets (around at least one star) — that we find in the Universe today.

A more detailed discussion of these events is best divided into two parts: the first 10^{-6} second and the evolution from age 10^{-6} second forward. Only during the first microsecond were conditions so severe that any treatment of them is unqualifiedly speculative and, no doubt, will be subject to major revisions in the future. In contrast, by age 10^{-6} second the content of the Universe and the physical laws governing it are believed to have been the same as they are today. Consequently, discussion of that period is much less in question, though uncertainties do remain.

Theoretical Background

Before discussing the Universe's birth, a few remarks are necessary concerning the theories upon which this description is based. One theory has already been introduced — the General Theory of Relativity — though to be fully applicable to the Universe's first instant of existence Einstein's formulation of relativity needs to be modified. In particular, Einstein's theory needs to be replaced by a quantum theory of gravitation. Unfortunately, no one has yet succeeded in working out such a theory in detail, and this constitutes the greatest source of uncertainty in the description below.

Another theory that we need goes by the poetic name of *quantum chromodynamics* or QCD. According to this theory, protons and neutrons are not really elementary, but are composed of more fundamental particles called *quarks* (see the beginning of the Introduction to Part One).

A third theoretical component of the description is the *grand unified theories* or GUTs (plural, because there are many grand unified theories, none of which is at present universally accepted). With GUTs, physicists seek to find relationships between the forces of nature and unify them into one. For example, the theories predict that at temperatures above approximately 10^{15} K two of the standard forces — the weak and the electromagnetic forces — become unified into the *electro-weak force*. At the same time, electrons and neutrinos lose their identity and behave in identical ways. They are then referred to collectively as *leptons* and they interact by the electro-weak force. Above approximately 10^{27} K, the electro-weak and nuclear forces become unified into the *grand unified force*. Finally, above approximately 10^{32} K, the grand unified force and gravitation are expected to become unified into a single *superforce*.

The concept of a theory that unifies the forces and, simultaneously, describes the fundamental nature of the elementary particles is not new. For example, after completing his General Theory of Relativity, Einstein attempted to unify gravity with the other forces. However, it was not until the late 1960s that real progress in this regard was made. The first to succeed in unifying the electromagnetic and weak forces were Steven Weinberg, Sheldon Lee Glashow (both of the United States) and Abdus Salam (of Pakistan), and in 1979 they were awarded the Nobel prize in physics for this work. Then in the 1970s, Howard Georgi (of the United States), Glashow, and others began working out theories (namely GUTs) unifying the electro-weak force with the nuclear force, an effort that continues today. No one has yet succeeded in unifying the grand unified force with gravity, in part because there is no quantum theory of gravitation. Therefore, the earlier reference to the existence of a single superforce above approximately 10^{32} K needs to be accepted with particular caution.

The application of GUTs to the early Universe described below was first made in 1980 by Alan H. Guth of the Massachusetts Institute of Technology and has been developed further by him and others since then. Guth called his theory the *inflationary model* of the Big Bang because it predicts an enormously fast initial expansion of the Universe, during which most of the matter that is still with us today came into existence.

First Microsecond

In the discussion of the birth of the Universe that begins here, the focus is on the sequence of transitions mentioned earlier. These transitions, in a step-by-step fashion, are believed to have marked the evolution from the initial extreme conditions to the more familiar ones of today (see figure 1.22).

Age 10^{-43} second. Age 10^{-43} second is the earliest point in time about which physics has anything to say. Neither matter and energy nor the laws of physics as we know them today existed then. Space–time was so severely curved that black holes were continually forming and bursting. They were microscopic in size, with diameters 10^{20} times smaller than a proton, and they lasted on the average about 10^{-43} second. Many black holes may have come into existence and then burst; we don't know. Along with the creation and destruction of black holes, space and time were also continually being broken and reconnected. Thus, space and time were not continuous. They were "quantized" analogous to the way today's photons and other elementary particles are quantized. Under these conditions even the concept of causality, namely such words as *earlier* and *later*, had little meaning.

At that early instant of time, the temperature was approximately 10^{32} K and the force that governed the black holes was the superforce introduced earlier. This force constituted a unification of our present-day gravitational, weak, electromagnetic, and nuclear forces. Sometime between ages 10^{-43} and 10^{-35} second, this era of black hole creation and destruction ended. The black holes that existed then were expanding, the temperature was dropping, and the superforce split into gravity and the grand unified force.

Stop and consider. Space–time was continually being broken and reconnected between ages 10^{-43} and 10^{-35} second. Does reference to these times make much sense when time itself was discontinuous and causality had little meaning? There is at present no clear answer. There may well have existed an infinite stretch of time before 10^{-43} second. Or that point in time may have been preceded by earlier cycles of oscillating universes. Or any number of conditions that, with our limited understanding, we cannot even conceive of may have existed then. We just don't know.

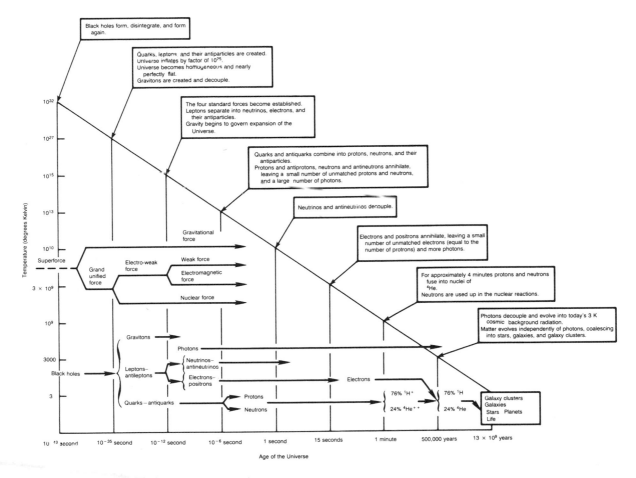

Figure 1.22. Key events during the birth of the Universe according to the inflationary model of the Big Bang.

Age 10^{-35} second. By age 10^{-35} second, the temperature had fallen to about 10^{27} K, and the grand unified force separated into the nuclear and electro-weak forces. Within each of the black holes, this transition was accompanied by the spontaneous and rapid creation of quarks and leptons, as well as antiquarks and antileptons. We may compare this creation to the spontaneous formation of raindrops in the present Earth atmosphere, when water vapor cools below the condensation point, though the energies involved were far more extreme. According to GUTs, the quarks, leptons, and their antiparticles came into existence because the vacuum of that time contained energy (unlike today, when a vacuum contains no energy and, hence, cannot produce matter). In the language of physics, the creation of the particles constituted a phase transition from pure energy into matter.

The rapid creation of quarks, leptons, and their antiparticles led to enormous pressures, which forced the black holes to expand at superfast rates. This was the expansion that Alan Guth called *inflation*. Even though each of the black holes, along with the disconnected regions of space-time associated with them—called *domains*—expanded extremely quickly, they moved away from each other still more rapidly. Consequently, they never became connected and, in the course of time, got more and more out of touch. We do not know what became of any of these domains except one. It continued to expand and evolved into the Universe in which we live, as described in this book. The other domains may have become annihilated. Or they may still exist as separate universes, forever inaccessible to us. We shall never know.

Age 10^{-12} second. By about age 10^{-12} second, the temperature had dropped to 10^{15} K and the last of the unified forces divided. The electro-weak force separated into the electromagnetic and weak forces. Thus, all four of the forces that we experience today had become established.

Quarks, leptons, and their antiparticles were now no longer created spontaneously, and the epoch of inflationary expansion ended. Inflation had increased the size of the Universe by roughly a factor of 10^{25} from its embryonic state at age 10^{-35} second. Furthermore, the distribution of the particles throughout the Universe was nearly perfectly homogeneous, a condition still observed today on the largest scale.

Even though inflation had come to an end, the Universe continued to expand simply because of the momentum it had acquired during inflation. However, the expansion was not unopposed. The global attractive force of gravity—that is, the force of gravity exerted by *all* matter and energy in the Universe—began to slow the expansion. As a result the large-scale evolution of the Universe followed—and still follows today—one of the three curves derived by Alexander Friedmann. As already noted, we do not know which of the three curves it was, and GUTs give us no answer either. We only know that the cosmic expansion rate at that early time must have been extremely delicately balanced against the global gravitational attraction to ensure that the Universe would coast to its present size. Any deviation from this finely tuned balance would have become greatly magnified with time. If the expansion rate had been only a fraction greater, the Universe would be totally dispersed today. Had it been slightly smaller, the Universe would long ago have recollapsed into its initial, highly condensed state. Put differently, our Universe is nearly perfectly flat, as recognized in the previous section.

With the separation of the electro-weak force into the weak and electromagnetic forces, the leptons and antileptons no longer remained single species of particles. They separated into electrons and positrons (electron antiparticles), and neutrinos and antineutrinos. The electrons and positrons interacted by the electromagnetic force, while the neutrinos and antineutrinos interacted by the weak force. No corresponding change occurred in the case of quarks and antiquarks.

In addition to these particles and antiparticles, which are characterized by having mass, there also existed massless particles, namely gravitons and photons. The gravitons had been created during the Universe's rapid inflationary acceleration, and they continued to be created now, though at an ever diminishing rate (see the beginning of the Introduction to Part One and box I.1 for a brief discussion of gravitons).

The particles that filled the Universe at age 10^{-12} second were in a chaotic state. There was no structure or order. The particles were flying around at high speed in all directions, continually colliding and interacting violently with each other. Upon colliding, the particles and their antiparticles annihilated each other, creating high-energy photons, which, in turn, quickly decayed and changed back into matter and antimatter particles.

The only particles that did not participate in these collisions were the gravitons. They passed right through the other particles without interacting directly. They were decoupled from the rest of the matter and the photons. This happened because they interacted only by the gravitational force, which is the weakest of the four forces. However, despite being decoupled, the gravitons still felt the global gravitational attraction of the rest of the matter and energy in the Universe. Consequently they traveled along curved paths as dictated by the geometry of space–time and, presumably, they still do so today. Should we ever succeed in detecting them, they would constitute the most ancient direct evidence of the birth of the Universe.

Age 10^{-6} second. At age 10^{-6} second the temperature dropped below 10^{13} K and the quarks and antiquarks combined into protons and antiprotons as well as neutrons and antineutrons. Simultaneously, the protons and neutrons collided with the antiprotons and antineutrons and became annihilated, producing a huge number of photons. At the existing temperature, the photons were no longer energetic enough to create either protons and antiprotons, neutrons and antineutrons, or quarks and antiquarks. Thus, the back-and-forth reactions between heavy particle–antiparticle pairs and photons came to an end. The heavy particles and their antiparticles annihilated each other without being recreated by the photons. (This was not true of the electron–positron pairs. They continued their back-and-forth reactions with the photons for about another 15 seconds; see below.)

Not all of the protons and neutrons were annihilated, however. At age 10^{-35} second, slightly more quarks had been created than antiquarks — approximately one unmatched quark for every billion quark–antiquark pairs. These unmatched quarks combined into protons and neutrons that did not find antiprotons and antineutrons with which to annihilate. Hence, they survived. These excess protons and neutrons (together with excess electrons; see below) constituted the matter that eventually evolved into the stars, galaxies, and clusters of galaxies that mark the Universe today. Had matter and antimatter particles been created in exactly equal numbers, they would have completely destroyed each other and left the Universe in a rather impoverished state, with photons, neutrinos, and gravitons as its only content.*

*We happen to live in a Universe made of matter consisting of protons, neutrons, and electrons. In addition there are neutrinos and antineutrinos (which may have a tiny mass), photons, and gravitons (the last two are massless). Generally, there are no antiprotons, antineutrons, and positrons in our present-day Universe. However, during such violent events as supernova explosions, galaxy explosions, and Earth-based experiments with high-energy particle accelerators, particle–antiparticle

To summarize, the Universe at age 10^{-6} second was in a state of rapid expansion. It contained protons, neutrons, electrons and positrons (both in nearly equal numbers), neutrinos and antineutrinos, photons, and gravitons. The forces governing the interactions among the particles were the nuclear, electromagnetic, weak, and gravitational forces.

All the particles were in continuous chaotic motion, incessantly colliding and interacting with each other. The only exception was the gravitons, which were decoupled from the rest. The electrons and positrons were still continually annihilating each other and creating photons. The photons, in turn, quickly decayed and changed back into electrons and positrons.

None of the collisions produced nuclei of helium or heavier elements, nor did protons and electrons combine to form atoms. At the high prevailing temperatures, the collisions were still too violent, by many orders of magnitude, for that. Had nuclei of heavy elements or atoms formed, they would have been instantaneously broken up into their component particles by further collisions.

After the First Microsecond

Space and time, matter and energy, and the four forces all came into existence during our Universe's first microsecond of existence. Let us now examine what happened next.

Age 1 second. By age 1 second, the temperature had dropped to about 10^{10} K. As well as causing this drop in temperature, the cosmic expansion had reduced the neutrinos' energies and changed their interaction properties such that they ceased to feel the presence of the other particles. Thus, neutrinos and antineutrinos became the second species of particles to decouple from the rest. Like the gravitons, they still were part of the Universe and subject to its global gravitational attraction. Presumably, these primordial neutrinos are still racing through the Universe today, though they have not yet been detected. Should they be detected in the future, they will be the second oldest evidence from the Big Bang. In fact, there may be so many of them that their combined mass closes the Universe, as noted earlier.

Age 15 seconds. As the Universe aged and expanded further, its radiation and matter continued to cool. By age 15 seconds, the temperature had fallen below 3 billion degrees Kelvin and the photons no longer had the energy to produce electron–positron pairs. The electrons and positrons experienced the same fate that had befallen the protons and neutrons earlier. They annihilated each other without being recreated by the photons. The only exception was the small excess number of electrons with which the Universe was born. Thus, the

pairs are created—electron–positron pairs, proton–antiproton pairs, neutron–antineutron pairs, and still other, more exotic pairs. When these particles collide with their respective antiparticles (electrons with positrons, protons with antiprotons, and so forth) they annihilate each other and produce high-energy photons, just as during the first microsecond of the Big Bang. Furthermore, these high-energy photons are observed to decay into particle–antiparticle pairs. The antiparticles have the same masses as their respective particles. However, the positron is positively, and the antiproton is negatively, charged. The antineutron has zero charge, just like the neutron.

last vestige of primordial antimatter (with the exception of antineutrinos) disappeared. The Universe now consisted of equal numbers of electrons and protons, of neutrons, neutrinos, antineutrinos, photons, and gravitons. The neutrinos, antineutrinos, and gravitons were decoupled from the rest.

Age 1 minute. The time between ages 1 and roughly 5 minutes was the epoch of primordial nucleosynthesis.* During this interval, the temperature dropped from about 1.3 billion to 600 million degrees Kelvin (which brought it into the range of 40 to 80 times the central temperature of the Sun). Collisions between the particles became much less violent than they had been earlier. Consequently, collisions between protons and neutrons produced stable nuclei that were not torn apart again by further collisions. In particular, protons and neutrons fused to form nuclei of hydrogen-2 ($^2H^+$, also called deuteron), helium-3 ($^3He^{++}$) and helium-4 ($^4He^{++}$):†

$$
\begin{aligned}
p^+ &+ n &&\rightarrow {}^2H^+ &&+ \gamma \\
p^+ &+ p^+ &&\rightarrow {}^2H^+ &&+ e^+ + \nu \\
p^+ &+ {}^2H^+ &&\rightarrow {}^3He^{++} &&+ \gamma \\
{}^3He^{++} &+ {}^3He^{++} &&\rightarrow {}^4He^{++} &&+ 2p^+ + \gamma
\end{aligned}
$$

(The same nuclear reactions, with the exception of the first one, occur today in the interiors of the Sun and of many other stars. In fact, they are the major source of the stars' radiant energy [see chapter 3].)

By the time the Universe was five minutes old, the temperature had dropped below 600 million degrees Kelvin and nucleosynthesis came to an end. All the neutrons had become used up in the nuclear reactions. The matter now consisted of roughly 76% hydrogen (^1H), 24% helium-4 (^4He), and traces of deuterium (^2H) and helium-3 (^3He), all in fully ionized form. (Percentages refer to mass.) Virtually no heavier elements had been manufactured. The primordial abundance of the elements remained unchanged until the birth of the first generations of stars hundreds of millions of years later. Only then did the buildup towards the heavier elements — carbon, nitrogen, oxygen, the metals, and the rest — continue.

With continued expansion the temperature kept falling further — to 10 million degrees, 1 million degrees, and lower. Although these temperatures were below the threshold for nuclear reactions, they were easily high enough to keep the matter fully ionized. Hydrogen still existed as free protons and free electrons. As soon as a proton captured an electron to form a neutral hydrogen atom, it was torn apart again into a proton and an electron either by a photon or by collision with some other particle. The same was true of helium. It existed in the form of free helium nuclei and free electrons.

Nucleosynthesis refers to the production of elements heavier than ^1H by nuclear reactions.

†A deuteron consists of a proton and neutron tightly held together by the nuclear force. It is the nucleus of deuterium, the isotope hydrogen-2 (^2H). The symbols p^+, $^2H^+$, $^3He^{++}$, and $^4He^{++}$ refer to the nuclei of hydrogen-1, hydrogen-2, helium-3, and helium-4. The right-hand superscripts indicate that these *nuclei* are singly or doubly charged. The symbols ^1H, ^2H, ^3He, ^4He (without the right-hand superscripts) refer to the *neutral atoms* hydrogen-1 and hydrogen-2 (deuterium) and helium-3 and -4. (For further details, see the Introduction to Part One.)

All the particles—except the neutrinos and gravitons—were still colliding with each other incessantly. In particular, the photons were unable to travel very far before striking an electron, a proton, or a helium nucleus. This made the Universe opaque. It resembled thick fog, except that it was much hotter.* Because of the constant collisions, the photons and the material particles shared their energies of motion; and because temperature is a measure of this energy, photons and matter shared a common temperature.

Age 500,000 years. Up to roughly 500,000 years, the state of the Universe was as just described. By then the temperature had fallen to approximately 3000 K. As a consequence, the energy of the photons had become so greatly reduced that they were less and less able to keep the nuclei and electrons from combining. Soon they were totally unable to do so and matter changed into neutral atomic form. This also changed the appearance of matter from opaque to transparent, and the photons did what the gravitons and neutrinos had done earlier. They ceased to interact with the matter and the Universe became transparent, as it still is today.

The decoupling of the photons from the matter is the last transition included in this description of the birth of the Universe. From here on, photons and matter evolved independently, each according to its own physical laws. The matter evolved into the structures we see all around us today—galaxy clusters, galaxies, stars, planets, and life. The photons evolved very little. Like the gravitons and neutrinos, the photons remained part of the Universe and continued to feel the global gravitational attraction. However, they merely partook in the cosmic expansion. They did not acquire any structure. Their energies were reduced more and more, their wavelengths became longer and longer, and the temperature characterizing them kept falling. Today we observe them as the 3 K cosmic background radiation, a faded remnant from the birth of the Universe.

One final comment is in order. As noted earlier, the 3 K cosmic background radiation observed today is nearly perfectly isotropic. For example, if we measure the intensity of this radiation from one direction and then from the opposite direction, we find that the two results are identical to roughly one part in 10^4. This is rather surprising. The two opposite locations were approximately 10^7 LY apart when the radiation that we measure today was emitted, at age 500,000 years (the age of photon decoupling). In 500,000 years no signal—not even light, the fastest mode of communication—could have traveled 10^7 LY to "instruct" the two places to emit the same intensity background radiation. Hence, these two locations should not have been causally linked. Yet they emitted radiation of nearly identical intensity.

*Fog consists of microscopic water droplets. It is opaque because light rays passing through continually collide with the droplets and are scattered. They cannot follow a straight path. The same was true of the photons in the early Universe—they kept colliding with and were scattered by electrons, protons, deuterons, and helium nuclei. It was also true of the gravitons and neutrinos before they decoupled from the other particles.

Apparently, the isotropy had been established much earlier, at age 10^{-35} second, when the Universe was much more compact and, despite its young age, all of its parts were in communication and causally linked. During the subsequent inflationary expansion, when its size increased by 10^{25} orders of magnitude, the various parts of the Universe got causally out of touch with each other. However, they all inherited the same conditions. That is why at age 500,000 years the intensity of the cosmic background radiation was virtually the same everywhere and why it still is so today. It is also the reason why the distribution of matter at that time was nearly the same everywhere and why, on the largest scale, the distribution of clusters of galaxies is still homogeneous and isotropic today.

Exercises

1 What is meant by "large-scale structure of the Universe"?

2 Using the fact that light travels 300,000 km per second and the number of seconds per year, calculate how far light travels in one year. Compare your answer with the value of one light-year given in the text.

3 (a) Listen to a car or train passing you at high speed. You'll notice a distinct change in the sound from high to low pitch. Explain this change in terms of changes in the wavelength and frequency of the sound waves as they arrive at your ear.
(b) Consider the following experiment: A friend throws tennis balls toward you at a constant rate (say, one ball every second). Catch the balls while first running toward and then away from your friend. How do the rates at which you catch the balls differ from the rate at which your friend throws them?
(c) How does this experiment relate to your answer to (a) of this exercise?
(d) How does it relate to observations of stars and galaxies?

4 Compare the meanings of *homogeneous, heterogeneous, isotropic,* and *anisotropic,* and find examples from everyday life illustrating these terms.

5 Look up yesterday's maximum and minimum temperatures reported in your local newspaper. If you live in the United States, the temperatures are probably given in degrees Fahrenheit. Convert them to degrees Celsius and Kelvin.

6 Throw a ball up vertically. Observe how gravity decelerates the ball, momentarily stops it, and then accelerates it downward.
(a) How does this simple experiment correspond to the evolution of the closed universe? How is it

an oversimplification of that evolution?
(b) How would you have to change the experiment to make it correspond to the evolutions of the flat and the open universes?

7 What is meant by the "missing mass of the Universe"? Is the mass really missing?

8 What would the age of today's Universe be if $H = 30$ km/(sec MLY) and if (a) the Universe's rate of expansion has been constant since the Big Bang and (b) the rate of expansion has slowed down due to the Universe's self-gravity. (Assume that the Universe can be represented by a Friedmann model.)

9 Inflate a spherical balloon. Cover its surface with dots to represent superclusters of galaxies. Continue to inflate the balloon and observe how the distances between the dots increase.
(a) Is there any central dot?
(b) Is there an edge?
(c) How can you use this simple experiment to model the evolution of the closed Universe? In this model, which characteristics of the balloon correspond to time; to space?

10 According to the Big Bang theory, why does the primordial abundance of the elements consist mainly of hydrogen and helium, and virtually of no elements heavier than helium?

11 You probably have experienced a whiteout, either in a snowstorm or heavy fog, and know that it is easy to lose your balance or get lost under that condition.
(a) Why is this so?
(b) How does the condition of a whiteout correspond to conditions in the early Universe, before the decoupling of radiation and matter?

Suggestions for Further Reading

Davies, P. 1985. "Relics of Creation." *Sky and Telescope*, February, 112–15. About progress in understanding nature's fundamental forces at the subnuclear level and how this enables cosmologists to study the first few moments of the Universe's existence.

DeWitt, B. S. 1983. "Quantum Gravity." *Scientific American*, December, 112–29. The author explains that in a quantum-mechanical theory of gravitation, space–time would be subject to continual fluctuations, and even the distinction between past and future might become blurred.

Dicus, D. A., et al. 1983. "The Future of the Universe." *Scientific American*, March, 90–101. A forecast for the expanding Universe through the year 10^{100}.

Dyson, F. J. 1954. "What Is Heat?" *Scientific American*, September, 58–63. The author defines heat as "disordered energy" and explains what this statement means. He also describes the range of temperatures found in nature, entropy, and other concepts from the science of thermodynamics.

Einstein, A. 1961. *Relativity: The Special and the General Theory; a Popular Exposition*. New York: Crown Publishers. An easy-to-read little book about the theory of relativity, revealing the clarity of thought and deep physical intuition that distinguished Einstein.

Gale, G. 1981. "The Anthropic Principle." *Scientific American*, December, 154–71. Traditionally it has been argued that life evolved on Earth because conditions in the Universe were favorable. The anthropic principle argues the reverse: The presence of life may "explain" the conditions in the Universe.

Gamow, G. 1956. "The Evolutionary Universe." *Scientific American*, September, 136–54. A lucid account of the geometrical interpretation of the space–time structure and evolution of the Universe.

Gore, R. 1983. "The Once and Future Universe." *National Geographic*, June, 704–49. A well-illustrated summary of our current understanding of the structure and evolution of stars, nebulae, galaxies, quasars, and the Universe, including brief descriptions of some of the modern instruments used in the study of these objects.

Gott, J. R., III. 1983. "Gravitational Lenses." *American Scientist*, March–April, 150–57.

Gregory, S. A., and L. A. Thompson. 1982. "Superclusters and Voids in the Distribution of Galaxies." *Scientific American*, March, 106–14.

Guth, A. H., and P. J. Steinhardt. 1984. "The Inflationary Universe." *Scientific American*, May, 116–28.

Hawking, S. W. 1984. "The Edge of Spacetime." *American Scientist*, July–August, 355–59. This article, written by one of the most imaginative cosmologists of our time, addresses the question, "What happened at the beginning of the expansion of the Universe?"

Hodge, P. 1984. "The Cosmic Distance Scale." *American Scientist*, September–October, 474–82. Two distinct approaches to the measurement of cosmic distances have produced two different values for the size of the Universe.

Hoskin, M. 1986. "William Herschel and the Making of Modern Astronomy." *Scientific American*, February, 106–12.

Hoyle, F. 1956. "The Steady-State Universe." *Scientific American*, September, 157–66.

Rowan-Robinson, M. 1985. *The Cosmological Distance Ladder: Distance and Time in the Universe*. San Francisco: W. H. Freeman and Co. In this important book, the author critiques the methods used to establish the cosmological distance scale. He disagrees with the derivations of the two most-often quoted values of the Hubble constant, 15 and 30 km/sec per million light years, and suggests instead the value of 20 km/sec per million light years with an uncertainty of about $\pm 20\%$.

Schneider, S. E., and Y. Terzian. 1984. "Between Galaxies." *American Scientist*, November–December, 574–81. The authors discuss how the search for elusive "missing matter" is changing some of the most basic ideas about the nature of the Universe.

Schramm, D. N. 1983. "The Early Universe and High-Energy Physics." *Physics Today*, April, 27–33. The author discusses what the new particle field theories predict about the early history of the Universe, and he explains that comparisons of these predictions with astronomical data may offer the only possible tests of these theories.

Silk, J., A. S. Szalay, and Y. B. Zel'dovich. 1983. "The Large-Scale Structure of the Universe." *Scientific American*, October, 72–80.

Sulak, L. R. 1982. "Waiting for the Proton to Decay." *American Scientist*, November–December, 616–25. In the grand unified theories of elementary particle structure the proton is not stable, but is predicted to decay on a time scale of about 10^{30} years. This article describes experiments designed to detect this decay.

Trefil, J. S. 1983. "How the Universe Began." *Smithsonian*, May, 32–51. "How the Universe Will End." *Smithsonian*, June, 72–83. A readable, two-part account of the recently developed grand unified theories of particle interactions and the predictions that follow from them concerning the origin and end of the Universe. The second article suggests that the Universe's future will be determined by the "missing mass" that is thought to exist in galaxies and galaxy clusters.

———. 1980. "Einstein's Theory of General Relativity Is Put to the Test." *Smithsonian*, April, 74–83. A summary of recent and still-in-progress experiments designed to test various predictions of Einstein's General Theory of Relativity, including the possibility that gravity is weakening over time.

Weisskopf, V. F. 1983. "The Origin of the Universe." *American Scientist*, September–October, 473–80. A respected scientist's discussion of recent theoretical developments linking cosmology and particle physics.

Wilczek, F. 1980. "The Cosmic Asymmetry Between Matter and Antimatter." *Scientific American*, December, 82–90. Evidence from cosmology and particle physics suggests an explanation of why the Universe today consists almost entirely of matter.

Wilson, R. W. 1979. "The Cosmic Microwave Background Radiation." *Science* 205: 866–74. Wilson's Nobel prize lecture on the discovery and status of understanding of the cosmic background radiation, delivered on December 8, 1978, in Stockholm.

Figure 2.1. Galaxies NGC 5194 and 5195. The larger of these two galaxies, NGC 5194 (type Sc), also known as the Whirlpool Galaxy, is one of the most magnificent spirals in the sky. It is connected by an extension of one of its arms to its smaller companion, an irregular galaxy.

Galaxies

THE PREVIOUS CHAPTER ended with a description of the birth of the Universe and its evolution to age one-half million years. At that time the temperature had dropped to 3000 K and the primordial photons had stopped interacting with the matter. Both the photons and the matter were smoothly distributed throughout space, but only the photons have remained so. During the intervening 13 billion years the matter acquired the structures and forms that surround us near and far—from atoms, molecules, and dust grains to stars, galaxies, and galaxy clusters.

How did matter come to organize itself in this astounding way? How, in addition, did it evolve into life on a tiny planet called Earth? These are among the central questions challenging scientists today.

THE FIRST STRUCTURES IN THE UNIVERSE

When matter set out on its independent evolution it consisted of approximately 76% hydrogen and 24% helium by mass. It was in gaseous form and smoothly distributed, and its initial, nearly infinite density had been lowered by the cosmic expansion to about 10^{-21} g/cm^3. However, this smoothness was not quite perfect; it contained weak, random fluctuations in density.

Such fluctuations are a common phenomenon and occur today everywhere in our environment. For instance, from a distance the surface of a lake or the ocean may appear to be smooth, but there are almost always small ripples or waves on the surface. Even on a perfectly windless day, when there are no perceptible waves, fluctuations on the molecular level still exist, so that at any instant the water molecules are concentrated a bit more in some places than others. Within fractions of a second the concentrations shift to different locations, and they continue to do so in an ever-changing, random pattern. Likewise, when water freezes and the molecules become firmly bonded into a hexagonal pattern (see the Introduction to Part One), the resulting crystal structure is not perfect. Microscopic variations exist in the density and structure of the ice, and the molecules oscillate back and forth perpetually.

Random fluctuations in density cannot be avoided in nature. When the gaseous matter and photons decoupled in the early Universe, weak random fluctuations in the density of matter existed everywhere, extending over a wide range of scales. From the present-day clumping of matter into superclusters of galaxies, we may assume that the largest fluctuations stretched over regions of about 10^{15} to 10^{16} $M_\odot$.* Superimposed on them were fluctuations over smaller regions, and they contained still smaller fluctuations. This state of alternating high and low concentrations of matter in the early Universe is analogous to today's ocean waves. They, too, consist of a hierarchy of structures—very long waves that have smaller ones of varying lengths superimposed on them.

Initially the fluctuations were not very strong. Nowhere did the density vary from its average by more than a fraction of a percent. This follows from the near isotropy of today's cosmic background radiation. Careful measurements of the intensity of this radiation from different directions in the sky show variations of

*The symbol $M_\odot$ (read "solar mass") is a common one in astronomy and refers to a mass equal to that of the Sun.

only about one part in ten thousand. The initial fluctuations in the matter density are expected to have been comparably small, because until decoupling, the matter and photons were interacting strongly.

Formation of Primordial Clouds

Upon decoupling, the primordial photons retained their almost perfectly smooth distribution, but the matter did not. Wherever there existed small enhancements in density, the attractive force of gravity counteracted and tended to slow the expanding matter. At first it succeeded in this only over the largest scales. Over smaller scales, the pressure exerted by the matter's constituent particles was too strong to permit gravitational contraction. As a result, regions extending over approximately 10^{15} to 10^{16} $M_\odot$ of matter expanded more and more slowly and eventually began to contract. This led to the formation of huge primordial clouds of hydrogen and helium—the first large-scale structures in the evolving Universe (see figure 2.2). Between these clouds, voids developed that were as large or larger than the clouds themselves. (Note that

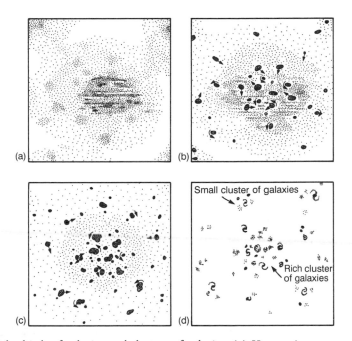

Figure 2.2. Model of the birth of galaxies and clusters of galaxies. (*a*) Huge primordial clouds of hydrogen and helium condense from density fluctuations present in the young Universe. (*b*) Local gravitational forces break up the primordial clouds into gaseous fragments. In the process, the fragments acquire large velocities relative to each other and many begin to rotate. (*c*) The fragments collide and merge into larger, more massive fragments. The first stars are forming and the fragments turn into galaxies. (*d*) Within roughly 2 to 3 billion years the birth of galaxies is largely completed. The original primordial clouds have evolved into superclusters of galaxies, each containing millions of member galaxies.

whenever matter gets compressed, the pressure rises and opposes the compressional forces. Under the conditions existing in the early Universe, the pressure was successful in opposing gravity on all but the largest scales. The concept of pressure is discussed in box 3.2).

Although there is no direct observational knowledge about the formation of the primordial clouds, it seems reasonable to assume that their contraction did not proceed smoothly and uniformly, and few if any of the primordial clouds contracted into spheres. As first suggested by Yokov B. Zel'dovich, professor of astrophysics at Moscow University, the compression of the matter generally tended to occur more rapidly along one of the three spatial axes than along the other two. As a result, the first large-scale primordial clouds are believed to have taken on the shapes of huge, flattened pancakes. Where these pancakes intersected, long, thin filaments formed. Thus, the structure and distribution of the primordial clouds was probably highly irregular and interconnected. Three-dimensional maps of the present-day distribution of superclusters of galaxies, which evolved from the primordial clouds, reflect this (see figure 2.3).

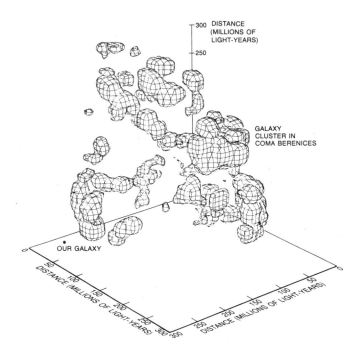

Figure 2.3. Three-dimensional map of the distribution of bright galaxies within 250 million LY of the Milky Way, as seen in the northern sky. The map outlines the central regions of the superclusters of galaxies, and it illustrates the interconnectedness of the distribution of matter on the largest scale. The map is based on a survey of 1801 galaxies. It was constructed by Carlos Frenk and Simon White of the University of California at Berkeley from a map prepared by Marc Davis and his colleagues at the Center for Astrophysics of the Harvard College Observatory and the Smithsonian Astrophysical Observatory. From J. Silk, A. S. Szalay, and Y. B. Zel'dovich, "The Large-Scale Structure of the Universe," *Scientific American*, October 1983, 78. Used by permission.

While the primordial clouds were forming, the Universe continued to expand. By contracting the gas into clouds, gravity succeeded in *locally* counteracting the cosmic expansion. However, *globally* the expansion of the Universe was not affected. The clouds participated in the cosmic expansion by moving farther and farther away from each other, a phenomenon still observed today among the superclusters of galaxies.

At present astronomers do not understand why the primordial clouds condensed over regions as large as 10^{15} to 10^{16} $M_\odot$ (values deduced from the masses contained in today's superclusters of galaxies). Simple theoretical estimates give masses that are one to two orders of magnitude smaller. For example, if the maximum initial density fluctuations of the matter, as allowed by the variations in the intensity of today's cosmic background radiation, are used in the mathematical modeling of the Universe's early evolution, structures of at most 10^{14} $M_\odot$ are obtained.

So how did the primordial clouds of hydrogen and helium acquire their large masses? At present, astronomers take very seriously the suggestion that the neutrinos created during the Big Bang were responsible, provided that they possess a tiny mass (see the Introduction to part 1). As noted in chapter 1, the primordial neutrinos decoupled from the photons and the rest of the matter when the Universe was only about one second old. From then on, they followed their own evolutionary course. Their initial density distribution, too, was marked by small fluctuations that, under the influence of gravity, began to grow with time. However, because this growth had already begun at age one second, when the Universe was still much more compact than at age 500,000 years (when the matter and photons decoupled), the force of gravity was stronger and more time was available to amplify the fluctuations. Therefore, the regions over which the neutrinos became concentrated were considerably larger than the structures of 10^{14} $M_\odot$ obtained for the evolution of the matter alone. These large regions constituted gravitational potential wells into which the matter was pulled, upon its decoupling from the photons. In this manner, the primordial clouds of hydrogen and helium, encompassing 10^{15} to 10^{16} $M_\odot$, are thought to have formed.

Though astronomers are still debating whether neutrinos indeed played a role in the formation of the first large-scale structures of the Universe, it is a plausible hypothesis and, in broad terms, the theoretical consequences are consistent with observations. Furthermore, neutrinos may constitute the dark matter that stellar and galactic motions indicate is present within galaxies and in intergalactic space, as noted in chapter 1. Another hypothesis is that this dark matter might also consist, at least in part, of black holes or some other, even more esoteric, form. We just don't know. Surely, a number of surprises still await us in our search to understand the formation of the Universe's first structures of large scale.

Breakup of Primordial Clouds

As the primordial clouds formed, they inherited all the weak, small-scale density fluctuations that had characterized the matter when it decoupled from the photons. Gradually, the local gravitational contractions that began on the largest scales affected the small-scale density fluctuations as well. In many regions where the density within the clouds was greater than the average, gravity managed to contract the gas further and to intensify the concentration of matter. This increased the local gravitational force still more and led to the breakup of the primordial clouds into large numbers of cloud fragments.

At present, we do not understand the details of this breakup. A good guess is that the fragments acquired masses in the range of 10^5 to 10^8 $M_{\odot}$ and, possibly, even more. The smaller ones were probably far more numerous than the larger ones. Furthermore, the fragments were strongly concentrated toward the centers of the primordial cloud concentrations where the density of matter was greatest and, hence, gravity acted most strongly.

The formation of the cloud fragments may be compared to the condensation of water vapor into rain droplets in the clouds of the Earth's atmosphere. The major differences between the two kinds of condensations are in scale, time, and the forces involved. The primordial cloud fragments extended over masses of millions of suns, their formation took hundreds of millions of years, and the participating force was gravity. In contrast, rain droplets weigh a fraction of a gram, they condense spontaneously, and the force responsible is the electric force, which pulls water molecules together.

The primordial clouds of hydrogen and helium were the first structures of large scale to form in the evolving Universe. Their breakup into fragments created second-order structures on the next smaller scale. Next, the fragments broke up further into stars and interstellar clouds, producing third-order structures on a still smaller scale.

BIRTH OF GALAXIES AND CLUSTERS OF GALAXIES

The breakup of the primordial clouds into fragments was chaotic and violent. At first the local gravitational forces, which produced the fragments, were rather weak and the contractions were slow. As the matter became more clumped, the gravitational forces increased in strength and the gaseous material fell together faster and faster. Eventually it reached supersonic speeds. Shock waves, like those from supersonic jets but much stronger, were created and thrust their way through the clouds. They acted like the blades of a giant mixer, agitating the matter and giving it rapid, turbulent motions. Whirlpools were stirred up, extending over thousands and millions of solar masses of gas, and windstorms began to blow with velocities of tens and hundreds of thousands of kilometers per hour. The fragments were accelerated to large velocities relative to each other and many acquired rapid rotational motions.

This violent activity occurred during the first few hundred million years, when the Universe was much more compact than it is today. Consequently the fragments had very little elbow room, particularly because many of them had large velocities. They started to collide, but not in a simple way like two tennis balls that bounce off each other. The collisions were inelastic and complicated. Fragments of gas containing masses equal to millions of suns were plowing into each other at thousands of kilometers per hour. On impact the gas of each fragment tried to maintain its own momentum and to penetrate into the oncoming fragment. The ensuing struggle compressed and heated the gas. New shock waves were created that spread through both fragments and caused further turbulence and compression. The friction between the oppositely directed streams of gas gradually slowed the motions and the fragments merged into one larger fragment.

Quite often this sequence of events was not permitted to run its full course. Before one merger was completed, collisions occurred with other fragments, creating further compression, heating, shock waves, turbulence, and merging. In the course of dozens to hundreds of collisions and mergers, the masses of the surviving fragments increased manyfold and their numbers decreased correspondingly.

Birth of the First Stars

The shock waves created by the collisions compressed the gas of the fragments. This happened to such a degree that huge regions, involving up to millions of solar masses and scattered throughout the fragments, became gravitationally unstable and collapsed into compact clouds within which the first generations of stars began to form. In some fragments star formation occurred in rapid bursts and most of the gas was quickly converted into stars. In others, which possessed a great deal of internal turbulence and motion, the collapse into stars was less efficient and much of the matter remained in gaseous form.

The birth of the first generations of stars had two far-reaching consequences. Up to now the fragments had been rather cold and dark objects,* except for regions where collisions and shock waves had temporarily heated the gas. This changed rather quickly when the newly formed stars began to pour forth their radiant energy and illuminate the gaseous matter surviving in their vicinity. In a semiregular pattern, along the paths where shock waves had compressed the gas and induced gravitational collapse, light after light turned on as stars and groups of stars were born. For the first time since decoupling from the primordial photons, the matter of the Universe lit up and became visible.

The second consequence of star formation resulted from the accompanying change in the density of matter. The gas that collapsed into stars was no longer spread out over vast regions of space as a diffuse and amorphous mass. It became concentrated into relatively small volumes. The stars' diameters were up to a million times smaller than the distances separating them. Consequently, when two fragments collided the stars did not generally hit each other; they were too compact for that.† They merely interacted by the long-range gravitational force and altered each other's motions until, after some time, their orbits became intermingled and the two groups of stars merged into one larger stellar system (see figure 2.4).

---

*Before the birth of stars, the fragments were cold and dark because they radiated away their heat energy very efficiently in the infrared and radio parts of the electromagnetic spectrum. This kind of heat loss is crucial for the formation of stars, as will be seen in chapter 3.

†A difference by a factor of 10^6 between the sizes and separations of the stars in galaxies is comparable to scattering a handful of sand so that the average distance between one sand grain and the next is one to two kilometers. If two batches of sand, dispersed to this extent, were thrown toward each other, the probability that two grains collide would be vanishingly small.

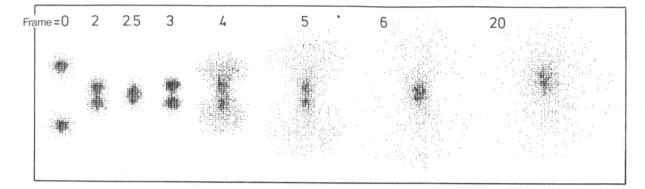

Figure 2.4. Computer simulation of the merger of two elliptical galaxies, leading to the formation of a daughter galaxy that is much more extended than either of the original ones. In nature, such a merger takes hundreds of millions of years. In this simulation, each galaxy consists of 1000 stars, though in reality elliptical galaxies contain millions to many billions of stars. From Joseph Silk, *The Big Bang* (San Francisco: W. H. Freeman & Co., 1980), 171. Used by permission.

Birth of Galaxies

In contrast to the stars, the uncondensed gaseous components of two colliding fragments interacted as they always had. They plowed into each other, became compressed and heated, shock waves and turbulence were created, and new

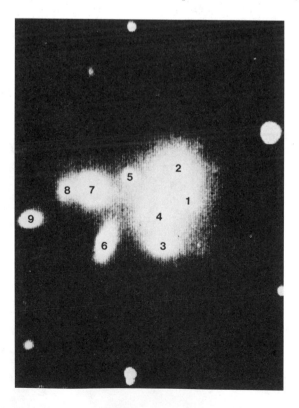

Figure 2.5. Nine small galaxies in the process of coalescing into a single galaxy. The motions of the nine galaxies suggest that they are the nuclei of once independent galaxies that collided and, due to their mutual gravity, are now merging into a single system. Within roughly a billion years, the nine galaxies are expected to be no longer identifiable as separate entities and the coalescence will have produced a new giant elliptical galaxy.

instabilities formed, leading to more star formation. In some collisions the gaseous components stayed with the stars. In other collisions the gas was cleanly swept away from the stars. Thus, a broad spectrum of end products resulted, from stellar systems consisting of only stars to clouds comprised of mostly gas.

As this sequence of collisions and mergers of fragments and of conversion of gas into stars progressed, the surviving systems looked more and more like star clusters and galaxies (figure 2.5). The fragments that were most successful in capturing and assimilating others grew into large galaxies with masses of 10^{11} $M_\odot$ and more. Those that managed to retain their gaseous material became spiral galaxies, like our own or the Andromeda galaxy (see figure 1.8). Fragments that lost their gas or converted most of it into stars, evolved into giant ellipticals, like galaxy M 87 in the Virgo cluster (see figure 1.12). Fragments that grew less successfully became dwarf ellipticals and globular star clusters. Finally, fragments that collided with or were captured into close orbit about another more massive galaxy experienced strong tidal perturbations and ended up with rather distorted shapes (figures 2.6 and 2.7). They turned into irregular galaxies, such as the Large and Small Magellanic Clouds (see figure 1.5).

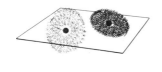

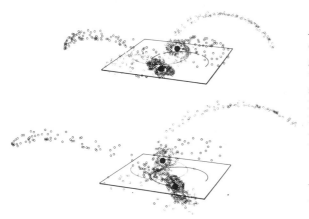

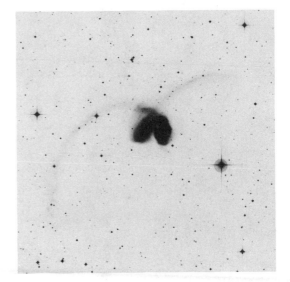

Figure 2.6. (Left) A sequence of five frames from a computer-generated motion picture, simulating the gravitational distortions resulting from the close encounter of two galaxies. The initial conditions were chosen so as to reproduce the tidal distortions observed in galaxies NGC 4038 and 4039 (also known as the Antennae, figure 2.7). From George O. Abell, *Exploration of the Universe*, 4th ed. (Philadelphia: Saunders College Pub., 1982), 618. Used by permission.

Figure 2.7. (Above) The Antennae, a pair of spiral galaxies, NGC 4038 and 4039, seen shortly after a very close encounter.

Figure 2.8. Stephan's quintet. This interesting group of five galaxies was named after their discoverer, a nineteenth-century astronomer observing from Marseilles, France. Four of the galaxies are in close proximity of each other, as indicated by the luminous bridges of gas connecting them and by their nearly identical redshifts. They are the compact center of a small cluster of galaxies at a distance of approximately 250 million LY. The fifth galaxy (an elliptical), in the lower right part of the photograph, is a foreground galaxy and not associated with the others.

Most of the newly formed galaxies found themselves in the vicinity of others and were bound to each other by the force of gravity. This clustering was particularly pronounced near the central regions of what used to be the primordial clouds (figure 2.8). Today that is where the rich clusters of galaxies with thousands of members are found, such as the Virgo and Coma clusters. Farther out the galaxies clustered into sparser groupings with only a handful to a few dozen members. Our Local Group is an example of such a small cluster. Most of these outlying galaxy clusters are weakly tied by gravity to a rich central cluster. Together they form the superclusters of galaxies, which we visited on our imaginary journey in chapter 1.

Note that this description of the origin of galaxies and galaxy clusters merely represents some of the current thinking of astronomers. It is far from the final word on the subject. The description is largely based on theoretical deductions from observations of cloud collisions that presently happen on much smaller scales in our galaxy, on observations of collisions and near-encounters among some nearby galaxies (see figures 2.5 and 2.7), and on computer simulations (see figures 2.4 and 2.6). It is not based on observations of the actual events that produced the galaxies and galaxy clusters.

In fact, Valérie de Lapparent, Margaret J. Geller, and John P. Huchra of the Harvard-Smithsonian Center for Astrophysics have suggested a scenario of galaxy cluster formation quite different from the one described here. From a study of some 1100 galaxies distributed in a three-dimensional wedge of sky 6° × 117° and extending to a distance of about 650 million LY (see figure 2.9), they deduce that galaxies are not grouped into distinct superclusters with voids separating them from each other. Instead, they find that galaxy clusters are much more interconnected and appear to have a shell-like or bubble-like distribution. The bubbles are roughly spherical or elliptical, with typical diameters of 150 million LY, though the largest bubble in the survey is twice that size. The galaxies are located predominantly on the surfaces of the bubbles, while the interiors of the bubbles are largely void of matter. Huchra suggests that "these bubbles fill the Universe just like suds filling the kitchen sink."*

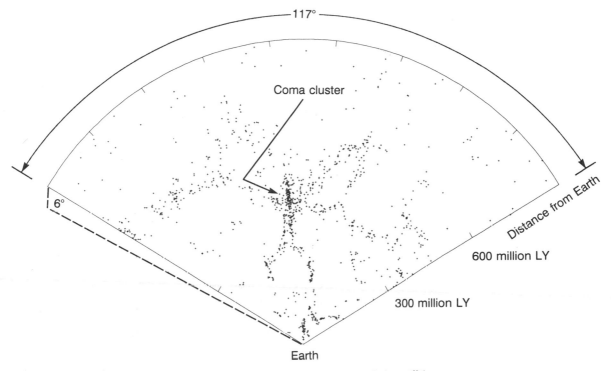

Figure 2.9. Distribution of nearly 1100 galaxies in a three-dimensional slice (6° by 117°) of the Universe in the direction of the Coma cluster, as compiled by Valérie de Lapparent, Margaret J. Geller, and John P. Huchra. The diagram suggests that the large-scale distribution of galaxies in the Universe is bubble-like. The Coma cluster of galaxies is the richest supercluster system in this survey and appears to lie at the intersection of several bubbles.

Science and the Citizen section, *Scientific American*, March 1986, pp. 59, 62.

If Huchra and his colleagues are right, then gravity may not have been the dominant force in initiating the formation of galaxies and galaxy clusters, as is traditionally assumed. The dominant force may have been shockwaves that resulted from explosions of enormous magnitudes within a few hundred million years after the Big Bang. What caused the explosions is not known, but according to one theory they were supernova explosions of early generations of massive stars. The shockwaves pushed the primordial matter apart locally and created the observed bubble-like voids. Where the shockwaves from neighboring bubbles collided, the matter became compressed (probably unevenly), cloud fragments resulted, and galaxies began to form, perhaps in the manner discussed in this section.

To decide how galaxies and galaxy clusters really formed—by one of the scenarios described in this chapter or, perhaps, by yet another one—observations like those by Huchra and his colleagues need to be extended over larger parts of the sky. Furthermore, galaxies 10 and more billion LY distant need to be observed in order to see them in their formative stages. However, light from such distant objects is, with few exceptions, too diffuse and faint to be detected with the present generation of telescopes. This should change when the Hubble Space Telescope, with a 2.4-meter mirror, is launched into Earth orbit aboard a space shuttle, probably some time in 1988 or 1989. (The launch was originally scheduled for 1985, rescheduled for 1986, and then delayed further by the explosion of the space shuttle *Challenger* on January 28, 1986.) Because this telescope will be located far above the atmosphere, objects 50 times fainter than are presently observable with ground-based optical telescopes should come within our view, allowing us to see and to study first hand some of those long-ago events that created the large-scale structures in the Universe.

THE STRUCTURE OF GALAXIES

Let us now turn from the topic of galaxy birth to that of galaxy structure. We begin with the normal galaxies, namely those that have been able to reach some semblance of equilibrium and do not experience major perturbations from either outside or within.

The most widely used scheme for classifying the normal galaxies is one introduced by Edwin Hubble, who lined them up along a tuning fork pattern according to their structural appearance (see figure 2.10). The handle of the fork is occupied by the *elliptical galaxies*, starting with the perfectly spherical ones and ending with the highly flattened ones. They are labeled E0 through E7, depending on their degree of flattening. The point where the prongs of the fork divide is marked by the *S0 galaxies*. They resemble the spirals except that they contain no gas or dust. Along the upper fork lie the *spiral galaxies* in order of tightness of the winding of their arms, from the most tightly wound (type Sa) to the most loosely wound (Sc). Along the lower fork are situated the *barred spirals*. They are like the ordinary spirals, except that they have a luminous bar running through their nucleus. This scheme of classifying galaxies, the *Hubble sequence*, is summarized in figure 2.10. It is further illustrated by figures 1.6, 1.8, 1.9, 1.12, 2.1, 2.11, and 2.13–2.15. Table 2.1 summarizes the range of physical characteristics of the galaxies, including the irregulars.

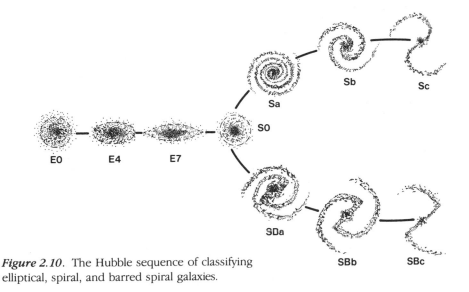

Figure 2.10. The Hubble sequence of classifying elliptical, spiral, and barred spiral galaxies.

Elliptical Galaxies

As table 2.1 indicates, the elliptical galaxies (figure 2.11) possess by far the greatest range in mass, size, and luminosity. The smallest of them resemble the globular star clusters, with approximately 1 million stars. The largest ones contain up to 10^{13} stars, nearly one hundred times more than there are in the Milky Way. Elliptical galaxies contain very little gas or dust. They either converted

Figure 2.11. Galaxy M 86, an elliptical of type E3. This galaxy does not have quite the perfectly spherical shape of M 87 (see figure 1.12); hence the designation E3.

Table 2.1. Range of Physical Characteristics of Elliptical, Spiral, and Irregular Galaxies

	Ellipticals	Spirals	Irregulars
Mass (solar masses)	10^6 to 10^{13}	10^9 to 10^{12}	10^8 to 10^{11}
Diameter (thousands of LY)	1 to 500	20 to 200	5 to 30
Radiant energy output (solar units)	10^6 to 10^{11}	10^8 to 2×10^{10}	10^7 to 2×10^9
Content	Old stars, little or no gas and dust	Old and young stars, much gas and dust	Old and young stars, much gas, dust may be present or absent

NOTE: Adapted from G. Abell, *Exploration of the Universe*, 4th ed., (Saunders College Publ., 1982), 615.

most of the gas and dust into stars or lost that material during encounters with other galaxies, probably during their formative years. As a result, star formation has ceased in them long ago, and the only stars present today are old ones.

Spiral Galaxies

Of all the galaxies, the spirals are of greatest interest to us. First, our galaxy—the Milky Way—is one of them. Second, they contain much gas and dust, from which stars are still forming today. By studying how this is happening we can learn about the origin of the Sun and the planets, including Earth.

Spiral galaxies span a considerably narrower range in mass, size, and luminosity than the ellipticals. The Milky Way is one of the largest with approximately 150 billion stars, a diameter of more than 100,000 light-years, and a luminosity roughly 10 billion times the Sun's. The structure of spirals is much more complicated than that of ellipticals (see figure 2.12). It consists of two main

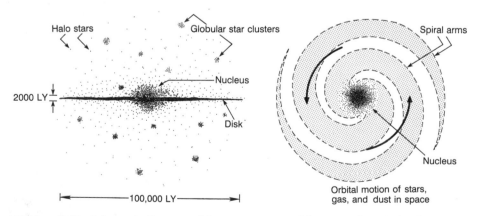

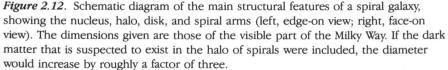

Figure 2.12. Schematic diagram of the main structural features of a spiral galaxy, showing the nucleus, halo, disk, and spiral arms (left, edge-on view; right, face-on view). The dimensions given are those of the visible part of the Milky Way. If the dark matter that is suspected to exist in the halo of spirals were included, the diameter would increase by roughly a factor of three.

Figure 2.13. Galaxy NGC 4565. This galaxy, seen edge-on, is of type Sb. Its distribution of gas, dust, and stars is similar to ours.

components, a spherical and a disk component. The *spherical component* possesses only old stars, many of which are grouped into globular star clusters (see figure 1.3). (The meanings of "old" and "new" stars are discussed in chapter 3.) Both the individual stars and globular star clusters show a strong concentration toward the galactic center, the *nucleus*; but they also extend in a sparser distribution outward to great distances, the *halo*. In addition, the halo of the spherical component contains a great amount of dark matter that extends far beyond the visible stars and star clusters. In fact, the motions of stars and interstellar clouds in the outlying regions of spiral galaxies (see below) indicate that this dark halo may constitute the major part of the masses of these galaxies. The *disk component* of most spirals is rather flat, with a ratio of thickness to diameter comparable to two or three 12-inch phonograph records stacked together (figure 2.13). The disk contains much gas and dust, clumped into large clouds, as well as many young stars. The brightest and youngest of the stars lie

Figure 2.14. Galaxy M 81. This beautiful galaxy, with a pronounced central bulge and graceful arms dotted with bright emission nebulae, is of type Sb. Dark, light-absorbing dust lanes are clearly distinguishable along the arms. The Milky Way, viewed from afar, is believed to resemble this galaxy.

mostly along the *spiral arms*, where they have recently been born and where they act as luminous markers defining the spiral pattern (figures 2.14 and 2.1).

The flatness of the disk component is due to gravity and the disk's rotation about the galactic center. Gravity concentrates the matter of the disk into the flat, relatively thin distribution, and rotation gives rise to centrifugal forces, which prevent the matter from falling toward the galactic center. The rotation however, is not uniform like that of a record. The galactic nucleus with its great concentration of mass acts as a gravitational hub, keeping the stars, gas, and dust of the disk in orbit. Consequently, the inner parts of the disk rotate more quickly than the outer parts, much as the Sun's inner planets revolve more quickly than the planets farther out and for the same physical reason. However, toward the outer regions of the disk, the rotational velocities do not drop off as they do in the case of the planets circling the Sun. They don't because of the presence of the halo, which, as noted above, contains a great deal of mass (in contrast to the Solar System, in which nearly all of the mass is concentrated in the Sun).

The arms of spiral galaxies do not rotate rigidly along with the disk. If they did, the differential rotation of the disk would long ago have wound them up into an indistinct blur. Instead, the arms—that is, the very bright, young stars that define the spiral pattern, along with clouds of gas and dust that give rise to them—are constantly being renewed.

This renewal is believed to take place as follows. When the interstellar gas and dust, in their orbits about the center of the galaxy, enter a spiral arm they collide with matter already in the arm. This compresses the gas and dust into huge clouds of many thousands of solar masses. Still other clouds form when shock waves from supernova explosions (see below and chapter 3) expand into interstellar gas and dust and compress them. In many of the clouds thus formed, the force of gravity becomes sufficiently strong to collapse some of the gas and dust into stars (for details on star formation see chapter 3). This happens in much the same way as it did billions of years earlier in the compact gas clouds of the protogalactic fragments, except on a much smaller scale. Usually hundreds of stars are born simultaneously and for some time they orbit around the galactic nucleus as loosely bound groups. They are called *open star clusters*, in distinction to the much older and richer globular star clusters. Figures 1.2, 3.3, 3.6, and 3.7 show open star clusters in various stages of their evolution.

Among the newly formed stars, a few are always much more massive than the Sun. Such stars use up their nuclear fuel at a furious rate. Within just a few million years, and before their motions have even carried them through the arm in which they were born, they literally burn themselves out. While they last, these massive stars shine up to a million times more brightly than the Sun and mark the location of the spiral arms. In the end, they explode as supernovae and throw much of their matter back into space. During such explosions, enormous amounts of energy are released and shock waves are created, which contribute to the formation of new generations of stars and the maintenance of the spiral arms.

Barred Spiral Galaxies

A third group of galaxies recognized in the Hubble sequence are the barred spiral galaxies (figure 2.15). They resemble normal spirals, except that their

Figure 2.15. The barred spiral galaxy NGC 1300, of type SBb.

arms start not from the nucleus but from the ends of a luminous bar of stars, gas, and dust, which passes right through the galactic center. Computer simulations show that a disk of orbiting stars, gas, and dust quite readily assumes a bar-like structure, suggesting that, possibly, all spiral galaxies periodically develop bars and then lose them again.

S0 Galaxies

The S0 galaxies, which lie at the origin of the fork in the Hubble sequence, are similar to the spirals as far as the spatial distribution of stars goes, but they contain little gas or dust. Hence, they have neither spiral arms nor clouds in which stars are forming. Very likely S0 galaxies were once normal spiral galaxies, but lost their gas during past collisions with other galaxies. This hypothesis is supported by the fact that they are most abundant in the central regions of rich clusters of galaxies, where collisions capable of sweeping away the gas and dust occur most often.

Irregular Galaxies

The irregular galaxies are galaxies that do not readily fit into the Hubble sequence. They come in the most unusual and contorted shapes (figures 2.16, 2.17). Many of them are close satellites of larger galaxies and their distorted

Figure 2.16. The Large Magellanic Cloud, one of the two companion galaxies of the Milky Way. Although this galaxy is classified as an irregular, its outer parts are distorted into what appear to be short spiral arms. Some astronomers have suggested that this galaxy might be evolving from an irregular into a barred spiral. (See also figure 1.5.)

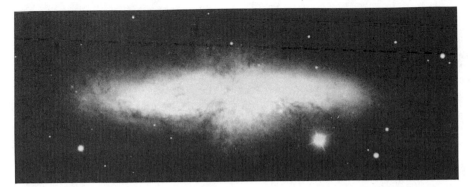

Figure 2.17. Galaxy M 82. Though this galaxy is classified as an irregular, it probably is a disk galaxy (possibly a spiral) seen edge-on that a few million years ago experienced an explosion of enormous magnitude at its center. The explosion accelerated millions of solar masses of gas to velocities of up to 500 km/sec and produced the chaotic structure seen in the galaxy's central regions. Incidentally, Galaxy M 82 lies in the vicinity of the spiral galaxy M 81 (see figure 2.14) and is connected to it by a faint bridge of gas, the result of a close encounter between the two galaxies about 100 million years ago.

shapes are caused by tidal forces, much like the Earth's oceans are distorted by the Moon's tidal force. Others undergo internal upheavals from explosions of gigantic magnitude. Still others either are colliding or have recently collided with another galaxy and are suffering from this experience.

Active Galaxies

A final group of galaxies, unknown in Hubble's time and hence not included in his classification, are the radio galaxies, Seyfert galaxies, and the quasars. Collectively they are known as *active galaxies* because all three are characterized by unusually energetic activities.

Figure 2.18. Radio telescope at Bad Münstereifel-Effelsberg, near Bonn, Federal Republic of Germany. The dish of this telescope is 100 meters across, making it the world's largest steerable radio antenna. Radio telescopes allow astronomers to extend their observations to the long-wavelength part of the electromagnetic spectrum (millimeters to many centimeters).

The *radio galaxies* emit much of their energy as radio waves (figure 2.18). In many cases the radio emission comes from pairs of sources placed hundreds of thousands to millions of light-years apart on opposite sides of the galactic nuclei. Exceptionally powerful explosions in the galactic nuclei must have ejected matter and thus created these double radio sources.

The previous chapter introduced two of the most interesting radio galaxies — M 87 and Centaurus A (figure 1.10). M 87, seen in figure 2.19, is the giant elliptical galaxy near the center of the Virgo cluster. The faint, lumpy jet shooting out from its nucleus is probably the result of a series of explosions. The radio emission of this galaxy is strongly concentrated toward the core of the galaxy and gradually tapers off farther out. Radio contour maps show that the

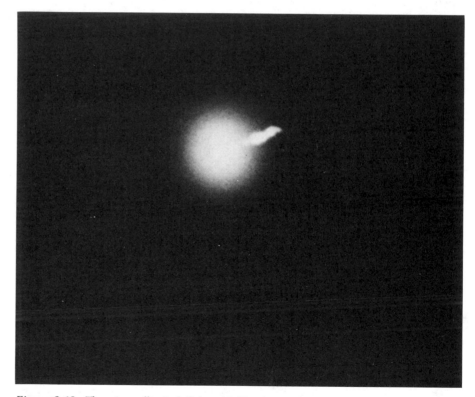

Figure 2.19. The giant elliptical Galaxy M 87 (of type E0, also known as Virgo A) in the Virgo cluster (see also figure 1.12).

tapering-off is elongated in opposite directions from the galactic nucleus and is aligned along the axis defined by the jet. This suggests that two jets are involved, which is only vaguely evident from the optical photographs. It is believed that most of the radio waves emitted by galaxies like M 87 and Centaurus A are *synchrotron radiation*, which originates from electrons accelerated to nearly the speed of light and forced by magnetic fields to travel in tight spiral paths.

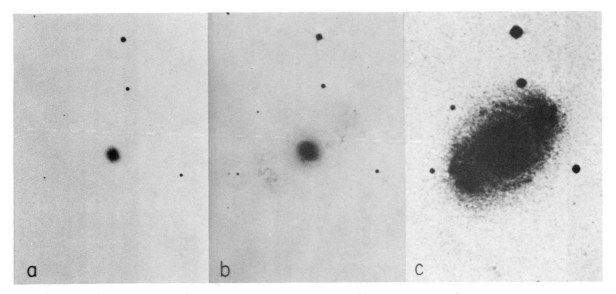

Figure 2.20. A sequence of three negative prints of progressively longer exposures of Seyfert galaxy NGC 4151. In the **(a)** photograph, the galaxy's nucleus is visible as a compact, star-like object. In the other two photographs, the galaxy's spiral structure can be seen.

The *Seyfert galaxies** are intrinsically much brighter than ordinary galaxies, with an energy output exceeding that of the giant elliptical galaxies by at least a factor of ten to a hundred (figure 2.20). The energy is emitted not only in the optical part of the spectrum, but also as infrared, ultraviolet, and X radiation. Often these galaxies are powerful sources of radio waves as well. Most of this energy originates from a very compact region, measuring only a few light-weeks or light-months across.

Quasars are another group of objects that many astronomers today suspect to be galaxies also — or, more accurately, the central nuclei of galaxies. They have the point-like appearance of stars, yet their spectra are clearly nonstellar. In the early 1960s, when they were first discovered, astronomers classified them as star-like or *quasi*-stell*ar* and contracted these two words into *quasar*. In many respects, their energy output resembles that of Seyfert galaxies, though most of them are brighter still. They a lie at very great distances where, if they are indeed galaxies, only their compact central nuclei would tend to show up on photographs (figure 2.21).

Initially, astronomers were quite uncertain about the nature of quasars, not only because of their star-like appearance, but also because of the unusual features in their spectra. They contain emission and absorption lines never seen in astronomical objects before. In 1963, three years after the discovery of the first quasar, Maarten Schmidt of the Hale Observatory recognized that these

*The Seyfert galaxies are named after Carl K. Seyfert who, as a postdoctoral fellow at the Mount Wilson Observatory in California, described their properties in 1943.

Figure 2.21. Quasar 3C 273. This quasar is the brightest one visible from Earth, with a recession velocity of 15% of the speed of light and a corresponding distance of 3 billion LY. Radio observations indicate that it consists of two sources, the fainter of which shows up on photographs as a jet, reminiscent of that of M 87.

spectral features originate in the UV part of the electromagnetic spectrum and are redshifted into the optical range. They appear so unusual because UV radiation is normally absorbed by the Earth's atmosphere and, hence, in the early 1960s astronomers were not familiar with its spectral features. In the case of the quasars, the redshift moves the UV radiation into the optical range where, like light from the Sun, it readily penetrates to the ground. For example, for quasar 3C 273, whose light Schmidt analyzed, the redshift amounts to 0.158, and for the quasar with the largest redshift discovered to date, it is 4.01. In comparison, stars in our galaxy have redshifts of only about 10^{-4}, with some being positive and others negative, and clusters of galaxies typically have redshifts in the range of 0.01 and 0.2.*

For two decades astronomers have debated what the source of the quasars' unusually large redshifts could be. Are they caused by intense gravitational fields, by very high recession velocities, or by yet other processes that we presently do not understand? If they are due to high recession velocities, are those velocities related to the cosmic expansion as are the recession velocities

---

*Astronomers refer to the *redshift* of a star or a galaxy by the letter z and define it as the ratio $(\lambda - \lambda_0)/\lambda_0$, where λ_0 is the wavelength of the radiation as it is emitted by the source and λ is the redshifted wavelength as observed on Earth. Redshifts of 4.01, 0.2, 0.01, and 10^{-4} correspond to recession velocities of 92%, 18%, 1%, and 0.01% of the speed of light, respectively.

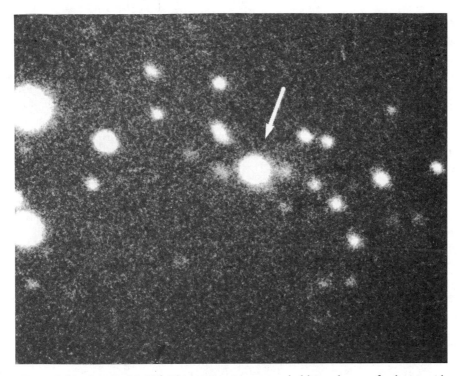

Figure 2.22. Quasar 3C 206. This quasar is surrounded by a cluster of galaxies with about 200 members. The quasar and galaxies have similar redshifts (z = 0.203 to 0.206, corresponding to a distance of about 3.7 billion LY), strongly suggesting that the quasar is a member of the cluster.

of distant clusters of galaxies? Or are they locally produced velocities? For most astronomers the controversy has now been resolved by the discovery that some of the quasars are members of clusters of normal galaxies (figure 2.22) and others show absorption lines originating from intervening intergalactic gas. These discoveries leave little doubt that the quasars' redshifts are due to large recession velocities and the expansion of the Universe. The recession velocities measured to date range from 5 to 92% of the speed of light, placing the quasars at distances from 1 to 12 billion light-years (providing that Hubble's velocity–distance relationship applies to the quasars and our notions of the large-scale structure of the Universe are correct). Thus, the most distant quasars emitted their radiation when the Universe was only about 1 billion years old.

Note, however, that a minority of astronomers still questions that the currently available data on quasars proves beyond the shadow of a doubt their cosmological nature. They suggest that quasars may be nearby objects, whose large redshifts have nothing to do with the cosmic expansion and whose origin and physical structure we do not presently understand. Therefore, they prefer not to think of quasars as the luminous nuclei of galaxies, but restrict their definition of quasars to *intense, point-like sources of light (and, in some cases, of radio waves, X rays, and gamma rays) that are characterized by large redshifts of the lines in their visible spectra.*

Black Holes: Engines of Active Galaxies

What could the energy source of quasars, Seyfert galaxies, and radio galaxies be that produces such powerful explosions and creates a brightness of up to a thousand times that of our galaxy, all from a volume not much bigger than the Solar System? One rather intriguing answer, which many astronomers are coming to accept, is that the sources are gigantic black holes, with masses of millions or even billions of suns, at the centers of the active galaxies. The energy is produced when gas clouds and stars, in the course of their orbital motions about a galaxy, pass too close to the black hole (see figure 2.23). Intense tidal forces tear the gas and stars apart. The gases, which may be at tens or hundreds of millions of degrees Kelvin, are then pulled inward to form a thin *accretion disk* around the black hole. Gradually they spiral closer and closer to the black hole and, eventually, are irretrievably drawn into it. During the final descent, the matter is accelerated to nearly the speed of light and enormous amounts of energy are released, far more than is available from nuclear reactions or any other brown source. The outbursts are repeated as new gas and stars approach the black hole and succumb to its inescapable gravitational pull.

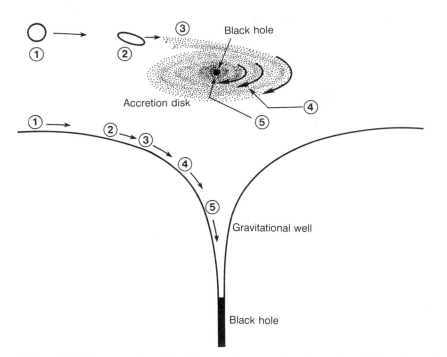

Figure 2.23. Proposed model of the engine powering the active galaxies. According to this model, the engine is a gigantic black hole lying at the centers of quasars, Seyfert galaxies, and radio galaxies. Stars (1) that pass too close to the black hole are tidally torn apart (2), spread out into an accretion disk (3), and finally are drawn into the bottomless gravitational well of the black hole (4, 5), releasing enormous amounts of energy. If the Sun, for example, were to meet this fate, the energy released would be 100 times more than it radiates in the course of its main-sequence lifetime of 10 billion years (see chapter 3).

The black hole model explains another feature observed in many active galaxies, namely the jets of gas and radio lobes. These features could result from the high pressures created by heating and compression when the gas spirals through the accretion disk toward the black hole. The pressures become so great that they squirt part of the gas outward at nearly the speed of light in opposite directions along the disk's spin axis (the directions of least resistance). Besides creating jets and radio lobes, this kind of ejection would also lead to the emission of gamma rays, X rays, and synchrotron radiation (due to relativistic electrons of the jets interacting with magnetic fields surrounding the black hole), which are observed from near the cores of many active galaxies.

It has been suggested that the sequence of events just described, which spells the demise of entire stars, is the engine powering more than just the active galaxies. Possibly, even ordinary galaxies, including our own, possess black holes at their centers and periodically flare up as matter falls into them.

BUILDUP OF HEAVY ELEMENTS

When the galaxies formed, matter consisted of approximately 76% hydrogen (by mass), 24% helium-4, traces of deuterium and helium-3, and virtually nothing else. By the time the Solar System was born, 4.5 billion years ago, the amount of hydrogen in our Galaxy had been reduced by about 2 to 3%, helium-4 had increased slightly, and roughly 1 to 2% of the matter had been converted into heavier elements—carbon, nitrogen, oxygen, neon, magnesium, silicon, iron, and the rest. Similar changes had occurred in all other spiral, barred spiral, S0, and most irregular galaxies. In elliptical galaxies and in the halo of the spiral galaxies, the changes were considerably less. Only one one-hundredth to one-tenth of 1% of their matter had been changed into heavy elements.

Significance of Heavy Element Buildup

Is such a small change as 1 to 2% buildup of heavy elements in the spiral galaxies significant? The answer is yes. First, it reveals a great deal about the formation and early evolution of the galaxies. Second, and more important for us, the heavy elements made a planet like Earth and life on it possible. The core of our planet consists largely of iron and nickel; its mantle and crust are made of silicon, oxygen, magnesium, aluminum, and many other heavy elements; our present atmosphere contains mostly nitrogen and oxygen; and life is based on the chemistry of carbon. Thus, the buildup of the heavy elements, even though it amounted to only 1 to 2%, was crucial for our existence.

The dependence of our planet's structure and of life's development on the heavy element buildup is discussed in later chapters. Here let us focus on what this buildup tells us about the making of the galaxies. Look at a few more facts. First, the buildup of heavy elements in galaxies like our own has not progressed much since several billion years before the formation of the Solar System. For example, stars 2 or 3 billion years older than the Sun have roughly the same composition as the Sun or stars currently forming. Second, stars older than about 8 to 10 billion years—the stars in the halos and nuclei of spiral galaxies, those in globular star clusters, and in elliptical galaxies—do have a measurably lower heavy element content. Third, stars that lie toward the inner part of the disks of spiral galaxies show a noticeably greater abundance of heavy elements than those lying further out.

Finally, heavy elements are manufactured only in the cores of stars much more massive than the Sun (to be discussed in some detail in chapter 3). This manufacturing process starts with the fusion of hydrogen into helium-4, much as it did during primordial nucleosynthesis (see chapter 1). Helium is then fused into carbon and oxygen, which subsequently undergo further fusion and are converted into still heavier elements. These nuclear reactions are the stars' source of energy. After some hundreds of thousands to millions of years of furious nuclear activity, the massive stars become structurally unstable and explode as supernovae, throwing much of the freshly synthesized elements back into space. There the elements intermingle with the primordial hydrogen and helium and gradually change the composition of interstellar matter.

Early Star Formation

These facts suggest that soon after the first fragments condensed out of the primordial clouds and began colliding and coalescing into protogalaxies, massive stars must have been forming in great numbers. They rapidly raced through their evolutions, like engines out of control, manufacturing heavy elements, exploding, and tossing their products back into space. The primordial gas thus became enriched with heavy elements. New generations of stars formed from the enriched gas and the cycle was repeated.

In protogalactic fragments that evolved into globular star clusters, elliptical galaxies, and the spherical components of spiral galaxies, the gas was used up rather quickly. Or, it was lost due to collisions between the galaxies. Very few stars were formed after the initial burst of star formation and nucleosynthesis ended after only very little hydrogen had become fused into heavier elements. In fragments that evolved into the disk components of spiral galaxies, much of the gas remained in uncondensed form and the cycle of stellar birth, nucleosynthesis, and supernova explosion continued for many generations. The heavy element content of the gas rose and within just a few billion years amounted to approximately 1 to 2% by mass. The rapid formation of massive stars then came to an end and the synthesis of heavy elements was reduced to a negligible level.

This theory of early star formation is further supported by the concentration of heavy elements toward the central regions of all galaxies. Any collapsing cloud of gas always develops the highest density near its center. The same must have been true during the collapse of the fragments and the growth of protogalaxies. Hence, the formation of massive stars, nucleosynthesis, and supernova activity may be expected to have been greatest there and to have produced the heavy element distribution observed in galaxies today.

To summarize, the buildup of heavy elements by massive stars was at its peak when the galaxies were forming. By the time the Universe was just a few billion years old, the rapid rate of star formation had come to an end. In spiral and irregular galaxies, star formation has continued to the present epoch, but at a much reduced rate. This suggests that, in comparison to today, the young galaxies must have been exceedingly bright, particularly toward their centers where the formation of massive stars and supernova explosions were concentrated. It may be that the early bursts of star formation and supernova activity created the giant black holes that are suspected to mark the centers of galaxies. We cannot be sure at present. But as new generations of telescopes are built that can extend our view into the Universe's distant past (in particular, space-based telescopes), we may be able to test these theories.

MILKY WAY GALAXY

When we look overhead on a clear night, we observe the faint glow of the Milky Way stretching from horizon to horizon. Away from the interfering lights and pollution of cities, observers with good eyesight can see approximately six thousand stars (3000 per hemisphere). The light of billions more, too faint to be seen individually, merges to form the silvery band of the Milky Way. This band is the disk of our galaxy and we are located right in the middle of it. In addition to stars, it consists of a patchwork of bright luminous clouds of gas interspersed with regions of dark, light-absorbing dust. We see it edge-on from within, just as we see some other galaxies edge-on from outside. A panoramic view of our galaxy is shown in figure 2.24. To appreciate this mosaic of photographs fully, remember that the photographs were taken from within the disk of the Galaxy, some 30,000 LY from its center, and constitute a full-circle view.

Overall Structure

Living right within the galactic disk and being surrounded by huge clouds of gas and dust has made it very difficult for astronomers to deduce the true structure of our galaxy. In the direction of the Milky Way disk, we literally cannot see the Galaxy for the clouds, particularly when we view it in the optical and UV parts of the spectrum. Only in directions perpendicular to the disk do we have a reasonably clear field of view.

The limitations on visibility are not nearly so severe in the radio spectrum. Infrared and radio waves, which are of longer wavelengths than optical and UV radiation, penetrate the interstellar gas and dust much more readily. Therefore, it became one of the tasks of radio astronomers to map the spiral structure of our galaxy. They focused their attention on clouds of neutral hydrogen gas that are concentrated along the arms and emit a characteristic radio signal at a wavelength of 21 cm. By observing this signal as it arrives from all directions and by noting its Doppler shifts due to the differential rotation of the gas about the galactic center, they deduced the size and the spiral layout of our galaxy (figure 2.25). This layout closely resembles the multiple-arm structures of some other large spiral galaxies, such as the Andromeda galaxy and M 81, and it marks the Milky Way as a galaxy of type Sb.

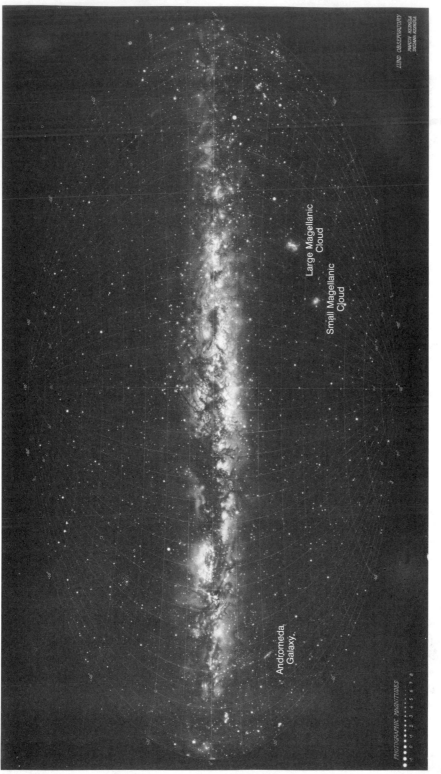

Figure 2.24. Panoramic view of the Milky Way. The coordinates shown are galactic longitude and latitude, where 0° longitude is the great semicircle in the direction of the Galaxy's poles and center (as viewed from Earth), and 0° latitude is the great circle that lies in the Galaxy's central plane (also called the galactic equator).

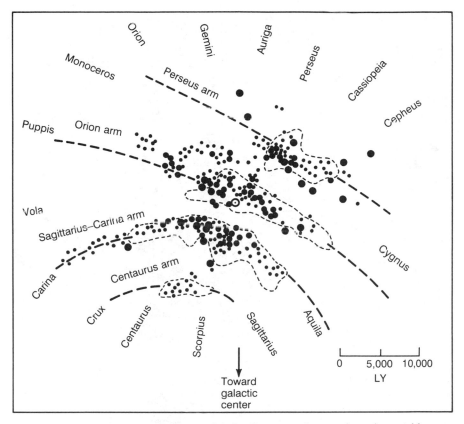

Figure 2.25. The spiral arm features of the Milky Way galaxy in the solar neighborhood. Four arms are shown — the Perseus, Orion, Sagittarius–Carina, and Centaurus arms. The Sun (⊙) is located on the inner edge of the Orion arm. The dots mark the distribution of open star clusters (•) and brightly illuminated clouds (●). The dashed lines enclose regions with the highest concentration of bright and recently formed stars. The directions of some of the constellations in neighboring regions in the sky are indicated along the periphery of the diagram.

Optical astronomers have also studied the size and the general distribution of matter in the Galaxy. Because they are so limited in seeing in the direction of the galactic plane, one of their most successful methods has been to determine the distances to the globular star clusters. Many of these clusters lie far above and below the galactic plane where there is little obscuring gas and dust. From the clusters' nearly spherical distribution about the galactic center and our off-center location relative to them, optical astronomers confirmed the results from the 21 cm observations.

According to both methods of observation, the visible part of the Milky Way (both in the optical and radio parts of the electromagnetic spectrum) is approximately 100,000 LY across, contains close to 150 billion stars, and has a mass of about 140 billion $M_\odot$. However, if the dark matter, which is suspected to be the dominant component of the Galaxy's halo, is included, the diameter of the Milky Way is at least 300,000 LY and its total mass is in excess of $10^{12} M_\odot$.

Figure 2.26. The Great Nebula in the constellation Orion, the brightest nebula visible from Earth. It lies at a distance of 1300 LY and is heated to incandescence by a number of young, hot, massive, and very luminous stars embedded in it. It is situated in Orion's Sword, roughly three degrees below the Belt, and is part of a much larger, dark cloud complex of gas and dust in which new stars are believed to be currently forming (see also figure 3.3).

Orion Arm

The Solar System is located some 30,000 LY from the galactic center at the inner edge of one of the spiral arms, known as the *Orion arm*. The two nearest neighboring arms are at distances of about 6000 LY each. They are the Sagittarius–Carina arm, which lies in the direction of the galactic center, and the Perseus arm, which lies in the opposite direction (see figure 2.25).

Because of its closeness, the Orion arm appears to wrap itself almost halfway around us in the night sky, from the constellations Monoceros and Orion to Gemini, Auriga, Perseus, Cassiopeia, Cepheus, and Cygnus. It contains most of the brightest stars that we see overhead—Sirius, Vega, Capella, Rigel, and Betelgeuse are just a few examples (see table 3.1 for data on these stars). It also contains many of the bright and dark nebulae that are visible through telescopes and that show up so spectacularly on photographs—such as the Orion and Eta Carinae nebulae (figure 2.26 and cover).

Nucleus

When the inhabitants of the middle to lower latitudes of the Northern Hemisphere look towards the southern horizon on a summer night, in the direction of the constellation Sagittarius, they see a gradual broadening in the band of the Milky Way. This broadening represents the outer regions of the central bulge of our Galaxy. The whole of it can only be seen from the Southern Hemisphere of

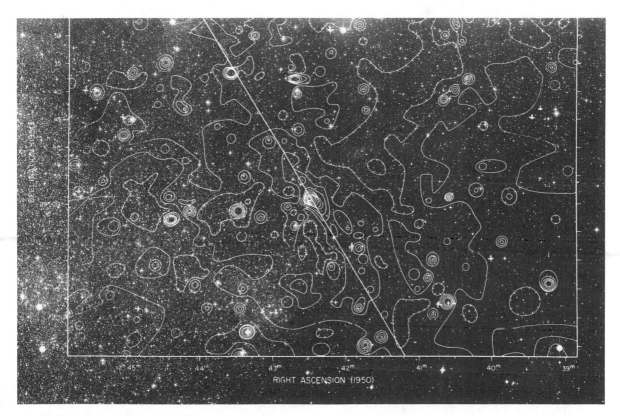

Figure 2.27. Photograph in the direction of the center of the Milky Way, with superimposed infrared map. The concentration of contour lines near the middle of the map signifies Sagittarius A West. The slanted line represents the central plane of the Galaxy's disk.

the globe and, with its high concentration of stars and nebulae, is easily the most striking visible feature of the Milky Way.

The clouds of gas and dust lie so thick in the direction of the Galaxy's nucleus that its center cannot be seen from Earth in optical light. Radio and infrared radiations do penetrate, however, and reveal a most interesting structure. As the contour map in figure 2.27 shows, the central region of the Galaxy contains several distinct infrared emission sources. The most intense of them is no bigger than Jupiter's orbit, yet the rapid motions of hydrogen gas in its vicinity indicate that it holds at least 4 million solar masses. It is called Sagittarius A West and signifies the center of the Galaxy.

The region surrounding Sagittarius A West shows ample evidence of a history of periodic, violent explosions. It is a strong source of synchrotron radiation, which indicates the presence of a mechanism capable of accelerating electrons to relativistic speeds in a magnetic field. It emits X rays and gamma rays, the most energetic kinds of radiation in the electromagnetic spectrum. The gamma rays have wavelengths that indicate they are produced by electron–positron pair annihilation, implying that there must be a source of positrons (the antiparticles of electrons, see the discussion of the birth of the Universe, age 10^{-12} second, chapter 1). The region also contains a number of IR sources. Furthermore, there are large streams of gas flowing rapidly outward from the center of the Galaxy. It is all reminiscent of quasars, Seyfert, and radio galaxies. Is Sagittarius A West a black hole, like those that are suspected to form the centers of the active galaxies, but on a smaller scale? The evidence is mounting for an affirmative answer. Matter passing through an accretion disk and falling into a black hole would explain the energetic activities taking place at the center of the Milky Way. Also, as far as we know, only a black hole can pack several million solar masses into a volume smaller than the Solar System.

Halo and Disk

The stars and globular star clusters of the halo of the Milky Way trace out large, elongated orbits about the galactic nucleus. These orbits are randomly oriented in space and give the halo its spherical symmetry. In contrast, the stars, open star clusters, and gas and dust of the disk trace out nearly circular orbits. They all lie very nearly in the same plane and give the disk its flat distribution.

As noted earlier, the matter of the disk does not rotate like a solid wheel. It rotates differentially, with the stars and the gas closer to the nucleus orbiting more rapidly than those farther out. The Solar System, which belongs to the disk and which lies at approximately 30,000 light-years from the galactic center, orbits with a speed of roughly 220 km/sec. This carries us around the Galaxy once every 250 million years. Despite this great orbital speed (equivalent to 800,000 km/hr) we have no sensation of motion. The distances to the stars are too great for any obvious positional changes to occur among the stars within the span of a human life. Still, careful observations through telescopes show that many of the nearby stars change their positions by up to a few seconds of arc every year.* These positional changes are mostly due to small random motions

*A second of arc is a measure of the spread of an angle. Sixty seconds of arc make up one minute of arc, and sixty minutes of arc make up one degree. There are 90 degrees in a right angle and 360

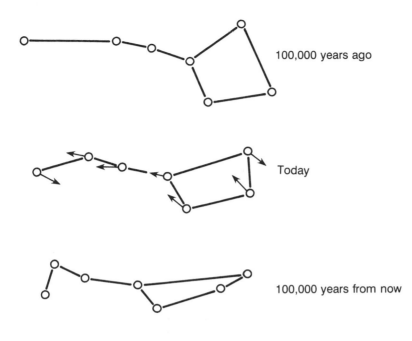

100,000 years ago

Today

100,000 years from now

Figure 2.28. The Big Dipper through time. The arrows (center) indicate the motions of the stars. Adapted from Robert D. Chapman, *Discovering Astronomy* (San Francisco: W. H. Freeman & Co., 1978), 328.

superimposed on the circular orbital velocities. To a smaller extent they are due to the differential orbital motions. If we were to return to Earth 100,000 years from now, these small yearly changes in the positions of the stars would have added up to rather noticeable changes and, for example, most of the constellations would have become distorted beyond recognition (see figure 2.28).

We may regard the 250 million years of the Sun's orbital journey about the galactic center as one *galactic year*. By human standards this is an exceedingly long span of time. The last time we were at our present position in the Galaxy, the dinosaurs were just beginning to become the dominant land animals on Earth. Mammals did not yet exist. The continents were converging into one single supercontinent, *Pangaea*, and bore no resemblance to their present shapes. Two galactic years ago, the earliest vertebrates (jawless fishes) swam in the sea, and the land was still devoid of plant and animal life. Four galactic years ago, life was single-celled and microscopic. The origin of life itself dates back about 16 galactic years, and the age of the Earth and the rest of the Solar System is approximately 18 galactic years. Thus, in terms of our orbital motion around the Galaxy, the Earth and life on it are still very young.

degrees in a complete circle. As viewed from Earth, the disks of the Moon and the Sun are approximately 30 minutes, or 1800 seconds, of arc across. Hence, one second of arc is a very small angle indeed.

To recapitulate, the weak density fluctuations that characterized the Universe during its early evolution were the source of a new cosmic order. They led to the contraction of matter into superclusters of galaxies, which are the primary structural units of the cosmos. The superclusters measure, on the average, 100 million light-years across and are interspersed by voids of comparable size. Each supercluster contains millions of member galaxies and each galaxy contains millions to billions of stars. The Milky Way is a member of a small cluster, the Local Group, which is an outlying member of the Virgo supercluster of galaxies.

The creation of order did not stop with the formation of galaxies and stars. In the case of the Solar System, for example, it continued on progressively smaller scales with the making of planets and the origin and evolution of life on one of them. In Part Two (chapters 6–10) I will explore how this evolution began with random chemical reactions in the primordial soup and, in the course of some 4 billion years, produced the enormous diversity of life — from bacteria to protists, fungi, plants, and animals — that inhabits the Earth today.

Exercises

1 Name a few everyday examples supporting the statement that small random fluctuations in the physical properties of matter (such as in density, in temperature, and in the sizes of objects) cannot be avoided.

2 Waves of moderate energy (such as sound waves, waves induced in a taut rope, and water waves of moderate amplitude in deep water) pass through each other with little interaction. Confirm this by inducing waves at both ends of a rope or by throwing two rocks into a lake. Then watch the two waves approach and pass through each other. (*Note*: Waves of large energy, such as shock waves caused by jets flying faster than sound or breakers at a shallow beach, behave differently. They interact strongly when they collide, causing local turbulence and losing much of their energy through friction.)

3 (*a*) Turn on two garden hoses and point the streams of water at each other (the water should flow slowly). Slightly vary the directions and the flows of the two streams. Describe the interaction between the two streams.
(*b*) What does the interaction between the two streams of water tell you about the possible consequences of collisions of gas fragments in the early Universe? (*Note*: Liquid water is nearly incompressible, while gaseous matter can be compressed much more readily.)

4 Discuss the spatial distributions of globular star clusters, open star clusters, gas and dust clouds, and young stars in spiral galaxies.

5 Why does galactic rotation not "wind up" and destroy the spiral arms of our Galaxy and those of other spiral galaxies?

6 S0 galaxies resemble spirals except that they contain no, or very little, gas and dust. Explain this characteristic, using the fact that S0 galaxies are found predominantly near the central regions of rich clusters of galaxies.

7 In 1983 astronomers observed a supernova in Quasar QSO 1059 + 730. What can be deduced from this observation about the nature of this quasar?

8 Why do globular star clusters and elliptical galaxies contain only "old" stars, that is, stars with a much lower percentage of heavy elements than are found in the Sun and other disk stars of spiral galaxies?

9 On a clear night, and preferably on a mountain top and away from city lights, look at the starry sky overhead and follow the faint band of the Milky Way from horizon to horizon.
(*a*) Depending on the time of the year, hour of the night, and viewing conditions, locate the constellations Sagittarius, Aquila, Cygnus, Cepheus, Cassiopeia, Perseus, Auriga, Gemini, Orion, and Monoceros, all of which lie in or close to the disk of the Milky Way.
(*b*) Find the direction toward the Galactic center (ℓ, galactic longitude, $= 0°$), the direction in which the Solar System travels in its orbit around the Galactic center ($\ell = 90°$), and the direction opposite to the Galactic center ($\ell = 180°$).
(*c*) In analogy to our geographic coordinate system,

define and locate the two Galactic poles and the Galactic equator.

(This exercise requires a star chart, listed in the bibliography for this chapter. Star charts can also be found in most large dictionaries and in most introductory astronomy texts, several of which are listed in the bibliography for the Introduction to Part One.)

10 (a) Draw a face-on diagram of the Milky Way, including the Galactic nucleus, the Perseus, Orion, and Sagittarius–Carina arms, and the location of the Sun.

(b) By arrows indicate the motion of the Sun as well as the motions of disk stars closer to and more distant from the Galactic center than the Sun.

(c) How will the positions of the Sun and its neighboring stars change relative to each other in the course of time?

11 Locate as many of the following astronomical objects on the night sky as the season and observing conditions allow. Use a star chart and binoculars.

Object	Type	Constellation	Comments
Andromeda galaxy (M 31)	Spiral galaxy	Andromeda	Visible with the unaided eye as a faint, hazy spot of light.
h and Chi Persei	Open star clusters	Perseus	Two beautiful clusters, each about 45′ across and visible with the unaided eye; located midway between the stars Mirfak (α Persei) and Cih (γ Cassiopeiae).
M 34	Open star cluster	Perseus	Located midway between the stars Algol and Almach, just visible with the unaided eye.
Pleiades (Seven Sisters	Open star cluster	Taurus	Six or seven stars can be seen with the unaided eye, many more with binoculars or a small telescope.
Orion Nebula (M 42)	Nebula	Orion	Located in Orion's Sword and visible with the unaided eye.
Praesepe (Beehive, M 44)	Open star cluster	Cancer	A large, scattered cluster, almost resolved by the unaided eye.
"Great Cluster" in Hercules (M 13)	Globular star cluster	Hercules	A beautifully symmetric cluster, just visible with the unaided eye.
Lagoon Nebula (M 8)	Nebula	Sagittarius	A poorly defined nebulosity, with dark patches and stars, visible with the unaided eye.

Suggestions for Further Reading

Bahcall, J. N., and **L. Spitzer, Jr.** 1982. "The Space Telescope." *Scientific American*, July, 40–51.

Blandford, R. D., M. C. Begelman, and **M. J. Rees.** 1982. "Cosmic Jets." *Scientific American*, May, 124–42. A discussion of the violent activities at the center of many galaxies, manifested in the production of narrow, focused streams of ionized gas.

Blitz, L., M. Fich, and **S. Kulkarni.** 1983. "The New Milky Way." *Science* 220: 1233–40. A revised model of the Milky Way as deduced from the distribution of atomic hydrogen. The model shows that the disk of the Galaxy is about twice as extended as was previously thought and that beyond the Sun the gas is concentrated in large-scale, coherent spiral arms indicative of a regular four-armed spiral pattern.

Boer, K. S. de, and **B. D. Savage.** 1982. "The Coronas of Galaxies." *Scientific American*, August, 54–62. Satellite observations indicate that the disks of our and other spiral galaxies are surrounded by enormous envelopes of hot gas.

Bok, B. J. 1981. "The Milky Way Galaxy." *Scientific American*, March, 92–120. An article about the newly emerging picture of the structure and content of the Milky Way, by an eloquent lecturer and writer.

Burns, J. O., and R. M. Price. 1983. "Centaurus A: the Nearest Active Galaxy." *Scientific American*, November, 56–66.

Geballe, T. R. 1979. "The Central Parsec of the Galaxy." *Scientific American*, July, 60–70.

Habing, H. J., and G. Neugebauer. 1984. "The Infrared Sky." *Scientific American*, November, 48–57. Comets, stars, interstellar gas and dust clouds, and galaxies as seen with the *Infrared Astronomical Satellite (IRAS)*.

Hodge, P. W. 1981. "The Andromeda Galaxy." *Scientific American*, January, 92–101.

Hutchings, J. B. 1985. "Observational Evidence for Black Holes." *American Scientist*, January–February, 52–59. Recent observational research that tries to establish whether quasars, the cores of galaxies, and certain X-ray-emitting binary stars are powered by black holes.

Margon, B. 1983. "The Origin of the Cosmic X-Ray Background." *Scientific American*, January, 104–19. One source of the cosmic X-ray background may be a multitude of distant quasars.

Mathewson, D. 1985. "The Clouds of Magellan." *Scientific American*, April, 106–14. The turbulent history of our nearest galactic neighbors, including close encounters with each other and with the Milky Way.

Meier, D. L., and R. A. Sunyaev. 1979. "Primeval Galaxies." *Scientific American*, November, 130–44. The first galaxies to form after the Big Bang have not been seen, but the characteristics of older galaxies suggest what younger ones might have been like.

Mewaldt, R. A., E. C. Stone, and M. E. Wiedenbeck. 1982. "Samples of the Milky Way." *Scientific American*, December, 108–21. The authors explain that atomic nuclei found in cosmic rays are clues to the isotopic composition of other parts of the Galaxy.

Osmer, P. S. 1982. "Quasars as Probes of the Distant and Early Universe." *Scientific American*, February, 126–38.

Peebles, P. J. E. 1984. "The Origin of Galaxies and Clusters of Galaxies." *Science* 224: 1385–91. Some of the competing theories and unresolved problems concerning galaxy formation. For the advanced student.

Readhead, A. C. S. 1982. "Radio Astronomy by Very-Long-Baseline Interferometry." *Scientific American*, June, 52–61. Observations made simultaneously by radio telescopes thousands of miles apart can be combined with the aid of atomic clocks to yield the highest resolution ever achieved.

Rubin, V. C. 1983. "Dark Matter in Spiral Galaxies." *Scientific American*, June, 96–108. The newly emerging picture of the structure of spiral galaxies, indicating that much of their matter emits no light and that this matter is not concentrated near their centers.

Sandage, A. 1961. *The Hubble Atlas of Galaxies*. Washington, D.C.: Carnegie Institution of Washington. A beautiful and comprehensive set of photographs defining the Hubble classification of galaxies.

Strom, S. E., and K. M. Strom. 1979. "The Evolution of Disk Galaxies." *Scientific American*, April, 72–82. Through collisions or near encounters, spiral galaxies can evolve into S0 galaxies, consisting of smooth disks without spiral arms. Such evolutions occur most likely in rich clusters of galaxies.

Star Charts

Astronomy and *Sky and Telescope*, monthly popular astronomy magazines, have star and planetary charts in each issue, including discussions of what to look for in the night sky.

Chandler, D. 1977. *The Night Sky*. Cambridge, Mass.: Sky Publishing Corporation. A compact, easy-to-use star chart for the range 38°–50° north latitude.

Muirden J. 1985. *Astronomy with a Small Telescope*. London: George Philip and Son, Ltd. A very useful book for the amateur astronomer who wants to observe the night sky with the unaided eye, a pair of binoculars, or a small telescope.

Norton, A. P. 1978. *Norton's Star Atlas and Reference Handbook (Epoch 1950)*. 17th ed. Edinburgh: Gall and Inglis. A comprehensive set of charts of stars, clusters, and nebulae; descriptive lists of objects observable through small telescopes; and definitions of astronomical terminology. For the serious amateur astronomer.

Philips' Planisphere (for Latitude 42° N). 1982. London: George Philip and Son, Ltd. A compact, easy-to-use chart.

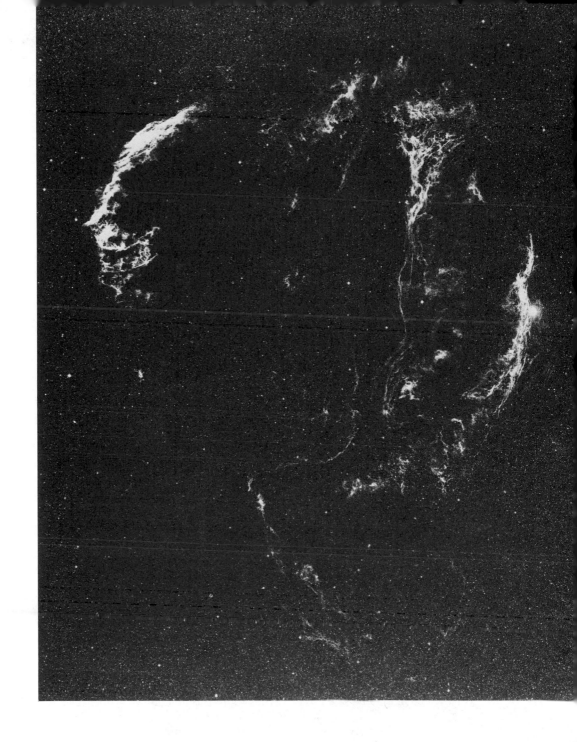

3

Stars and Heavy Element
Synthesis

THIS STORY OF THE EVOLUTION of matter will now gradually progress from structures of cosmic dimension to ones closer in size to human experience. The previous chapter was devoted to the origin of galaxies and galaxy clusters, whose sizes range from thousands to roughly 100 million light-years and whose masses lie in the range from 10^6 to 10^{15} solar masses. In this chapter the discussion focuses on the structure and evolution of the stars, most of which have sizes of up to a few hundred or thousand Earth diameters, although some stars, namely the red supergiants, may be 10^5 times as large as the Earth, and some unusual stars, such as neutron stars, are smaller than the Earth.

The masses of stars are a few ten thousand to a few ten million times that of the Earth. The Sun is quite an ordinary star, with a radius and mass that are neither unusually large nor unusually small:

$$1\ R_\odot = 700{,}000 \text{ km}$$
$$= 110 \text{ Earth radii}$$
$$1\ M_\odot = 2 \times 10^{30} \text{ kg}$$
$$= 330{,}000 \text{ Earth masses}$$

(As noted earlier, the symbol $\odot$ stands for the Sun. $R_\odot$ and $M_\odot$ refer to the Sun's radius and mass. $L_\odot$ refers to the Sun's luminosity. Radius of Earth [equatorial] = 6380 km, mass of Earth = 6.0×10^{24} kg.)

In general, stars distinguish themselves from all other objects in the Universe today by producing their own energy through nuclear reactions and pouring that energy forth into space as electromagnetic radiation. That is what makes them shine. The only exceptions are such stars as white dwarfs and neutron stars (see below) that have exhausted their nuclear energy supply. They shine by radiating stored-up heat energy. The Sun's rate of energy output is called *one solar luminosity* ($L_\odot$). The Earth intercepts only about 4.6×10^{-10} of this amount, which is the solar energy that heats our planet, provides the energy for photosynthesis, and gives us light (see box 3.1).

Most stars are very stable and neither expand nor contract perceptibly in the course of a human life span, although there are exceptions. Stars are not solid or liquid, like the bulk of the Earth—they are gaseous. However, unlike the gases of our planet's atmosphere, the gaseous matter of stars is ionized, with exceptions occurring only in the cooler, outermost layers (ionization is defined in the Introduction to Part One). The force that holds stars together is gravity, and it tends to shape them into nearly perfect spheres. We may, therefore, define stars as *giant, self-luminous, gaseous spheres.**

←   *Figure 3.1*. The Cygnus Loop. This tenuous and rather delicate-looking shell of gas and dust is the ejecta from a supernova that exploded about 30,000 years ago. It lies at a distance of 2500 LY from Earth, measures roughly 130 LY across, and is expanding at an average rate of 116 km/sec. Initially, the rate of expansion was much greater, but the ejecta steadily decelerated as it plowed into the surrounding interstellar medium, compressing it and carrying it along.   .

*According to this definition, planets, such as Venus and Mars, are not stars because they do not have their own nuclear sources of energy. They shine by reflecting the light of the Sun. The same is true of the Moon, the asteroids, the comets, and all the other smaller bodies in the Solar System.

THE HERTZSPRUNG–RUSSELL DIAGRAM

When astronomers focus their telescopes on a given star, they intercept its light and obtain information about the star's surface temperature and luminosity. Stars that are green to bluish-white have surface temperatures hotter than the Sun's. Yellow stars like the Sun have surface temperatures of about 6000 K. Orange to red stars are 5000 K and cooler at their surfaces.

The amount of light that reaches Earth from a star depends both upon its luminosity and its distance: the nearer the star of a given luminosity, the brighter it is. The luminosity is the actual rate, independent of distance, by which a star pours forth radiant energy in all directions. This is the quantity to consider when studying stars. It allows meaningful comparisons to be made between them, even when they lie at different distances.[*]

In the 1910s, Ejnar Hertzsprung (1873–1967) of Denmark and Henry Norris Russell (1877–1957) of the United States recognized that a plot of the luminosities of stars versus their surface temperatures, as shown in figure 3.2, produces a very interesting pattern. Most stars fall along a fairly narrow band that stretches from the upper left corner of the diagram to the lower right. This band is called the *main sequence*. The stars at the upper left end of the main sequence have surface temperatures of up to 50,000 K and luminosities 10^6 times that of the Sun. Those at the lower right end are at about 3000 K and 1000 times fainter than the Sun. The Sun is located on the main sequence just below its middle.

---

[*]Observational astronomers commonly measure the brightness of stars in units of *magnitude*. This unit was introduced by antiquity's greatest astronomer, Hipparchus (born in Nicaea in Bithynia, fl. 146–127 B.C.). He divided the stars into six equally spaced brightness categories from one (brightest) to six (just barely visible with the unaided eye). With some modification, these categories have become the modern magnitude scale. It is a logarithmic scale because, very approximately, the human eye responds to physical stimuli logarithmically. More precisely, the difference in magnitude between two stars (m_1 and m_2) is related to the difference in their luminosities (L_1 and L_2) by the equation $m_2 - m_1 = 2.5 \log_{10}(L_1/L_2)$. *Absolute magnitude*, which is shown in some of the diagrams of this chapter, is the magnitude a star has if it is observed from a distance of 10 parsecs or 32.6 light-years. The *parsec* is the unit of interstellar and intergalactic distance measurement used by astronomers. It is the distance at which the radius of the Earth's orbit about the Sun subtends one second of arc. More specifically, 1 parsec = 3.26 LY.

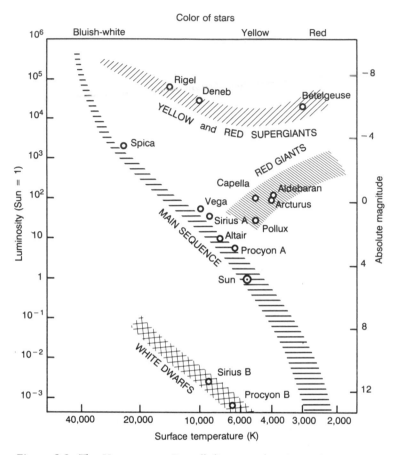

Figure 3.2. The Hertzsprung–Russell diagram, showing main sequence stars, red giants, yellow and red supergiants, and white dwarfs. The diagram also shows the twelve brightest stars visible from the Northern Hemisphere, the Sun, and the white dwarf companions of Sirius and Procyon.

In addition to main sequence stars, the diagram contains several other prominent groupings of stars. Along the right side and above the main sequence lie the *red giants*. As their name implies, they are very large stars and radiate predominantly in the red. Their surface temperatures range from about 2000 K to 6000 K and they are very luminous. To the right of the upper main sequence and extending clear across the diagram lie the *supergiants*. They are even larger and more luminous than the red giants. The lower part of the diagram is occupied by the *white dwarfs*. They are very compact in size and appear bluish-white to yellow. They are stars of low luminosity with surface temperatures somewhat hotter than the Sun's. Clearly, these names — red giant, supergiant, and white dwarf — were given to the stars during an age more

Table 3.1. The Twelve Brightest Stars Visible from the Northern Hemisphere
(Listed in Order of Decreasing Brightness).

Star	Constellation	Type of star	Distance (LY)	Mass $M_\odot$	Radius $(R_\odot)$	Surface temperature (K)	Intrinsic luminosity $(L_\odot)$
Sirius A	Canis Major	Main sequence	9	2.3	2.5	9,000	35
Arcturus	Boötes	Red giant	36	4.5	20	4,000	100
Vega	Lyra	Main sequence	26	3	2.5	10,000	50
Capella	Auriga	Red giant	46	3.5	13	5,000	100
Rigel	Orion	Blue supergiant	800	20	36	15,000	60,000
Procyon A	Canis Minor	Main sequence	11	1.5	2	6,500	6
Betelgeuse	Orion	Red supergiant	600	20	500	3,000	20,000
Altair	Aquila	Main sequence	16	2	1.5	8,000	10
Aldebaran	Taurus	Red giant	70	5	20	4,000	100
Spica	Virgo	Main sequence	260	15	3	25,000	2,000
Pollux	Gemini	Red giant	35	4	8	5,000	30
Deneb	Cygnus	Yellow supergiant	1600	14	60	10,000	30,000

Note: The masses, radii, and luminosities are given in solar units. For example, Sirius A has 2.3 solar masses, is 2½ times the size of the Sun, and is intrinsically 35 times brighter than the Sun.

romantic than ours. Today contractions such as pulsar and quasar or simply letters and numbers are used to refer to new astronomical discoveries.

The diagram invented by Hertzsprung and Russell neatly separates stars of different types into distinct groupings according to just two easily measured parameters. It plays a fundamental role in astronomers' attempts to understand the physical nature of stars and is referred to as the *Hertzsprung–Russell diagram*, or *HR diagram*.

Of the twelve brightest stars (besides the Sun) visible from the Northern Hemisphere, five are main sequence stars, four are red giants, and three are supergiants. None of them is a lower main sequence star or a white dwarf, although both Sirius and Procyon have faint white dwarf companions, which are visible only through very large telescopes. Table 3.1 lists the twelve stars, including the constellations in which they are located, their distances, and a few of their physical characteristics.

There must be some underlying reason why stars fall into the distinct groupings that characterize the HR diagram. Nature does not produce this kind of order by accident. It turns out that the groupings are in part due to physical characteristics that the stars acquire at birth, such as mass and composition. Mostly, however, they are due to structural changes stars experience in the course of their lives. These structural changes affect their energy output and surface temperatures and, thereby, show up in the HR diagram.

By studying the pattern of the HR diagram, astronomers have learned a great deal about the evolution of the stars. This seems paradoxical, for humans have been observing stars with the purpose of understanding their physical structures for only a few hundred years—a time span far shorter than the millions and billions of years it takes stars to evolve perceptibly. How could we possibly have learned about events that take so long? Remember that the HR diagram contains information gathered from a great many stars. The collective nature of this information allows astronomers to deduce the evolution of individual stars. How this is done may be seen best by way of a simple analogy.

Suppose a visitor from outer space arrives on Earth and wishes to learn about us. For reasons beyond his control, he can stay for only one hour—far shorter than the average life span of humans. He lands his ship at a busy American shopping mall and, in the course of his one-hour stay, hurriedly videotapes the shoppers. After departure, he scrutinizes the tape. He notices that the majority of the people are about the same size, but some are quite small and a few are so tiny that they need to be carried or pushed around in carriages. Still others, although of normal size, walk bent over with the aid of canes. Being intelligent, our visitor soon realizes that he is observing an aging process. People are born very tiny; they grow up and spend most of their lives as active adults; and eventually they become old. Because the number of old people is relatively small, he concludes that in the end they die.

Astronomers are in the same position with regard to the stars as the visitor from outer space is with regard to people. In a relatively short span of time astronomers have observed hundreds of thousands of stars. They have measured their luminosities and surface temperatures, and they have taken detailed spectra of their light. By careful scrutiny of this information, and by relying on physical theory and computer modeling, they have deduced the following story about their evolution:

Stars are born within large interstellar clouds of gas and dust. They settle as young adults on the main sequence, where they spend most of their lives. They then evolve into red giants or supergiants, and eventually they die. They die either quietly or explosively. Those that die quietly end up as white dwarfs, growing dimmer and dimmer with time. Those that die explosively throw much of their matter back into space, leaving behind a collapsed object just a few kilometers across. The explosions are the supernovae mentioned in chapter 2 and the collapsed remnants are neutron stars or black holes.

Thus, stars do not occupy forever the same location in the HR diagram but follow evolutionary paths. The details of the paths depend on the stars' masses and compositions. In figure 3.14 such paths are shown for the Sun and for a star of $25\,M_\odot$, including the years they spend in their major evolutionary stages. The Sun will spend by far the greatest part of its adult life—10 billion years—on the main sequence, and it will die quietly as a white dwarf. In contrast, the $25\,M_\odot$ star will race through its entire evolution in just a few million years and end its life as a supernova. The remainder of this chapter is devoted to filling in details of the story of the origin and evolution of stars as observed today in the disk of our Galaxy.

THE BIRTH OF STARS

The birthplaces of stars in our and other spiral galaxies are the huge interstellar clouds of gas and dust that are mainly strung out along the spiral arms. Typically, these clouds range from a few thousand to several ten thousand solar masses and are roughly 10–50 light-years across, though a few are much larger. The most impressive one visible from Earth is the great cloud in the constellation Orion (figure 3.3). It is an enormous complex of gas and dust, measuring nearly 100 light-years across and filling most of the area outlined by the constellation. It contains at least 100,000 solar masses of gas and dust in various stages of compression. Much of the material is cold, dark, and impenetrable to optical and ultraviolet radiation. However, in a relatively small region on the side facing us and surrounding Orion's Sword, a group of very bright stars is heating the gas and dust to incandescence and creating the beautiful Orion nebula. The four brightest stars, known as the Trapezium, are not much older than a million years. Buried deep within the dark expanse of gas and dust are hundreds to thousands of compact, dense regions that astronomers call *clumps*. Some of these clumps are collapsing under their own gravity and 10,000 to a million years hence will have become stars. The brightest and most massive of the newly born stars (like the Trapezium stars) will race through their lives within a few million years and explode as supernovae, spewing freshly synthesized heavy elements back into space and sending shock waves through the cloud. Others will live for hundreds of millions to billions of years, traveling numerous times in nearly circular orbits around the center of the Galaxy.

Formation of Interstellar Clouds

Cloud complexes like the one in Orion form when powerful external forces act on interstellar gas and dust and compress them. The origins of these external forces are at present not well understood. It seems that they arise when interstellar gas and dust, in the course of their orbital motions about the galactic center, plow into a spiral arm. They probably also arise when the shock front from one or more nearby supernova explosions expands into the gas and dust. Both of these mechanisms of compression are rather complicated and involve a number of interlinked events.

While the gas and dust are still in the uncompressed, dispersed state, they are quite hot. Their average density amounts to about one atom per cubic centimeter and their temperature is maintained at roughly 10,000 K by steady bombardment from cosmic rays.* The high temperature creates a pressure large enough (see box 3.2) to resist successfully gravity's attempt at compression. This is analogous to the way the pressure of the air in an inflated tire or balloon offers resistance to compression, particularly on a hot day.

When interstellar gas and dust become compressed the situation changes drastically. The individual gas and dust particles start emitting infrared (IR) and

Cosmic rays are streams of atomic nuclei—mostly protons and nuclei of helium, plus a sprinkling of nuclei of heavier elements—that travel at nearly the speed of light through interstellar and intergalactic space. The Earth is continuously bombarded by them. They are believed to originate during supernova explosions and when matter falls onto a neutron star or into a black hole.

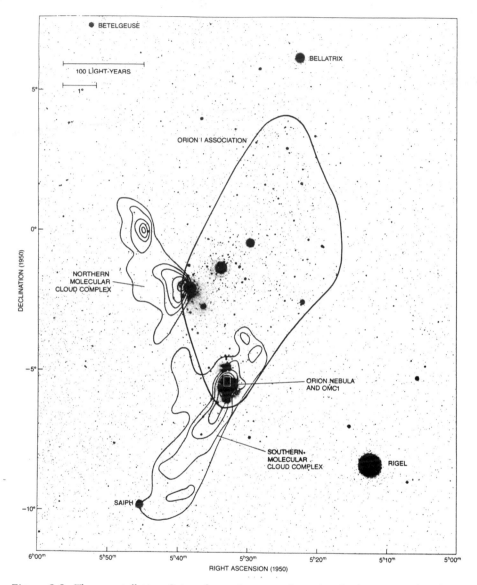

Figure 3.3. The constellation Orion shown in a negative print. In the area outlined by the constellation and at a distance of roughly 1300 LY from Earth lies an enormous cloud complex of interstellar molecules and dust grains. The contour lines superimposed on the photograph show the strength of the IR radiation emitted by the carbon monoxide of the cloud complex. The southern component of the cloud complex includes the Orion Nebula, outlined by the small rectangle (shown in greater detail in figure 2.26). The region also contains a very young open star cluster of about a thousand members, known as the Orion I Association (indicated by a heavy line).

radio waves. They do this because the compression squeezes them closer together, which makes them collide more often and, during each collision, some of their energy of motion is converted into IR and radio waves. The long wavelengths of this radiation allow it to escape from the cloud and to carry away

BOX 3.2.
The Concept of Pressure

IN physics, the concept of pressure refers to a force that is applied over an area. If the force (F) is applied at an angle to the area (A), only the component of the force perpendicular to the area ($F_\perp$) enters into the calculation of the pressure (P):

$$P = \frac{F_\perp}{A}$$

(see figure A) If the force is applied perpendicularly to the area, as in all of the examples below, then the full value of F contributes to the pressure because $F_\perp = F$.

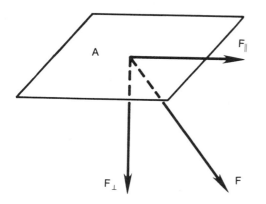

Figure A. Geometric decomposition of a force, F, acting at some angle over an area, A. The force may be thought of as consisting of two components: a component perpendicular to the area, $F_\perp$, and a component parallel to the area, $F_\parallel$.

A few examples will illustrate the concept of pressure. When we stand or walk on level ground, the pressure we create on the ground equals the force with which gravity pulls us downward divided by the surface area over which the force is distributed, in this case the surface area of the soles of the shoes in contact with the ground. Thus, a 160-pound man standing on both feet creates a pressure of 2.5 lb/in² on the ground, assuming that the surface of contact equals 64 square inches.

$$P = \frac{F_\perp}{A}$$

$$= \frac{160 \text{ lb}}{64 \text{ in}^2}$$

$$= 2.5 \text{ lb/in}^2.$$

If the man stands on just one leg, the pressure is doubled because the area over which the force is distributed is halved. If the man stands on skis, with a total surface area of, say, 350 in², the resulting pressure is slightly less than 0.5 lb/in². This pressure is low enough to allow him to walk through deep snow without sinking down very far. In contrast, a 120-pound woman wearing high-heeled shoes creates a pressure of nearly 500 lb/in² when she puts all of her weight on one heel (assuming the heel's surface area equals ¼ in²), which is sufficient pressure to damage floors.

The last example illustrates that for a given force the pressure can be made very large by reducing the area over which the force is applied. People exploit this fact when they pierce an object with a needle or cut it with a knife. In these cases the area—the point of the needle or the cutting edge of the knife—is very small.

Pressure of Gases

The concept of pressure applies not only to solid objects pressing against an area, but also to gases. For example, the recommended air pressure in car tires today is 25–30 lb/in². And *the pressure of the Earth's atmosphere at sea level is approximately 14.7 lb/in²* (= *1 atm**). This means that the column of air above every square inch of the Earth's sea level presses down with a force equal to 14.7 pounds. Interestingly, the atmosphere exerts this pressure not only on level surfaces, but on all surfaces no matter how they are oriented. For example, at sea level every square inch of our bodies is pressed upon by a force equal to 14.7 pounds. The reason is that air is a fluid that can flow in any direction.

*British units have been used in this discussion because few people in the United States are familiar with the metric units of pressure, dynes/cm² or Newtons/m², but most are aware of how many pounds per square inch of pressure are in a properly inflated car tire. In the rest of this box, and in the text, pressure is expressed in units of *atmospheres* (abbreviated atm). One atm is defined to equal the pressure of the Earth's atmosphere at sea level and a temperature of 15° C:

$$1 \text{ atmosphere (atm)} = 14.7 \text{ lb/in}^2$$
$$= 1.01 \times 10^6 \text{ dynes/cm}^2$$
$$= 1.01 \times 10^5 \text{ Newtons/m}^2$$

BOX 3.2 continued

The pressure of the Earth's atmosphere equals 14.7 atm only at sea level. At higher elevations, it diminishes because the greater the height, the less atmosphere there is above to press down. For example, on top of Mt. Mitchell, the highest point in the eastern United States with an elevation of 2037 m, the atmospheric pressure is 0.781 atm (11.5 lb/in^2). On top of Mt. McKinley, 6194 m, it is 0.453 atm (6.7 lb/in^2). On top of Mt. Everest, 8848 m, it is a mere 0.310 atm (4.6 lb/in^2). And at a height of 100 km it is 3.0 × 10^{-7} atm (4.4 × 10^{-6} lb/in^2).

The pressure of a gas exerts outwardly directed forces that expand the gas unless it is confined. In the cases of a balloon or car tire, the confinement is due to the enclosing rubber walls. The tension in the rubber walls creates an inward force per unit area that exactly balances and cancels the outwardly directed force of the pressure of the gas. If the walls are punctured, then the outwardly directed force becomes unopposed (assuming the puncture is large), the gas escapes, and the balloon or tire deflates.

In the case of the Earth's atmosphere, the confinement is due to gravity. To understand the meaning of this statement, consider the Earth's atmospheric layer between 2000 m and 2500 m, assuming that it is in equilibrium (that is, no winds are blowing). At 2000 m

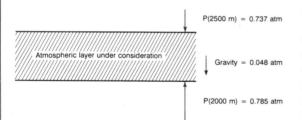

Figure B. Schematic illustration of the forces acting on each unit area of an atmospheric layer between 2000 m and 2500 m elevation: P(2000 m), P(2500 m), and gravity.

the atmospheric pressure is 0.785 atm and at 2500 m it is 0.737 atm, as shown in figure B. These numbers mean that the layer experiences a force per unit area of 0.785 atm from below and of 0.737 atm from above. The difference between these two pressures, namely 0.048 atm, represents a force per unit area that pushes on the layer in the upward direction (upward

because the force from below is greater). However, the layer is not lifted off into space because gravity pulls down on it with an equal force per unit area. Thus, the upward force (due to pressure) and the downward force (due to gravity) cancel and the layer is moved neither upward nor downward.

There is another way to express these ideas. The difference in pressure between the bottom and top of the layer divided by its thickness is an estimate of the *pressure gradient* of the atmosphere between 2000 m and 2500 m.* This pressure gradient exerts an upward force on the layer. There also is a downward force, namely the force of the Earth's gravity that acts on the mass contained in the layer. The two forces are equal in magnitude, but oppositely directed. Hence, they cancel each other and the layer moves neither upward nor downward. We say that the layer is in *hydrostatic equilibrium*. The same considerations apply to the gaseous layers at all other heights in the Earth's atmosphere as well as to the layers in stellar interiors, provided that they are in hydrostatic equilibrium.

What happens when the gaseous layers of the Earth's atmosphere, or of stellar interiors, are not in hydrostatic equilibrium? The pressure gradients are then not balanced by opposing forces and, as a result, gas flows occur. In the case of our planet's atmosphere, we observe these gas flows as winds, storms, and hurricanes that, along with variations in temperature and precipitation, constitute our weather.

This raises another question: What brings about the unbalanced pressure gradients in our atmosphere? The chief cause is the heating of the atmosphere by the Sun and the variations of this heating from one location to another: more in the equatorial regions and above the continents than in the polar regions and above water, snow, and ice. Furthermore, there is a steady loss of heat from the atmosphere due

---

*The *pressure gradient is a measure of how rapidly the pressure changes with height*. In the case of the atmospheric layer in this example,

$$\frac{P(2500 \text{ m}) - P(2000 \text{ m})}{500 \text{ m}}$$

is an estimate of the pressure gradient, where P stands for the pressure at a given height. The division by the thickness of the layer is necessary to cancel out the fact that if the layer were thicker (or narrower), there would be a proportionately greater (or smaller) pressure difference. The narrower the layer, the closer the estimate is to the true value of the pressure gradient.

BOX 3.2 continued

to the emission of infrared radiation, which also varies from one location to another. These differences in heating and cooling create differences in the pressure of the atmosphere from one location to another or, in meteorological terms, they create highs and lows. The highs and lows are responsible for the unbalanced pressure gradients in our atmosphere and, hence, for winds, storms, and hurricanes.

We may regard these atmospheric currents as nature's attempt to erase all pressure gradients except those that are balanced by gravity. However, unequal heating and cooling continually create new highs and lows. Therefore, equilibrium is never achieved globally in our planet's atmosphere and winds, storms, and hurricanes keep blowing.

Pressure, Temperature, and Volume

Now let us examine the relationship between the pressure of a gas and two other physical parameters—the temperature and volume of the gas. For gases such as air or those in the interior of the Sun and other similar stars, this relationship can be expressed by the *equation of state of ideal gases*:

$$P = \text{constant } \frac{T}{V},$$

where P, T, and V stand for pressure, temperature, and volume of the gas and the constant depends on the particular units used.

We all have some familiarity with the physics expressed by this equation. It states, for example, that if the temperature of a gas increases, then the pressure increases also (assuming for the time being that the volume remains unchanged). We experience this when driving on a hot day. The tires warm up and, as a result, the pressure of the air in the tires increases. Another example is the high pressure regions created by the heating of our planet's atmosphere. If the temperature of the gas decreases, then the pressure decreases also (again, assuming that the volume remains unchanged). This situation arises, for instance, when the temperature of interstellar gas clouds is lowered by the emission of IR and radio waves. The pressure is then lowered also, which aids in the compression of the gas and its collapse into stars.

The equation of state also tells how the pressure of a gas responds to a change in volume. If the volume of a gas decreases, the pressure increases (assuming that the temperature remains constant). This happens, for instance, when we pump up a bicycle tire with a hand pump. We compress the gas in the pump (that is, we decrease its volume). This increases the pressure, which, in turn, forces the gas through the valve into the tire. Likewise, if the volume of the gas increases, the pressure decreases (again, assuming that the temperature is kept constant).

In general, the temperature and volume of a gas may both change simultaneously. The equation of state still applies and tells the resulting pressure.

energy. This lowers the temperature and the pressure. The lowering of the pressure reduces the resistance to compression, and contraction continues. This, in turn, increases the rate of emission of IR and radio waves, and the whole process of cooling, lowering of pressure, and further contraction accelerates. The result is the formation of a large interstellar cloud with an average density of about 1000 atoms per cubic centimeter, a temperature of 100 K, and nearly total opaqueness (called *opacity* by astronomers) to all short-wavelength radiation. This large opacity makes the clouds appear so dark and ominous.

Interstellar Molecules

In addition to the steps just described, one other significant physical process occurs during the formation of interstellar clouds. The increasing rate with which atoms collide with each other and with the dust particles induces chemical reactions among them and they start combining into molecules.

Molecules are microscopic structures consisting of atoms held together by the electric force, as described in the Introduction to Part One and illustrated by a few examples in figures 3.4 and 3.5. The low temperature and the large opacity of the cloud assures that the molecules are not destroyed, as they otherwise would be, by collisions with each other or by bombardment from ultraviolet (UV) radiation and cosmic rays. The molecules formed most abundantly are those for which the atomic raw materials — hydrogen, carbon, nitrogen, oxygen, silicon, sulfur, etc. — are amply available. Most prominent among interstellar molecules are molecular hydrogen (H_2), the radicals hydroxyl (OH) and cyanogen (CN), water (H_2O), ammonia (NH_3), hydrogen sulfide (H_2S), carbon monoxide (CO), silicon oxide (SiO), and many organic molecules. The organic molecules found in interstellar space include methane (CH_4), the radical CH, hydrogen cyanide (HCN), formaldehyde (H_2CO), methanol (CH_3OH, popularly known as "wood alcohol"), formic acid (CHOOH, the acid that gives ant bites and nettles their sting), and many more. (A *radical* is a molecular fragment. For instance, the radical OH is a fragment of H_2O and the radical CH is a fragment of any of a number of organic molecules, such as CH_4 and CH_3OH. Organic molecules consist of carbon and hydrogen atoms, though other kinds of atoms may be present also.)

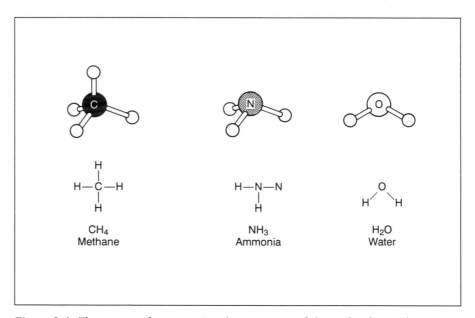

Figure 3.4. Three ways of representing the structures of the molecules methane, ammonia, and water. In the three-dimensional models (top) the spheres stand for the atoms carbon (C), nitrogen (N), oxygen (O), and hydrogen (H). The bars represent the chemical bonds that hold the atoms together in the molecular structures. (The bars are only simplistic representations of real bonds, which consist of pairs of electrons orbiting about pairs of atoms. For instance, methane has four such electron pairs bonding the central carbon atom with the four hydrogen atoms.) The two-dimensional models (middle) are simpler and more commonly used than the three-dimensional models. The symbolic notations (bottom) CH_4, NH_3, and H_2O are the simplest and the most commonly used representations.

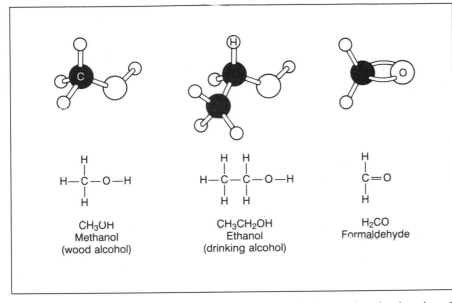

Figure 3.5. Three ways of representing the organic molecules methanol, ethanol, and formaldehyde, all three of which are produced both biologically on Earth and abiologically in interstellar clouds. (The spheres and bars have the same meaning as in figure 3.4.) In formaldehyde the carbon and oxygen atoms are joined by a double bond, indicating that two pairs of electrons orbit about these two atoms, holding them together.

The fact that the interstellar clouds harbor gigantic and very active chemical laboratories is a rather new discovery. Most of the IR and radio emissions that refrigerate the clouds originate from molecules, and every molecule radiates at characteristic wavelengths.* Up to the late 1960s, astronomers had been able to detect only the emission from the chemical radicals CH, CN, and OH. However, since then the technology of radio reception has become sufficiently sophisticated that the signals from many additional molecules can be distinguished. Interstellar ammonia and water were discovered in 1968, formaldehyde in 1969, and carbon monoxide and methanol in 1970. By the mid–1980s more than 60 molecular species had been found, most of which were organic molecules. Today the search continues for ever more complex interstellar molecules, in particular for such life-related ones as amino acids, sugars, and the nitrogenous bases.

New Stars

Consider again the compression of interstellar gas and dust into clouds. This compression is accompanied by much violence, rapid streaming of gas and dust, turbulent motions, and the creation of shock waves. As the compression

---

*When interstellar molecules collide with each other, with atoms, or with dust particles, some of their energy of motion goes into rotational and vibrational motions. Rotating and vibrating molecules possess energy due to these motions. They give up these energies and return to the quiet state (the lowest energy state) by emitting IR and radio waves of characteristic wavelengths.

Figure 3.6. The Trifid nebula. The name of this beautiful nebula means "cleft into three" and refers to the three dark, radial dust lanes that divide it into three sections. The luminous component of the nebula consists mostly of hydrogen gas that is heated to incandescence by a central bright, hot star. The star is a member of an open star cluster roughly 7 million years old.

continues, the force of gravity gradually becomes dominant and pulls the matter closer together. Finally, the cloud assumes a rather squashed and uneven shape, with some parts much more compressed than others. It all is reminiscent of an epoch billions of years earlier when the primordial clouds became unstable and broke up into protogalactic fragments—though those earlier events occurred on a much larger scale.

In most parts of the interstellar cloud, the compressive force of gravity sooner or later is balanced by the expansive forces of pressure and rotation. However, there usually exist small, clumpy regions in which the gas and dust have become much more compressed than elsewhere. In these regions, gravity remains dominant and contraction continues. The temperature drops to as low as 5 to 10 K, the pressure is reduced correspondingly, and the matter collapses into the very compact clumps mentioned earlier. These clumps are created along semiregular patterns throughout the cloud, first here, then there, and they can be seen as pitch-black splotches in photographs of luminous clouds, such as the Eta Carinae and Trifid nebulae (see cover and figure 3.6). Sir William Herschel called them *Löcher im Himmel* ("holes in the heavens"), which quite accurately describes their visual appearance.

In the central regions of some of these clumps, the gravitational collapse eventually releases energy so rapidly that the emission of infrared and radio waves can no longer carry it all away. As a result, energy accumulates in the form of heat and the temperature begins to climb to 1000 K, 10,000 K, and still higher. The dust particles and molecules break up into individual atoms, the atoms become stripped of their electrons, and soon all the matter at the centers of the clumps is in fully ionized form. The resulting atomic nuclei and electrons fly around in a helter-skelter fashion and at very high speeds, continually colliding and bouncing off each other. Simultaneously, with the rise in temperature and the increase in the concentration of matter, the pressure rises. It is able to oppose gravity more and more effectively and gradually slows down the contraction.

The rise in temperature just described is greatest in the central regions of the clumps. This rise continues, but at a reduced rate because of the slowdown in contraction. When the central temperature of the clumps surpasses approximately 5 million degrees, the protons collide so violently with each other that they come within the range of each other's nuclear forces and begin to fuse into helium. Nuclear burning is now being induced and large amounts of energy are released. The centers of the clumps have contracted into self-luminous, gaseous spheres and a new generation of stars has been born.

A newly formed star is not immediately visible. For some time it remains shrouded in thick, opaque layers of uncondensed gas and dust. The first clue to the outside world that a star has actually been born comes in the form of strong infrared radiation. The radiation originates from the star as ordinary optical and UV light, but as it filters through the surrounding dust it becomes altered toward the infrared. Within a few thousand years, the most massive and luminous of the stars vaporize the cocoon-like envelopes of dust within which they were born and their intense radiation then escapes undiminished. It pushes the surrounding gas and dust of the cloud outward (by radiation pressure), heats the material to incandescence, and literally burns a hole into the cloud. That is how the bright interstellar nebulae, such as the one surrounding Orion's Sword or the Eta Carinae and Trifid nebulae, are created.

Depending on the conditions under which a clump forms — its mass, rate of rotation, disturbances from shock waves, exposure to radiation from nearby luminous stars, amount of dust compared to gas — it may give birth to a single star, a multiple star system (double, triple, or quadruple), or a star surrounded by planets. Altogether several hundred stars are usually formed from the clumps of a single interstellar cloud. (Actually, in most interstellar clouds of gas and dust, many more clumps than that start forming, but for reasons not well understood most do not complete the collapse into stars. Instead, they eventually become

Figure 3.7. The Pleiades, or Seven Sisters. These relatively young, hot stars are the brightest members of an open star cluster containing roughly 300 stars. Most of the stars occupy a nearly spherical volume 30 LY across and 400 LY from Earth, in the constellation Taurus.

dispersed.) For hundreds of millions to a few billions of years, the newly formed stars orbit around the center of the Galaxy as an open star cluster, such as the Pleiades (figure 3.7) and the double cluster h and Chi Persei (figure 1.2).

Presumably the Sun was also born as a member of an open star cluster, roughly 4.5 billion years ago. However, we do not know today which of our stellar neighbors are the Sun's brothers and sisters. These stars have long ago ceased to travel through space as an identifiable group. Open star clusters are loosely bound by gravity and within a dozen or so orbits about the Galaxy become dispersed beyond recognition by the tidal forces of the Galaxy.

THE MAIN SEQUENCE

A newly formed star adjusts its structure, by slight contractions and expansions, until at every point throughout its interior gravity is exactly balanced by the pressure gradient. At the same time it assumes a temperature profile, from its hot interior to its relatively cool surface, such that energy flows outward at precisely the rate at which it is generated in the core by the fusion of hydrogen (see box 3.3). Once a star has reached equilibrium in pressure and temperature, it is stable and changes its structure only very slowly.

BOX 3.3.
Heat Flow

HEAT is a form of energy. It is the energy a body possesses due to the kinetic motions of its molecules, atoms, or other constituent particles (such as the electrons and atomic nuclei in stellar interiors). The faster the kinetic motions, the greater is the heat energy of an object and the higher is its temperature, as discussed in box 1.2.

The heat energy of a body is not an intrinsic, unchanging quality of that body. It may increase or decrease. For example, the heat energy of a body is increased by the compression of that body or by nuclear and chemical reactions that release energy (see boxes 3.2 and 3.4). The heat energy of a body is decreased by expansion of that body or by nuclear and chemical reactions that absorb energy. Another way of changing the heat energy of a body is by the flow of heat from one place to another. This flow of heat is the topic discussed here.

It is a common experience of daily life that heat energy flows from places of high temperature to places of low temperature. The flow of heat generally occurs by three mechanisms: radiation, convection, and conduction. In *radiative heat flow*, heat energy is carried by electromagnetic radiation. If the radiating body is at a low temperature, such as a human body, the radiation lies mainly in the IR part of the spectrum. If the radiating body is very hot, such as the surface of the Sun or other stars, the radiation lies mainly in the visible or in a still shorter wavelength part of the spectrum. In *convective heat flow*, heat energy is carried by a fluid, such as a gas or a liquid. Finally, in *conductive heat flow*, heat energy is passed from molecule to molecule, atom to atom, or electron to electron of the conducting body. Conduction by electrons occurs mainly in metals, which contain large numbers of free electrons capable of picking up heat energy at one place and depositing it at another place.*

*One of the characteristic features of metals is that they contain electrons that are not firmly attached to atoms. These free electrons are able to move about with little resistance, which makes metals good conductors of heat. It also makes them good conductors of electricity. In contrast, the electrons of most other substances in our terrestrial environment—for example, wood, plastic, glass—are not free. They are firmly attached to atoms, making such substances poor conductors of heat and electricity.

We continually experience the flow of heat by one or more of the three mechanisms just described, as the following list illustrates.

Radiative Heat Flow

Radiant energy given off by the Sun and other stars.

Infrared radiation given off by heaters, lamps, fires, our own bodies.

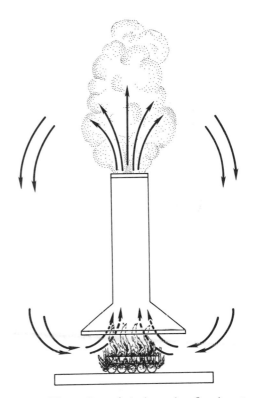

Figure A. The cycling of air through a fireplace is an example of convection. The convective cycle is driven by forces that result from the large temperature difference between the bottom and the top of the chimney. Air heated by the fire expands, becomes less dense, and rises by the force of buoyancy. Rising hot air coming out of the chimney cools by radiating off its heat and mixing with the surrounding cold air. As a consequence, it contracts, becomes denser, and sinks. The same mechanism drives convective currents in stellar layers whose temperature decreases rapidly with height.

BOX 3.3 continued

Convective Heat Flow

Hot air coming from a blower (such as a hair drier) or rising from a heater or a fire (see figure A).

Transport of heat or cold by ocean and air currents.

Transport of heat by the blood stream from the cores of our bodies to the tissues below the skin.

Conductive Heat Flow

The heating (or cooling) of the handle of a metal spoon in a hot (or iced) cup of tea.

The loss of heat through the walls of a stone house.

The flow of heat through the tissues of our bodies from the core to the skin (a process that takes place simultaneously with convective heat transport by the blood).

As stated above, heat always flows from hot to cold. Because temperature differences exist everywhere in our environment—in our bodies, in homes, car engines, the ground and atmosphere—the flow of heat is a common phenomenon.

The same phenomenon occurs in stars. They possess large temperature differences between their cores and their surfaces—from many millions of degrees to thousands of degrees. Hence, heat energy flows continuously from a star's core, where it is generated by nuclear reactions, to its surface, where it is radiated off into space.

As in our immediate environment, the flow of heat through a star's interior may occur by radiative, convective, and conductive transport. Radiative heat transport operates in all stars. It does so because in the hot interior the photons (that is, electromagnetic radiation) that are inevitably present are more energetic than the photons in the outer, cooler layers. Hence, the photons that diffuse from the interior outward carry more energy than those that diffuse inward. The result is a net radiative transport of heat energy toward the star's surface.

Convective and conductive heat transports are of significance in stars only under special circumstances. Conduction of heat is important in stars that are very compact, such as white dwarfs and the cores of red supergiants. Their matter contains free electrons, as in metals, that can travel over great distances and, thereby, transport heat energy. Convective heat trans-

port occurs only if the temperature drops very rapidly from the interior to the surface. In that case, hot gases from the interior rise and deposit their heat in cooler layers further out. The cooled-off gases then sink, heat up, and rise again (see figure A).

In the solar interior, the heat flow from the center outward occurs by radiation. However, in approximately the outer one-third of the Sun, where the temperature drops very rapidly, convective transport takes over. The presence of convection in the Sun's surface layers can be seen in photographs of the Sun taken through special filters. They show a granular pattern, which is caused by convection (see figure 3.11). The centers of the granular cells are brighter and their boundaries are dimmer. In the bright areas, hot gases from the interior arrive at the surface. From there they stream horizontally toward the boundaries of the cells, radiating their heat content off into space and cooling. At the boundaries, the cooled-off and, hence, dimmed gases sink toward the interior and the cycle is repeated.

In many stars, yet another mechanism of heat transport occurs: heat loss by neutrinos. Some of the neutrinos are the by-product of certain nuclear reactions, such as those of hydrogen burning. Others come into existence spontaneously under extreme temperature and density conditions, such as are found in white dwarfs and the cores of red supergiants. All such neutrinos carry energy. Because neutrinos are very elusive particles and tend not to interact with matter, they carry their energy directly off into space. In the case of main sequence stars, this energy loss does not amount to very much compared to the energy loss by radiation and convection. However, in the case of white dwarfs and red supergiants, it is rather substantial and greatly speeds up the evolution of those stars.

The flow of heat energy from stellar interiors to the surfaces is crucial for the evolution of the stars. It is as crucial for them as it is, for example, for the proper running of car engines or the maintenance of our own bodies. Without the efficient expulsion of heat, car engines and human bodies would quickly heat up past tolerable levels and destroy themselves. Similarly, if stars were unable to get rid of the energies released by their nuclear reactions, they, too, would heat up, expand, and destroy themselves.

There is, however, a difference in the time scales involved over which human bodies and car engines on one hand and stars on the other rid themselves of their heat. Human bodies and car engines expel their heat energies rapidly (that is why we feel cold so quickly when we step outside without adequate clothing in cold weather). In contrast, stars require considerably more time to transport their energy to the surface. In the case of the Sun it takes roughly 10 million years. That means the energy of the radiation we receive today from the Sun was produced by nuclear reactions at the center of our star approximately 10 million years ago, long before our species walked on Earth.

In the HR diagram, the star is now approaching the main sequence from the right (that is, from the low-temperature region). During this approach it still experiences periodic instabilities in its surface layers, which evidence themselves as surface eruptions, accompanied by mass ejection and intermittent bursts in its light output. Stars passing through this phase are called *T Tauri stars*, after the prototype variable star T in the constellation Taurus.

The violent surface eruptions cease once the star reaches the main sequence. It stays in this extremely stable evolutionary phase as long as hydrogen remains in its core for fusion into helium and release of energy. How long that takes depends on the star's mass. A $50\,M_\odot$ uses up its hydrogen so swiftly that in just 1 million years its main sequence life is over. The Sun, with its smaller mass, burns its hydrogen more conservatively and stays in this evolutionary stage for roughly 10 billion years. A star less massive than the Sun lasts still longer. This large difference in the rate of nuclear fuel consumption and energy release is what makes massive stars so luminous and low-mass stars so relatively faint.

In table 3.2, the lifetimes and other physical characteristics of main sequence stars of different masses are compared. The masses listed range from 0.1 to $50\,M_\odot$, the approximate mass range of most stars observed in nature. Stars much more massive than $50\,M_\odot$ are not stable and rather rapidly shed their excess mass in the form of strong stellar winds. Objects below approximately $0.1\,M_\odot$ never reach sufficiently high temperatures for the ignition of hydrogen. In the absence of a nuclear energy source these objects are not self-luminous and, hence, are not stars. The planet Jupiter, with just under $0.001\,M_\odot$ is an example of such a low-mass object. Had Jupiter been born with one hundred times more mass, it would be a faint star (in the bottom right-hand corner of the main sequence) and, together with the Sun, would form a double star system.

The Sun

The variation of temperature and pressure of a main sequence star, from the center to the surface, depends only on its mass and composition. Figure 3.8 shows the distribution of these and a few other parameters for the Sun, which is approximately halfway through its main sequence life. The Sun's center has a temperature of 16 million degrees and its density is 160 times that of water. Furthermore, its central pressure equals about 300 billion earth atmospheres. These are large values and they testify to the severity of conditions in stellar interiors. Temperature, pressure, and density all decrease from the Sun's center

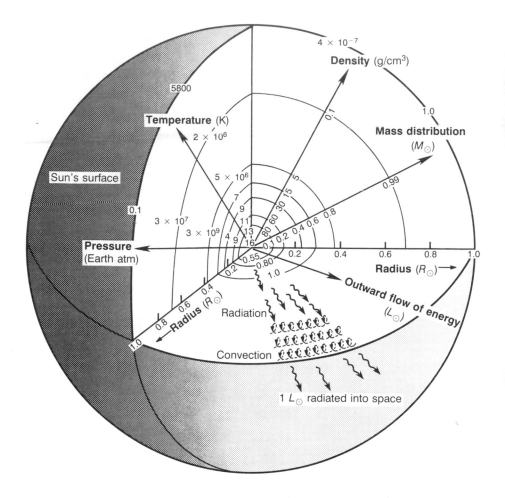

Figure 3.8. Diagram illustrating the internal structure of the Sun, as derived by computer calculations.

outward. The rate of this decrease is very gradual in the interior and very sharp in the outer layers. For instance, in the outer one percent of its total mass, the Sun's temperature drops from 2×10^6 K to 5800 K, the pressure decreases from 3×10^7 earth atmospheres to about one-tenth of an earth atmosphere, and its radius goes from 0.69 to 1.0 $R_\odot$. The enormous spatial extent of the Sun's outermost layers is responsible for its low surface density of a mere 4×10^{-7} g/cm^3, which is approximately 3000 times less than the sea level density of the Earth's atmosphere.

Note the strong concentration of nuclear energy generation toward the Sun's center, as indicated by the increase in the outward flow of energy in figure 3.8: the central 10% of the Sun's mass produces 55% of its luminosity; the next 10% produces another 25%; and beyond 60% in mass from the center, the luminosity increases no further. The outflowing energy is generated by nuclear reactions. The flow itself is due to the steep drop of temperature from 16 million degrees at the Sun's center to a mere 5800 K at the surface. Throughout roughly the inner two-thirds of the Sun this energy flux is carried by photons, which slowly filter outward much as light filters through dense fog or through a cloud here on Earth. In the outer third of the solar surface the energy is transported mainly by convective motions of the gases, much as heat energy is carried by gases up the chimney of a fireplace (see box 3.3). As the energy arrives at the solar surface, it is radiated off into space, making the Sun shine.

The surface of the Sun, as well as of other stars, is not well defined, for stars do not possess sharp outer edges. Gases of extremely low density expand for hundreds of thousands of kilometers into space. In the case of the Sun, these outermost layers contain the chromosphere and the corona. When astronomers speak of the surface of a star, they do not mean those tenuous outermost gases. They are referring to the layer that emits the light that we see. This surface layer

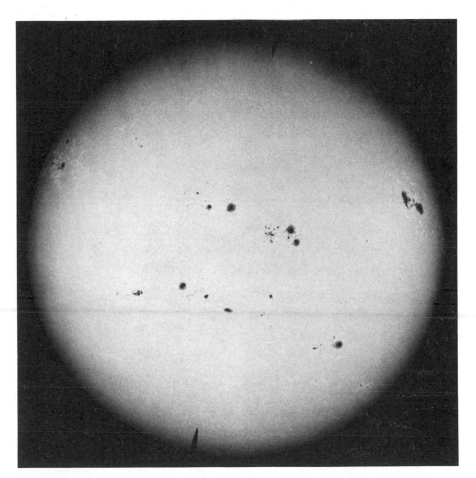

Figure 3.9. The Sun, photographed near sunspot maximum on July 13, 1937. Sunspots are the most conspicuous features of the Sun's photosphere. They are regions in which the gases are up to 1500 K cooler than those of the surroundings. The number of spots varies over an 11-year cycle (next maximum is expected to occur in 1991).

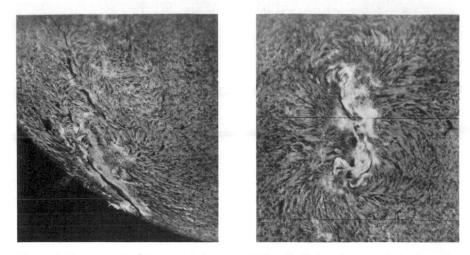

Figure 3.10. A group of *sunspots*, photographed in the light of the Hα line of hydrogen.

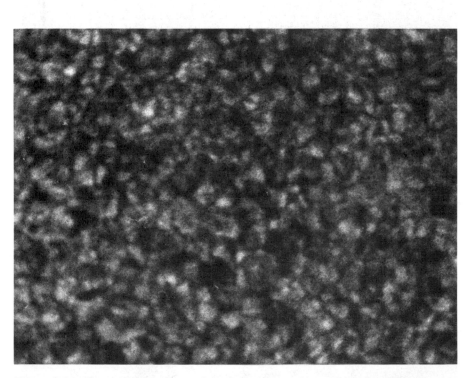

Figure 3.11. The granular pattern of the solar photosphere. The *solar photosphere* is the Sun's visible surface. Its granular appearance arises from the convective motions of gases by which energy is transported to the Sun's surface. Regions in which convective currents arrive at the surface from below are hotter and brighter, while regions in which convective currents move downward are cooler and darker.

Figure 3.12. The total eclipse of the Sun of June 8, 1918, photographed at Green River, Wyoming. As the Moon moves in front of the Sun, it blocks out the light from the solar photosphere and makes it possible to photograph the Sun's extended halo of very tenuous and hot gas (2 million K and higher), known as the *corona*.

is known as the *photosphere* (figures 3.9–3.11) and is spatially rather narrow and well defined. If we were to go below it, we would find the matter becoming opaque rapidly, like dense fog. None of the radiation present there can reach us directly because it is absorbed by the overlying gases. In contrast, the layers above the photosphere—the *chromosphere* and *corona* (figure 3.12) in the case of the Sun—are transparent and the radiation emitted from the photosphere readily passes through them and into space. Very little radiation is emitted from the chromosphere and corona, and without special instruments they cannot be seen.

Open Star Clusters

Next let us consider the HR diagram in figure 3.13. It is constructed from the data of several open star clusters. This diagram demonstrates the effect of stellar aging. When the clusters were born, all of their stars lay along the main sequence. Today, each of the graphs of the different clusters has its own distinct

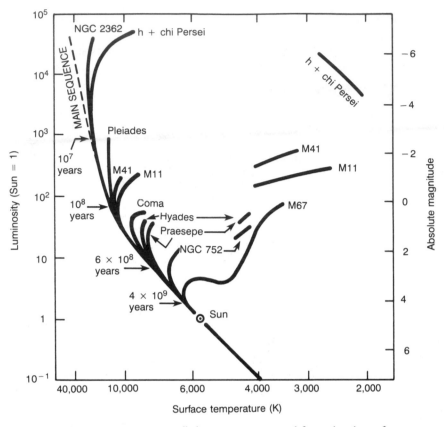

Figure 3.13. Hertzsprung–Russell diagram constructed from the data of ten open star clusters and showing the effect of stellar aging. The ages of all ten star clusters can be deduced from the age scale given to the left of the main sequence.

termination point on the main sequence. In the case of the two youngest clusters, NGC 2362 and h and Chi Persei, this point lies far to the left. However, even they no longer include any of the most massive stars that evolve through their main sequence lives in fewer than 10 million years. These stars have already burned all of the hydrogen in their cores and evolved into yellow and red supergiants or exploded as supernovae. In contrast, less massive stars, which stay on the main sequence longer, are still there; and thus we may conclude that these two clusters are roughly 10 million years old. In the case of the Hyades cluster, only stars with main sequence lives of 600 million years and longer are still on the main sequence; the others have evolved off it. Hence, this cluster has an age of about 600 million years. Cluster M 67 is one of the oldest known open star clusters in the Galaxy. Its main sequence terminates at the point corresponding to an age of 4 billion years, which is nearly as old as the Solar System. M 67 must be an extremely tightly bound system to have lasted so long, for most open star clusters disperse within 1 to 2 billion years.

Core Hydrogen Burning

We may regard the main sequence as a way station in the evolution of stars. The fusion of hydrogen into helium in the stars' cores temporarily arrests or slows down the evolutionary changes of the stars' surface features and makes them bunch up along the main sequence. The burning of helium into heavier elements affects the evolution of the stars similarly (see the next section). It produces surface conditions that concentrate those stars into the red giant and supergiant regions in the HR diagram. Likewise, the cooling of stars after they have exhausted their nuclear fuel creates yet another way station, namely the white dwarf region.

Why is the slowing down of the evolution along the main sequence not similar for all stars? Why do those with 10, 25, or 50 times more mass than the Sun complete their main sequence lives 1000, 2000, and 10,000 times faster than solar mass stars? The reason is that the core temperatures of the massive stars are higher than those of the less massive stars (see table 3.2) and that hydrogen burning is extremely temperature sensitive. It is much more rapid at high temperatures than at low temperatures. Consequently, the massive stars convert their hydrogen into helium much more rapidly than the smaller-mass stars. This means that the massive stars are not only brighter, but they also run out of hydrogen fuel in their cores much sooner and have shorter main sequence lifetimes than the less massive stars.

Interestingly, the mechanisms by which hydrogen is converted into helium tends to be different for the high- and low-mass stars. Main sequence stars more massive than about $2 M_\odot$ burn their hydrogen mainly by the *carbon cycle*, in which carbon-12 (^{12}C) is used as a catalyst. Main sequence stars less massive than about $2 M_\odot$ burn their hydrogen mainly by the *proton–proton chain*. The details of these two and other nuclear reaction mechanisms of importance in stellar evolution are discussed in box 3.4.

The Sun has been on the main sequence for roughly 4.5 billion years and will remain in this evolutionary phase for about another 5.5 billion years. It burns its hydrogen relatively slowly and mainly by the proton–proton chain. If the Sun

Table 3.2. Comparison of Physical Characteristics of Main Sequence Stars.

Mass $(M_\odot)$	Radius $(R_\odot)$	Luminosity $(L_\odot)$	Surface temperature (K)	Central temperature (10^6 K)	Central density (g/cm^3)	Main sequence lifetime (years)
0.1	0.15	0.001	2,500	5	200	10^{12}
1	1	1	5,800	16	160	10^{10}
5	3	500	20,000	25	25	10^8
10	5	10,000	25,000	30	10	10^7
25	10	100,000	35,000	40	5	5×10^6
50	20	500,000	45,000	45	3	10^6

used up its hydrogen more rapidly, say as rapidly as a 10 $M_\odot$ star, it would have passed through its entire life long ago and as a main sequence star would have been much brighter and hotter than it actually is. In that case it would be very difficult to imagine how life on Earth could have come into existence or, if it had, how it could have survived and evolved. This is particularly true of land life, which is directly exposed to the solar radiation.

During the Sun's main sequence life up to now, its surface temperature has remained nearly constant and its luminosity has increased by about 30%. Considering the length of time involved, this is a remarkably small change and attests to the stability of the Sun's structure.

POST–MAIN SEQUENCE EVOLUTION

While stars are on the main sequence, only their cores experience major evolutionary changes — the conversion of hydrogen into helium. The outer layers remain largely unaffected. This situation ends rather abruptly when all of the hydrogen in the central 10–15% of a star's mass has been converted into helium. The star's main sequence life comes to a close and, in the absence of central nuclear energy generation, the stellar interior does what it did earlier. It contracts and releases gravitational energy, bringing about a number of significant physical changes:

1. About half of the energy released by the contraction heats the star's interior and the temperature rises. As a consequence, the hydrogen just exterior to the helium core burns in a shell.

2. The rest of the energy released by contraction plus the energy generated by the newly created hydrogen shell source diffuse slowly outward and inflate the star's outermost layers. Thus, even though the core of the star contracts, its outer layers expand.

3. The expansion lowers the star's surface temperature and the color of its radiation shifts toward the red.

4. Finally, as the temperature at the center of the contracting core approaches 100 million degrees, helium ignites and fuses into carbon and oxygen. In the case of stars with masses similar to the Sun's, the ignition of helium occurs rather suddenly and violently, and it is known as the *helium flash*. In the case of more massive stars, the ignition of helium occurs more gradually. The star now has two nuclear energy sources: helium burning in the core and hydrogen burning in a thin, adjacent shell. Together, the energy released by these two sources brings the contraction of the core almost to a halt and the outer layers stop expanding. Equilibrium is reestablished between the rate of nuclear energy release in the interior and the rate of energy loss (mainly by radiation) from the star's surface.

Red Giants and Supergiants

The physical events just described are the beginning of the star's post–main sequence evolution. This evolution is characterized by tracks in the HR diagram that may be rather complex depending on the star's mass, as illustrated in figure 3.14 for a 1 $M_\odot$ and a 25 $M_\odot$ star. A fuller description is in the following pages.

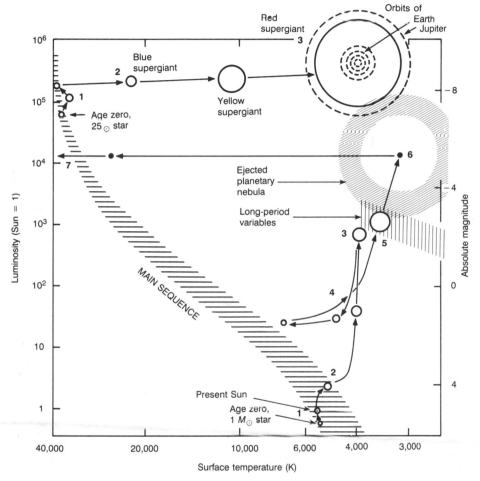

Figure 3.14. Evolutionary tracks of 1 $M_\odot$ (Sun) and 25 $M_\odot$ stars in the Hertzsprung–Russell diagram. The major stages, indicated by numbers in the figure, are:

1 $M_\odot$ star

1. Main sequence (present Sun). H burns in core. Age 4.5 billion years.
2. End of main sequence. Core H burning ceases, H burning shifts to shell, He core contracts, H-rich envelope expands. Age 10 billion years.
3. Helium flash. He ignites explosively in core.
4. Red giant. He burns in core, H burns in surrounding shell.
5. Evolved red giant, long-period variable. Very compact C–O core, He and H burn in shells, highly extended H-rich envelope.

Figure 3.14 continued.

6. Ejection of H-rich envelope as a planetary nebula. Nuclear reactions are extinguished. Age 14 billion years.

7. Upon ejection of planetary nebula, stellar remnant moves to the left in the HR diagram and its surface temperature exceeds 100,000 K. Then the star cools and fades; within approximately 100 million years of the ejection of the planetary nebula it becomes a white dwarf (not shown in this figure).

25 $M_\odot$ star

1. End of main sequence. Core H burning ceases. Age 5 million years.

2. Beginning of supergiant phase. He, C, O, and Si burn in successive stages in the core and surrounding shells. Outer H-rich envelope expands. Star becomes a blue, yellow, and then a red supergiant.

3. Red supergiant, with enormously expanded H-rich envelope. Star explodes as a supernova about 500,000 years after the end of its main sequence of life.

Note that the stars are not drawn to scale. Structural details of the evolution of the 1 $M_\odot$ star are found in figure 3.16. (The track of the 25 $M_\odot$ star is taken from Weaver 1980.)

Figure 3.15. Sirius, also known as Alpha Canis Majoris or the Dog Star. This star is a double star system, consisting of Sirius A and B. Sirius A is by far the brighter of the two stars and usually the only one seen. It is, in fact, the brightest star in the sky and, along with its companion, constitutes the sixth nearest stellar system to Earth, at a distance of 8.7 LY. Sirius A is a main sequence star of about 2.3 $M_\odot$, while Sirius B is a white dwarf of about 1 $M_\odot$.

Stars below approximately 8 $M_\odot$ become *red giants*. During this evolutionary phase they usually shed enough mass, by a steady stellar wind and by sudden ejection, that eventually they become less massive than 1.4 $M_\odot$ and end their lives as *white dwarfs* (figure 3.15). Stars above approximately 8 $M_\odot$ become *supergiants*. They also lose mass, but generally not enough to become reduced below 1.4 $M_\odot$, which is the upper limit for white dwarfs (more about white dwarfs and their upper mass limit below). These more massive stars undergo complicated nuclear evolutions and end their lives as *supernovae*.

Long-period Variables and Planetary Nebulae

The final evolution of a star like the Sun, which is typical of stars that become white dwarfs, is illustrated in figure 3.16. When the helium in the core of such a star has been depleted by conversion into carbon and oxygen, the central parts of the star contract again. Gravitational energy is released and the temperature rises. Helium starts burning in a thin shell adjacent to the carbon–oxygen core, and hydrogen continues to burn (in a shell) further out. The

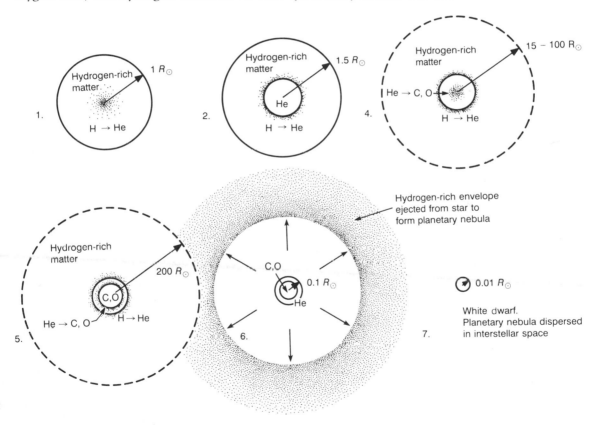

Figure 3.16. Schematic illustration of the evolution of a 1 $M_\odot$ star from the main sequence to the white dwarf stage. The numbers refer to the stages indicated and defined in figure 3.14. Dots represent regions of nuclear burning. Stellar radii are not drawn to scale; therefore, the outer surfaces of the two red giants are indicated as dashed circles.

hydrogen shell source produces helium, and the helium shell source below it converts the helium into carbon and oxygen. Thus, the two shell sources create an ever-growing carbon–oxygen core (note the core grows in mass, though in size it slowly contracts). The temperature never becomes high enough for the nuclear ignition of carbon or oxygen.

As the helium and hydrogen shell sources continue to burn, and the carbon–oxygen core continues to grow in mass and to contract, the rate of nuclear energy generation speeds up and the flow of radiation outward increases. Gradually the radiation flow becomes so intense that it creates instabilities in the remaining hydrogen-rich surface layers. Periodically the pressure gradient exceeds gravity and then falls below the value of gravity. As a result, the star begins to pulsate. Over a period of months to a year, the surface layers slowly expand and then contract again in repeating cycles. Such a pulsating star is known as a *long-period variable*. It is rather cool and luminous, as indicated by its location in the HR diagram. It is also characterized by a steady loss of mass from its surface in the form of an outflowing wind.*

In the course of time, the pulsations of a long-period variable grow stronger until eventually expansion is no longer followed by contraction. The pressure gradient created by the outward flowing radiation becomes sufficiently strong to push the entire hydrogen-rich envelope off into space. The ejection is helped along by the fact that the envelope is already quite extended and, hence, relatively weakly bound by gravity. Furthermore, the ejected matter cools to the point where the ions and electrons combine into atoms and molecules, which releases energy and contributes to the outward push. Such an ejected stellar envelope is known as a *planetary nebula*. This name was coined by early astronomers who, through their low-power telescopes, found that planetary nebulae resemble both interstellar gaseous nebulae and the Sun's outer planets (although planetary nebulae bear no relationship to planets, and they have no appreciable motions in the sky as do planets). Subsequent analysis of the spectra of their light confirmed that they are indeed gaseous. There are more than one thousand planetary nebulae in our part of the Milky Way. Approximately 4 billion years have elapsed between the end of the star's main sequence life and the ejection of the planetary nebula.

A planetary nebula typically contains about 0.2 $M_\odot$ of hydrogen-rich matter, with some enrichment of helium and other products of the parent star's nuclear history, and it expands at 10–20 km/sec (about 50,000 km/hr or 30,000 mi/hr). Initially it is brightly illuminated by the intense radiation from the central star as, for instance, in the case of the Ring nebula shown in figure 3.17. However, as the

---

*All stars with surface temperatures like the Sun's and cooler lose mass from a hot corona surrounding them. The corona consists of a very sparse expanse of gas and is created by energy released from convective and magnetic activities in the surface layers of these stars. The wind emanating from the Sun is extremely weak and carries away only about 0.0001 $M_\odot$ in the course of the Sun's entire main sequence lifetime. Still, the solar wind creates noticeable effects in the Earth's environment. It passes Earth with a speed of about 450 km/sec. By injecting protons, electrons, and other charged particles into the Earth's magnetic field, it causes auroras (northern and southern lights); and when it strikes the skin of a spacecraft, X rays are created which can harm the astronauts. In red giants and supergiants, which are much more luminous and have more extended coronae than the Sun, the stellar winds are much stronger and may carry away up to 1 $M_\odot$ in just a few million years.

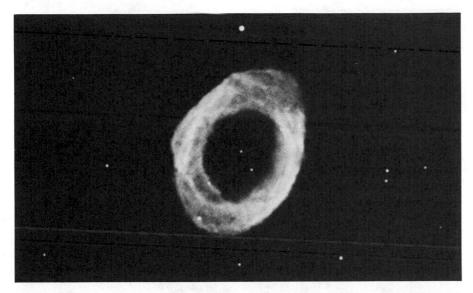

Figure 3.17. The Ring nebula, a planetary nebula in the constellation Lyra. This nebular shell of gas is expanding at roughly 20 km/sec radially away from the star seen near its center and from which it was ejected approximately 5000 years ago. The star is exceptionally compact and hot and, like all central stars of planetary nebulae, is evolving toward the white dwarf stage. The nebula lies at approximately 2000 LY from us and has a radius of about one-third of a LY.

star fades (see below), the nebula fades also, and within a few ten thousand years it can no longer be seen. The expansion continues until the ejected matter is completely dispersed and intermingled with the gas and dust of interstellar space.

The star that remains after the ejection is compact and luminous. As the exposed interior of a former red giant, it has a very high surface temperature, which initially may exceed 100,000 K. In the HR diagram it lies far to the left. Despite the high temperature at the surface, conditions in the interior are no longer favorable for nuclear reactions. The burning of hydrogen has become extinguished during the ejection, and the burning of helium does not last much longer either because of its proximity to the star's surface. For some time the star continues to shine very brightly by radiating away its stored-up heat energy. Gradually the star cools and grows fainter and fainter. At the same time it contracts, but only slightly because it is already quite compact. Little else happens to the star and over a period of about 100 million years it evolves into a white dwarf.

White Dwarfs and Black Dwarfs

Even though white dwarfs do not possess any nuclear energy sources and shine by radiating stored-up heat energy, they are rather interesting objects. They typically have a mass of about 0.5 to 1 $M_\odot$ and a size comparable to that of the Earth, making them very compact. Their average densities range from about 10^5 to 10^7 g/cm^3, which means that a volume the size of a fingertip contains from

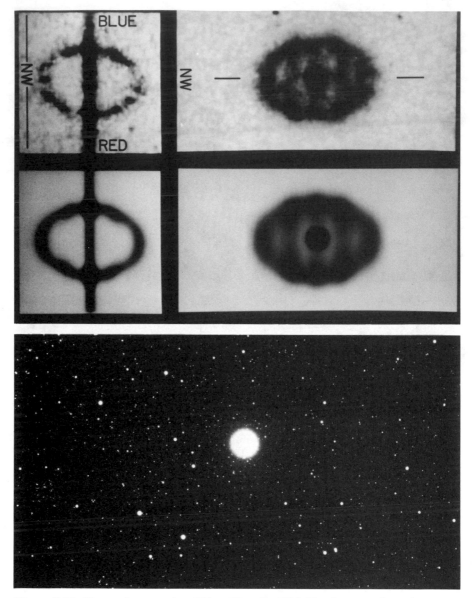

Figure 3.18. Three photographs of Nova Herculis 1934. The first two show the star before and during the outburst in 1934. The third was taken in 1972 and shows a magnified view of the star and the expanding ejecta.

0.1 to 10 tons of matter. At such high densities the electrons (which are part of the matter along with nuclei of helium, carbon, oxygen, and a few other heavier elements) touch each other and cannot be compressed further. Hence, gravity is no longer able to contract such stars. In the course of billions of years the white dwarfs cool and fade to a point where they can no longer be seen, even through telescopes. They have then become *black dwarfs*—dead cinders of once living stars.

A white dwarf is incompressible, however, only if its mass is less than approximately 1.4 $M_\odot$, the *Chandrasekhar mass limit* (named after Subrahmanyan Chandrasekhar, an astrophysicist from India who since 1937 has been at the University of Chicago and who in 1984, along with William A. Fowler of the California Institute of Technology, received the Nobel Prize in Physics for the fundamental contributions he has made to our understanding of stellar evolution). If the mass of the white dwarf is greater than 1.4 $M_\odot$, gravity is able to overwhelm the pressure and collapse the star further, which leads to a supernova explosion (see discussion of the evolution of a 25 $M_\odot$ star, next section).

Novae

Occasionally exceptions arise to the rule that white dwarfs merely cool and fade. If a white dwarf has a close stellar companion that loses hydrogen-rich matter, the white dwarf may become a *nova* "new star" (figure 3.18). A nova outburst is illustrated in figure 3.19 in four stages:

1. A white dwarf (carbon–oxygen core, surrounded by a helium layer) and a red star orbit each other in close proximity. The red star has an extended hydrogen-rich envelope, which keeps expanding.

2. Much of the hydrogen-rich matter the red star loses is attracted gravitationally by the white dwarf. The matter spirals through an accretion disk and falls

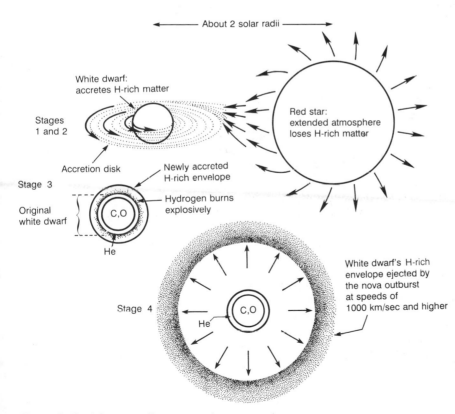

Figure 3.19. Schematic illustration of a nova outburst.

onto the white dwarf. A hydrogen-rich envelope builds up on the white dwarf.

3. When the envelope has grown to approximately $10^{-4} M_\odot$, the temperature and density of its base become severe enough for the fusion of hydrogen into helium. Because of the extreme densities existing in the white dwarf, the nuclear burning occurs explosively and a great deal of energy is released in a short time.

4. The nuclear energy released is sufficient to eject most of the accreted matter at speeds of 1000 km/sec and higher (3.6 million km/hr and higher) and to cause the star's brightness to increase suddenly by a factor of several million. This sudden brightening makes the star visible to great distances, which is why such an explosion is called a nova.

Nova outbursts are repeated periodically as long as the companion star loses hydrogen-rich matter and the white dwarf accretes it. Once this mass exchange ceases, the nova outbursts come to an end and the white dwarf resumes its evolution of cooling and fading. (Note that novae are the result of explosions in a white dwarf's surface layers. Supernovae are much more energetic and involve the entire star.)

FINAL EVOLUTION OF MASSIVE STARS

The more massive stars do not end their lives quietly as white dwarfs nor does their nuclear history terminate with the burning of helium. Considerable excitement awaits them during their final years, culminating in a superenergetic and catastrophic explosion that destroys the normal structure of the star and expels much of its matter into interstellar space at very high speeds.

Consider a computer simulation of the final evolution of a $25 M_\odot$ star to see how these events come about. When the star has completed its main sequence life, it starts burning helium in the core and hydrogen in a surrounding shell. Up to now the evolution has been much like that of the less massive star discussed in the previous section, except that the burning of hydrogen in the $25 M_\odot$ star took place mainly by the carbon cycle and the luminosity was higher, as shown by the star's evolutionary track in the HR diagram of figure 3.14. Upon depletion of helium in the core, gravitational contraction releases energy once again, raises the temperature, and greatly expands the star's outer layers, making it a supergiant.

Heavy Element Synthesis

Now the evolution begins to deviate sharply from that of the less massive star. Helium starts burning in a shell next to the carbon–oxygen core, hydrogen continues to burn in a shell further out, and soon carbon ignites at the center. The synthesis of the heavy elements has now begun in earnest. When the carbon in the core becomes exhausted, contraction raises the temperature to the point where oxygen burns in the core and carbon in an adjacent shell. Simultaneously, helium and hydrogen continue to burn in shells further out.

BOX 3.4.
Nuclear Reactions: The Energy Source of Stars

NUCLEAR reactions convert atomic nuclei of one kind into nuclei of another kind. The forces participating in these conversions are the nuclear and the weak forces, with the latter coming into play only when neutrinos are present.

Nuclear reactions are to be distinguished from chemical reactions. Chemical reactions never affect the nuclei of atoms. They affect the electrons surrounding the nuclei and either create or break the chemical bonds that hold atoms together in molecules, as discussed further in the Introduction to part 2. The only force participating in chemical reactions is the electromagnetic force. Examples of chemical reactions are the formation of molecules in interstellar space, the burning of coal and other fossil fuel, and the digestion of food. Chemical reactions liberate considerably less energy than most nuclear reactions. For example, if the Sun derived its energy from the burning of coal, it would shine at its present rate for only about 5000 years (assuming it were made entirely of coal and burned all of it). In reality, the Sun will shine for well over 10 billion years by fusion reactions.

Nuclear reactions may be divided into two major classes: fusion and fission reactions. In *fusion reactions*, nuclei of lighter elements are fused into nuclei of heavier elements. *Fission reactions* are the reverse of fusion reactions. In them, nuclei of heavy elements are split into nuclei of lighter elements.

Stars obtain their energy by fusion reactions. These reactions liberate enormous quantities of energy that comes from a slight lowering in mass of the nuclei undergoing fusion. For example, in the fusion of hydrogen into helium-4, the mass of the initial hydrogen is lowered by roughly 0.7% compared with the mass of the resulting helium-4. If we call the reduction in mass Δmass (read "delta mass"; the symbol Δ is the letter D in the Greek alphabet and stands for "difference"), the energy liberated is given by Einstein's equation

$$\text{energy} = \Delta\text{mass} \times c^2,$$

where c stands for the speed of light. The amount of mass that is converted into energy in fusion reactions is very little (a mere 0.7% in this example), but because of the factor c^2 in Einstein's equation, the energy liberated is very large (note that $c^2 = 9 \times 10^{16}$ m^2/sec^2). To appreciate just how much energy is released, note that the fusion of 1 kg of hydrogen into helium-4 yields as much energy as the chemical burning of 25,000 tons of coal.

Fusion reactions that build up nuclei of elements beyond iron do not liberate energy; they require energy.[*] Hence, in general the elements beyond iron are formed only during such violent events as supernova explosions, which are able to provide the required energy. The energy that thereby becomes locked up in the elements beyond iron may be released by fission reactions as, for example, by the fission of uranium into simpler products (such as lanthanum, barium, krypton, and bromine) in atomic power generators.

Fusion of ^{1}H into ^{4}He

By far the most energetic of all the fusion reactions occurring in stars is the conversion of hydrogen into helium. There are two major routes by which this conversion may proceed. In stars of about 2 $M_\odot$ and less, including the Sun, whose core temperatures are below 20 million K, it proceeds mainly by the *proton–proton chain*. In stars more massive than about 2 $M_\odot$, whose core temperatures are above 20 million K, it proceeds mainly by the *carbon cycle*:

The Proton–Proton Chain
(5 million K and above)

$$\text{p} + \text{p} \rightarrow {}^2\text{H} + e^+ + \nu$$
$$\text{p} + {}^2\text{H} \rightarrow {}^3\text{He} + \gamma$$
$${}^3\text{He} + {}^3\text{He} \rightarrow {}^4\text{He} + 2\text{p} + \gamma$$
$$+ \text{ energy}$$

(*Note*: The first two reactions must occur twice in order to yield the two ^{3}He nuclei required by the third reaction.)

The Carbon Cycle
(20 million K and above)

$$^{12}\text{C} + \text{p} \rightarrow {}^{13}\text{N} + \gamma$$
$$^{13}\text{N} \rightarrow {}^{13}\text{C} + e^+ + \nu$$
$$^{13}\text{C} + \text{p} \rightarrow {}^{14}\text{N} + \gamma$$
$$^{14}\text{N} + \text{p} \rightarrow {}^{15}\text{O} + \gamma$$
$$^{15}\text{O} \rightarrow {}^{15}\text{N} + e^+ + \nu$$
$$^{15}\text{N} + \text{p} \rightarrow {}^{12}\text{C} + {}^4\text{He}$$
$$+ \text{ energy}$$

[*]The Periodic Table of the Elements, page 23, shows which elements are lighter and which are heavier than iron.

BOX 3.4 *continued*

The net result of the reactions of both the p–p chain and the carbon cycle is the conversion of nuclei of hydrogen into nuclei of helium-4 along with the release of large amounts of energy: $4 \, p \rightarrow {}^4He +$ energy.

The positrons (e^+) produced by the reactions eventually collide with electrons (e^-) and are annihilated, giving off additional energy. The neutrinos escape from the star and carry away some energy. The photons become rapidly absorbed by the stellar matter and deposit whatever energy they carry as heat.

The unusual feature of the carbon cycle is the way nuclei of ${}^{12}C$ participate in it. In the course of the cycle, these nuclei change into nuclei of ${}^{13}N$, ${}^{13}C$, ${}^{14}N$, ${}^{15}O$, ${}^{15}N$, and, in the final reaction step, back into ${}^{12}C$. Thus, nuclei of ${}^{12}C$ act as catalysts.* They cycle through the reactions, temporarily changing into nuclei of other elements and facilitating the reactions.

During the late stages of a massive star's evolution, when the carbon cycle (which occurs then in a shell beyond the star's core; see text) is interrupted by a supernova explosion, some of the nuclei are in the form ${}^{14}N$ and, along with the rest of the envelope material, are ejected into space. This constitutes the main source of nitrogen in the Universe, an element that is very abundant in our atmosphere and is essential for all earth-bound life.

Temperature dependence. The next question is, "Why do the p–p chain and the carbon cycle, as well as other fusion reactions, take place only at the high temperatures found in stellar interiors?" The answer is that the nuclear force, which plays the key role in all fusion reactions, has an extremely short range, namely about 10^{-13} cm (see tables I.1 and I.2). Unless the collisions among the nuclei bring them within that distance of each other, they will not feel each other's nuclear forces and nuclear reactions will not

occur. It is not easy, however, to get nuclei to within 10^{-13} cm of each other because of the repulsive force exerted by their positive electric charges. This force increases the closer two nuclei come to each other. It also increases the more charge the nuclei carry. Only at temperatures of several million K and higher do nuclei fly around sufficiently rapidly and collide sufficiently violently to overcome their mutual repulsions and come within the range of each other's nuclear forces.

Because protons carry only one positive electric charge, the minimum temperature required for the p–p chain is the relatively low temperature of about 5 million K. In the case of the carbon cycle, in which protons react with nuclei that carry six or seven positive charges, the temperature must be at least 20 million K before the reactions occur at appreciable rates. And for most other fusion reactions taking place in stars, some of which are discussed below, the temperatures required are higher still.

Except for scale and the forces involved, the requirement of high temperatures to induce nuclear reactions is analogous to the requirement of a minimum temperature for burning wood. At room temperature, wood is quite stable, just as nuclei are quite stable at sufficiently low temperatures. However, if the temperature rises above approximately 500° C, then the molecules of wood oscillate and knock into each other so violently that chemical reactions are induced, unlocking the molecules' stored-up energies and creating the phenomenon we call fire.

Reaction rates. There is one final point to consider regarding the p–p chain and the carbon cycle, namely the rates at which these reactions take place. The rapidity with which fusion reactions occur depends in part on the violence and frequency with which the nuclei collide with each other. The higher the temperature, the more violent and frequent are the collisions and, hence, the more rapid is the nuclear burning.

Temperature is, however, not the only factor influencing the reaction rates. Another factor is the probability that a nuclear reaction occurs once two nuclei have come within the range of each other's nuclear forces. It turns out that the presence of the weak force tends to reduce the reaction probability considerably. This effect is particularly pronounced in the first reaction of the p–p chain (the emission of a neutrino

---

Catalysts are particles that participate in nuclear or chemical reactions and speed them up without themselves undergoing any permanent changes. The catalysts of nuclear reactions are aways atomic nuclei as, for example, the nuclei of ${}^{12}C$ in the carbon cycle. The catalysts of chemical reactions are usually molecules or chemical compounds. As will be discussed in Part Two of this book, all of life's chemical processes depend on the participation of catalysts, such as the enzymes.

BOX 3.4 continued

indicates the presence of the weak force). This reaction is a kind of bottleneck in the p–p chain. Protons must collide many times before two of them react and fuse into a ^2H nucleus. Once this has happened, the rest of the reaction sequence to ^4He goes relatively quickly. In contrast to the p–p chain, the carbon cycle is not much slowed down by the weak force, even though this force participates in the second and fifth reaction steps.

Fusion of ^4He into ^{12}C and ^{16}O

The fusion of hydrogen into helium in stellar evolution is followed by the fusion of helium into carbon and oxygen. The fusion into carbon occurs by the *triple-alpha reaction*. This reaction is so called because in it three nuclei of ^4He—which are also known as *alpha particles*—collide and fuse into a nucleus of ^{12}C:

$$^4He + {}^4He + {}^4He \rightarrow {}^{12}C + energy$$
(100 million K and above)

The fusion into oxygen occurs by the addition of alpha particles to nuclei of ^{12}C:

$$^{12}C + {}^4He \rightarrow {}^{16}O + energy$$
(100 million K and above)

Because this reaction requires the presence of nuclei of ^{12}C, it takes place only after the triple-alpha reactions have built up a supply of carbon nuclei.

The production of carbon and oxygen by these reactions requires much higher temperatures than the earlier fusion of hydrogen into helium. The reason is simple. Each alpha particle carries two positive electric charges and the nuclei of ^{12}C carry six charges. Furthermore, in the triple-alpha reactions, three alpha particles must be brought almost simultaneously within the range of each other's nuclear forces. Consequently, the electric forces of repulsion that need to be overcome are considerably greater than those in the reactions of the p–p chain and the carbon cycle, and carbon and oxygen are manufactured only at temperatures of 100 million K or higher. That means that these reactions do not occur while stars are on the main sequence. Only after their cores

have contracted further and their central temperatures have risen to 100 million K or higher, do those reactions take place.

Fusion of ^{12}C and ^{16}O into Heavier Elements

The next major phase of nuclear burning consists of the fusion of ^{12}C and ^{16}O into still heavier nuclei.

Carbon Fusion
(600 million K and above)

$$^{12}C + {}^{12}C \rightarrow {}^{24}Mg + \gamma$$
$$\rightarrow {}^{23}Na + p$$
$$\rightarrow {}^{20}Ne + {}^4He$$
$$+ energy$$

Oxygen Fusion
(1 billion K and above)

$$^{16}O + {}^{16}O \rightarrow {}^{32}S + \gamma$$
$$\rightarrow {}^{31}P + p$$
$$\rightarrow {}^{31}S + n$$
$$\rightarrow {}^{28}Si + {}^4He$$
$$+ energy$$

Because the electric repulsive forces are even greater in these reactions than they were in the earlier reactions, the minimum temperatures required are greater also. They are approximately 600 million and 1 billion K for the carbon and the oxygen reactions, respectively. This means that they can occur only after the stellar core has contracted and heated up still further than during the helium-burning phase.

As the equations indicate, several different products result during carbon and oxygen burning: ^{20}Ne, ^{23}Na, ^{24}Mg, ^{28}Si, ^{31}P, ^{31}S, and ^{32}S, as well as protons, neutrons, and alpha particles (^4He). The protons, neutrons, and alpha particles do not last long. As rapidly as they form, they react either with each other or with the heavier nuclei and produce still additional nuclear species and isotopes. The result is a broad spectrum of nuclear products.

The nuclear reactions discussed so far were induced by violent collisions between nuclei. Upon completion of oxygen burning, further nuclear reactions are induced by photons, which are inevitably present in stellar interiors. The temperatures are now in excess of 2 billion degrees, and the photons are so

BOX 3.4 *continued*

energetic that they knock protons, neutrons, and alpha particles out of the heavier nuclei. These photon-induced fission reactions are rapidly followed by fusion reactions in which the protons, neutrons, and alpha particles react with each other and with heavier nuclei, much as they did during the burning of carbon and oxygen earlier. However, now, because of the higher temperatures, they push the buildup of the heavier nuclei all the way to the formation of iron, nickel, and other elements of similar atomic weight. These *iron peak nuclei* (see figure 3.22) are more stable than the lighter nuclei and are, therefore, relatively immune to the breakdown by photons. They accumulate, which explains their relatively high abundance in nature.

With the formation of the iron peak elements, fusion reactions and the release of nuclear energy come to a close in stellar evolution. The buildup of still heavier elements requires energy and occurs only during such violent events as supernova explosions.

Summary

A summary of the most dominant nuclear reactions in stars follows.

hydrogen burning $\quad\quad ^1H \rightarrow {}^4He, {}^{14}N$ (byproduct of the carbon cycle)

helium burning $\quad\quad\quad ^4He \rightarrow {}^{12}C, {}^{16}O$

carbon burning $\quad\quad\quad ^{12}C \rightarrow {}^{20}Ne, {}^{24}Mg$

oxygen burning $\quad\quad\quad ^{16}O \rightarrow {}^{28}Si, {}^{32}S$

photon-induced $\quad ^{28}Si, {}^{32}S \rightarrow {}^{56}Fe$
burning

Because these reactions are dominant, their nuclear products — ^4He, ^{12}C, ^{14}N, ^{16}O, ^{20}Ne, ^{24}Mg, ^{28}Si, ^{32}S, and ^{56}Fe — are the most abundant elements in the matter ejected during supernova explosions.

The matter ejected during supernova explosions intermingles with and enriches the gas and dust of interstellar space and, thereby, becomes part of the raw material for the birth of new generations of stars. Some 4.5 billion years ago, the Sun was formed from such enriched matter and that is why the elements listed above are so abundant in the solar matter, as indicated by the graph of the relative abundance of the elements in the Sun (figure 3.22). It is also the reason why, with the exception of the chemically inert gases helium and neon, these elements are among the most abundant ones on Earth.

Step by step the star synthesizes the heavy elements. The ashes from the burning of one element become the fuel for the successive nuclear reactions: hydrogen fuses into helium, helium into carbon and oxygen, and these two elements fuse into neon, sodium, magnesium, silicon, phosphorus, and sulfur (see box 3.4 for details). The energy liberated by these reactions keeps the star shining and retards the gravitational contraction of its interior which (see below) would otherwise occur very rapidly.

Throughout the carbon and oxygen burning phases, neutrinos are created in great abundance in the core of the star. They come into existence spontaneously because of the extreme temperatures and densities that exist there — approximately 0.6 to 1.0×10^9 K and 10^5 to 10^7 g/cm^3. As noted before, neutrinos are very elusive and as quickly as they are created they dart off into space, hardly ever interacting with the intervening layers of the star. With them they carry energy. This loss of energy by neutrino emission greatly increases the speed with which the star evolves from here on. It causes the star's core to contract much more quickly than it would without neutrino emission. The contraction, in turn, raises the density and temperature of the core, and carbon as well as oxygen burn faster and faster. Carbon burning in the core is so intense that it is completed in 600 years. Oxygen burning is even more extreme and is over in just one-half year. By stellar standards, these are very fast evolution times indeed.

In the next phase of nuclear burning, the elements produced by the fusion of oxygen — mainly ^{28}Si and ^{32}S — are ignited, and in a series of reactions they are converted into iron and other elements of similar weight (again, see box 3.4). The structure of the star now consists of a rapidly growing iron core (growing in mass, but shrinking in size), surrounded by successive shells in which the burning of sulfur, silicon, oxygen, carbon, helium, and hydrogen are the major sources of nuclear energy, as shown in figure 3.20.

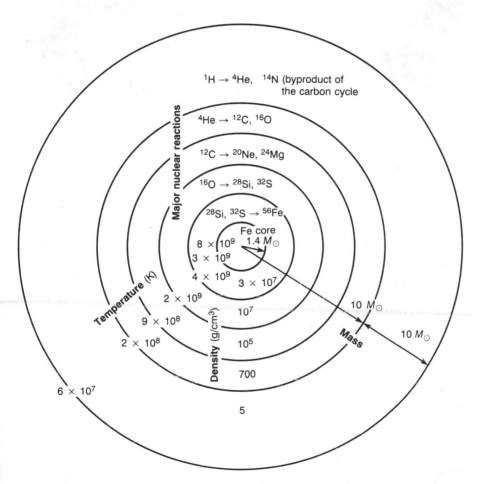

Figure 3.20. Structure of a 20 $M_\odot$ star just before implosion of its iron core and explosion of its outer parts as a supernova. The diagram shows the major nuclear reactions and the distribution of temperature, density, and mass. (The shells are not drawn to scale.)

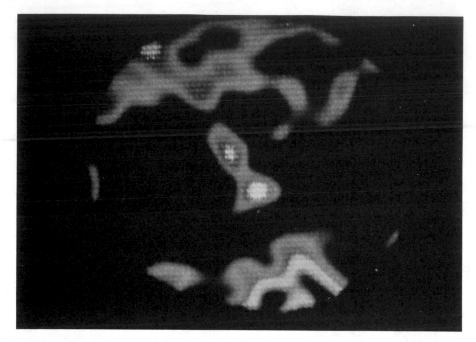

Figure 3.21. Betelgeuse, a red supergiant and the brightest star in the constellation Orion. This photograph was obtained by a special technique (called *speckle interferometry*) designed to overcome the problem of twinkling, which is caused by the Earth's atmosphere. It shows several enormous starspots, each roughly as large as the Earth's orbit around the Sun. The star's total size is approximately 1.5 times greater than the orbit of Mars.

The ongoing gravitational contraction of the star's central regions raises their density to approximately 10^9 g/cm^3. This means that the central one-solar mass of the star is compressed into a sphere ten times smaller than the Earth. Along with the contraction of the core, the steady outflow of large amounts of radiant energy toward the surface greatly expands the hydrogen-rich outer layers. If the star were at the location of the Sun, the outer layers would easily engulf the inner part of the Solar System to beyond the planet Mars. Simultaneously, steady loss of mass in the form of a strong stellar wind reduces the star's mass to perhaps as low as 20 $M_\odot$. Because of its enormous size, the surface layers of the star are now quite cool and it radiates largely in the red. It has become a red supergiant (figure 3.21) and in the HR diagram it lies in the upper right-hand corner. Approximately 500,000 years have elapsed since the star left the main sequence.

Core Collapse

With the formation of iron in the star's core, the traditional sequence of successive stages of nuclear burning comes to a halt. The reason is that iron cannot undergo nuclear reactions *and* release energy; it requires energy for fusion into still heavier elements. Hence, the iron core of the star resembles a white dwarf: its density is high, the electrons touch each other and resist contraction, and it possesses no nuclear energy source. However, the white dwarf configuration of the iron core does not last very long. The neutrino emission has speeded up the evolution to such a feverish pace that within just one day of silicon burning the iron core grows beyond the Chandrasekhar limit of $1.4\,M_\odot$ and the structural stability that has sustained the star so far breaks down. The star is coming face to face with its final destiny.

As soon as the iron core grows larger than $1.4\,M_\odot$, gravity becomes so strong that the electrons are no longer able to resist contraction and the core begins to collapse. In just a few milliseconds the following sequence of events runs its course:

1. The collapse crunches the iron nuclei together so violently that they break up into their constituent elementary particles — protons and neutrons.

2. The electrons present in the collapsing core are forced into the protons and produce neutrons. The collapsing core now consists entirely of neutrons. The equation of the reaction between electrons and protons is

$$e^- + p^+ \rightarrow n + \nu.$$

3. With the disappearance of the electrons, virtually all resistance to gravity vanishes and the collapse becomes "free-fall," reaching speeds of up to 10% the speed of light.

4. As the density of the collapsing matter approaches 10^{14} g/cm^3, the neutrons begin to touch each other and strenuously resist further compression (just as the electrons did earlier at a lower density). As suddenly as it began, the collapse stops. It stops with a bounce, followed by a brief period of sharp oscillations, somewhat akin to the ringing of a bell when it is struck by its clapper. After the oscillations have damped out, the collapsed core reaches a new equilibrium configuration about 20 km across, corresponding to the size of an average American town. The core has become a neutron star.

The collapse of the star's core from white dwarf dimensions into a neutron star takes just a few milliseconds because of the enormously strong gravitational forces that exist when $1.4\,M_\odot$ of matter are compressed into a volume smaller than the Earth. Furthermore, with the removal of the electrons, the pressure virtually vanishes and gravity becomes nearly unopposed, so it is able to collapse the core at an extremely fast rate.

Supernova Explosion

The collapse of the stellar core goes far too quickly for the envelope to be immediately affected. It is simply left behind. However, that period of grace does not last long. The collapse releases colossal amounts of energy, easily exceeding the energy emitted by 100 stars like the Sun during their entire main sequence lives. That energy has to go somewhere. A lot of it goes into breaking up the iron nuclei into their constituent protons and neutrons and into squeezing the electrons into the protons. Some of it is carried away by neutrinos, mainly during the bounce and the ensuing oscillations. The remainder of

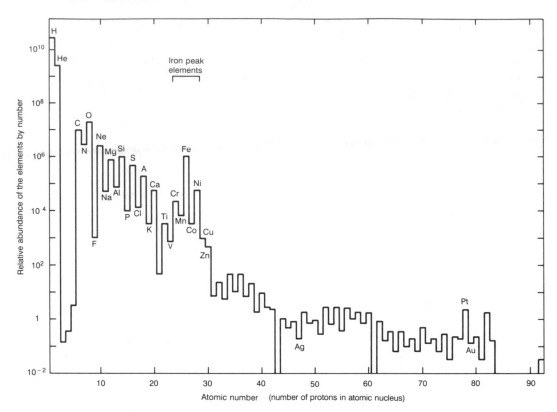

Figure 3.22. Relative abundance of the elements in the Sun. The abundance is by number, with silicon arbitrarily set at 10^6. For instance, for every 10^6 silicon atoms, the solar material contains 3.0×10^{10} hydrogen atoms, 2.5×10^9 helium atoms, and 10^7 carbon atoms. The 26 most abundant elements are indicated by their chemical symbols (see the Periodic Table of the Elements, table I.3). Silver (Ag), platinum (Pt), and gold (Au) are shown as well. Gold is an uncommon element: for every 10^6 silicon atoms, there exist only 0.13 gold atoms. Converting the abundance by number into abundance by mass gives 73.6% hydrogen, 24.8% helium, and 1.6% other elements.

Data: Robert W. Noyes, *The Sun, Our Star* (Cambridge and London: Harvard University Press, 1982), Table 2.1.

the energy goes into the creation of a superstrong shock wave, which runs radially outward toward the star's surface layers and affects them in a number of irreversible and far-reaching ways:

1. All along its way toward the surface, the shock wave loses energy by friction to the matter of the envelope, thus heating it and causing brief bursts of explosive nuclear burning in the silicon and oxygen layers. Both of these elements fuse into iron. The iron, in turn, undergoes further nuclear reactions and is converted into still heavier elements, such as copper, zinc, silver, and gold (see figure 3.22 and box 3.4). The energy required by these fusion reactions is supplied by the shock wave.

2. When the shock wave reaches the surface of the star, it has a whiplash effect on the outermost layers. The nuclei and electrons that make up these layers

are accelerated to nearly the speed of light and sent flying off into space. For millions and, possibly, billions of years they keep speeding through interstellar and intergalactic space until they are absorbed by gaseous clouds and other objects. They are one source of the cosmic rays that bombard the Earth continuously.

3. The heating produced by the shock wave plus the energy released by the brief bursts of nuclear reactions raise the pressure throughout the envelope to explosive conditions. The star now resembles a hydrogen bomb that has just gone off, except that it is bigger by far and more energetic. It is a gaseous sphere almost as large as Jupiter's orbit and contains approximately twenty times the mass of the Sun. At first slowly and then faster and faster the star's envelope gathers momentum. Layer after layer begins to roll and thunder outward, each driven by its own internal pressure and the push from the layers below. Within a few weeks, speeds exceeding 5000 km/sec (corresponding to 20 million km/hr or 10 million mi/hr) are reached and all the original, unburned hydrogen plus the freshly synthesized elements beyond the collapsed core—from helium to carbon, oxygen, neon, sodium, magnesium, silicon, phosphorus, sulfur, iron, and beyond—are ejected into space. The star is exploding as a supernova.

Outside observers, such as astronomers on Earth, who watch a supernova go off are much more limited in what they see. They cannot see the implosion of the core, for it is hidden deep within the star's envelope. Nor can they see the shock wave running through the envelope. They first become aware that something unusual and violent is happening when the shock wave arrives at the surface and the whiplash effect tosses the star's outermost layers into space at relativistic speeds. At that instant the star suddenly begins to brighten and within a day or two it outshines the rest of the galaxy in which it is situated (figure 3.23). Almost simultaneously with the brightening, the Doppler shift of the star's light reveals that the envelope is gathering speed and expanding. Soon it becomes apparent that the expansion is explosive and will end in the expulsion of enormous amounts of matter at speeds of thousands of kilometers per second. The large luminosity lasts for a few weeks and then gradually fades.

Figure 3.23. A supernova in galaxy NGC 7331. The left-hand photograph was taken before the supernova explosion, and the right-hand one was taken during the supernova's maximum brightness (1959). The other stars seen in these photographs (the sharp, circular images) are all foreground stars within our own galaxy, but despite their relative nearness they are much fainter than the supernova.

However, the ejected envelope can be seen for hundreds of thousands of years. It keeps expanding into the surrounding interstellar gas and dust, pushing them out of the way and compressing them (figure 3.1), and, perhaps, inducing the formation of yet another generation of stars. Gradually the expansion slows and the ejected matter mixes with the interstellar medium. Eventually the mixing becomes so thorough that the supernova ejecta can no longer be recognized, except for the presence of freshly synthesized heavy elements.

Black Holes

This description of supernovae focused on a star with an original mass of $25\,M_\odot$. If the star is much more massive, the mass of the imploding core may exceed $3\,M_\odot$. In that case the neutrons are unable to bring the collapse to a halt. They are squeezed into each other like the electrons and protons were earlier, and the collapse of the core continues undiminished. When its size becomes less than about 10 km across, the gravitational force at its surface is so strong that nothing can escape from it—neither material particles nor electromagnetic radiation—and anything that falls into it is irretrievably lost. The core has become a *black hole*. However, gravitational energy is still released during the collapse and leads to a supernova explosion and the ejection of the envelope, as in the case of the $25\,M_\odot$ star.

Black holes do not interact with the outside world except by gravity. Should some future human astronauts ever travel among the stars and suddenly feel that their spaceship is strongly attracted towards a point in space that appears to contain nothing, they are being pulled into a black hole. If they have the misfortune of insufficient engine power to climb back out, they will be doomed. The enormous gravitational forces suck them at ever-increasing speeds into the bottomless well. Tidal forces rip the spaceship apart and disintegrate everything into atomic form. The atoms are broken up into electrons and nuclei and, with the speed of light, these particles cross the boundary of the black hole. Except for a slight increase in the mass of the hole, the spaceship and occupants have vanished from the Universe. For the astronauts and their ship the event is over in seconds.

Observers on a sister ship, at a safe distance from the black hole, see the catastrophe as if it were happening in slow motion. For many hours they continue to receive electromagnetic signals of the descent and destruction of the ship and its unfortunate occupants. That radiation arrives more and more redshifted, because it has to climb out of an ever-deepening gravitational well. At first it is shifted from the optical to the infrared and then to the radio part of the spectrum. Eventually the wavelengths become stretched so much that the observers' receivers are inadequate for reception. For them, too, the event is now over. (See chapter 1 for additional discussion of the gravitational redshift and of the difference in time perception between observers in gravitational fields of differing strengths.)

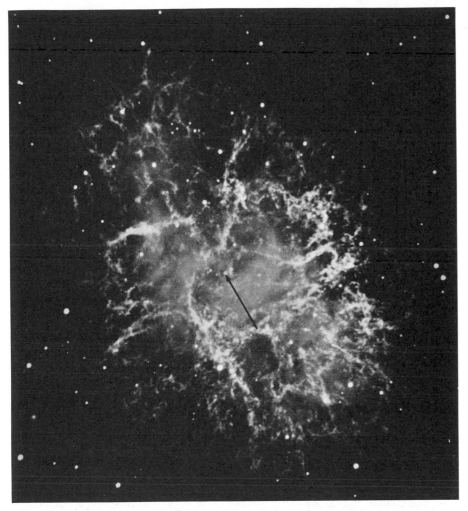

Figure 3.24. The Crab nebula. This nebula lies at a distance of about 6000 LY and consists of the gaseous matter ejected by the supernova observed in 1054 A.D. The energy of these emissions is believed to come from the stellar remnant of the explosion, the Crab pulsar (indicated by an arrow).

Historical Supernovae

At the present epoch, supernovae are rather rare events. Only seven are known to have occurred in our Galaxy during recorded history. Of these only the supernovae of 1006, 1572, and 1604 A.D. were recorded by European scholars. References to the others (the earliest dating from 185 A.D.) are found in Chinese, Japanese, and Korean chronicles.* The most famous historical supernova is one

*Observations of galaxies like our own have shown that on the average one supernova explodes in them every 100 years. The last one observed in the Milky Way was almost 400 years ago. Hence, we may be tempted to think that soon we should be treated to the spectacle of another one. This may

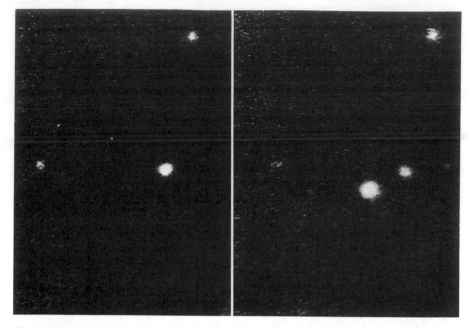

Figure 3.25. The Crab pulsar. This pair of photographs illustrates the on–off optical pulsation of the pulsar that lies embedded in the Crab nebula and that is the stellar remnant of the Crab supernova explosion.

that occurred in 1054 A.D. in the constellation Taurus and was recorded by the Chinese and Japanese.

Despite a distance of about 6500 LY, the supernova of 1054 A.D. was brighter than Venus for weeks before fading from view. At the location in the sky where the supernova exploded, modern astronomers have found a highly contorted and agitated nebula, the Crab nebula (figure 3.24), and a very hot and compact star. The Crab nebula is the stellar envelope ejected by the explosion, and it still is expanding at about 1500 km/sec. It radiates in all wavelengths from γ and X rays to UV rays, optical light, and IR and radio waves, attesting to the violence and turbulence still present today, more than 900 years after the outburst. The compact star is a neutron star — the collapsed remnant of the parent star's core.

Interestingly, the neutron star in the center of the Crab nebula sends out pulses of light and radio waves with startling regularity 30 times every second. Many of the neutron stars resulting from supernova explosions do that and, hence, they are known as *pulsars* (figure 3.25). The pulses are believed to be caused by a lighthouse effect. The neutron star rotates and, at the same time, strong magnetic fields on its surface focus the light and radio waves into a narrow beam. During every rotation, the beam sweeps out an arc of 360°, much like the rotating light beam from a lighthouse. As the beam crosses our line of

be the case, but we cannot be sure. The explosion of supernovae is a random phenomenon and there is no certain way of predicting when the next one will take place. It may happen in one year or in one or more centuries.

sight we see a pulse. Because the pulsing of neutron stars is observed in the frequency range from about once every few seconds to several hundred times per second, we may assume that neutron stars rotate at those rates.* The most rapidly pulsing pulsar discovered to date is Pulsar PSR 1937 + 214, which pulses 642 times per second.

Our galaxy, like all others, may well be populated by millions of neutron stars and black holes. We don't know for sure how many there are because after some time neutron stars stop pulsing and can no longer be detected. Furthermore, black holes cannot be seen at all, except when matter (as from a nearby companion star) falls into one and emits X rays and other electromagnetic radiation. Most of the neutron stars and black holes are believed to date back to the formative years of our galaxy, when nearly all of the heavy elements present today were manufactured and when supernova activity must have been at its peak.

The birth, evolution, and death of the stars illustrate perhaps better than any other phenomena taking place in the Universe today the interplay between the four fundamental forces of nature. The gravitational force pulls the interstellar gas and dust together into clouds. Emission of IR and radio waves (electromagnetic force) contributes to the contraction by removing heat, and it allows the further collapse of the gas and dust into clumps. Nuclear reactions (nuclear force) supply the energy necessary to bring the collapse to a halt and to transform the clumps into stars. Emission of light (again, the electromagnetic force) carries away the energy liberated by the nuclear reactions and makes the stars shine. Neutrino emission (weak force) speeds up the final evolution of the massive stars. In the end, gravity asserts itself and the stars, or parts of them, collapse into white dwarfs, neutron stars, or black holes. None of the four forces acts independently of the others. Like the different sections of an orchestra, they perform together, influencing each other and taking turns in controlling the evolution.

While this evolution runs its course, conditions arise that stand in stark contrast to each other. For instance, in following the life cycle of stars we have encountered cold, dark clouds with temperatures as low as ten degrees and hot stellar interiors with temperatures of tens of millions and billions of degrees. In interstellar space densities may be as low as 10^{-24} g/cm^3, while those of neutron stars are 10^{14} g/cm^3 and those of black holes are higher still. Nuclear reactions require that nuclei come to within 10^{-13} cm of each other; the size of a star may be as large as 10^{13} cm. The low-mass stars stay on the main sequence for 10 billion years and longer, while the final stages of nuclear burning in massive stars are completed in years or days and the collapse of their cores is over in a fraction of a second.

---

*Normal stars rotate much more slowly than neutron stars. For instance, the Sun and other stars like it rotate approximately once every 20 to 40 days. More massive stars generally rotate more rapidly, but never faster than about once every two or three days. When the core of such a star collapses, its rate of rotation greatly increases, for the same reason that skaters spin much more quickly when they bring their extended arms close to their bodies. A physicist would say that this speeding up of the rotation is evidence of the conservation of angular momentum.

Ultimately, these contrasting conditions result from the properties of five elementary particles—the electron, proton, neutron, neutrino, and photon—and the four fundamental forces by which they interact. Stars may be regarded as laboratories that allow the study of these particles and their interaction properties under conditions far broader than is possible on Earth. In particular, stars permit scientists to study the synthesis of heavy elements as well as the triggering of interstellar cloud collapse, both of which were preconditions for the formation of the Solar System and life on Earth.

Exercises

1 Locate on the night sky as many of the stars listed in table 3.1 as the season and observing conditions allow. Estimate the stars' relative brightness and order them accordingly, from bright to faint. Compare your order with that of table 3.1. (This exercise requires a star chart, listed in the bibliography in chapter 2.)

2 Fill a glass with water, place an index card over it, invert the filled glass while pressing the card against the glass, and let go of the card. Unless the card is warped or bent or the glass is uncommonly tall, the water will stay in the inverted glass. Why? (*Hint*: Compare the two kinds of pressures pushing on the card from above and below.)

3 (*a*) The boiling point of water diminishes with increasing elevation. For example, on the tops of Mt. Mitchell (2037 m), Mt. McKinley (6194 m), and Mt. Everest (8848 m), it is 93°, 79°, and 70° C, respectively. Why is this so?
(*b*) On what principle does a pressure cooker work?

4 (*a*) Describe the physical conditions under which molecules form in interstellar space.
(*b*) How do such conditions arise?
(*c*) Why are H_2, CO, OH, NH_3, CN, H_2O, H_2S, SiO, HCN, CH_3OH, and H_2CO among the most abundant molecules found in interstellar space?

5 Using one of the molecule kits listed in the bibliography in chapter 6, build three-dimensional models of the molecules listed in figures 3.4 and 3.5.

6 Distinguish between fusion and fission reactions, and give a few examples of each.

7 Discuss the importance of nucleosynthesis for the origin and evolution of life on Earth. Make reference both to the nucleosynthesis occurring presently in the Sun and to the nucleosynthesis that occurred in the Galaxy's early generations of massive stars.

8 (*a*) Describe the three mechanisms of energy transport in stellar interiors.
(*b*) Name a few examples other than those listed in box 3.3 illustrating the application of these mechanisms in everyday life.

9 Why do stars, in the course of their evolution, bunch up along the main sequence?

10 Why are massive stars much more short-lived than low-mass stars?

11 What would happen if:
(*a*) the temperature of the Sun's core were doubled?
(*b*) the temperature of the Sun's core were halved?
(*c*) all of the hydrogen in the central 15% of the Sun's interior were helium?
(*d*) all of the matter of the Sun were iron?
(*e*) the mass of the Sun were increased by 50% and all of its matter were iron?

12 Discuss the major differences among:
(*a*) main sequence stars, red giants, supergiants, and white dwarfs.
(*b*) stellar wind, long-period variable stars, planetary nebulae, novae, and supernovae.
(*c*) neutron stars and black holes.
(*d*) pulsars and quasars.

Suggestions for Further Reading

Bethe, H. A., and **G. Brown**. 1985. "How a Supernova Explodes." *Scientific American*, May, 60–68. The article focuses on the supernova explosions of massive stars (8 $M_\odot$ and greater), which give rise to Type II supernovae. It briefly touches upon Type I supernovae, which occur in less massive stars. (Also see the articles below by Kirshner and by Wheeler and Nomoto.)

Blitz, L. 1982. "Giant Molecular-Cloud Complexes in the Galaxy." *Scientific American*, April, 84–94. The sizes, content, physical conditions, evolution, and galactic distribution of interstellar cloud complexes, including star formation that takes place in them.

Boss, A. P. 1985. "Collapse and Formation of Stars." *Scientific American*, January, 40–45. Computer modeling of star formation.

Dyson, F. J. 1971. "Energy in the Universe." *Scientific American*, September, 50–59. An interesting article about the genesis and flow of energy both here on Earth and in stars, galaxies, and the Universe at large.

Greenberg, J. M. 1984. "The Structure and Evolution of Interstellar Grains." *Scientific American*, June, 124–35.

Kafatos, M., and **A. G. Michalitsianos**. 1984. "Symbiotic Stars." *Scientific American*, July, 84–94. Binary star systems that consist of a red giant and a nearby compact companion star. Mass transfer from the red giant to the compact star may give rise to an accretion disk, nova outbursts, and ejection of highly directional jets of matter.

King, I. R. 1985. "Globular Clusters." *Scientific American*, June, 78–88.

Kirshner, R. P. 1976. "Supernovas in Other Galaxies." *Scientific American*, December, 88–101.

Lada, C. J. 1982. "Energetic Outflows from Young Stars." *Scientific American*, July, 82–93.

Leibacher, J. W., **R. W. Noyes**, **J. Toomre**, and **R. K. Ulrich**. 1985. "Helioseismology." *Scientific American*, September, 48–57. Oscillations of the Sun, observed on its surface, hold clues to the structure, composition, and dynamics of the Sun's interior.

Mathis, J. S., **B. D. Savage**, and **J. P. Cassinelli**. 1984. "A Superluminous Object in the Large Cloud of Magellan." *Scientific American*, August, 52–60. A giant nebula in this small galaxy holds an object that is 50 million times brighter than the Sun. If it is one body, it is far more massive than any known star.

Noyes, R. W. 1982. *The Sun, Our Star*. Cambridge, Mass.: Harvard University Press. A nonmathematical, up-to-date account of current understanding of the physical structure of the Sun, including chapters on solar observations, the Sun's impact on the Earth's climate, and solar energy.

Penrose, R. 1972. "Black Holes." *Scientific American*, May, 38–46.

Philip, A. G. D., and **L. C. Green**. 1978. "The H-R Diagram as an Astronomical Tool." *Sky and Telescope*, May, 395–98. Highlights from a symposium of the International Astronomical Union on "The HR Diagram," held in memory of Henry Norris Russell in 1977 in Washington, D.C.

Scoville, N., and **J. S. Young**. 1984. "Molecular Clouds, Star Formation, and Galactic Structure." *Scientific American*, April, 42–53. Radio observations show that the giant molecular cloud complexes where stars are born are distributed in various ways in spiral galaxies, perhaps accounting for the variation in the clouds' optical appearance.

Seward, F. D., **P. Gorenstein**, and **W. H. Tucker**. 1985. "Young Supernova Remnants." *Scientific American*, August, 88–96. Space observations of the expanding shells from recent supernova explosions reveal X-ray and radio emissions, shock fronts, and clumping of the ejected matter.

Stephenson, F. R., and **D. H. Clark**. 1976. "Historical Supernovas." *Scientific American*, June, 100–107. The seven supernovae that were seen between 185 A.D. and 1604 A.D.

Thorne, K. S. 1967. "Gravitational Collapse." *Scientific American*, November, 88–98. In the case of burnt-out stars of critical mass, star clusters, and perhaps galaxies, gravity may overwhelm all other forces and crush matter out of existence to form black holes.

Weisberg, J. M., **J. H. Taylor**, and **L. A. Fowler**. 1981. "Gravitational Waves from an Orbiting Pulsar." *Scientific American*, October, 74–82. Pulsar PSR 1913 + 16 is a member of a binary system whose second member is an ordinary star. The orbital radius of the pulsar is slowly shrinking, which is interpreted as being due to the emission of gravitational waves.

Wheeler, J. C., and **K. Nomoto.** 1985. "How Stars Explode." *American Scientist*, May–June, 240–47. Two classes of supernovae are observed: Type I and Type II. Relying on observations of their light curves and remnants and on computer modeling, the authors describe the physical mechanisms that appear to give rise to each class.

Williams, R. E. 1981. "The Shells of Novas." *Scientific American*, April, 120–31.

Wilson, O. C., A. H. Vaughan, and **D. Mihalas.** 1981. "The Activity Cycles of Stars." *Scientific American*, February, 104–19. Variations in activity similar to the Sun's 11-year sunspot cycle have been followed and studied in 91 nearby stars.

Wolfson, R. 1983. "The Active Solar Corona." *Scientific American*, February, 104–19.

Wynn-Williams, G. 1981. "The Newest Stars in Orion." *Scientific American*, August, 46–55.

Additional References

Weaver, T. A. 1980. "The Evolution of Massive Stars and the Origin of the Elements." *Energy and Technology Review*, February, 1.

Figure 4.1. Saturn and its ring system, photographed by *Voyager 2* from a distance of 21 million km. Four of Saturn's satellites are visible. Below the planet, from left to right, are Tethys (which casts a dark shadow on Saturn's disk), Dione, and Rhea. Mimas is barely visible against the planet's disk, below the rings and above and to the left of Tethys.

4

The Solar System: Orbital Layout

NEAR THE INNER EDGE of the Orion Arm and approximately 30,000 LY from the center of the Milky Way, the Sun speeds silently through interstellar space at nearly 1 million kilometers per hour. Opposing gravitational and centrifugal forces keep it firmly locked into a nearly circular orbit around the massive, central bulge of the Galaxy, as noted in chapter 2. Every 250 million years the Sun completes one circuit. It has done so since its birth some 4.5 billion years ago and, barring some near encounter or collision with another star, will do so in the future.

From a distance of a few light-years or more the Sun appears to be just another star among the 150 billion populating the Milky Way. However, when viewed from nearby it is a very special star. It is the gravitational hub of a planetary system that includes Earth — our home and shelter in the vastness of space — and, through its steady outpouring of radiation, provides the energy that sustains terrestrial life.

It is, therefore, not surprising that since the dawn of history humans have been fascinated by the Solar System. Our ancestors observed the daily rising and setting of the Sun, the waxing and waning of the Moon, and the wandering of the planets across the firmament. Periodically they were startled and frightened by the appearance of a comet or the falling of meteorites, and they ascribed plagues and other calamities to these events. In their myths and religions they deified the Sun and the planets and they explained their motions by ingenious and, sometimes, fanciful theories.

Today we are still fascinated by the Solar System and employ science to probe its secrets. We use telescopes to observe the Sun, the planets, and the lesser bodies; we bounce radar waves off their surfaces; we send space probes to them; and in 1969, we first set foot on the Moon. As a result of the space missions, we have extensive photographic records of the surface features of the planets and their satellites, excepting only Neptune and Pluto. We know about their magnetic fields. We possess Moon rocks. We have on-site analyses of the surface compositions of Venus and Mars. We are beginning to understand the structures of the planets and their satellites. And, with the aid of high-speed computers, we are developing new theories of the origin and evolution of the Solar System. The goal of this chapter and the next is to describe the orbital distribution and structures of the member bodies of the Solar System, and to make you familiar with some of the current theoretical efforts concerning its origin and evolution. We shall begin with a brief summary of the historical background to our contemporary perspective.

EMERGENCE OF THE HELIOCENTRIC VIEW OF THE SOLAR SYSTEM

Most ancient scholars thought that the Earth was the center of the Universe, with all other celestial objects revolving around it. However, an early exception was Aristarchus of Samos (c. 310–230 B.C.), who advocated a heliocentric Solar System. The ancients clearly distinguished between the *wandering stars* or *planets* and the *fixed stars*. The planets included the Sun, the Moon, Mercury, Venus, Mars, Jupiter, and Saturn, all of which, in the course of days to weeks, can be seen to wander across the firmament.* The fixed stars were the objects still called stars today and that do not appear to partake in the wandering motions. Today we know that the planets and fixed stars differ not only in their apparent motions across the firmament, but are fundamentally different objects. The planets belong to the Solar System and shine by reflecting sunlight. In contrast, the fixed stars (that is, those in our galaxy) lie at distances from a few to tens of thousands of light-years, and they shine by radiating energy released by nuclear reactions in their interiors.

The commonly held view among the ancients (figure 4.2) was that the seven planets as well as the fixed stars are carried on crystalline spheres around the Earth once every day and that the friction between the spheres gives rise to the "music of the spheres." This *geocentric* or *Earth-centered* view was formalized in the second century A.D. by the great Egyptian astronomer and mathematician, Ptolemy, in his book the *Almagest* (derived from the Arabic, meaning "The Great Treatise"). He stated that all the planets move along small circular paths, called epicycles, whose centers in turn move along larger circles around the stationary Earth. This system allowed him to approximate the motions of the planets across the sky, including the retrograde motions of the outer three.

Copernicus and Kepler

Ptolemy's geocentric view prevailed for more than 1300 years, until the early part of the sixteenth century. By then a new intellectual spirit had taken hold in Europe and many traditional teachings came under critical scrutiny. The mood was one of adventure and of looking beyond what was known or had been done before. Explorers sailed the seven seas and merchants expanded their trade to the far corners of the globe. This new adventurism also spread to astronomy. Growing discrepancies were noticed between the Ptolemaic predictions of the positions of the planets and their true positions in the sky, and a few astronomers dared to acknowledge that something fundamental was wrong with the geocentric model. The first major step in resolving the problem was taken by the Polish astronomer Nicolaus Copernicus, who in his celebrated work *De revolutionibus orbium coelestium* ("On the Revolutions of the Celestial Spheres") proposed a new and much simpler model of the Solar System. He placed the Sun at the center of the system and let the six known planets, including Earth, move in circular orbits around it. The Moon lost its status as a planet and became a satellite of Earth. This is the *Sun-centered* or *heliocentric* model of the Solar System.

Although Copernicus' heliocentric model came closer to reality than the geocentric model, it still did not quite correctly predict the motions of the planets. This became particularly evident in the case of the planet Mars, whose positions were very accurately measured by the Danish astronomer Tycho

---

*The ancients attributed mystical meanings to the planets and named most of them after their gods:

Mercury	Roman god of commerce and messenger of the gods
Venus	Roman goddess of love
Mars	Roman god of war
Jupiter	chief Roman god
Saturn	Roman god of agriculture

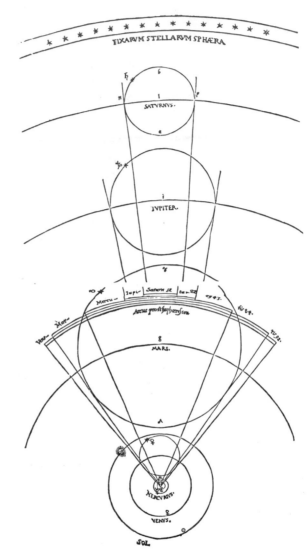

Figure 4.2. The orbital layout of the Solar System according to the theory of the ancients in the western world, showing the order of the planets, the relative sizes of their orbits, and epicycles. The Latin text is translated (Kepler 1981):

In the center is the Earth, which alone is motionless.

The innermost small orbit round the Earth represents the Sphere of the Moon, of which the motion is monthly.

The next round that is the orbit of Mercury; it is followed by that of Venus and after it is the Sphere of the Sun, which all go round in an annual revolution.

The orbits of the other three, the superior planets, Mars, Jupiter, and Saturn, as well as the Sphere of the Fixed Stars, are indicated by arcs, which anyone can complete by describing the whole of them about the Earth as the center. The orbit of Mars makes a turn in two years. That of Jupiter requires 12 years, as nearly as possible, and that of Saturn about 30 years. The Fixed Stars complete a period in 49,000 years, according to the tenets of the Alfonsine tables.

The amounts of the equations for each of them (except the Moon) produced by the epicycles on the concentric circle at their mean distances are shown by the arcs intercepted by straight lines drawn from the Earth and touching each of the epicycles, that number of degrees being added.

Brahe (1546–1601) over a period of some 20 years. Brahe attempted to resolve the disagreement between theory and observations, but without success.

Brahe's assistant, the German mathematician Johannes Kepler (1571–1630), succeeded with the task. Kepler found that he could achieve agreement if he made two assumptions: One, the planets do not move in circular orbits, as proposed by Copernicus, but in elliptical orbits; and, two, the Sun lies not at the center of the orbits, but at one of the foci of the ellipses. He also noted that the planets travel faster when they are near the Sun and slower when they are more distant. Finally, he discovered a simple relationship between the planets' distances from the Sun and their orbital periods. Kepler summarized these results in his famous *three laws of planetary motion*, which to this day form the basis of our understanding of celestial orbital motions (they are described in more detail in box 4.1). They apply to the motions of planets around the Sun, as well as to the motions of satellites, including artificial ones, around planets. They also apply to the motions of double stars and double galaxies around each other.

These accomplishments by Copernicus, Brahe, and Kepler contributed greatly to the intellectual optimism of their times. They had proved that it was possible to gain valuable insights into natural phenomena by making careful observations and measurements and by using them in the construction of new theoretical models. No longer was it necessary to accept ancient dogmas on faith. We must rank this development — reliance on observation and experimentation combined with theoretical reasoning — among the great revolutionary breakthroughs in the intellectual history of humankind. It was, in fact, the beginning of modern science.

Galileo and Newton

The heliocentric model of the Solar System was met initially with much opposition, as Galileo's trial by the Inquisition amply demonstrates. Only in the succeeding centuries, after a series of remarkable discoveries confirmed the model's validity, did it become universally accepted. Many scientists contributed to this success, but two of them stand out particularly. They are the two giants of seventeenth century science, Galileo Galilei and Isaac Newton.

Galileo was the first to observe celestial bodies through the then newly invented telescope. He noticed that Venus has phases just like the Moon and correctly explained them as being due to that planet's motion around the Sun, during which increasing and decreasing portions of its illuminated side face toward us. When he trained his telescope toward Jupiter, he saw four small bright objects orbiting around that planet with periods of between two and seventeen days. They are Jupiter's four largest satellites, which today we call the Galilean satellites — Io, Europa, Ganymede, and Callisto.* Galileo fully realized

*It is possible that Galileo was not the first to observe the satellites of Jupiter. This honor may go to Gan De (about 365 B.C.), one of the earliest Chinese astronomers. He is quoted in *The Kaiyuan Treatise on Astrology*, which was compiled between 718 and 726 A.D., as having said (Hughes 1982): "In the year of chan yan . . . , Jupiter was in Zi, it rose in the morning and went under in the evening together with the lunar mansions Xunü, Xü and Wei. It was very large and bright. Apparently, there was a small reddish star appended to its side. This is called 'an alliance'," The term *alliance* probably indicates that Gan De thought the reddish star to be a subsidiary or satellite of Jupiter. We do not know for certain that what Gan De saw was one of Jupiter's satellites. But if it was, it may have been Ganymede, Jupiter's largest and brightest satellite.

BOX 4.1.
Kepler's Laws of Planetary Motion

KEPLER formulated his three laws of planetary motion by carefully analyzing Tycho Brahe's data of the planets' positions in the sky. He began with this work in 1600 when he was hired by Brahe, who was then the imperial mathematician at the court of Emperor Rudolf II in Prague, and was asked to find a theoretical explanation for the orbit of Mars. Upon Brahe's death one year later, Kepler succeeded him in his court appointment and continued for some twenty years with the studies of planetary orbits. At first he experimented with traditional epicycles and tried to fit various combinations of circles to the orbit of Mars, but without obtaining satisfactory agreement. He then used "oval" curves and, finally, ellipses, which he soon realized can be made to agree quite accurately with the observations. By 1609 he had worked out his first two laws of planetary motion, which he published under the title *Astronomia nova* ("New Astronomy"). He published the third law in 1619 in his *De harmonices mundi* ("Harmonies of the World"). The three laws are:

Law I: Each planet revolves around the Sun in an elliptical path, with the Sun occupying one of the foci of the ellipse.

Law II: The straight line joining the Sun and a planet sweeps out equal areas in equal intervals of time.

Law III: The squares of the planets' orbital periods are proportional to the cubes of the semimajor axes of their orbits.

Kepler's *first law* states that the planets' orbits are ellipses. An *ellipse* may be constructed on paper by taking a piece of string of a certain length, attaching it firmly to two fixed points, and tracing out a line, keeping the string taut (figure A).

In the mathematical analysis of ellipses, the length of the string is equated to $2a$, where a is the length of the *semimajor axis* (see figure A). The letter b is used to indicate the length of the *semiminor axis*. The two points to which the string is attached are said to be the *foci* of the ellipse and their distances from the

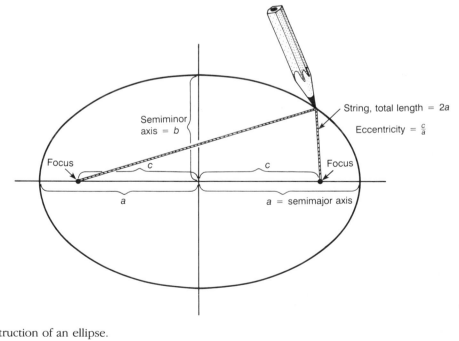

Figure A. Construction of an ellipse.

BOX 4.1 continued

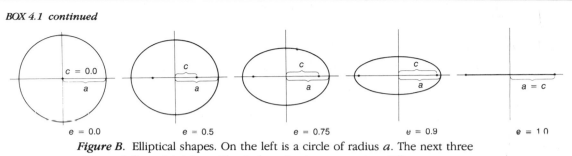

Figure B. Elliptical shapes. On the left is a circle of radius a. The next three ellipses (left to right) have identical semimajor axes, a, but different eccentricities. The straight line segment on the right has length $2a$.

center of the ellipse are labeled c. The ratio c/a is defined as the *eccentricity*, e. For elliptical orbits, e has a value between zero and one.

Differently shaped ellipses result if the two foci are moved closer together or farther apart (that is, by changing c) but the length of the string (that is, $2a$) remains the same. Changing the length of the string (the length of $2a$) changes the size of the ellipse. If the two foci overlap, a circle is obtained. If the two foci are so far apart that the distance between them is equal to the length of the string, a straight line results. Between these two extreme cases lies the full range of elliptical shapes, as illustrated in figure B.

The orbits of most planets have very small eccentricities and, hence, are nearly circular. The orbits of most asteroids have eccentricities in the range from 0.1 to 0.3 and, hence, deviate slightly from circles. Finally, the orbits of most comets that visit the inner

regions of the Solar System have eccentricities between 0.5 and nearly 1.0 and, hence, are noticeably elongated. In all cases, the Sun occupies one of the two foci of the body's orbit; the other focus is unoccupied and without any physical significance.

Kepler's *second law* quantifies the fact that an orbiting body travels more slowly far from the Sun and more rapidly close to it. Such changes in orbital speed are yet another consequence of the *conservation of angular momentum*. The speeds of an orbiting planetary body are related to the corresponding distances from the Sun in such a way that the line joining the Sun and the planetary body sweeps out equal areas during equal intervals of time. In other words, if the orbital travel times from position 1 to position 2 ($t_{1,2}$) and from position 3 to position 4 ($t_{3,4}$) are the same, then $\text{Area}_{1,2}$ equals $\text{Area}_{3,4}$ (figure C).

Figure C. Conservation of angular momentum. When the orbiting body is close to the Sun, its orbital speed is fast; when it is far from the Sun, its orbital speed is slow. If the orbital travel time from point 1 to point 2 ($t_{1,2}$) amd from point 3 to point 4 ($t_{3,4}$) are the same, then, according to Kepler's second law, $\text{Area}_{1,2} = \text{Area}_{3,4}$.

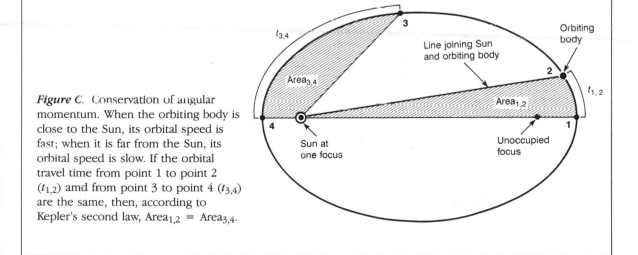

BOX 4.1   continued

Kepler's *third law*, which is also known as the harmonic law, is demonstrated by the entries in the following table. The planets' orbital periods are given in Earth years and their orbits' semimajor axes are in astronomical units. The use of these units makes the proportionality between the square of a planet's orbital period, P, and the cube of the planet's semimajor axis, a, an equality: $P^2 = a^3$ (see columns 4 and 5 of the table):

	P (years)	a (AU)	P^2	a^3
Mercury	0.240 85	0.387 10	0.058 01	0.058 01
Venus	0.615 21	0.723 33	0.378 5	0.378 5
Earth	1.000 0	1.000 0	1.000	1.000
Mars	1.880 9	1.523 7	3.538	3.538
Ceres	4.602	2.766	21.2	21.2
Jupiter	11.862	5.202 8	140.7	140.8
. . .				
Comet Halley	76.1	17.8	5800.	5600.

The small differences between some of the entries in the final two columns are due to inaccuracies in the measurements of a and P and to the fact that the Sun is not the only body in the Solar System exerting a gravitational pull.

the importance of this latter discovery. He had found a new satellite system—a miniature version of the Solar System. Apparently, systems in which one or several objects revolve around a larger central object are common in nature, with the Earth–Moon system being yet another example.

Perhaps the most spectacular sight for Galileo was the surface of the Moon. There he saw mountains, craters, valleys, and large dark areas, which he thought were seas. Obviously, that body is not the perfectly round and smooth sphere envisioned by the ancients. In many ways it resembles the Earth. Nor is the Sun a perfect body. On its surface, Galileo found dark spots, which today we call sunspots. These blemishes had been observed before, but they were dismissed as being either illusions in the Earth's atmosphere or shadows cast by intervening planets. From the motions of the sunspots across the solar disk, Galileo deduced that our star rotates with a period of slightly less than one month. Galileo described these and other telescopic discoveries in his book *Sidereus nuncius* ("The Starry Messenger").

Besides his astronomical discoveries, Galileo made numerous other major contributions to the development of modern science. In particular, through his experiments and mathematical analysis of the acceleration of material bodies, he paved the way for the work of Isaac Newton. Newton was convinced that a fundamental principle governs the motions of all objects and he set out to find it. This eventually led him to realize that whenever a body experiences a *change in motion*—when it is either speeded up or slowed down or when the direction of its motion is changed—a force is acting on it. He expressed his idea in three laws, which he published under the title *Philosophiae naturalis principia mathematica*, known as the *Principia*. With these three laws (more fully described in box 4.2) Newton laid the foundation of modern physics.

In the course of his work on the laws of motion, Newton wondered whether the force responsible for the falling of objects on Earth might not also be the force that keeps the Moon in orbit about the Earth. Legend has it that this idea

BOX 4.2.
Newton's Three Laws of Motion

SINCE the days of the ancient Greeks, scholars have endeavored to understand the motions of material bodies. Until the sixteenth century, however, these efforts were based strictly on geometrical interpretations of movement, without any recourse to experimentation, and therefore they were largely unsuccessful. Then, starting with Galileo, the French philosopher René Descartes (1596–1650), and the two English scientists Sir Christopher Wren (1632–1723) and Robert Hooke (1635–1703), physical reasoning based on observations and experiments was brought to bear on the problem. Each of these men acquired some understanding of motion, but Isaac Newton was the first to understand it thoroughly and to describe it quantitatively. He published his results in the form of three laws in the introductory part of his *Principia* (first edition published in 1687):

> Law I: Every body continues in its state of rest, or uniform motion in a right line, unless it is compelled to change that state by forces impressed upon it.

> Law II: The change of motion is proportional to the motive force impressed; and is made in the direction of the right line in which that force is impressed.

> Law III: To every action there is always opposed an equal reaction: or, the mutual actions of two bodies upon each other are aways equal, and directed to contrary parts.

All three of these laws are based on the concept of *force*. More specifically, they are concerned with the net force that acts upon a body. *Net force* refers to *the sum of all the forces* that act on the body. It is important to keep this in mind because in general bodies are exposed simultaneously to a number of different forces, often pointing in a number of different directions.

The *first law* describes the response of a body when *there is no net force* exerted upon it: if the body is initially at rest, it will remain at rest; if the body is initially in motion, it will continue with this motion without any change in speed or in the direction of the motion.

We observe the application of the first law while parking or driving a car. When the car is parked and the engine is off, no net force acts on the car and it remains at rest. If we drive the car on a straight, level road and press down on the gas pedal just enough so that the torque of the engine produces exactly the forward push required to overcome the frictional drag, then no net force is acting on the car either (the forward push and frictional drag cancel) and we will drive along at constant speed.

The *second law* describes the case when *there is a net force* exerted upon a body. The effect of the net force is a change in the motion of the body. In particular, this change in motion is proportional to the applied net force and it points in the direction of that force.

The change in motion may be a change in the speed of the body, a change in the direction in which the body is moving, or both. For instance, if the net force is applied in the direction in which the body is moving, the body's speed increases. This would be the case if, in the car example above, we stepped harder on the gas pedal. If the net force is applied in the direction opposite to that in which the body is moving, the body's speed decreases. This would be the consequence if we let up somewhat on the gas pedal or if we let up entirely on the gas and pressed on the brake pedal. Finally, if the net force is applied at an angle to the direction in which the body is moving, the direction of motion will change. This would be the case if we drive the car around a curve. All three of these changes in motion are called *accelerations*, including the slowing down (which is a negative acceleration) and the change in the direction of motion.

As just described, the *direction* of the acceleration of a body depends on the direction in which the net force is applied, but the *magnitude* of the acceleration depends on the strength of the applied force. In fact, the magnitude of the acceleration is proportional to the net applied force. This means that if the net force applied to an object is doubled or tripled, then the acceleration is also doubled or tripled.

In his formulation of the second law, Newton referred only to the proportionality between acceleration and applied force. However, the acceleration also depends on the mass of the body. For example, for a given force, a small mass experiences a much greater acceleration than does a large mass. If we use the car

BOX 4.2 continued

example once more, this statement expresses the common experience that an engine of a given horsepower accelerates a light sports car much more rapidly than a large family car.

With the introduction of the concept of mass, the proportionality between acceleration and applied force can be expressed by the equation

$$F = ma,$$

where F stands for the applied force, a for the acceleration, and m for the mass of the body. Newton's second law is commonly presented in this form today.

Incidentally, Newton's second law also expresses the content of the first law. In the first law, $F = 0$. According to the above equation, this means that $ma = 0$, which, in turn, implies that $a = 0$. But zero acceleration means that the body's speed and direction of motion remain unchanged, as predicted by the first law.

The *third law* states that a force cannot exist by itself, but is always accompanied by a second force. The two forces—called action and reaction—are of equal magnitude and point in opposite directions.

Further, the law indicates that the two forces never act on the same body, but aways on two, as the following examples illustrate.

We experience the presence of the two forces when we jump off a small rowboat: our bodies are accelerated in one direction, while the boat is accelerated in the opposite direction. Still other examples in which the presence of the two forces is readily apparent are the firing of a rocket engine (the rocket goes forward, exhaust gases backward), and the shooting of a gun (bullet flies forward, gun recoils). These examples also make it amply clear that the two forces never act on just one body, but always on two: person and boat, rocket and exhaust gases, gun and bullet.

In many situations we are not aware of the presence of both forces as, for example, when we drive a car on an asphalt road. We notice that there is a forward push on the car, but we do not usually notice that there is also a backward push on the road. We don't because the road is in firm contact with the Earth and the resulting backward acceleration is too small to be observed. However, when we accelerate the car on a gravel road, we do notice the backward push because gravel is thrown out in the backward direction.

first occurred to him while he was watching an apple fall from a tree. He thought that, perhaps, the same force is also responsible for the planets' motions around the Sun as well as the Galilean satellites' motions around Jupiter. To test these hypotheses, he substituted the orbital data of the planets and satellites into Kepler's third law. He found that all of the motions could be explained, including the motions of falling objects on Earth, if he assumed that the force (F) varies with the inverse square of the distance (R) separating the bodies: $F \propto 1/R^2$. He found further that the force is also directly proportional to the product of the masses of the bodies (for example, the masses of the Earth and the Moon, the Sun and a planet, M_1 and M_2): $F \propto M_1 M_2$. By combining the two proportionalities, Newton derived a mathematical expression for the *universal force of gravity*:*

$$F \propto \frac{M_1 M_2}{R^2}.$$

(See discussion of gravitational force, Introduction to Part One). *Universal* means that this force applies not only to the effect of the Earth on falling apples and the Moon or of the Sun on the planets, but to the attraction between all matter and energy in the Universe. With this discovery, Newton demonstrated for the first time that there exists but one "design" in nature and that the scientific method allows us to discover it.

*The term *gravity* is derived from the Latin *gravitas*, meaning "heaviness" or "weight."

MODERN PERSPECTIVE OF THE SOLAR SYSTEM

During the three centuries following the pioneering work of Copernicus, Brahe, Kepler, Galileo, and Newton, knowledge of the Solar System grew at a steadily accelerating pace.

Three New Planets

In 1781, the German–English astronomer and musician Sir William Herschel noted, in the course of a systematic survey of the sky, that the image of a faint star was larger than is typical for stars. He observed the object for several weeks and found that it moved relative to the fixed background stars. At first he thought he had discovered a comet. Several months later, with more positional data on hand, the object's orbit was computed. It turned out not to be the elongated orbit of most comets (more about them later), but a nearly circular path around the Sun at a distance of about nineteen times that of the Earth. This is a distance greater than Saturn's and, judging from the object's brightness, meant that it was much bigger than a comet. Herschel had discovered a new planet. He named it *Georgium Sidus* (meaning "Star of George") in honor of George III, England's reigning king, but the name was later changed by J. E. Bode (see below) to *Uranus*, after the Greek god of the heavens (figure 4.3). Incidentally, Herschel was not the first to sight Uranus. The planet had been plotted on charts of the sky on at least seventeen previous occasions since the year 1690 without recognition of its true nature.

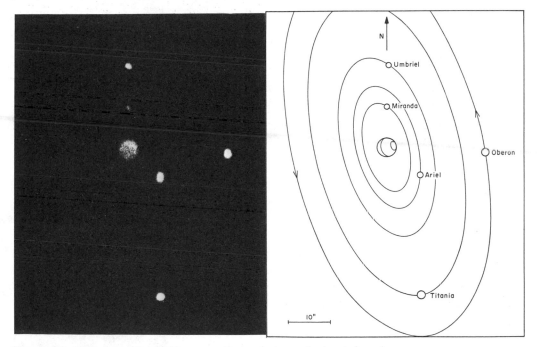

Figure 4.3. Uranus and its five large satellites, photographed at infrared wavelengths from the Mauna Kea Observatory, Hawaii.

Figure 4.4. Neptune and its two known satellites, Triton and Nereid. Triton is visible between the upper two diffraction spikes of Neptune's overexposed image, while Nereid is marked by an arrow. A long exposure was used to capture the image of Nereid, which is a body only 300 km across and extremely faint.

By 1820, it had become clear that the orbit of Uranus is not exactly an ellipse. Over the years, its path in the sky deviated more and more from that predicted by theory, even after the perturbing effects of Jupiter and Saturn were taken into account. It appeared that an unknown and still more distant planet was pulling on Uranus. The question was where in the sky to find the unknown planet. Tentative positions were computed in 1841 by John Couch Adams (1819–1892), an undergraduate at Cambridge University in England, and, four years later, by Urbain Jean Joseph Leverrier (1811–1877) in France. A search was conducted for the unknown planet at the Cambridge and Berlin observatories. It was finally discovered in 1846 by Johann Gottfried Galle (1812–1910), a young German astronomer at the Berlin Observatory, within one degree of the position predicted by Leverrier. It was christened *Neptune*, in honor of the Roman god of the sea (figure 4.4).

As in the case of Uranus, Neptune seems to have been sighted before it was identified as the Sun's eighth planet. Galileo's notebook shows that on several occasions in 1613 he had seen in the vicinity of Jupiter what he thought was a faint star that moved, and recent calculations indicate that it may well have been Neptune.

Early in the twentieth century, it became evident that Neptune alone could not account for the full perturbation that was observed in Uranus's orbit. There had to be yet another planet — *Planet X*, as it became known — that exerted a gravitational tug. Two American astronomers, William Henry Pickering (1858–1938) and Percival Lowell (1855–1916), estimated that Planet X should lie somewhere in the constellation Gemini and that its mass should be between that of the Earth and Neptune. An intense search ensued. Lowell looked for the planet from 1906 until his death, at the observatory he built in Arizona for the purpose of studying the planets. Success finally came in 1930 to Clyde Tombaugh, a 22-year-old Kansas farm boy and assistant at the Lowell Observatory. For many months he photographed sections of the night sky and compared photographic plates taken a few days apart. His aim was to see if any of the stellar images showed displacement and, thereby, revealed itself as the missing planet. The work was tedious and time consuming. Then, on February 18, 1930, Tombaugh noted that the image of what he thought was a faint star had moved the expected amount on photographs taken six nights apart earlier in January. He had discovered the Sun's ninth planet. It lay within 6 degrees of the position predicted by Lowell. The new planet was called *Pluto,* after the Greek god of the underworld. It was exceedingly faint and not sighted visually until 1950.

There is an ironic twist to Pluto's discovery that became evident only 48 years later. In 1978, James Christy, an astronomer at the U.S. Naval Observatory, noted that on high-quality photographs Pluto's image has a bump and that this bump moves systematically around the planet. The bump turned out to be a satellite. Its orbit is so close to Pluto that on photographs the two bodies are not fully resolved, but merge into a single bumpy image (figure 4.5). The satellite has a

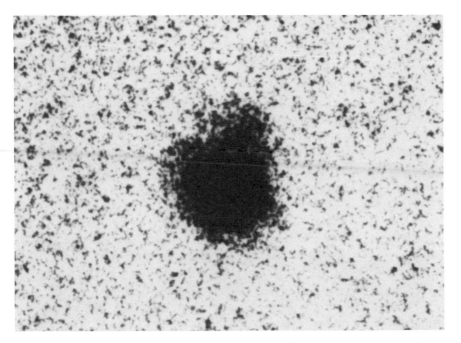

Figure 4.5. The discovery photograph of Pluto's satellite Charon, taken on July 2, 1978, with the 1.54-m reflector of the U.S. Naval Observatory. Charon orbits Pluto at a distance of a mere 19,700 km once every 6.4 days and appears in this photograph as a bump on the upper edge of Pluto's image.

period of 6.4 days, the same as Pluto's rotation period, and it circles Pluto at a distance of about 20,000 km (equal to three earth radii). Christy named the satellite *Charon*, for the boatman of Greek mythology who ferries the souls of the dead to Hades, the underworld and Pluto's domain. Charon's orbital period and distance imply—through Kepler's third law—that Pluto's mass is only 0.002 earth masses, which is considerably less than the mass assumed by Lowell in making his prediction of Pluto's position. It is, in fact, such a small mass that it cannot possibly be responsible for the observed residual perturbation in Uranus's orbit. Therefore, Pluto's discovery on the basis of Lowell's prediction was pure coincidence. It was, however, a lucky coincidence: without it we might still not be aware of Pluto's existence.

If Pluto is not Planet X, perhaps there exists a larger and as yet undiscovered planet at a still greater distance from the Sun that is the true source of the residual perturbation of Uranus's orbit. The case for this hypothesis is strengthened by small and unaccounted for discrepencies in the orbits of the other planets, from Jupiter to Neptune. In fact, the case for a tenth planet is so strong that astronomers in the United States and Europe are now actively searching for it. They do not presently know where exactly in the sky it is located, if, in fact, it does exist; but they estimate that it has a mass between two and five earth masses and an orbit beyond Pluto's, at about 50 to 100 astronomical units* from the Sun.

The Planets' Orbital Characteristics

Before continuing with the description of the Sun's planetary system, a summary of the planets' orbital characteristics is in order (see figures 4.6, 4.7, and table 4.1):

1. All planets revolve around the Sun in nearly the same plane. Relative to the Earth's orbital plane—also known as the *plane of the ecliptic*—the other planets' orbits are inclined by only three degrees or less. The only exceptions are Mercury and Pluto, whose orbits are inclined by 7 and 17 degrees, respectively.

2. The orbits of all but two of the planets are nearly circular. Again, the two exceptions are Mercury and Pluto, whose orbits have eccentricities of 0.21 and 0.25, respectively (see box 4.1 for a definition of eccentricity).

3. All planets orbit the Sun in the same direction, namely from west to east. This direction is called *prograde*, in distinction to *retrograde*, which refers to the opposite direction.

4. With two exceptions, the planets rotate in the same direction in which they orbit. That is why, for example, on Earth the Sun rises in the east and sets in the west. Only Venus and Uranus are different. Both rotate in the retrograde

---

*The *astronomical unit*, or AU, is a measure of length equal to the semimajor axis (see box 4.1) of the Earth's orbit about the Sun:

$$1 \text{ AU} = 1.50 \times 10^8 \text{ km};$$
$$1 \text{ LY} = 63,000 \text{ AU}.$$

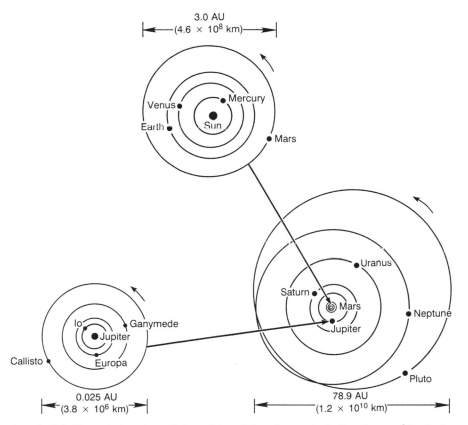

Figure 4.6. The relative sizes of the orbits of the planets, including those of Jupiter's Galilean satellites. The Sun is not drawn to scale (it is 80 times smaller than Mercury's orbit).

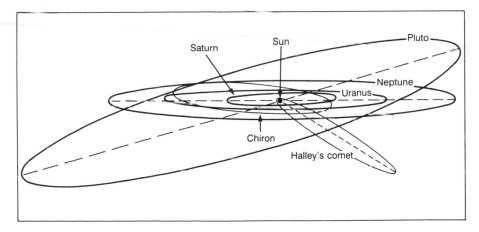

Figure 4.7. Diagram of the orbits of Saturn, Uranus, Neptune, and Pluto; the asteroid 2060 Chiron; and Halley's comet; as viewed from 5° above the ecliptic. Pluto, Chiron, and Halley's comet have noticeably elongated and inclined orbits.

Table 4.1. Orbital and Physical Data of the Sun's Planets.

	Orbital characteristics					Physical characteristics				
	Semi-major axis[a] (AU)[b]	Sidereal period[c] (years)	Average orbital velocity (km/sec)	Eccen-tricity[d]	Inclina-tion to ecliptic[e] (degrees)	Mass (earth masses)[f]	Equato-rial radius (earth radii)[g]	Average density (g/cm³)	Sidereal rotation period[c]	Inclina-tion of equator to orbit
Mercury	0.39	0.24	48.	0.21	7	0.056	0.38	5.4	58.7 days	0°
Venus	0.72	0.62	35.	0.007	3	0.82	0.95	5.3	243 days (retro-grade)	2°
Earth	1.00	1.00004	30.	0.02	—	1.00	1.00	5.5	23.9345 hours	23°44
Mars	1.52	1.88	24.	0.09	2	0.11	0.53	3.9	24.6 hours	24°
Jupiter	5.20	11.9	13.	0.05	1	318.	11.2	1.3	9.8 hours	3°
Saturn	9.55	29.5	9.6	0.06	2	95.	9.5	0.69	10.2 hours	29°
Uranus	19.2	84.0	6.8	0.05	1	14.5	4.0	1.2	17.3 hours (retro-grade)	82°
Neptune	30.1	165.	5.4	0.009	2	17.	3.8	1.7	ca. 16 hours	29°
Pluto	39.4	248.	4.7	0.25	17	ca. 0.002	ca. 0.24	ca. 0.9	6.4 days	≥ 50°

Source: Data from Beatty, et. al., *The New Solar System.*

[a]Semimajor axis: half the major axis of a planet's elliptic orbit (see box 4.1).

[b]Astronomical unit (AU): semimajor axis of Earth's orbit; equal to 1.5×10^8 km.

[c]Sidereal period: period of rotation or revolution relative to the fixed stars.

[d]Eccentricity: measure of deviation of ellipse from a circle; the eccentricity of a circle is zero, that of a straight line segment is one (see box 4.1).

[e]Ecliptic: plane of Earth's orbit.

[f]Earth mass: 6.0×10^{24} kg

[g]Earth radius (equatorial): 6,380 km the west.

direction. However, while Venus' equator is inclined by only 2° to its orbit, Uranus' equator is tilted almost perpendicularly to its orbit.

5. The planets orbital distances from the Sun are so regular that they can be approximated by a very simple numerical "law," discovered in 1781 by Johann Daniel Titius (1729–1796) of the University of Wittenberg in Germany. Titius found that taking the sequence of nine numbers 0, 3, 6, 12, . . . , 384, adding 4 to each, and dividing the sums by 10, yields approximately the distances from the Sun to Mercury, Venus, and so forth in astronomical units (with the single exception of Neptune), as shown in the table on the following page. This law is named after its discoverer and after Johann Elert Bode (1747–1826) who, as director of the Berlin Observatory, published and popularized it.

Symbol	Planet	Distance from the sun according to the Bode-Titius Law (AU)	Actual average distance from the Sun (AU)
☿	Mercury	(0 + 4)/10 = 0.4	0.39
♀	Venus	(3 + 4)/10 = 0.7	0.72
⊕	Earth	(6 + 4)/10 − 1.0	1.00
♂	Mars	(12 + 4)/10 = 1.6	1.52
	Asteroids	(24 + 4)/10 = 2.8	2.70
♃	Jupiter	(48 + 4)/10 = 5.2	5.20
♄	Saturn	(96 + 4)/10 = 10.0	9.55
♅	Uranus	(192 + 4)/10 = 19.6	19.2
♆	Neptune		30.1
♇	Pluto	(384 + 4)/10 = 38.8	39.4

Asteroids

At the time of the discovery of the Bode–Titius law, the space between Mars and Jupiter was marked by a glaring gap. The law predicted a planet at a distance of 2.8 AU, but none was known to exist. The problem was solved 20 years later in 1801, when a small body approximately one-third the size of the Moon was sighted orbiting the Sun at almost the exact distance predicted. It was named *Ceres*, after the Roman goddess of agriculture. Soon other similar bodies were found, most of them in the gap between Mars and Jupiter, and today we know of thousands of them. Most are less than one kilometer across and their total mass is no more than one-tenth the mass of our Moon. These bodies are known as the *asteroids* or *minor planets*.

The word *asteroid* comes from the Greek and means "starlike," although these bodies are not like stars at all. Most of them are rocky objects, and all but the largest ones are irregularly shaped and heavily marked by collisions they have suffered over the eons. Some appear to be double or multiple bodies and are thought to be the broken-up pieces of formerly larger bodies. All asteroids are believed to be leftover debris from the formation of the Solar System. A later discussion will show why they never managed to condense into one single planetary mass.

Most asteroids orbit the Sun in a broad belt, the *Asteroid Belt*, at an average distance from the Sun of 2.7 AU (see table 4.2). However, not all asteroids are so confined. Some have very elliptical orbits that extend inward past the orbits of Mars and Earth. They are the *Amor* (Mars-crossing) asteroids and *Apollo* (Earth-crossing) asteroids. Asteroid Icarus, which is a chunk of rock 700 m across, has an orbit that carries it to within 0.19 AU of the Sun (in comparison,

Table 4.2. Orbital and Physical Data of the Moon, the Galilean Satellites, Titan, Triton, and Some of the Small Bodies in the Solar System.

	Total number of known satellites or bodies		Orbital characteristics				Physical characteristics		
			Average distance from central body	Sidereal period	Eccentric-ity	Inclination to planet's equator (degrees)	Mass (kg)	Radius (km)	Average density (g/cm^3)
Earth	1	Moon	384,400 km	27.3 days	0.055	23	7.4×10^{22}	1740	3.3
Mars	2	Phobos	9380 km	0.3 days	0.02	1	9.6×10^{15}	14×10	1.9
		Deimos	23,500 km	1.3 days	0.002	2	2.0×10^{15}	8×6	2.1
Asteroids	about 2000	Vesta	2.4 AU	3.6 years	0.09	7	2.0×10^{20}	263	3.0
		Iris	2.4 AU	3.7 years	0.23	6	1.5×10^{19}	105	3.0
		Hebe	2.4 AU	3.8 years	0.20	15	2.0×10^{19}	98	5.0
		Eunomia	2.6 AU	4.3 years	0.19	12	4.0×10^{19}	124	5.0
		Juno	2.7 AU	4.4 years	0.26	13	2.0×10^{19}	134	2.0
		Ceres	2.8 AU	4.6 years	0.08	11	1.0×10^{21}	487	2.0
		Pallas	2.8 AU	4.6 years	0.24	35	2.0×10^{20}	269	2.0
		Psyche	2.9 AU	5.0 years	0.14	3	4.0×10^{19}	119	5.5
		Hygiea	3.2 AU	5.6 years	0.10	4	6.0×10^{19}	211	1.5
		Davida	3.2 AU	5.7 years	0.18	16	3.0×10^{19}	159	1.5
Jupiter	16 (4 in retrograde orbits), Ring System	Io	413,000 km	1.8 days	0.00	0	8.9×10^{22}	1820	3.6
		Europa	671,000 km	3.6 days	0.00	0.5	4.9×10^{22}	1560	3.0
		Ganymede	1,070,000 km	7.2 days	0.00	0.2	1.5×10^{23}	2640	1.9
		Callisto	1,880,000 km	16.7 days	0.01	0.3	1.1×10^{23}	2410	1.8
Saturn	23 (1 in retrograde orbit), Ring System	Mimas	186,000 km	0.9 day	0.02	2	4.5×10^{19}	196	1.4
		Enceladus	238,000 km	1.4 days	0.00	0	8.4×10^{19}	250	1.2
		Tethys	295,000 km	1.9 days	0.00	1	7.6×10^{20}	530	1.2
		Dione	377,000 km	2.7 days	0.00	0	1.1×10^{21}	560	1.4
		Rhea	527,000 km	4.5 days	0.00	0.4	2.5×10^{21}	765	1.3
		Titan	1,220,000 km	15.9 days	0.03	0.3	1.4×10^{23}	2580	1.9
		Iapetus	3,560,000 km	79.3 days	0.03	15	1.9×10^{21}	730	1.2
Uranus	15, Ring System	Miranda	130,000 km	1.4 days	0.00	3	7.5×10^{19}	240	1.3
		Ariel	191,000 km	2.5 days	0.00	0	1.4×10^{21}	580	1.7
		Umbriel	266,000 km	4.1 days	0.00	0	1.3×10^{21}	595	1.4
		Titania	436,000 km	8.7 days	0.00	0	3.5×10^{21}	805	1.6
		Oberon	583,000 km	13.4 days	0.00	0	2.9×10^{21}	775	1.5
Neptune	2, Ring System	Triton	355,000 km	5.9 days (retrograde)	0.00	20	3.4×10^{22}	1750	1.5
		Nereid	5,560,000 km	365 days	0.75	28	2.8×10^{19}	150	2.0
Pluto	1	Charon	19,700 km	6.4 days (retrograde)	0.0	86	1.6×10^{21}	ca. 750	ca. 0.9
Comets	about 100 about 500 about 200 billion	short-period long-period Oort cloud	<35 AU >35 AU 10^3–10^5 AU	<200 years >200 years 10^4–10^7 years	0.1 to 0.97 (?) (?)	mostly <30 all values all values	10^9–10^{17}	0.1–25	1–2

Sources: Data from Beatty, et al., 1982; Allen 1973; Brown, R. H. and D. P. Cruikshank 1985; Stone and Miner 1986.

Note: The table includes data for all satellites and asteroids with masses greater than 10^{-4} times the mass of the Moon. The satellites of Mars and comets are included also, even though their masses are less.

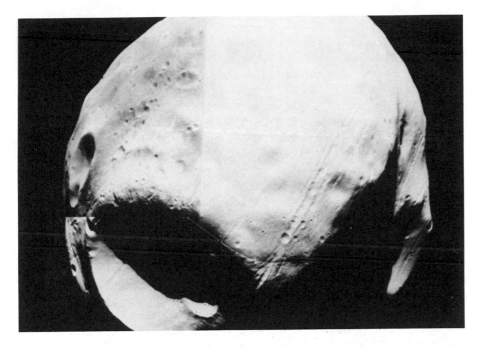

Figure 4.8. Phobos, one of the two asteroid-like moons of Mars. This potato-shaped satellite measures roughly 14 × 10 km and revolves about Mars in a nearly circular orbit at a distance of 9380 km. The crater near the bottom of Phobos (called Stickney) is evidence of the frequent violent collisions among planetary bodies that characterized the early history of the Solar System.

Mercury orbits the Sun at roughly twice that distance) and outward past the orbit of Mars. Other asteroids stray outward to beyond the orbits of Saturn and Uranus as, for example, Chiron (or Asteroid 2060), discovered in 1977. Still others, the *Trojan Asteroids* (most of which are thought to be icy rather than rocky objects), are firmly locked into orbits with the same period as Jupiter's, with one group traveling 60 degrees ahead and the other group 60 degrees behind Jupiter, as seen from the Sun's position.

Many of the smaller satellites of the planets are thought to be asteroids that were captured by the planets (see table 4.2). These small satellites have masses, sizes, and compositions like asteroids. For example, the two moons of Mars, Phobos (figure 4.8) and Deimos, have dimensions of only 14 × 10 km and 8 × 6 km, respectively. They are irregularly shaped and photographs reveal surfaces marred by many large and small impact craters, indicating countless collisions in the past with other rocky debris in interplanetary space. Eleven of Jupiter's 16 known satellites are less than 100 km across and the outer four of them are in retrograde orbits. Twelve of Saturn's 23 known satellites are also quite small (100 km in size or smaller), and so are 10 of Uranus's 15 known satellites. No doubt, as astronomers continue their explorations of the outer Solar System, they will discover many additional small asteroid-like satellites circling the giant planets.

Comets

Far beyond the planets, in the dark and cold outer fringes of the Solar System, many billions of small, primordial bodies of rock and ice are suspected to circle the Sun. They constitute the *Oort Cloud* of dormant cometary bodies, which, like the asteroids, are thought to be leftover debris from the formation of the Solar System. This cloud is believed to extend up to 100,000 AU outward from the Sun, which is more than a third of the way to the Alpha Centauri system (the nearest stars, at a distance of 270,000 AU; see chapter 1).

The existence of this cloud of cometary bodies was first proposed in 1950 by the Dutch astronomer Jan Oort, in order to explain the frequency of appearance of comets in the inner part of the Solar System. According to Oort, the occasional passage of nearby stars perturbs the orbits of some of these bodies and deflects them toward the inner part of the Solar System (see box 4.3). There, bombardment by radiation and charged particles from the Sun brings them to "life." Within about 3 AU of the Sun their ices start vaporizing, and gas and dust particles stream outward to form an envelope—the *coma* and *hydrogen cloud*—around the central body, the *nucleus* (see figure 4.9). Pressure exerted by the solar radiation and solar wind drives the gases and dust many millions of kilometers through interplanetary space to form the cometary *tail*. Due to the way it is formed, the tail always points away from the Sun. The main features of the components of comets are summarized in table 4.3.

BOX 4.3.
Nemesis: Companion Star of the Sun?

IN 1984 a group of scientists proposed that the gravitational perturbations that regularly deflect comets from the Oort Cloud toward the inner part of the Solar System may, in fact, arise from a small and very faint companion star of the Sun, with an orbital period of 26 million years. They made this proposal because of the recent realization that during the past 250 million years life on Earth has experienced episodes of mass extinctions roughly every 26 million years. It is thought that some of these extinctions may have been caused by comets that collided with Earth. Such collisions would have thrown enormous quantities of dust into the upper atmosphere and prevented sunlight from reaching the ground. This, in turn, would have lowered the average ground temperature over much of the globe, slowed or stopped plant growth, and severely interrupted the food chain (see box 9.4 for a discussion of the mass extinction 65 million years ago). Perhaps these periodic comet showers are triggered every time the postulated companion star—which has been named *Nemesis*, after the Greek goddess of doom—is on its closest approach to the Sun in what may be a highly elliptical orbit. (But the point of closest approach would be still far beyond Pluto's orbit.)

Few scientists are convinced that the case for a companion star of the Sun is very strong. If the star's period is 26 million years, the major axis of its orbit would be 2.8 LY, which is a goodly fraction of the average distance between stars in the solar neighborhood. This means that Nemesis's orbit would be exposed to strong tidal forces from other stars and from the Galaxy as a whole, and that the orbit would probably not be stable for much more than about a billion years (which is much less than the age of the Solar System). Still the possible existence of Nemesis cannot be entirely ruled out. For example, the difficulties with orbital stability would be eased if the star had been captured by the Sun during the past few hundred million years (an improbable event) or if it originally had a smaller orbit, for which the tidal disturbances were much smaller. In any case, several astronomers are now actively searching for Nemesis.

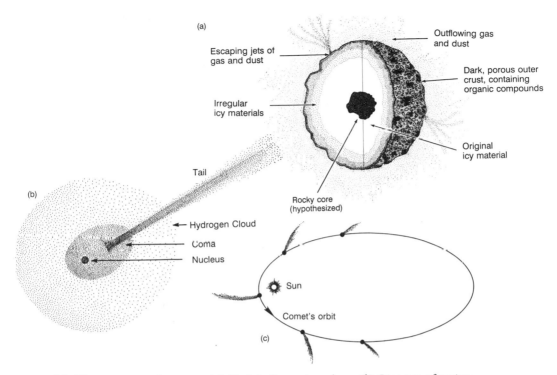

Figure 4.9. The structure of comets. (*a*) Model of comet nucleus; (*b*) Diagram of major components of comets; (*c*) Shape and growth of tail of typical comet as it passes near the Sun.

Table 4.3. Features of Components of Comets.

Component	Composition	Size
Nucleus: central solid body of comet	rocks and ices of water (H_2O), methane (CH_4), and ammonia (NH_3), source of all the dust and gas of the comet's coma, hydrogen cloud, and tail	100 m–25 km irregularly shaped; *mass*: 10^9–10^{17} kg (10^6–10^{14} tons)
Coma: envelope of gas and dust surrounding nucleus	*gas*: neutral atoms and molecules rich in hydrogen, carbon, oxygen, nitrogen, sulfur, and various metals (H, C, O, S, OH, CH, CN, CO, CS, NH, NH_2, HCN, CH_3CN, Na, Fe, K, Ca, V, Cr, Mn, Co, Ni, Cu) *dust*: microscopic silicate minerals (similar to terrestrial sandy dust but less compact)	10^5–10^6 km (approx. the size of the Sun)
Hydrogen Cloud: enormous nearly spherical envelope	atomic hydrogen, derived from the disintegration (by sunlight) of OH and other hydrogen-rich radicals of the coma	up to many million km (much larger than the Sun)
Tail: elongated structure of dust and gas	*dust*: primarily silicate minerals *gas*: charged particles (e^-, CO^+, CO_2^+, H_2O^+, OH^+, CH^+, CN^+, N_2^+, C^+, Ca^+)	*length*: up to 1 AU

Sunlight reflected from the gas and dust particles produces the image that we commonly have of comets — the long, sweeping tail, extending out from the bright, globular coma. We obtain this image of comets mostly from photographs. With the unaided eye comets can usually be seen only as inconspicuous, faintly glowing objects, right before sunrise or after sunset. However, occasionally comets do become spectacular. Then they arouse people's interest and imagination, and even strike awe and fear in some of them. That was the effect Comet Halley had during its apparition in 1910.

About four comets visit the inner part of the Solar System every year. Most of them come from far beyond the orbit of Pluto, approach the Sun in highly elongated orbits, whip around it, and then recede again. Hundreds of thousands of years may go by before they return once again. There are, however, a number of comets that by near encounters with a planet were placed into orbits closer to the Sun. If their orbital periods are less than 200 years (corresponding to about 35 AU for the length of the semimajor axis of their orbits), they are called *short-period comets*. If their periods are greater than 200 years, they are called *long-period comets*.

Comet Halley (named after the British astronomer and mathematician Edmond Halley [1656–1742]) is a short-period comet. It completes an orbit every 75 to 80 years, approaching the Sun to within 0.59 AU and receding to 35 AU. This comet has been sighted regularly since 240 B.C., when Chinese records first make mention of it. Its most recent approach to the inner part of the Solar System occurred in 1985–86, when it was extensively investigated and photographed by five unmanned spacecraft (see box 4.4).

Every time a comet passes through the inner part of the Solar System it loses some of its constituent gases, dust, and rocks. After hundreds to thousands of such passages, nothing is left but a fragile rocky body that will eventually break up due to the Sun's tidal forces. The rocky rubble that results continues to travel along the comet's original orbit. In the course of time, the combined gravitational effects of the Sun and the planets disperse the cometary debris further and further until it isspread over the entire orbit. This debris is the source of shooting stars or meteors (see chapter 5).

BOX 4.4.
Comet Halley, 1985–86

FOR most nonscientists, the 1985–86 apparition of Comet Halley turned out to be disappointing. Even in the Southern Hemisphere, where viewing the comet was optimal, most people who looked for it were not sure whether they had actually seen it. This disappointment was in stark contrast to the excitement Comet Halley had created in 1910, when its coma became as bright as Polaris, the North Star, and its tail stretched more than halfway across the sky. On May 6 of that year, the Earth encountered rocky debris from the comet, which caused a meteor shower, and on May 19 and 20 the Earth passed either through or very close to the comet's tail.

The differences in the visual displays of Halley's comet between 1910 and 1985–86 (figure A) were mainly due to differences in the relative positions of the comet, Earth, and Sun during these two apparitions. For example, in 1910 the closest approach between Comet Halley and Earth was 0.14 AU and it occurred one month after the comet's perihelion passage, when the coma and tail were still substantial and very bright. In 1986 the closest approach was 0.42 AU and it occurred two months after perihelion passage, when the coma and tail had already begun to fade.

In contrast, professional astronomers were excited by the unprecedented wealth of data, including the first photographs of a cometary nucleus ever taken, that were sent back to Earth from five spacecraft: *Vega 1* and *2* from the Soviet Union, *Suisei* and *Sakigake* from Japan, and *Giotto* from the European Space Agency. Besides television cameras, these spacecraft carried instruments for measuring the energy and chemical composition of ions and neutral gas particles, the rate of impacts by dust particles, the mass and chemical composition of individual dust particles, the strength of magnetic and electric fields, the brightness of the coma at different wavelengths,

Figure A. Comet Halley photographed during its 1985–86 apparition.

BOX 4.4 continued

and the turbulent interaction between the outer coma and the solar wind.

Sending spacecraft to Halley's comet was a scientific adventure of the highest order and a feat that probably even the most optimistic observers of 1910 would not have thought possible. For example, *Giotto*, which weighed 520 kg and was equipped for the most ambitious scientific program of the five probes, was launched on July 2, 1985, atop an Ariane rocket, and sent into an orbit that for the first two months carried it just slightly beyond that of Earth. It then began falling sunward to roughly halfway between the orbits of Earth and Venus, while its forward momentum carried it more than three-quarters of the way around the Sun toward its distant rendezvous with the comet (see figure B).

When *Giotto* was launched, Comet Halley was still beyond the asteroid belt, traveling at approximately 20 km/sec in a retrograde orbit toward the Sun (see figure 4.7). As the comet approached the inner part of the Solar System and the Sun's gravitational pull became stronger, it speeded up rapidly. Late in November, the comet crossed the orbit of Mars. On December 31 it crossed the orbit of Earth, traveling now at slightly more than 40 km/sec. On February 9 it passed perihelion at 0.59 AU from the Sun and a speed of 55 km/sec. It then began its long climb out of the Sun's gravitational well. Finally, on March 13, the comet and *Giotto* passed each other at a distance of 605 km, traveling in opposite directions at a relative speed of nearly 70 km/sec (about 250,000 km/hr, or 50 times the speed of a rifle bullet). The rendezvous point was 0.88 AU from the Sun and 0.99 AU from Earth.

Giotto's designers thought that the craft's journey so close to Halley's comet was going to be a suicide mission, in that cometary dust particles and chunks of debris were likely either to destroy the craft or at least incapacitate some of its instruments. Therefore, they shielded the spacecraft with an inner blanket of Kevlar (material used in bulletproof vests) surrounded by a sheet of aluminum. To be on the safe side, all data

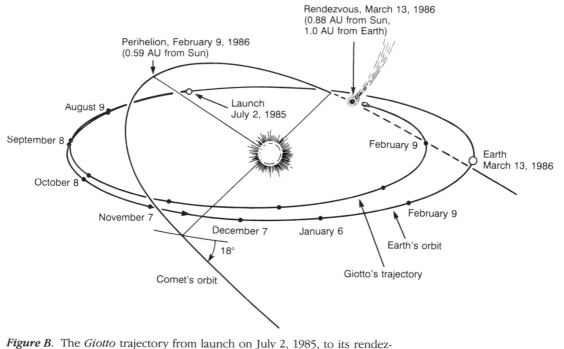

Figure B. The *Giotto* trajectory from launch on July 2, 1985, to its rendezvous with the nucleus of Halley's comet on March 13, 1986.

BOX 4.4 *continued*

were sent back to Earth in real time; the craft carried no tape recorders for storing information at a fast rate to be transmitted more slowly to Earth after the encounter, as is common with interplanetary probes.

The first indication that *Giotto* was nearing Comet Halley came roughly a day and a half before the rendezvous when, at a distance of 8 million km from the nucleus, the craft began encountering ionized gas particles from the comet, mostly H^+. At 4 million km away from the nucleus, the spacecraft passed through a region of magnetic turbulence, probably caused by interaction between the solar wind and cometary ions. At 3.5 million km, the ionized gas had thickened sufficiently to slow the solar wind by measurable amounts. At 1.1 million km and just 4 hours from closest approach, *Giotto* crossed the comet's bow wave, a 250,000 km wide region of ionized gases created by deflection of the solar wind around the inner part of the comet's coma (somewhat analogous to the bow wave in water in front of and around a moving boat). A variety of atoms, molecules, and radicals were detected, among them singly ionized oxygen (O^+), hydroxyl radicals (OH), carbon dioxide (CO_2), and water (H_2O).

At 280,000 km from the nucleus, *Giotto* encountered its first dust particles. They were microscopic in size, with masses of less than a microgram. Roughly 90 percent of them contained carbon compounds, while some of the rest contained iron. All of the early dust particles were held off by the craft's outer shield. However, at 8000 km distance, the dust particles were bigger and more concentrated, and they started to penetrate the craft's outer shield. In fact, the steady impact of cometary gas and dust at 50 times the speed of bullets slowed *Giotto* noticeably.

At 4300 km from the nucleus, instruments on *Giotto* detected a big decrease in the number of ions and a corresponding increase in neutral particles. The strength of the magnetic field, which tends to follow the motions of charged particles, began to diminish sharply as well. This was the point where purely cometary particles held off the last of the solar wind particles.

About 3 hours before closest approach, and still some 750,000 km distant, *Giotto* turned on its television camera and began taking pictures. They showed a peanut or potato-shaped nucleus, 15 km by 8 km in size, silhouetted against the bright background of the coma. They also showed two very bright jets of dust spurting outward from the nucleus.

The nucleus appears to be pitch-black. Its surface reflects only about 1 to 2 percent of the incident sunlight, making Halley the darkest known object orbiting the Sun. (In comparison, the Moon reflects 7 percent of the incident sunlight.) The dark color took most experts by surprise, for they had thought of cometary nuclei as being dirty snow or ice balls, and the presence of ices should raise the reflectivity. The concept of a dirty snow ball is probably still correct for the interior of Halley's nucleus. But the surface layers are depleted in ices. The ices that originally were present there have been vaporized and were driven off into interplanetary space during the comet's many close passages around the Sun. This repeated loss of volatiles probably produced a porous, rocky surface crust or loosely bound "rubble pile,"[*] as one scientist suggested, containing tarry material rich in organic compounds that give rise to the observed blackness.

Giotto took its last picture 1350 km from the nucleus, when all data reception ceased. Apparently, dust jolted the spacecraft and caused it to wobble violently. This knocked its narrow radio beam out of alignment with Earth and interrupted transmission. A few seconds later, intermittent signals were received again, as *Giotto*'s transmitter swung in and out of alignment. A little over half an hour later, the onboard damping device had steadied the craft sufficiently for continuous transmission of data to resume. Surprisingly, all of the craft's 10 instruments survived the incident, though six suffered some damage. The European Space Agency is now making plans to boost *Giotto* into a slightly altered orbit that would bring it by Earth in 1990 and then to use our planet's gravitational field for redirecting the craft toward another comet or asteroid.

The other space probes sent to Halley's Comet passed the nucleus a few days earlier than *Giotto* and at greater distances. Qualitatively they confirmed the findings of the *Giotto* mission, though there were differences. For example, the two *Vega* spacecraft encountered more dust than *Giotto*. Apparently, much of the gas and dust is emitted from the nucleus in jet-like bursts that vary in time and direction. Because the spacecraft followed courses that differed

Science News, March 22, 1986, p. 181.

BOX 4.4 continued

in space and time, they encountered different dust conditions.

In addition to emitting gas and dust in bursts and unevenly from its surface, Halley's nucleus is rotating. This was deduced from pictures sent back by *Suisei*, one of the Japanese craft, which indicate that the comet's extensive hydrogen cloud was growing and waning regularly. This is best explained by assuming that new gas is injected into the hydrogen cloud from a nucleus that releases the gas preferentially from one side and that also rotates. The rotation period was calculated to be 53 hours.

All of the data sent back to Earth from the five spacecraft were freely shared with the international scientific community. In fact, only through this shar-ing was *Giotto*'s close fly-by of the nucleus possible. Before any of the spacecraft had come close to Comet Halley, the position of the nucleus was known only to within several thousand kilometers within the haze of the coma's scattered light. However, by using infor-mation from several sources—precise measurements of the trajectories of each of the *Vega* spacecraft, obtained by Soviet and United States tracking stations, and position estimates of the comet's nucleus relative to the *Vegas*, deduced from the photographs sent back by these two spacecraft—European scientists were able to reduce *Giotto*'s targeting error to within about 100 km. Perhaps this spirit of international cooperation will remain Comet Halley's major legacy.

Exercises

1 (*a*) Name the seven planets or "wanderers" known to the ancients.
(*b*) Using a planet chart locate as many of these objects on the sky as conditions allow. (Monthly planet charts are included in each issue of such popular astronomy magazines as *Astronomy* and *Sky and Telescope*.)

2 Draw a diagram showing the Moon's orbit around Earth and the Moon's dark and sunlit hemispheres at several positions. Use this diagram to explain the Moon's phases. In particular,
(*a*) Why is the first-quarter Moon overhead at sunset?
(*b*) Why does the full Moon rise at sunset and set at sunrise?
(*c*) Why is the last-quarter Moon overhead at sunrise?
(*d*) Why do solar eclipses occur only when the Moon is new?
(Note that the directions of the Moon's orbital mo-tion and the Earth's rotation are the same.)

3 Mercury and Venus can only be seen in the evening and morning. Why?

4 Convert the Bode–Titius law into one that gives the planets' distances from the Sun in units of 5.2 AU, the Sun–Jupiter distance.

5 (*a*) From the values of the astronomical unit and the light-year given in the text, calculate how many astronomical units equal one light-year.
(*b*) Calculate in AU the distance from Earth to α Centauri (the distance equals 4.38 LY).

6 According to one theory, Planet X orbits the Sun beyond Pluto, at about 50 to 100 AU from the Sun. If the Bode–Titius law applies to this hypothetical planet, what would its orbital distance be?

7 (*a*) Using a planet chart, check which of the fol-lowing planetary objects are presently visible in the night sky: Jupiter, Saturn, Ceres, Vesta, and Pallas. Locate as many of these objects, including Jupiter's Galilean satellites, as conditions allow. (Use binocu-lars in the case of the asteroids and Jupiter's satel-lites.)
(*b*) If Jupiter is visible, record the positions of its Galilean satellites over the course of several weeks and determine their orbital periods.

8 (*a*) Where in their orbital journeys, and how, do comets acquire their comae, hydrogen clouds, and tails?
(*b*) Describe the composition and structure of each of these cometary components.

Suggestions for Further Reading

Beatty, J. K. 1986. "A Place Called Uranus." *Sky and Telescope*, April, 333–37. An article about *Voyager 2*'s close-up view of Uranus in January 1986.

Beatty, J. K., B. O'Leary, and A. Chaikin, eds. 1982. *The New Solar System*. 2d ed. Cambridge: Cambridge University Press. An in-depth, yet very readable, treatment of the Solar System, consisting of 20 chapters written by research scientists and covering planetary observations through the *Voyager* missions to Jupiter and Saturn.

Boss, A. P. 1986. "The Origin of the Moon." *Science*, 231: 341–45. The author suggests that the Moon originated as a result of a collision of the proto-Earth with a roughly Mars-sized body; the ejecta from this collision created a circumterrestrial disk of matter from which the Moon later accreted. The author also discusses the older fission, capture, and binary accretion theories for the Moon's origin and presents arguments why they are physically either impossible or highly improbable. For the advanced student.

Brandt, J. C., and M. B. Niedner, Jr. 1986. "The Structure of Comet Tails." *Scientific American*, January, 48–56.

Briggs, G., and F. Taylor. 1982. *The Cambridge Photographic Atlas of the Planets*. Cambridge and New York: Cambridge University Press.

Brown, R. H., and D. P. Cruikshank. 1985. "The Moons of Uranus, Neptune, and Pluto." *Scientific American*, July, 38–47. A preview of what *Voyager 2* might find when it flies by Uranus in January 1986 and Neptune in 1989.

Chaikin, A. 1986. "Voyager Among the Ice Worlds." *Sky and Telescope*, April, 338–43. The rings and satellites of Uranus as they were observed by *Voyager 2* during its January 1986 flyby.

Chapman, R. D., and J. C. Brandt. 1984. *The Comet Book: A Guide for the Return of Halley's Comet*. Boston: Jones and Bartlett Publishers. An excitingly written, easy-to-read book about all aspects of comets—their observations past and present, structure, composition, origin, eventual death, and legends about them.

Cohen, I. B. 1981. "Newton's Discovery of Gravity." *Scientific American*, March, 166–79.

———. 1955. "Isaac Newton." *Scientific American*, December, 73–80.

Gehrels, T. 1985. "Asteroids and Comets." *Physics Today*, February, 32–41.

Gingerich, O. 1982. "The Galileo Affair." *Scientific American*, August, 132–43. The article describes the prevailing pattern of thinking when Galileo defended the Copernican system, the mode of reasoning introduced by Galileo, and the new method of hypothesis testing that developed from it.

Gore, R. 1981. "*Voyager 1* at Saturn, Riddles of the Rings." *National Geographic*, July, 3–31. A well-illustrated article about the drama of *Voyager 1*'s flyby and observations of Saturn, its ring system, and its moons.

———. 1986. "Uranus. *Voyager* Visits a Dark Planet." *National Geographic*, August, 178–94.

Haberle, R. M. 1986. "The Climate of Mars." *Scientific American*, May, 54–62. The climate of Mars started out much like Earth's, with temperatures sufficiently high for water to flow on the planet's surface. But Mars evolved differently, and today the Martian surface is so cold that carbon dioxide freezes at its poles.

Hess, W., et al. 1969. "The Exploration of the Moon." *Scientific American*, October, 54–72.

Horowitz, N. H. 1977. "The Search for Life on Mars." *Scientific American*, November, 52–61. Experiments designed to detect life on Mars, carried out by the two *Viking* landers in 1976.

Ingersoll, A. P. 1981. "Jupiter and Saturn." *Scientific American*, December, 90–108. Description of two competing models of the Sun's two largest planets.

Johnson, T. V., and L. A. Soderblom. 1983. "Io." *Scientific American*, December, 56–67.

Lawless, J. G., C. E. Folsome, and K. A. Kvenvolden. 1972. "Organic Matter in Meteorites." *Scientific American*, June, 38–46.

Lewis, R. S., and E. Anders. 1983. "Interstellar Matter in Meteorites." *Scientific American*, August, 66–77. Carbonaceous chondrites incorporate material that originated outside the Solar System, including matter expelled by supernovae and other stars.

McCloskey, M. 1983. "Intuitive Physics." *Scientific American*, April, 122–30. Newton's laws are well known, but tests show many people think moving objects act otherwise.

Moore, P., and J. Mason. 1984. *The Return of Halley's Comet.* New York: W. W. Norton and Co. A well-written book, with excellent discussions of Comet Halley's 1910 and earlier apparitions, its orbital characteristics, and the spacecraft sent to meet it in 1986.

Owen, T. 1982. "Titan." *Scientific American*, February, 98–109. Titan's atmosphere, whose chemistry may resemble that of Earth before the origin of life.

Pollack, J. B., and J. N. Cuzzi. 1981. "Rings in the Solar System." *Scientific American*, November, 104–29. The structure and composition of the rings of Jupiter, Saturn, and Uranus, the processes that shape them, and their possible origins.

Press, F., and R. Siever. 1982. *Earth.* 3d ed. San Francisco: W. H. Freeman and Co. A college-level geology text.

Prinn, R. G. 1985. "The Volcanoes and Clouds of Venus." *Scientific American*, March, 46–53.

Schramm, D. N., and R. N. Clayton. 1978. "Did a Supernova Trigger the Formation of the Solar System?" *Scientific American*, October, 124–39. Unusual isotope abundances in carbonaceous chondrites offer strong observational evidence that the formation of the Solar System was triggered by a nearby supernova.

Schubert, G., and C. Covey. 1981. "The Atmosphere of Venus." *Scientific American*, July, 66–74.

Soderblom, L. A., and T. V. Johnson. 1982. "The Moons of Saturn." *Scientific American*, January, 100–116.

Weaver, K. F. 1986. "Meteorites: Invaders From Space." *National Geographic*, September, 390–418.

Wetherill, G. W. 1981. "The Formation of the Earth from Planetesimals." *Scientific American*, June, 162–74. Computer simulations show how the terrestrial planets might have formed. This article is one of the key sources for the last section of chapter 4.

Wood, J. A. 1970, "The Lunar Soil." *Scientific American*, August, 14–23.

Additional References

Allen, C. W. 1973. *Astrophysical Quantities*, 3rd ed. London: The Athlone Press.

Hughes, D. W. 1982. "Was Galileo 2000 Years Too Late?" *Nature* 296 (18 March): 199.

Kepler, J. 1981. *Mysterium Cosmographicum*, trans. A. M. Duncan. New York: Abaris Books.

Press, F., and R. Siever. 1978. *Earth*, 2nd ed. San Francisco: W. H. Freeman and Co.

Stone, E. C., and E. D. Miner. 1986. "The *Voyager 2* Encounter with the Uranian System." *Science* 233 (4 July): 39.

Erupting volcano on Io, photographed from a distance of 500,000 km by *Voyager I* during its historic encounter with Jupiter, March 4–5, 1979.

 The Solar System: Structure of Planetary Bodies and Origin

In chapter 4 we have primarily concerned ourselves with the orbital layout of the Solar System. In order to progress toward our goal of understanding the origin of the Sun and the planets, we must also learn about the physical makeup of the system's member bodies. The structure of the Sun was discussed in chapter 3. Now let us look at other bodies in the Solar System and, in the last section of this chapter, at some of the current theories of its origin and evolution.

THREE FAMILIES OF BODIES IN THE SUN'S PLANETARY SYSTEM

The bodies circling the Sun and constituting the planetary system range from dust and pebble-sized debris to planets with radii of thousands to tens of thousands of kilometers. The lower part of this range includes the asteroids, comets, many (but not all) of the satellites of the planets, the particles composing the rings of Jupiter, Saturn (figure 4.1), Uranus and Neptune, and the interplanetary dust and rock fragments that have resulted from collisions among the bodies. I am including among these *small bodies* of the Solar System those objects that have radii and masses up to approximately 1000 km and 1/1000 earth masses (a few times 10^{21} kg; figures 5.1, 5.2, 5.3). Many of them have orbits that are highly elliptical and inclined to the ecliptic. Together they form a family of bodies with a common history and with compositions that, on the average, vary systematically with distance from the Sun. They are believed to be the leftover debris from the original material out of which the planets and their large satellites were assembled. In the theory of Solar System formation, such small bodies are referred to as *planetesimals*.

A second family of Solar System bodies includes the terrestrial planets—Mercury, Venus, Earth, and Mars. In this book, the Moon, the Galilean satellites, Saturn's satellite Titan, and Neptune's satellite Triton are also included in this family. I refer to them as the *intermediate-sized planetary bodies*. They have radii between roughly 1500 km and 6500 km (figure 5.4); their masses range from about 1/200 to 1 earth mass; they have similar structures; and related physical processes shaped them. The orbits of most of them are nearly circular, with inclinations either close to the ecliptic or to the equatorial plane of the central body about which they revolve. The major exceptions are Mercury, as already noted, and Triton. Triton is in a retrograde orbit that is inclined by 20° to Neptune's equator. As in the case of the small bodies of the Solar System, the compositions of the intermediate-sized planetary bodies also vary systematically with distance from the central body about which they revolve.

In the past, Pluto has been included in the family of intermediate-sized planetary bodies; but, as this family is defined here, Pluto is a marginal case. Pluto's size is comparable to Europa, the smallest of the Galilean satellites; yet its mass is only 1/460 earth masses, which places it somewhere between a large asteroid and a small intermediate-sized planetary body. Furthermore, Pluto's

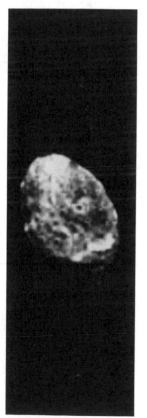

Figure 5.1. Hyperion, one of Saturn's smaller moons, photographed by *Voyager 2*. The photograph reveals a heavily cratered and battered surface. Hyperion is a mere 110 × 205 km in size, so its self-gravity is too weak to have shaped it into a spherical body.

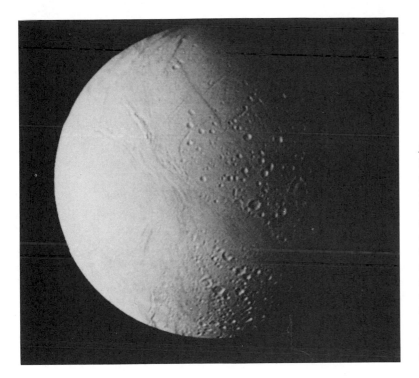

Figure 5.2. Dione, Saturn's fifth largest satellite, with a radius of roughly 560 km. Its density is about 1.4 g/cm³, indicating that it consists largely of ices intermingled with some rocks. Thousands of craters cover the satellite's surface. A fracture can be seen near the terminator in this photograph, suggesting that Dione has experienced some geologic evolution since its formation. (*Terminator* refers to the dividing line between the illuminated and dark parts of a satellite's or planet's disk.)

Figure 5.3. Miranda, one of Uranus's five large satellites (about 480 km in diameter), photographed by *Voyager 2* from a distance of 31,000 km. This satellite is one of the most unusual bodies in the Solar System. Its surface is characterized by a bizarre patchwork of ridges and grooves (a chevron near the center and oval "racetracks" near the upper-right and lower-left limbs), jagged cliffs, and crater-dotted regions resembling the lunar highlands. The grooves are up to several kilometers deep and the cliff visible at the lower right is 20 km high. No one knows what forces and events gave rise to this strange moon.

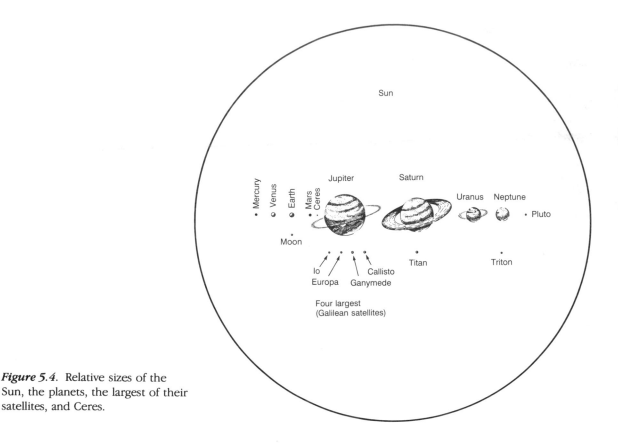

Figure 5.4. Relative sizes of the Sun, the planets, the largest of their satellites, and Ceres.

orbit is quite elliptical and inclined, similar to the orbits of asteroids and comets. Hence in this book Pluto (as well as its companion, Charon) is regarded as a member of the family of small bodies of the Solar System, even though in the tables it will be listed among the planets.

Of course, despite the division made here, there is no really natural or absolute borderline between the families of small and intermediate-sized planetary bodies. Together these bodies constitute a continuum from microscopic dust through small irregularly shaped boulders, to large, roundish bodies of rocks, metals, and volatiles. The choice of the borderline between the two families is quite arbitrary and mainly one of convenience.

The third family of Solar System bodies are the *giant* or *Jovian planets*—Jupiter (figure 5.33), Saturn (figure 4.1), Uranus, and Neptune. Their radii extend from roughly 25,000 to 70,000 km and their masses range from approximately 15 to more than 300 earth masses. Their structures are largely gaseous and liquid, and they have been formed by physical processes that were somewhat different from those of the smaller bodies. Their sizes, structures, and origins place these giant planets somewhere between the terrestrial planets and a small star, though none of them is even close to generating energy by nuclear fusion. Unlike the intermediate-sized planetary and small bodies in the Solar System, they do, however, radiate more energy than they receive from the Sun

Table 5.1. Viewing Data and Orbital Characteristics of the Four Meteor Showers Whose Orbits Are Represented in Figure 5.5.

Meteor shower	Date of maximum display	Normal period of visibility	Constellation in which meteor tracks appear	Period of original comet (years)	Velocity relative to Earth (km/sec)	Eccentricity of orbit	Inclination of orbit to ecliptic (degrees)
Quadrantids	Jan 3	Jan 2–4	Boötes	7	43	0.71	75
Perseids	Aug 12	July 29–Aug 18	Perseus	110 (retrograde)	61	0.96	65
Leonids	Nov 16	Nov 14–19	Leo	33 (retrograde)	72	0.92	17
Geminids	Dec 13	Dec 8–15	Gemini	1.6	37	0.90	26

Source: Adapted from Allen 1973, 158.

by processes that are at present not fully understood. (Summaries of the physical characteristics of members of the three families of Solar System bodies are found in tables 4.1–4.3 and 5.1–5.3.)

SMALL BODIES OF THE SOLAR SYSTEM

The best way to study the small bodies of the Solar System would be to send space probes to the Asteroid Belt and comets and to collect samples. That is what the United States did with its *Apollo* missions to the Moon. Similar missions to the asteroids and comets are technically feasible, but would be so expensive that to date they have not been funded. Until there is the opportunity to collect samples of asteroids and comets directly, scientists have to make do with the information they can obtain through remote sensing, flyby missions, and by collecting and analyzing whatever debris from these smaller bodies happens to collide with our planet and survives the fall to the ground.

Meteorites

As it turns out, many asteroids or fragments of asteroids, as well as the rocky remnants of past comets, are in elongated Earth-crossing orbits (figure 5.5), and every year Earth collides with and sweeps up roughly ten thousand tons of this interplanetary rubble. Most of these encounters are harmless, for the rubble is usually microscopic in size and burns up by friction as it passes through the Earth's upper atmosphere, briefly creating fiery trails across the sky at night. Such trails are popularly known as *shooting stars* and technically as *meteors*. Some meteors arrive in groups of thousands regularly every year and can give rather spectacular displays. Viewing data and orbital characteristics of a few *meteor showers* are given in table 5.1 and figure 5.5 on the next page.

Occasionally, large bodies weighing several kilograms or even tons enter the atmosphere. Typically they travel at speeds of a few times 10 km/sec (about 10^5 km/hr) and friction with the air sometimes produces so much heat that they may become brighter than the full Moon. Often they are sheared apart into two or more fragments. The largest of them, weighing thousands of tons or more, create sonic booms that can be heard for many miles. These large bodies do not

Table 5.2. Physical Characteristics of Meteorites

		Dominant characteristics	Density (g/cm³)	Elemental composition	Comments
Undifferentiated Meteorites[a]	Ordinary chondrites	Fine-grained earthy matrix containing two distinct components: (1) hard irregularly shaped *mineral inclusions* with high concentrations of Ca, Al, Ti, and (2) round or oval *mineral chondrules*[b] with high concentrations of Fe, Mg, Si.	about 3.7	Solar composition, except volatiles are underrepresented.	*Ca–Al–Ti inclusions* in chondrites are the most refractory constituents found in meteorites, meaning they were at one time exposed to high temperatures (1500 K or higher) that vaporized and drove off the volatile compounds. *Fe–Mg–Si chondrules* were once molten droplets, probably created during high-speed collisions of planetesimals.
	Carbonaceous chondrites	Similar to ordinary chondrites including the presence of Ca-Al-Ti inclusions, but chondrules are fewer in number or absent, and H_2O (2 to 20% by mass) and *organic compounds* (up to 5%) are present.	2.2–3.6		Organic compounds of carbonaceous chondrites include amino acids and tarlike chains of hydrocarbons. Among the amino acids, glycine, alanine, valine, proline, and glutamic acid are abundant, as well as amino acids not present in biologically synthesized protein (see the Introduction to Part Two and chapter 6). Both L- and D-amino acids are present, indicating an abiotic origin.
Differentiated Meteorites[a]	Achondrites	Similar to terrestrial and lunar igneous rocks.	3.0–3.5	Composition very different from solar due to process of differentiation.	Differentiation is believed to have occurred in parent bodies (that is, the planetesimals) of solar or ordinary chondritic composition and with diameters exceeding several ten kilometers.
	Stony irons	Roughly equal proportions of silicate minerals and Fe-Ni metals.	5.5–6.0		
	Irons	About 90% Fe, 5–10% Ni.	7.5–8.0		

[a]*Differentiation* in this context refers to the same process that separated the Earth early in its history into an iron–nickel core, a mantle of high-density silicate rocks, and an outer crust of low-density rocks.

[b]*Chondrite* and *chondrule* are derived from the Greek *chondros*, meaning "grain."

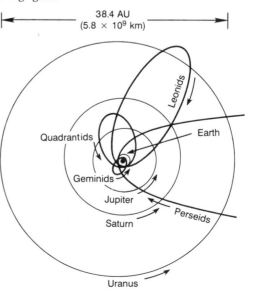

Figure 5.5. Diagram of the orbits of four meteor showers. The shapes (that is, eccentricities) and sizes of the orbits are drawn to scale. The inclinations of the orbits to the ecliptic are given in table 5.1.

Table 5.3. Characteristics of the Atmospheres of Five Intermediate-Sized Planetary Bodies.

	Composition of gases (fractions by number)	Surface pressure P_{Planet}/P_{Earth}	Average surface temperature (K)	(C)	Surface gravity (m/sec^2)
Venus	CO_2 (0.96) N_2 (0.035) H_2O (0.0001) SO_2 (0.0001)	90.	730	457	8.9
Earth	N_2 (0.77) O_2 (0.21) H_2O (0.01) Ar (0.0093) CO_2 (0.00033)	1.	288	15	9.8
Mars	CO_2 (0.95) N_2 (0.027) Ar (0.016) O_2 (0.0013) CO (0.0007) H_2O (0.0003)	0.007	218	−55	3.7
Titan	N_2 (0.80–0.95) CH_4 Organic Molecules Smog Particles	1.6	93	−180	1.4
Triton	CH_4	0.0001	(?)		0.9

Note: O_2 is molecular oxygen, SO_2 is sulfur dioxide.

burn up, but pass right through the atmosphere with very little deceleration. Most of these *meteorites** fall into the ocean, but about one-third of them do come down on land where they carve out craters that may be several meters to many tens of kilometers across.

Fortunately, impacts by meteorites are relatively rare today, and the probability of being struck by one is extremely low. Still, on every continent there exist a number of craters, such as the Barringer Crater near Winslow, Arizona (see figure 5.6), that give evidence of violent collisions between Earth and interplanetary bodies. Another example of a collision of Earth with an interplanetary body is the violent explosion that shook the Tunguska region of Siberia on the morning of June 30, 1908 (see figure 5.7). Apparently, a cometary fragment entered the Earth's atmosphere and exploded just above the ground. People as far as 750 kilometers away saw the event, and the energy released by the explosion scorched a 30-kilometer wide area and leveled trees like matchsticks.

Planetary scientists get greatly excited about new finds of meteorites because they are presently the only samples of asteroid and comet material. Because

*Interplanetary fragments that have actually fallen to the ground are known as *meteorites*, those that burn up in the Earth's atmosphere are called *meteors*, and the bodies still in space are termed *meteoroids*.

Figure 5.6. The Barringer Crater near Winslow, Arizona. The volcanic area of the San Francisco Peaks, near Flagstaff, Arizona, is visible on the horizon, at a distance of 70 km. The crater measures 1.2 km across, has a depth of 180 m, and its rim rises 60 m above the surrounding plain. It was formed about 25,000 years ago by the impact of an iron meteorite estimated to have weighed several thousand tons.

*Figure 5.*7. The Tunguska region of Siberia, showing the destruction wrought by the violent explosion of a meteorite on June 30, 1908. The photograph was taken a number of years after the explosion.

asteroids and comets are thought to be part of the original material from which the terrestrial planets and the cores of the Jovian planets were assembled, analysis of their chemical and structural makeup is expected to reveal much about the early history of the Solar System.

There exist two fundamentally distinct types of meteorites: meteorites that have changed very little since their formation and meteorites that have experienced major alterations. Meteorites that have changed very little are said to be *undifferentiated* or *primitive*, and they have compositions resembling that of the Sun, except for a paucity in the volatiles (such as the noble gases and compounds containing H, C, N, and O). Among these meteorites astronomers recognize two subspecies—the *ordinary chondrites* and the *carbonaceous chondrites*. Meteorites that have experienced major alterations are called *differentiated* meteorites. Among them astronomers recognize three subspecies—the *achondrites*,* the *stony-iron*, and the *iron* meteorites. The main features of these five kinds of meteorites are summarized in table 5.2.

Planetesimals

The composition of the meteorites indicates that they—or their parent bodies, the *planetesimals*—formed by the coalescence of fine- and coarse-grained mineral matter, metals, and gases of various kinds, including organic compounds (see figure 5.8). This coalescence probably took place in the *protoplanetary disk*, which theory suggests surrounded the nascent Sun and initially was of the same chemical composition as the material that went into the making of the Sun.

Two different mechanisms are believed to have been at work in the formation of planetesimals. One mechanism produced the undifferentiated planetesimals, while the other mechanism produced the differentiated planetesimals:

1. *Origin of the undifferentiated planetesimals (ordinary and carbonaceous chondrites).* The undifferentiated planetesimals were bodies that never grew much larger than a few ten kilometers across, which was too small for internal melting and differentiation by gravitational forces. However, these planetesimals acquired rather different compositions, depending on where in the protoplanetary disk they formed—compositions resembling those of ordinary chondrites in the inner part of the protoplanetary disk to compositions resembling those of carbonaceous chondrites in the middle and outer parts. These differences in composition resulted from the differential heating the protoplanetary disk experienced, with temperatures ranging from above 1500 K near the young or still forming Sun to below 100 K near the outer edge of the disk. The matter near the inner part of the disk, which was exposed to the highest temperatures, lost all but the most refractory minerals.† Only the calcium–aluminum–titanium (Ca–Al–Ti) inclusions found

---

*The chondrites and achondrites together are often referred to as *stony meteorites*.

†The term *refractory* refers to elements or compounds that vaporize only at high temperatures (generally above 500 K), such as the silicate minerals and metals. The term *volatile* means the opposite of refractory. It refers to elements or compounds that vaporize at low or moderate temperatures, such as the noble gases (He, Ne, Ar, Kr, Xe, Rn), H_2O, molecular nitrogen (N_2), carbon monoxide (CO), carbon dioxide (CO_2), and organic compounds.

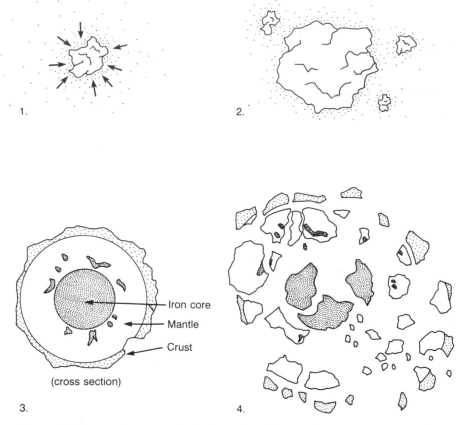

Figure 5.8. Four major stages in the formation history of stony, stony-iron, and iron meteorites.

1. Dust grains in the protoplanetary disk collided and coalesced first into pebble-sized and then into boulder-sized and larger objects.

2. Additional collisions with dust grains and with other larger objects led to the growth of planetesimals. The force holding planetesimals together was mainly gravity.

3. After a planetesimal had grown to several ten kilometers in size, heating by the decay of radioactive isotopes (such as ^{26}Al) melted the interior and led to differentiation. Iron, nickel, sulfur, and other heavy compounds sank toward the center to form a core. Lighter compounds (mostly silicate rocks and volatiles) rose to form a mantle and crust.

4. Violent collisions with other large planetesimals caused shattering and produced fragments containing stone, stone mixed with iron, and iron–nickel. The larger of these fragments are called asteroids when they are in orbit. When they fall to Earth, we call them stony, stony-iron, and iron meteorites.

in chondrites (see figure 5.9) and a few other silicates survived there. Matter more distant from the Sun, which was exposed to somewhat lower temperatures, was able to retain the less refractory minerals as well as some of the volatiles, including carbonaceous compounds. Finally, matter in the outer regions of the disk, which never became heated much at all, retained all elements (with the possible exception of the noble gases), including ices of water, ammonia, and methane.

2. *Origin of the differentiated planetesimals (achondrites, stony-iron, and iron meteorites)*. The bodies that gave rise to the differentiated meteorites grew initially by the same process that formed the undifferentiated planetesimals. However, in time they grew to much larger sizes and they became heated to the point of partial melting. The densest of the elements—mostly iron and nickel—sank toward the centers of the bodies, while the lighter materials—the silicates and volatiles—rose to the outer layers. Thus, the bodies became differentiated into iron-nickel cores surrounded by mantles

Figure 5.9. A large fragment from the Allende meteorite shower, which fell over a 10- × 50-km area near the small village of Pueblito de Allende in north-central Mexico, on February 8, 1969. The Allende meteorite fragments are classified as carbonaceous chondrites. The near end of this sample was broken off to expose unaltered meteoritic material, revealing mineral inclusions and chondrules (one to a few millimeters across) embedded in a matrix of fine-grained, dark gray, and relatively dense (3.67 g/cm^3) carbonaceous matter.

whose compositions varied from mixtures of iron and stone to pure stone. Subsequent collisions or rotational instabilities broke the differentiated bodies into fragments and created the achondrites and stony-iron and iron meteorites. (The same mechanism of differentiation operated in all of the terrestrial planets and larger satellites.)

The variation in composition acquired by the planetesimals due to the temperature gradient in the protoplanetary disk is still in evidence today among the asteroids and comets. For example, asteroids near the inner edge of the Asteroid Belt (at distances of about 2 AU from the Sun) generally consist of rocks (comparable to ordinary chondrites), rocks intermixed with iron (stony-iron meteorites), and iron–nickel (iron meteorites). The densities of the iron asteroids are 7.5 to 8.0 g/cm^3, while those of the rocky and stony-iron asteroids range from 3.0 to 6.0 g/cm^3, depending on the amount of iron they contain. Many of the asteroids in the middle of the belt (at about 3 AU from the Sun) are carbonaceous chondrites, with densities of 3 g/cm^3 or somewhat less, though there are still some stony and stony-iron as well as an occasional iron asteroid among them. In the outer belt and beyond (that is, beyond 4 AU), virtually all of the asteroids are of the carbonaceous type. Many are reddish in color, indicating increased admixtures of organic matter, including amino acids. Finally, the space beyond the planets is occupied by comets, which consist of rocks intermixed with ices of H_2O, NH_3, and CH_4 as well as many other volatiles. They have densities of about 1 to 2 g/cm^3.*

Deviations from this general pattern occur. Objects of cometary origin can be found close to the Sun, and stony or stony-iron asteroids exist beyond the Asteroid Belt. These exceptional cases are thought to be the result of orbital perturbations that the bodies experienced by past collisions among themselves, by gravitational interactions with Jupiter and the other planets, or, in the case of the comets, by gravitational interactions with passing stars.

What could have been the sources of heat that were responsible for the melting of the larger planetesimals? There is at present no sure answer to this question, though it is known that the gravitational energy released during the formation of the planetesimals was far too insignificant to have caused melting of bodies of only a few or even several hundred kilometers in size.

However, one source of heat that may have been very important in the melting of at least some of the planetesimals was the decay of the radioactive isotope aluminum-26 (^{26}Al) into magnesium-26 (^{26}Mg). The telltale sign of this decay is a slight excess of ^{26}Mg, relative to its abundance in solar material, in the Ca–Al–Ti inclusions of ordinary chondrites. Presumably, the excess ^{26}Mg was not incorporated into the inclusions as magnesium, but as ^{26}Al, which subsequently decayed into ^{26}Mg and released energy (see box 5.1 for details concerning the decay of unstable isotopes).

---

*Information about the composition of interplanetary bodies is obtained by analyzing the sunlight reflected from them. For example, asteroids that resemble ordinary chondrites and stony-iron meteorites are quite light in color; carbonaceous chondrites are very dark, mainly because of the tarry compounds they contain; and carbonaceous chondrites that are particularly rich in organic matter tend to be reddish, in addition to being dark, probably due to the presence of amino acids. In the case of comets, the analysis is made of the sunlight reflected from their comas and tails as they pass through the inner part of the Solar System.

The ^{26}Al must have come to the protoplanetary material within a few million years of being synthesized, as we can infer from this isotope's relatively short halflife of 720,000 years. Had ^{26}Al been synthesized much earlier, most of it would have decayed into ^{26}Mg and given off its energy before it ever became incorporated into planetesimals.

The only known sources of ^{26}Al are red giants, novae, and supernovae. We do not know which of these potential sources actually supplied the ^{26}Al to the protosolar material. However, Alastair G. W. Cameron of Harvard and James W. Truran of the University of Illinois made the interesting suggestion that if the ^{26}Al came from a supernova—perhaps one that exploded within a few dozen light-years from the interstellar gas cloud out of which the Solar System formed—the shock wave from the explosion may have been the trigger of the cloud's collapse.

Age of the Solar System

The decay products of long-lived unstable isotopes in meteorites allow us to deduce yet another important characteristic about the Solar System, namely its age. For example, potassium-40 (^{40}K) decays into argon-40 (^{40}Ar) with a halflife of 1.3×10^9 years. The amount of ^{40}Ar present in chondrites indicates that ^{40}K has been decaying in these meteorites for the past 4.5 to 4.6 billion years. Because the chondrites have experienced no changes in their composition other than by radioactive decay of unstable isotopes, this must be the age of these meteorites. It is also regarded to be the age of the Sun and the planets, for it is believed that all of these bodies formed together out of the same interstellar cloud. In addition, similar ages are obtained from the decay products present in some lunar rocks (figure 5.10) as well as from theoretical model studies of the Sun. (In comparison, minerals and rocks from geologically active planets like the Earth, in which the composition is continually altered physically and chemically, give widely varying ages, from more than 4 billion years to very recent ages.)

Figure 5.10. Lunar rock, collected by the *Apollo 17* astronauts. The age of this rock, as determined by comparing traces of the long-lived radioactive element rubidium-87 and its decay product strontium-87, is estimated to lie between 4.5 and 4.6 billion years. It is among the oldest known rocks of the Solar System and is believed to date back to the time of its formation.

BOX 5.1.
Decay of Radioactive Isotopes

IN 1896, the French physicist A. Henri Becquerel (1852–1908) performed a series of simple, yet significant experiments. He was intrigued by the fact that X rays, which had been discovered in the previous year by the German physicist Wilhelm C. Röntgen (1845–1923), are capable of inducing fluorescence in nearby materials.* He thought that there might exist some fundamental relation between fluorescence and the emission of X rays, and he set out to investigate this possibility.

Because some compounds of uranium fluoresce when exposed to sunlight, Becquerel began by concentrating on them. He placed samples of the compound on photographic plates wrapped in black paper and let sunlight shine on them. Upon developing the plates, he noticed that they were blackened. Then, after a succession of cloudy days, during which there was insufficient sunlight to induce fluorescence, he developed a photographic plate that had been left lying near some of the uranium. Much to his surprise, the plate turned out to be blackened, too, with the image of the uranium crystal clearly standing out. When he placed other, nonfluorescent compounds of uranium, including the metal itself, upon photographic plates, they were blackened as well. Clearly, the blackening had nothing to do with the fluorescence, but was a property of uranium itself.

During the subsequent years, Becquerel and others—among them Marie Curie (1867–1934), Pierre Curie (1859–1906), Ernest Rutherford (1871–1937), and Paul Villard (1860–1934)—discovered that many other substances have properties similar to that of uranium and emit "radiation." Three kinds of radiation were recognized and labeled *alpha-, beta-, and gamma-radiation* after the first three letters of the Greek alphabet. Within about a dozen years of Becquerel's first experiments with uranium, the alpha-radiation was identified as streams of ^{4}He nuclei (alpha-particles), beta-radiation as streams of elec-

trons or positrons (beta-particles), and gamma-radiation as short-wavelength electromagnetic radiation. Furthermore, it was discovered that these particles and electromagnetic radiation originate when certain elementary particles or nuclei change spontaneously into other kinds of particles by processes now called *radioactive decay* or *nuclear fission*. Thus, the modern nuclear age was born.

The simplest example of radioactive decay is the decay of free neutrons. These particles are not stable, but decay spontaneously into protons, with the simultaneous emission of electrons and neutrinos:

$$n \rightarrow p^+ + e^- + \nu.$$

Because this decay emits electrons, it is called *beta-decay*. Another example of radioactive decay is the spontaneous transformation of uranium-238 (^{238}U) into thorium-234 (^{234}Th), with the simultaneous emission of alpha-particles:

$$^{238}_{92}U \rightarrow ^{234}_{90}Th + ^4_2He.$$

This kind of decay is known as *alpha-decay*. Incidentally, the product of this decay, ^{234}Th, is itself radioactive and decays in a sequence of steps to lead-206 (^{206}Pb), which is stable.

The decays of neutrons and of ^{238}U take place at very different rates. For example, if we could contain a number of neutrons, we would find that after approximately 13 minutes half of them would have undergone beta-decay, leaving only half of the original number of neutrons. After another 13 minutes, half of those would have decayed, leaving only one-quarter of the original. And during every succeeding 13-minute interval, the number of neutrons remaining would again be reduced by a factor of one-half. The process stops only after all neutrons are gone and have been converted into protons, electrons, and neutrinos. If we started with a number of ^{238}U atoms, however, we would have to wait 4.5 billion years before half of them decayed into ^{234}Th. Yet another 4.5 billion years would go by before half of the remaining ones would decay, and so forth. The two time intervals—13 minutes and 4.5×10^9 years—are called the *halflives* of neutrons and ^{238}U, respectively. Every unstable elementary particle and unstable atomic nucleus has a characteristic mode of decay and halflife, and in table 5.4 these characteristics are summarized

---

Fluorescence refers to the emission of light by certain substances when they are exposed to some energy source. The most common example of this phenomenon is the fluorescent lamp, which emits light when an electric current (the energy source) is sent through the gas inside the glass tube. Other examples are the dials of many alarm clocks, which keep glowing for hours after they have been exposed to light.

BOX 5.1 *continued*

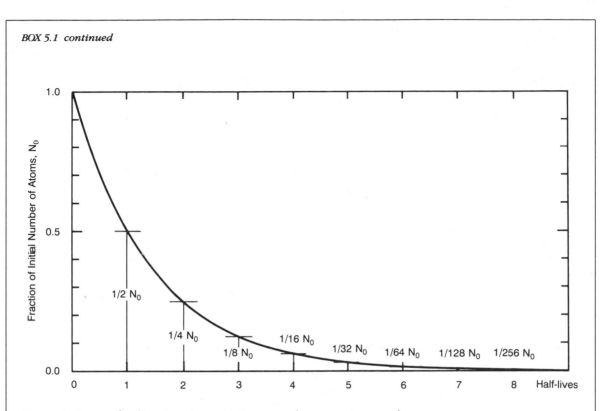

Figure A. Decay of radioactive atoms. At time zero, there is a given number of radioactive atoms, N_0. The atoms decay into their offspring products at rates such that after one halflife, half the N_0 atoms remain; after two halflives one-quarter of the N_0 atoms remain; and so forth.

for those radioactive isotopes that are of greatest interest to us.

Radioactive decay is illustrated in figure A. In particular, the figure shows the fraction of the initial number of radioactive atoms, N_0, remaining after one, two, three, and so forth halflives. The decay is a probabilistic phenomenon. It is not possible to predict exactly *which* atoms will decay during a given time interval. One can predict only *how many* will decay during a given time interval (provided that the number of atoms is large, so that the laws of probability apply). Initially, when there are still many radioactive atoms present, the curve drops steeply (that is, many atoms decay per unit time). Later when the number of radioactive atoms has diminished, the curve flattens (that is, fewer atoms decay per unit time). From the fraction of radioactive atoms remaining and a knowledge of the halflife one can calculate how long the decay has been going on.

The obvious question now is, "Why are some particles and nuclei—such as electrons, protons, alpha-particles, and the nuclei of ^{12}C and ^{16}O—stable, while others are not?" The answer is, in principle, quite simple. Stable particles and nuclei are in their lowest possible energy states and, therefore, last forever unless perturbed by outside forces (for example, by forces arising from collisions with other particles traveling at high speeds). Unstable particles, in contrast, are in higher energy states and it is only a matter of time—as indicated by their halflives—before they decay into lower, more stable energy states. The emission of electrons, positrons, alpha-particles, and gamma rays during such decays constitutes the energy difference between the initial, higher energy states (those of the unstable particles) and the final, lower energy states (those of the stable particles).

For example, the energy liberated by the decay of ^{26}Al into ^{26}Mg is thought to have caused the internal

BOX 5.1 continued

melting of the larger planetesimals during the formative stages of the Solar System. The decays of the long-lived radioactive isotopes ^{40}K, ^{232}Th, ^{235}U, and ^{238}U were and still are responsible for much of the heat content of the intermediate-sized planetary bodies. In the case of the Earth today, these sources of energy make up roughly 80% of the heat that flows from the crust to the surface (the geothermal energy); only 20% comes from stored-up, primordial heat.

Radioactive decay may also affect us in more direct ways than by contributing to the Earth's heat budget. The particles and gamma-rays emitted by radioactive materials may do damage to human tissues, cause cancer, and induce genetic mutations. But, when properly applied by trained medical personnel, they may also have beneficial uses, as in the treatment of cancer and other tumors.

The decay of radioactive isotopes has yet another important application. It allows geologists and archaeologists to deduce the ages of rocks and organic matter. For example, the decay of potassium-40 (^{40}K) into argon-40 (^{40}Ar) may be used to determine the ages of chondrites, which are meteorites that have not changed since their origin during the birth of the Solar System. This is done by assuming that when these meteorites formed the rock crystals contained only ^{40}K, but no ^{40}Ar. This is a reasonable assumption because ^{40}Ar is a noble gas, which does not readily combine with other elements and, hence, would not have become incorporated into meteorites in significant amounts except by the decay of ^{40}K. Next, the relative abundance of ^{40}K and ^{40}Ar is measured in chondrites today and, from the known rate of decay of the former into the latter, it is a simple matter to compute the chondrites' ages. The decay of the other long-lived radioactive isotopes listed in table 5.4 may

be used likewise in estimating the ages of chondrites and is useful for obtaining independent checks. The same isotopes may also be used to date terrestrial rocks. The ages so obtained refer only to the epochs when the rocks last experienced melting or other compositional changes, however; they do not tell us the age of the Earth. (The same is true if the method is applied to meteorites that have undergone compositional changes since their formation.)

The dating methods just described, which are based on the decay of long-lived radioactive isotopes, are useful only for estimating ages of hundreds of thousands of years or longer because over shorter time spans too few of the isotopes decay to yield dependable measurements. To estimate ages of shorter duration, radioactive isotopes that decay more rapidly must be used. One such short-lived radioactive isotope is carbon-14 (^{14}C). This isotope is continuously created in the Earth's upper atmosphere as a result of bombardment of nitrogen by neutrons. The newly synthesized ^{14}C reacts with atmospheric oxygen to form CO_2 and is subsequently incorporated into organic matter through photosynthesis. By this sequence of processes, approximately one out of every 10^{12} carbon atoms in living matter turns out to be ^{14}C rather than ^{12}C. Because ^{14}C is unstable, its presence in dead organic matter (matter that no longer takes up fresh ^{14}C from the food chain) diminishes with time. By comparing the relative abundance of ^{14}C in a piece of fossil bone, cloth, leather, wood, or the like with that in living matter, it is possible to obtain age estimates of such objects. This *radiocarbon dating method* works well for time intervals of up to 50,000 years. In objects older than that, too little ^{14}C remains for accurate measurements, and other dating techniques need to be used.

INTERMEDIATE-SIZED PLANETARY BODIES: INTERNAL STRUCTURE

The family of intermediate-sized planetary bodies includes Mercury, Venus, Earth, Mars, the Moon, the Galilean satellites, Titan, and Triton. These bodies have similar sizes, masses, and other characteristics, as noted earlier and shown in tables 4.1 and 4.2. They also have similar, though not identical, structures. The differences that do exist between them can be explained in terms of their masses, their different distances from the Sun or, in the case of the Galilean satellites, from Jupiter, and the kinds of materials that were available for their assembly at those distances. Let us examine the compositions and physical structures of these bodies, starting with the terrestrial planets.

The single most meaningful parameter concerning the composition of a planetary body is its density. A body made entirely of iron has a density of about 7.5 g/cm^3; one made mainly of silicate rocks has a density of about 3.0 to 3.5 g/cm^3; one made of rocks plus some admixture of volatiles has a density of between 2.0 and 3.0 g/cm^3; and one containing mostly ices has a density between 1.0 and 2.0 g/cm^3. (In comparison, basalt and granite, the major constituents of the Earth's rocky crust, have density ranges from 2.7 to 3.4 and 2.5 to 2.8 g/cm^3, respectively.)

Care must be taken in using density as an indicator of a celestial body's internal constitution. If a body's radius exceeds a few thousand kilometers, its self-gravity becomes so strong that even rocks and metals become appreciably compressed. In such cases it is not the actual density that is meaningful, but the body's *uncompressed* density, namely the density it would have if it were not compressed by self-gravity.

Terrestrial Planets

The distances from the Sun, radii, and actual and uncompressed densities of the terrestrial planets are shown in the following table.

	Distance From Sun (10^6 km)	Equatorial Radius (km)	Actual Density (g/cm^3)	Uncompressed Density (g/cm^3, approximate)
Mercury	58	2440	5.4	5.4
Venus	108	6050	5.3	4.3
Earth	150	6380	5.5	4.2
Mars	220	3400	3.9	3.8

The relatively high values of their uncompressed densities suggest that the terrestrial planets are not made of rocks alone, but must contain substantial amounts of iron and other high-density elements. Furthermore, there is a general trend of decreasing values with increasing distance from the Sun. Mercury, the planet closest to the Sun, has the highest uncompressed density and, hence, contains the greatest proportion of iron, in addition to rocks. Venus, Earth, and Mars have progressively lower densities and, hence, contain less iron and more rocks and volatiles. This is the same pattern observed in the case of the asteroids and comets, and it is yet another indication of the existence of a protoplanetary disk that became differentially heated. Near the Sun the temperatures rose to such high values that only the metals and most refractory minerals survived and became incorporated into the planets. With increasing distance from the Sun the temperatures stayed progressively lower and greater

Table 5.4. Decay Modes and Halflives of Some Important Radioactive Isotopes.

Radioisotope	Decay mode	Decay reaction	Halflife (years)	Applications or effects
Carbon-14	β-decay	$^{14}_{6}C \rightarrow \,^{14}_{7}N + e^- + \nu$	5730	Dating of organic matter.
Aluminum-26	β-decay	$^{26}_{13}Al \rightarrow \,^{26}_{12}Mg + e^+ + \nu$	7.2×10^5	Heating and melting of interiors of planetesimals.
Potassium-40	β-decay e-capture[a]	$^{40}_{19}K \rightarrow \,^{40}_{20}Ca + e^- + \nu$ $^{40}_{19}K + e^- \rightarrow \,^{40}_{18}Ar + \nu$	1.3×10^9	Heating and melting of interiors of the intermedi- ate-sized planetary bodies,
Rubidium-87	β-decay	$^{87}_{37}Rb \rightarrow \,^{87}_{38}Sr + e^- + \nu$	5.0×10^{10}	including Earth.
Thorium-232	α-decay	$^{232}_{90}Th \rightarrow \,^{228}_{88}Ra + \,^{4}_{2}He \;(\rightarrow \,^{208}_{82}Pb)^b$	1.4×10^{10}	Dating of rocks and meteorites.
Uranium-235	α-decay	$^{235}_{92}U \rightarrow \,^{231}_{90}Th + \,^{4}_{2}He \;(\rightarrow \,^{207}_{82}Pb)^b$	7.1×10^8	
Uranium-238	α-decay	$^{238}_{92}U \rightarrow \,^{234}_{90}Th + \,^{4}_{2}He \;(\rightarrow \,^{206}_{82}Pb)^b$	4.5×10^9	

[a]Electron-capture is a decay mode (also called inverse β-decay), in which one of the nuclear protons captures an orbital electron and changes into a neutron.
[b]The lead isotopes are the final products, reached by decay reactions beyond those shown explicitly.

amounts of volatiles survived, giving the more distant planets their lower densities.

The terrestrial planets are not homogeneous bodies of metals, rocks and volatiles. They are differentiated into high-density cores, mainly of iron and nickel, surrounded by rocky mantles of intermediate densities and surface crusts of low-density rocks and volatiles. These differentiations took place early in the planets' formation by mechanisms similar to those that differentiated the larger planetesimals, though the release of energy from gravitational coalescence and from the decay of long-lived radioactive isotopes was much more severe. Figure 5.11 depicts the internal structure of the Earth, which is typical (but not identical in details) of all terrestrial planets.

Moon

The Moon is an oddity among the intermediate-sized planetary bodies. Because it is close to Earth and, presumably, was formed nearby, its density might be expected to be similar to the uncompressed density of Earth, about 4.2 g/cm^3. Instead it is 3.3 g/cm^3, lower even than the density of Mars. This indicates that the Moon has either no iron core at all or only a very small one.

There are other anomalies about the Moon's makeup. Its rocks are depleted in the volatiles (H, C, N, O, and the noble gases) relative to Earth. For example, none of the lunar rocks brought back by the Apollo astronauts shows any trace of water, whereas terrestrial igneous rocks (like basalt and granite) generally contain between 0.2 and 0.5% water. The lunar rocks are also depleted in iron and all elements that tend to combine with iron—the "siderophiles" or "iron-loving" elements—such as nickel, cobalt, rubidium, iridium, and platinum.

Clearly, an explanation is required why the Moon has no substantial iron core and why its rocks are depleted in the iron and siderophiles that once must have been part of the material from which it formed. John A. O'Keefe, a NASA

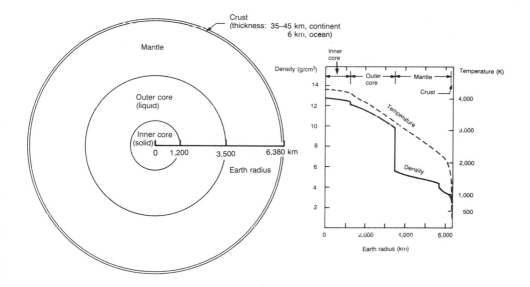

Figure 5.11. Model of the structure of the Earth. On the left are the major components of the Earth: the solid inner core and molten outer core, the mantle, and the crust (the latter is not drawn to scale). Both cores consist largely of iron and some nickel; the mantle mainly of iron and magnesium silicates; and the crust mainly of basaltic and granitic rocks. The graph on the right shows the approximate distribution of density (from 12.6 to 2.6 g/cm³) and temperature (from 4300 to 280 K). The density profile displays marked discontinuities at the edges between the inner and outer cores, between the outer core and the mantle, and within the mantle at a depth of about 670 km.

astronomer and geophysicist, gives a simple and rather compelling answer (personal communication 1979): "The Moon's iron and siderophiles are in the core of the Earth." He suggests that the material that today comprises the Moon was at one time part of a rapidly spinning Earth. After the Earth's iron and siderophiles had sunk toward its center to form the terrestrial iron core, the outer layers were torn off by the strong centrifugal forces resulting from the rapid rotation. Others have suggested that the Earth's outer layers were thrown into space by the impact of a very large (perhaps Mars-sized) planetesimal. However the layers were ejected, part of them—depleted in iron, siderophiles, and volatiles (presumably, many of the volatiles were outgassed and lost due to tidal friction and heating)—formed the Moon, and the rest escaped into interplanetary space.

These theories of the origin of the Moon by fission from Earth, originally proposed by Sir George H. Darwin (1845–1912, son of Charles Darwin), or by the impact of a large planetesimal, are not without some difficulties. In particular, they require that either the young Earth rotated unusually rapidly or that it was hit by an uncommonly large planetesimal. Hence, some planetary scientists prefer the *accretion* theory. According to it, the Moon accreted from material that was in Earth orbit from the outset, rather than part of the Earth. However, this theory cannot explain the Moon's many composition anomalies.

Therefore, in my opinion, further research will eventually prove either the fission or the impact theory to be the correct one. (The theory that the Moon formed somewhere else in the Solar System, perhaps beyond the orbit of Mars, and was captured by the Earth is now generally discounted.)

Galilean Satellites

The densities shown in table 4.2 for the Galilean satellites of Jupiter indicate that both Io and Europa have largely rocky structures and, possibly, small iron cores. Photographs show that Io's surface is rocky and covered by sulfur-containing compounds, while Europa is enveloped by a crust of water ice. The low densities of Ganymede and Callisto suggest that they contain substantial amounts of water (perhaps as much as 50% by mass), the rest being rock. They probably have rocky central cores that are overlain by thick mantles of water, either in liquid or solid (ice) forms. On top of the mantles float hard and rigid crusts of water ice with a thin covering of rock debris from meteorite impacts.

The decrease in density in the Galilean satellites with increasing distance from Jupiter follows the same pattern noted for the smaller bodies and the terrestrial planets. It suggests that Jupiter was at one time surrounded by its own protoplanetary—or protosatellitic—disk, that became heated as a result of the planet's formation. For reasons similar to the ones discussed above, the ensuing temperature gradient was probably responsible for the composition and density differences observed in the Galilean satellites today. Thus, Galileo's suggestion that Jupiter and its four giant satellites are a miniature Solar System was correct beyond the conditions he could envision.

Figure 5.12. Oblique view of intercrater plains on Mercury. Numerous large and small impact craters, many with central peaks, alternate with terrain of a gently rolling nature. This terrain is believed to be the result of planetwide surface melting during Mercury's final accretion phase.

Titan and Triton

The last two planetary bodies to be discussed are Saturn's Titan and Neptune's Triton. The data in table 4.2 show that Titan resembles Ganymede in size, distance from its central planet, and density. Hence, astronomers expect that its structure, too, is similar to Ganymede's, though they are not sure. But they do know that Titan has a substantial atmosphere, consisting of 80 to 95% molecular nitrogen, with the rest being mainly CH_4 and traces of H_2 and organic molecules. It is believed that these gases are the result of evaporation of ices of H_2O, CH_4, and NH_3 from the satellite's surface, followed by chemical reactions.

Our knowledge of Triton is even more limited than that of Titan, and it will remain so until *Voyager 2* makes its rendezvous with Neptune in 1989. We know that Triton is roughly of the same size as Europa, but its density is smaller. Consequently, ices of H_2O, CH_4, and NH_3 are probably major constituents of this satellite, too. The presence of CH_4 in Triton's thin atmosphere is further evidence of this supposition.

INTERMEDIATE-SIZED PLANETARY BODIES: SURFACE FEATURES

Descriptions of the surface features of the intermediate-sized planetary bodies require selectivity because two decades of space exploration have amassed far more information than can possibly be covered here. Hence, the surfaces of each of the bodies will not be described separately. Instead, they will be compared with each other, paying attention to four topics: (1) impact craters, (2) geological activities, (3) presence of water, and (4) atmospheres. This comparative coverage is consistent with grouping the intermediate-sized planetary bodies into a single family and with the modern trend of regarding the forces that shaped them as part of a continuum.

Impact Craters

The most heavily cratered intermediate-sized planetary bodies are the Moon, Mercury, Callisto, and Ganymede, as well as the southern hemisphere of Mars. Earth, Venus, the northern hemisphere of Mars, and Europa show relatively few impact craters. Io has none at all. Titan and Triton may be cratered similarly to Ganymede and Callisto, but at present they are not well enough explored for us to be sure.

On Callisto, Ganymede, and the highlands of the Moon and Mercury, the crater density is close to the saturation point, with craters edge-to-edge or overlapping and larger craters having smaller ones within them. Some of these impact craters are truly gigantic as figures 5.12–5.15 and figure 5.28 demonstrate. Others are much smaller, and lunar rocks show that they exist even on the microscopic level (see figure 5.16). Very little geological or weathering activity has disturbed these heavily cratered regions, and therefore, the cratering record, much of which dates back nearly 4 billion years, has been preserved.

Figure 5.13. View of the Caloris Basin of Mercury, situated near the planet's terminator in this Mariner 10 mosaic. This enormous impact feature is 1300 km in diameter and was formed about 3.7 billion years ago. It is surrounded by a pattern of concentric rings of mountains, interspersed with lava plains. The name *Caloris* is derived from the Latin *calor,* meaning "heat," and refers to the fact that the basin lies near one of Mercury's "hot poles," which alternately face the Sun when the planet is at its closest distance from the Sun.

The Moon and the other heavily cratered bodies are, however, not without some evidence of past geological activity. For example, from about 3.3 to 2.5 billion years ago, lava flows filled in most of the large impact basins on the Moon and formed the dark colored and relatively smooth "seas" or *maria.* Similar, though somewhat more ancient, lava flows have created the intercrater plains on Mercury.

At the other extreme with regard to crater density lies Io, geologically the most active body in the Solar System. Its volcanic ejectae and lava flows erase craters as rapidly as they form. On Earth, geological activity is somewhat less severe but is still sufficiently dynamic that only a few impact features older than about 500 million years remain. The same is true of Europa, though on this body it is not the movement of rock that obscures impact craters, but the flow of ice. Mars is interesting. Its heavily cratered southern hemisphere also contains intercrater plains like those on the Moon and Mercury. However, its northern hemisphere is dominated by volcanoes and other features of geological activity, which have erased most of the older impact craters.

The collective record of impact craters on the intermediate-sized planetary bodies indicates that up to approximately 3.8 billion years ago all of them were

Figure 5.14. Oblique view of lunar craters Eratosthenes (right of center) and Copernicus (on horizon, at left), and of Sinus Aestuum (foreground) and Mare Imbrium (upper right). Eratosthenes has a diameter of about 60 km. It was formed approximately 3.4 to 3.2 billion years ago, probably by the impact of a large planetesimal. However, the absence of obvious ejection features, such as rayed ejection patterns, has prompted the suggestion that it may be a caldera, the collapsed center of a former volcano.

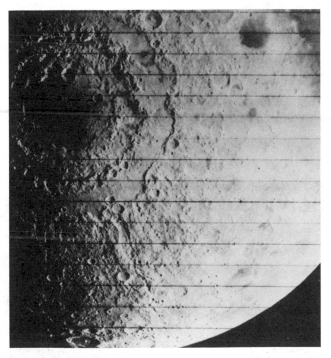

Figure 5.15. Mare Orientale, as viewed by *Lunar Orbiter 4* in 1967. This gigantic impact feature, with a central crater and surrounding multi-ring structures, was formed when an object perhaps 25 km across struck the Moon roughly 4 billion years ago. The outermost ring of mountains, named *Cordillera Mountains*, is nearly 900 km in diameter and rises about 6 km above the surrounding plains.

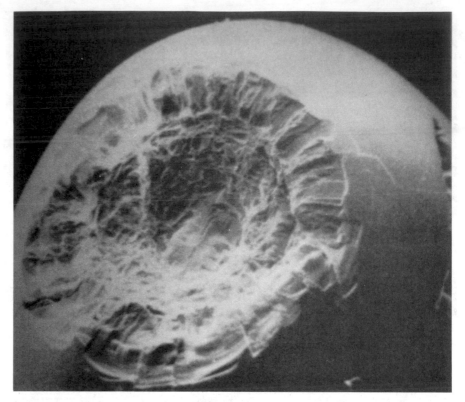

Figure 5.16. Microscopic crater on a spherule of glass, from a lunar rock sample. The crater is less than 1 mm in size and was carved out by the impact of a tiny particle of cosmic dust. The crater exhibits a multi-ring pattern of fractures similar to that found in lunar impact craters many kilometers in size.

heavily bombarded by large and small planetesimals (see figure 5.17). Most likely, this bombardment constituted the terminal phase of the accretion that created these bodies in the first place. From about 3.8 billion years forward, the bombardment declined rapidly; though even today it has not yet reached the zero point, as noted in the discussion of meteorite falls on Earth.

Geological Activity

Geological* activity on planetary bodies is caused by surface forces that are sufficiently strong to bring about movements in the crust. The forces may be in the horizontal or vertical directions or both, and they may be pushing, pulling, twisting, contracting, or lifting kinds of forces. They usually are the result of gravity, convective motions inside the body, weathering activities on the surface,

---

*The term *geological* is derived from the Greek root *geo-*, meaning "earth" or "land," and, strictly speaking, refers only to the Earth. In particular, it refers to changes in the Earth's surface features due to forces from within and from outside, as revealed by the rock record. However, with the development of the science of planetology in recent years, the term is now also used to refer to changes experienced by the other planetary bodies with solid surfaces.

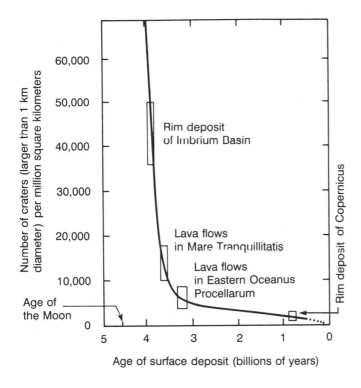

Figure 5.17. Density of impact craters at four different regions of the lunar surface plotted against the regions' ages. The slope of the curve drawn through the data rectangles shows that the rate of massive impact cratering diminished rapidly between 3.8 and 3.3 billion years ago. It was followed by a lower and much more slowly diminishing rate of impact cratering that continues to this date. (Adapted from Beatty et al. 1982, 39.)

and tidal interactions with another nearby body. The forces are most likely to produce movement if the surface crust is thin and weak, and if the interior is hot and in a molten or plastic state. The interior then acts as a lubricant that enhances movement. In contrast, if the surface crust is thick and strong and if there is little or no melting inside, the surface forces are unlikely to bring about movement. The body is then geologically inert.

The major sources of heat for melting the interior of the intermediate-sized planetary bodies are the gravitational energy that was released during their formation and the energy liberated since then by the decay of long-lived unstable isotopes, such as uranium-235 and -238, thorium-232, and potassium-40 (see box 5.1). Both of these heat sources are most effective in the largest bodies, for their surface-to-volume ratios are relatively small (surface $\propto$ radius^2, volume $\propto$ radius^3; hence, surface-to-volume ratio $\propto$ 1/radius), so that they lose their internal heat slowly. Also, much more heat was released early in the bodies' histories than, for example, today because the stored-up heat from their formation and the amount of unstable isotopes were greatest then.

Io. Io is one of the few bodies in the Solar System whose geological activity is mainly due to tidal heating. This heating is brought about by gravitational interactions with nearby Jupiter, which creates an appreciable tidal bulge on Io. As Io revolves about Jupiter and alternately approaches and recedes from it due to a slight eccentricity in its orbit (the eccentricity is maintained by gravitational interaction with Europa), the tides raise and lower different parts of Io. This causes friction by rock rubbing against rock and produces enough heat to melt part of Io's interior. Unbalanced forces, probably also due to the tides, open up cracks in Io's crust and allow the molten materials of the interior, consisting largely of sulfur and sulfur-containing compounds, to push to the surface. There they gush in geyserlike fashion into the open, creating enormous lava flows and huge volcanoes, giving the satellite its mottled appearance (figures 5.18, 5.19). Unquestionably, Io deserves its reputation as the Solar System's strangest and geologically most active planetary body.

Earth. After Io, the geologically most active intermediate-sized planetary body is Earth. This is not surprising. It is the most massive body among them. Thus it has the greatest amount of stored-up internal heat, is molten to a greater extent than the others (except for Io, because of its tidal heating, and, possibly, Venus), and its outer solid layer, the lithosphere, is relatively thin. In

Figure 5.18. Io, the innermost of Jupiter's four Galilean satellites. Nearly all features seen on Io appear to be linked to volcanic activity, including lava flows and volcanic caldera up to 200 km across. The doughnut-shaped structure near the center of this photograph is the plume of an erupting volcano.

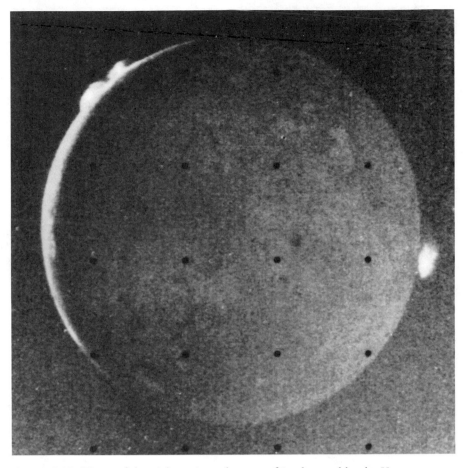

Figure 5.19. Three of the eight active volcanoes of Io, detected by the Voyager spacecrafts.

fact, the terrestrial lithosphere is a mere 100 km thick on the average, compared to a total radius of 6380 km.

The terrestrial lithosphere is composed of a number of distinct and rigid tectonic plates that float like rafts on the denser underlying asthenosphere. Forces within the Earth push and pull on the plates, slowly moving them about relative to each other. This pushing and pulling is called *plate tectonics* and is responsible for most volcanism (figure 5.20) and earthquakes, as well as for the building of mountains and the shaping of continents and ocean basins. (For a discussion of terrestrial plate tectonics see box 9.2.)

In addition to plate tectonics, the surface of the Earth is continuously altered by weathering. Flowing water, glaciers, and wind erode the land, and rivers carry the resulting debris toward the oceans, where they deposit it as sediments. Over time, the layers of sediments become compressed and solidified as more sediments are piled on top of them. When subsequent uplifting and erosion expose these ancient deposits, their banded structures reveal their mode of

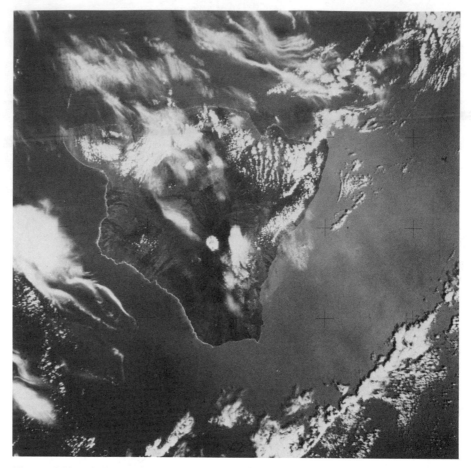

Figure 5.20. LANDSAT photograph of Mauna Kea (just above center in this photograph, 4205 m) and Mauna Loa (center, 4169 m), the two giant shield volcanoes of the Island of Hawaii. These two volcanoes rise roughly 9000 meters above the ocean floor and are 120 km wide at their base on the ocean floor. They were formed, along with the rest of the island, by the rising from the ocean floor of basaltic lavas and are probably the closest terrestrial analogues to the large volcanic mountains on Mars.

origin. Examples are the banded structures of the Canadian Rocky Mountains and the Grand Canyon.

Plate tectonics and weathering have been going on on Earth for at least 4 billion years. The oldest surviving rock formations of these processes are the *continental shields*, which are present on every continent and comprise about 10% of the Earth's surface (more about them in figure 7.3). These shields are the closest terrestrial analogues to the cratered highlands of the Moon, Mercury, and the southern hemisphere of Mars, though few craters remain.

Venus. Venus has nearly the same mass and density as Earth. Hence, we might expect that plate tectonics and volcanism were, and, perhaps, still are, also important in sculpturing its surface. Unfortunately, we do not yet know for certain if this is the case because the planet's surface remains hidden under a thick, opaque atmosphere. However, radar soundings bounced off Venus from Earth stations and orbiting spacecraft reveal a topography not unlike that of Earth and hence point to geological activity (figure 5.21). There are two massive highlands, Ishtar Terra and Aphrodite Terra, that resemble terrestrial continents. There also are several smaller islandlike regions and broad depressions comparable to the ocean basins on Earth. Most of these landforms are shallower than the contours on Earth, but not all. For example, Ishtar Terra contains a number of substantial mountain ranges. The highest of them, Maxwell Mons, rises 12,000 m above the planet's average radius and is comparable to the Himalayan Mountains.

Mars. Like Earth and Venus, Mars has had a complex and interesting geological history. Its southern hemisphere is heavily cratered and shows evidence of interspersed flooding by lava (see figures 5.22, 5.23). The northern hemisphere also contains craters and intercrater lava plains, but to a much smaller extent. Its most conspicuous features are enormous mountain ranges and canyon systems that dwarf their counterparts on Earth. The largest of the mountain ranges is a 6000-km-long uplift called the Tharsis Ridge that is punctuated by four gigantic volcanoes. The volcanoes bear a striking resemblance to the shield volcanoes on Earth, such as Mauna Kea and Mauna Loa of Hawaii. The largest of them, Olympus Mons (see figure 5.24), measures 600 km across at its base and rises 26,000 m above the surrounding terrain. To the east of the Tharsis Ridge lies a 4000-km-long system of interconnected canyons called Valles Marineris (figure 5.25). Some individual canyons are more than 200 km wide and 7000 m deep. They show modifications by landslides, wind erosion, and flowing water, mud, or, possibly, glaciers (though no liquid water exists on the Martian surface today; see below). In comparison, Earth's Grand Canyon is 450 km long, 30 km across at its widest point, and 2000 m deep.

The lava flows, mountains, and canyons of Mars indicate that, with regard to geological activity, this planet lies somewhere between the Earth and the Moon. After having been heavily cratered early in its history, its geology began to evolve like that of the Earth. However, before the entire planet was affected, its geological activity came to a halt. Most likely, the reasons were the relatively small mass and size of Mars, which, respectively, are one-tenth and one-half

Figure 5.21. The first television image of Venus's surface, made by the Soviet soft-landing spacecraft *Venera 9*, on October 22, 1975. The rocky landscape is strewn with boulders, most of them tens of centimeters in size.

those of Earth. Consequently, Mars was heated less initially; its crust became thicker than that of Earth; and its volcanism and tectonic activities stopped early in its history.

Europa. The only intermediate-sized planetary body besides Earth, Venus, and Mars that is suspected to have had recent and, maybe, current tectonic activity is Europa. The basis for this speculation is its lack of any large (over 100 km) and, hence, old craters. Furthermore, Europa's surface is marked by a vast tangle of light and dark streaks that are probably filled-in cracks and indicate the presence of tensional forces pushing and pulling on the crust (see figure 5.26). If tectonism does exist on Europa, the tectonic plates would not be

Figure 5.22. Mars, photographed by the *Viking 2* spacecraft from a distance of 418,000 km in 1976. Near the bottom of the photograph and just above the southern pole lies the Argyre Basin, covered here by thin frost. Near the middle of the planet's crescent and close to the terminator is the enormous canyon system of Valles Marineris. And near the top of the photograph lies the extinct volcano Ascraeus Mons (27,000 m), which is part of the Tharsis Ridge and responsible for the cloud streamers in its vicinity.

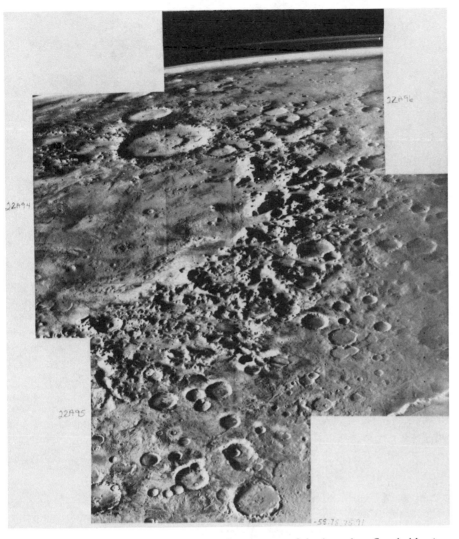

Figure 5.23. Oblique view across Argyre Planitia, one of the large lava-flooded basins of Mars. The basin, which was carved out by an asteroid impact, is bordered on the right by the old, cratered mountains of Chartium. Beyond it lies the crater Galle (210 km in diameter). Notice the thin haze of CO_2 above the Martian horizon.

rock, like those on Earth, but layers of water ice. The heat necessary for keeping the underlying rocky layers molten so that tectonic forces can produce crustal movement would largely come from tidal interactions with Jupiter, though to a much smaller degree than on Io.

Mercury, Moon, Ganymede, Callisto. The remaining intermediate-sized planetary bodies—Mercury, Earth's Moon, Ganymede (figure 5.27), and Callisto (figure 5.28)—are *one-plate planets* (too little is known of Titan and Triton to say anything about their geological activities). No large-scale tectonism deforms their surfaces today, nor is there much evidence of volcanism. However, the

Figure 5.24. Olympus Mons on Mars, the largest known volcano in the Solar System. Differences in crater densities around its flanks indicate that Olympus Mons grew over a considerable period of time.

presence of ancient intercrater lava plains on Mercury and the Moon does indicate that during the first half of their existence molten magma managed to penetrate from their interiors to the surface. Likewise, the rather complex system of grooves and ridges that crisscross Ganymede, with estimated ages of about 3.5 billion years, suggests that this body, too, has experienced some form of geological activity. It appears that by the time the Solar System was 2.5 billion years old, the crusts of all of these bodies began to harden into single, unyielding plates and their geological histories drew to a close.

Presence of Water

Water is present today on all intermediate-sized planetary bodies with the exception of Mercury, the Moon, and Io.

Mercury, Moon, Io. Mercury and Io have no water because of their proximities to the Sun and Jupiter, respectively, and the heating they experienced during their formation. The Moon is without water probably because of the way it formed.

Figure 5.25. View of the central section of Valles Marineris, the enormous canyon system just south of the Martian equator. Numerous huge landslides are visible on the north walls of the canyons, where they have created alcoves up to 100 km across (such as the landslide near the right edge and half way between top and bottom of this photo). Some of the larger landslides extend all the way across the canyon and ride part way up on the opposite wall.

Venus. Venus is also very dry. The little water it does possess exists as vapor in its atmosphere. However, at one time it may have had as much water as Earth, but lost most of it as a result of the evolution of its atmosphere. The evidence comes from the abnormally high ratio of deuterium to hydrogen (^{2}H:^{1}H) in the water that is still present. This discovery was made in 1978 with an instrumented capsule that descended through the Venusian atmosphere, after it was delivered to the planet by the United States spacecraft *Pioneer Venus Multiprobe*. The high ^{2}H:^{1}H ratio suggests that early in Venus's history UV radiation from the Sun broke the planet's water molecules into their atomic constituents, ^{1}H, ^{16}O, and an occasional ^{2}H. The oxygen reacted with other compounds and is now locked up in the crust of Venus. The heavy isotope of hydrogen, ^{2}H, also was retained, but most of the ^{1}H escaped into space because of its small mass. This raised the ^{2}H:^{1}H ratio in the remaining water to roughly 100 times its value on Earth.

Europa, Ganymede, Callisto, Titan, Triton. Of the intermediate-sized planetary bodies that have retained appreciable amounts of water, the greatest amount is found in the outermost ones — Europa, Ganymede, Callisto, Titan,

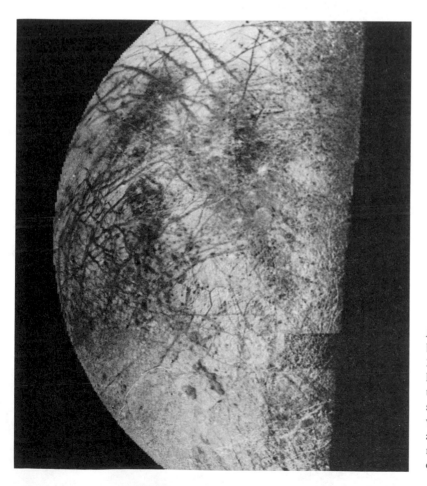

Figure 5.26. Europa, observed by *Voyager 2* from a distance of 241,000 km. The vast tangle of light and dark markings, some up to hundreds of kilometers across, suggest filled-in cracks. They have very little relief, and are primarily albedo markings (that is, they reflect sunlight with differing degrees of intensity).

Figure 5.27. Ganymede, the Galilean satellite that superficially resembles the Moon, with irregular dark regions on a brighter background. The prominent dark area in the lower left part of this photograph, called Galileo Regio, is about 3200 km across and heavily cratered. Like the other dark areas, it is older than the light regions, which exhibit a variety of features (such as complex systems of ridges and troughs) suggestive of past tectonic activity.

and Triton—because of the low temperatures that prevailed during and after their formation. In fact, water may constitute up to 50% of their masses, excepting only Europa. Differentiation early in the histories of these bodies forced the water out of their cores to form massive mantles surrounded by thin crusts. The surfaces are now at temperatures of 150 K or less and are frozen into hard, rigid ice that is capable of retaining the shapes of impact craters and other surface features for considerable lengths of time. For example, Ganymede and Callisto are heavily cratered, much like the rocky crusts of Mercury and the Moon, though the ice craters are considerably shallower (because ice slowly flows and deforms, as glaciers on Earth demonstrate).

Earth. A little more than two-thirds of the surface of Earth is covered by oceans. From this we might conclude that, concerning the abundance of water, Earth falls somewhere in the middle between those bodies that consist half of water and those that have none. That is a serious misconception. The water on the surface of the Earth adds up to approximately $2.3 \times 10^{-2}\%$ of the total mass of our planet. Several times this amount of water may be locked up in the rocks of the crust and mantle (basalt and granite contain between 0.2 and 0.5% water). But the total can hardly be more than a fraction of a percent of the

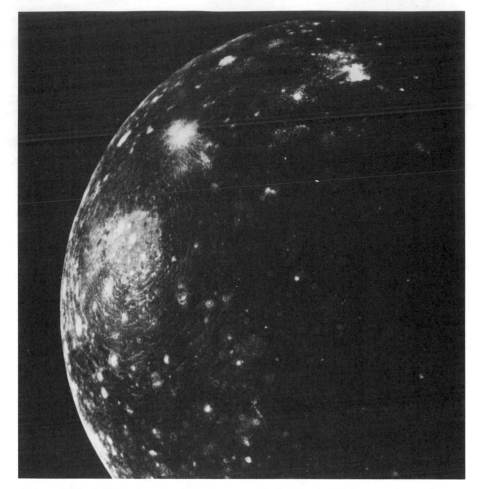

Figure 5.28. Callisto, the most heavily cratered of Jupiter's Galilean satellites. The prominent impact structure seen in this photograph, with an extensive ring system, is known as Valhalla and is similar in many respects to the large impact basins found on Mercury and the Moon. Valhalla's bright central area is about 300 km across, while the system of discontinuous, concentric rings extends radially outward some 1500 km.

total mass of the Earth.* Hence, with regard to water, Earth resembles barren Mercury much more than water-laden Ganymede or Callisto. Clearly, the planetesimals from which Earth was assembled had lost most of their water or never

---

*The mass of all of the water on the surface of the Earth adds up to approximately 1.4×10^{21} kg, and the total mass of the Earth is 6×10^{24} kg. The Earth's water is distributed approximately as follows (data from Press and Siever 1978, 150):

oceans	97.3%
glaciers and polar ice	2.1
groundwater	0.6
lakes and rivers	0.01
atmosphere	0.001
biosphere	4×10^{-5}

possessed much in the first place. Otherwise Earth should have ended up with roughly the same proportion of water as the outer planets.

Because the water loss was so great, we may assume that any number of minor changes in the protoplanetary disk—such as slightly more or less heating by the nascent Sun—could have significantly altered the amount of water that did end up on Earth, with important consequences for the environment and the origin and evolution of life.

Mars. Though Mars is more distant from the Sun than Earth, and hence should be much wetter, the only water apparent on its surface today is in thin layers of ice that cover its poles. These icecaps grow and wane with the seasons, subliming (that is, changing from the solid to the gaseous state without passing through the liquid state) during the summer and recondensing during the winter. If liquid water were present on the Martian surface, it would not be stable, given Mars's present temperature range of 150–290 K and a barometric pressure of a mere 7/1000 that of the sea level pressure on Earth. It would either boil or freeze, depending on the temperature.

Still, there is ample evidence that at one time liquid water flowed on Mars. For example, the cratered terrain is cut by countless *runoff channels* that bear some resemblance to terrestrial valleys eroded by flowing water (see figure 5.29). These channels start out small, have numerous tributaries, increase in size downstream, and are, on the average, a few tens of kilometers long. Most were probably cut between 3.0 and 3.5 billion years ago, shortly after the heavy initial cratering ended. Other kinds of channels, the *outflow channels*, have the markings of sudden flooding by huge quantities of water or other kinds of fluvial matter (see figure 5.30). These channels are much bigger than the runoff channels, frequently measuring tens of kilometers across, and they meander for hundreds of kilometers across the Martian terrain. They usually are flat-bottomed, often contain islands, and generally start abruptly and full-grown. It is as if huge reservoirs of water had suddenly burst their dams and emptied. In fact, there is some evidence for the previous existence of large standing bodies of water in some regions, in the form of sedimentary layering, but their relation to the outflow channels is ambiguous. Underground reservoirs of water, soil saturated with water that suddenly became unstable, and glaciers have also been suggested as likely sources of the flooding.

The outflow channels are younger than the runoff channels and probably were formed when the geological activity on Mars reached its peak, roughly 2.5 billion years ago. The geological activity was accompanied by outgassing and the release of large quantities of water and other volatiles, as occurred on Earth (see chapter 6). Much of the water was probably lost to space because of the low surface gravity of Mars. A small fraction is still present in the planet's polar icecaps. The rest is suspected to exist as permafrost or possibly in liquid form under the Martian surface.

Atmospheres

Of the intermediate-sized planetary bodies, only Venus, Titan, and Earth have substantial atmospheres, followed by Mars and Triton. The major gases in the

Figure 5.29. Ma'adim Vallis, a 600-km long runoff channel that runs through old cratered terrain on Mars. It contains long, branched tributaries upstream (lower right) and shorter tributaries resembling terrestrial box canyons downstream. Its gentle, meandering form leaves little doubt that it was carved out by a flowing medium, probably water.

atmospheres, the surface pressures, average surface temperatures (in degrees Kelvin and Celsius), and surface gravities of these bodies are listed in table 5.3 (page 201). In comparison, the other intermediate-sized planetary bodies have barren rocky or icy surfaces.

The sizes of planetary atmospheres are determined by two factors: (1) the sources that bring gases to the surface, and (2) the loss mechanisms that tend to remove them. The sources of atmospheric gases are tectonism and volcanism. They deliver severely heated, molten rock to the surface and expose them to low pressures, which allows volatiles trapped in the rocks to escape or outgas into the open. For example, H_2O, CO_2, and N_2 are the most common volatiles trapped in the surface and upper mantle rocks of the Earth, as the compositions of the clouds billowing forth from terrestrial volcanoes demonstrate. Hence, H_2O, CO_2, and N_2 were the major original constituents of our planet's atmosphere. A second source of atmospheric gases is evaporation of liquids or sublimation of ices from the planetary surfaces. The mechanisms by which planets lose atmospheric gases are (1) diffusion into interplanetary space if their surface gravities are small or the atmospheres are very hot, and (2) loss to the ground. The loss to the ground may occur either by condensation, such as the falling of rain, or by chemical reactions that bind the gases to the surface rocks.

The various gain and loss mechanisms of planetary atmospheres are interdependent and far too complex to be discussed here in detail. Briefly, Earth and Venus, for example, have substantial atmospheres because of the extensive geological activities they experienced throughout most of their pasts (figure 5.31). Furthermore, both planets are massive and have relatively strong surface

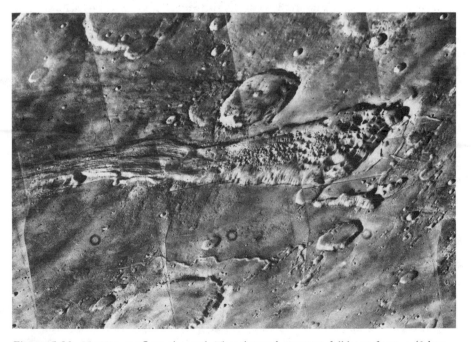

Figure 5.30. Martian outflow channel. The channel emerges full-born from a 40-km wide depression enclosing chaotic terrain.

gravities with which they retain their atmospheres. On Mars, by contrast, the early geographical activity was interrupted and, thereby, so was the growth of its atmosphere. Besides, Mars is less massive than Earth or Venus, with only about one-third their surface gravity. Hence, atmospheric gases escape to space more easily. Mercury is too hot and the other terrestrial planetary bodies have surface gravities that are too weak to have accumulated appreciable atmospheres. The one exception is Titan. Its surface gravity and temperature are comparable to those of Ganymede and Callisto, but unlike these two bodies it has a very thick atmosphere. In fact, above a given area of its surface, Titan's atmosphere exceeds that of Earth by a factor of ten. At present we do not know why this is so.

THE JOVIAN PLANETS

The Jovian planets — Jupiter, Saturn, Uranus, and Neptune — are the "gas giants" of the Sun's planetary system. They possess no solid surfaces and no ancient impact craters or deformations by geological activities mar their surfaces. Instead, thick atmospheres envelope them, whose top layers are characterized by gigantic and often colorful wind and storm patterns. The bottom layers of the atmospheres pass smoothly over into liquid mantles, and near their centers they possess cores of mainly rocks and iron.

Thus, the Jovian planets are quite unlike the intermediate-sized planetary bodies and deserve to be grouped into a family of their own. Their unique structures stem from their large masses and sizes in comparison to the other planetary bodies and from the volatiles they possess in great abundance. For example, Jupiter has roughly 300 times the mass and 10 times the size of the Earth, and Saturn is not much smaller, as table 4.1 indicates. Furthermore, the compositions of Jupiter and Saturn are dominated by hydrogen and helium, followed by carbon, nitrogen, and oxygen. Uranus and Neptune are less massive and smaller than Jupiter and Saturn, but they, too, consist largely of volatiles, though they lost much of their original hydrogen and helium. The large abundances of volatiles in these four planets follow from both their observed surface compositions and from their unusually low average densities (see table 4.1).

The combined mass of the Jovian planets adds up to 0.13% of the mass of the Sun and it exceeds the mass of the other planets, satellites, and smaller bodies of the Solar System by about a factor of 200. Thus, next to the Sun, the four Jovian planets are the Solar System's major components.

Core and Mantle Structure

The three-layered structures of the Jovian planets — gaseous atmospheres, liquid mantles, and solid or liquid cores — differ considerably from one planet to the next because of differences in their masses, compositions, and distances from the Sun.

The smallest difference in structure among the Jovian planets is found in the makeup of their cores. Model calculations indicate that the cores of all four planets consist mainly of silicates and iron, with small admixtures of other heavy

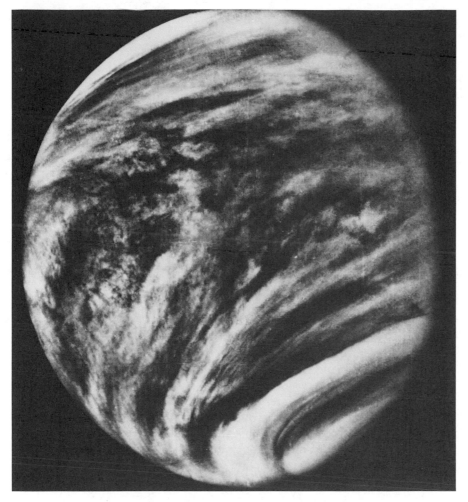

Figure 5.31. The cloud cover of Venus photographed in ultraviolet light.

elements. Clearly, some form of composition differentiation, similar to the one in the intermediate-sized planetary bodies, took place in these giant planets. Uranus and Neptune probably have the largest cores in proportion to their total sizes because they have lost the most volatiles. In contrast, Jupiter and Saturn have the highest core pressures and temperatures because they are the most massive. The estimated conditions at their centers, compared to the Sun's, are as follows:

	Jupiter	Saturn	Sun
Temperature (K)	25,000	20,000	16×10^6
Pressure (earth atmospheres)	8×10^7	5×10^7	3×10^{11}
Density (g/cm^3)	20	15	160

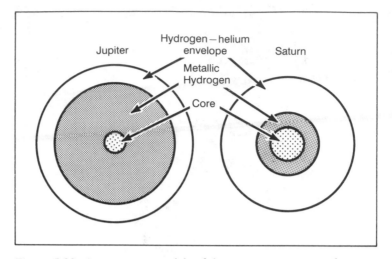

Figure 5.32. Approximate models of the interior structures of Jupiter and Saturn.

The mantles of the four planets are liquid, but beyond that they show a great deal of variation. The mantles of Jupiter and Saturn consist mainly of hydrogen and helium and have two-layered structures (see figure 5.32). In the inner layers, high pressures keep the hydrogen molecules stripped of some of their electrons, which produces metallic characteristics, such as high conductivity of heat and electricity. Hence, these layers are commonly referred to as the *metallic layers*. In the outer mantle layers, which are at lower pressures, the molecules are not stripped of electrons and, consequently, they are not metallic—they are just ordinary liquids. The mantles of Uranus and Neptune are believed to be single-layered and to consist chiefly of liquid water, methane, and ammonia, without any metallic characteristics.

Atmospheres

The greatest differences in structure among the four planets arise in their atmospheres. Jupiter's and Saturn's atmospheres are characterized by well-defined colored zones or belts that encircle the planets and rotate with them (see figures 4.1 and 5.33). These zones are remarkably stable and have not changed their latitudinal positions during the last 80 years. Close-up photographs sent back by the *Voyager 1* and *Voyager 2* spacecrafts show that the zones of both planets are in a general state of turmoil, with large and small eddies, fluctuating atmospheric motions, and elongated fast-moving gas streams abounding. The most prominent of these dynamic features have the appearance of dark or bright spots and are useful markers for tracking the planets' global circulation patterns. The atmospheric compositions are H_2 and He, roughly in solar proportions, followed by small admixtures of NH_3, CH_4, and, in the case of Jupiter, H_2O and various kinds of simple organic compounds such as C_2H_6 (ethane), C_2H_2 (acetylene), and HCN (hydrogen cyanide).

Figure 5.33. Jupiter, photographed by *Voyager 1* from a distance of nearly 33 million km. The planet's atmospheric band structure of alternating bright and dark clouds is clearly visible, as are the Great Red Spot, several large white ovals, and many smaller eddies and vortices.

The atmospheric zones of both Jupiter and Saturn are closely associated with alternating westerly and easterly winds. For example, if in one zone the atmospheric winds blow toward the east, those in the two adjacent zones blow toward the west. These winds are analogues of the jet streams and trade winds on Earth, which also flow in adjacent zones and in opposite directions.* Earth

*The Earth's trade winds are tropical and subtropical surface winds that on the average blow from east to west, with average speeds of about 20 km/hr. The Earth's jet streams are located further north and south of the equator than the trade winds and blow from west to east at heights between 10 and 50 km, reaching maximum speeds of 500 km/hr, though speeds of 100–300 km/hr are more typical.

has only two such wind zones (jet streams and trade winds) in each hemisphere, but Jupiter and Saturn have several and they are much wider. For example, the winds of their equatorial zones are approximately 30,000 km and 80,000 km wide and they blow eastward with speeds of up to 500 km/hr and 1800 km/hr, respectively. In the case of Saturn, this is two-thirds the local speed of sound.

The bright and dark spots on the surfaces of Jupiter and Saturn are eddies of circulating atmospheric gases, comparable to the rotating high- and low-pressure systems that determine the weather on Earth. The smaller eddies do not last long. Within days of their origin they usually are caught between oppositely flowing zonal wind currents, and the resulting shear forces tear them apart. Sometimes they are absorbed by clouds and may be disgorged hours or days later. Complex patterns of oscillating vortex motions are often set up in the wakes of these atmospheric disturbances. Eventually, the eddies and their fragments disappear altogether, only to be followed by new ones that suddenly appear.

The larger eddies last much longer. The most famous of them, the Great Red Spot in the southern hemisphere of Jupiter, has been observed since the invention of the telescope more than 300 years ago and, probably, has existed considerably longer than that (see figure 5.33). The three white ovals just to the south of the Great Red Spot were first observed in 1938; other large eddies have lasted for months or years. They are so stable because instead of being sheared apart, they roll along between the wind currents of adjacent zones.

In contrast to Jupiter and Saturn, very little is known about the atmospheric conditions of Uranus and Neptune. These two planets are too distant and faint to reveal much of their constitution to earth-bound observations. However, in January, 1986, *Voyager 2* passed Uranus and it is now on its way to Neptune, which it will reach in August of 1989.

Uranus is odd in that its equatorial plane as well as its ring system and the orbits of its 15 satellites are nearly perpendicular to the plane of the ecliptic. In contrast, the equatorial planes of all other planets are aligned much more closely with the ecliptic. In fact, at present Uranus's south pole points almost directly toward the Sun. Data sent back by *Voyager 2* indicate that Uranus rotates once every 17.3 hours. Its atmosphere contains hydrogen and helium in proportions similar to those of Jupiter and Saturn. It also contains small admixtures of methane and various other hydrocarbons. There is evidence of cloud structure and atmospheric winds, but *Voyager 2* detected no bands or other detailed surface characteristics. The planet is shrouded in a permanent smoglike haze, which hides most of the atmospheric features that lie below. The Uranian temperature is 50 K at a level where the pressure equals 0.1 earth atmospheres, and it increases with depth. Surprisingly, the temperature varies very little between the planet's sunlit and dark sides. This is further evidence of the existence of atmospheric circulations, which distribute the heat received from the Sun.

Neptune, the most distant and faintest of the Jovian planets, shows up as a greenish disk when viewed through earth-bound telescopes. Its known atmospheric constituents are hydrogen, helium, and methane. Virtually no surface markings are discernible from Earth. Few other details are presently known about this planet.

ORIGIN OF THE SOLAR SYSTEM

Now that we have discussed the orbital distribution and physical structures of the major components of the Solar System, let us turn to their origin. This is a topic of age-long interest and was one of the first to be examined during the early development of western science, more than 300 years ago. In 1644 René Descartes, the great French philosopher, suggested that at one time the Sun was surrounded by a large rotating disk of gas from which the planets and their satellites formed. One hundred years later, this *nebular theory* was challenged by the French naturalist Georges Louis Leclerc de Buffon (1707–1788). He proposed a *collision theory* according to which a massive comet or, as proposed later, a star passed close to the Sun and ripped out the material from which the planets and their satellites then assembled.

Neither of these two theories was based on sufficient facts or theoretical reasoning to be fully convincing, and during the succeeding centuries first one and then the other gained favor. The German philosopher Immanuel Kant (1724–1804) and the French mathematician Pierre Simon Laplace (1749–1827) supported the nebular theory. Laplace pointed out that a rotating disk of gas would slowly contract and develop rings of materials from which planetary bodies could form. Throughout much of the nineteenth century, this remained the accepted theory among most scientists. However, in the second half of that century, the British physicist James Clerk Maxwell (1831–1879) calculated that shear forces resulting from the differential rotation of the disk would prevent the matter from condensing into individual bodies. Thus, Buffon's collision theory was resurrected, and during the first few decades of the present century the English physicist Sir James H. Jeans (1877–1946) and others developed it further. Because a collision between two stars is an extremely rare event, they argued that our planetary system must be unique or at most one among a few.

Maxwell's critique of the nebular theory was based on the mistaken notion that stars and the interstellar material have the same composition as the Earth. Only in the late 1930s was it discovered that in fact hydrogen and helium are the major constituents of the Cosmos and that material like that of rocks adds up to no more than about 2%. From this new knowledge, the German physicist Carl Friedrich von Weizsäcker concluded that the primitive circumsolar disk must have been much more massive than are the planets today, and he demonstrated that Jeans's objections to the nebular theory were inapplicable. It was also realized that material ripped out of the Sun by a passing star would disperse or fall back onto the Sun, rather than condense into planetary bodies. These new theoretical insights led to the abandonment of the collision theory and the general acceptance of the nebular theory.

Since the 1940s, an increasing number of researchers have worked on the nebular theory of the origin of the Solar System, using the growing knowledge about interstellar clouds and star formation, building numerical models with the aid of high-speed computers, and, more recently, applying the wealth of new observational data gathered from space. The theory is still far from complete, with many rival ideas about some of the details. However, there is a general consensus today among the experts how, in broad terms, the Sun and its planetary system formed from an interstellar cloud of gas and dust (see figure 5.34).

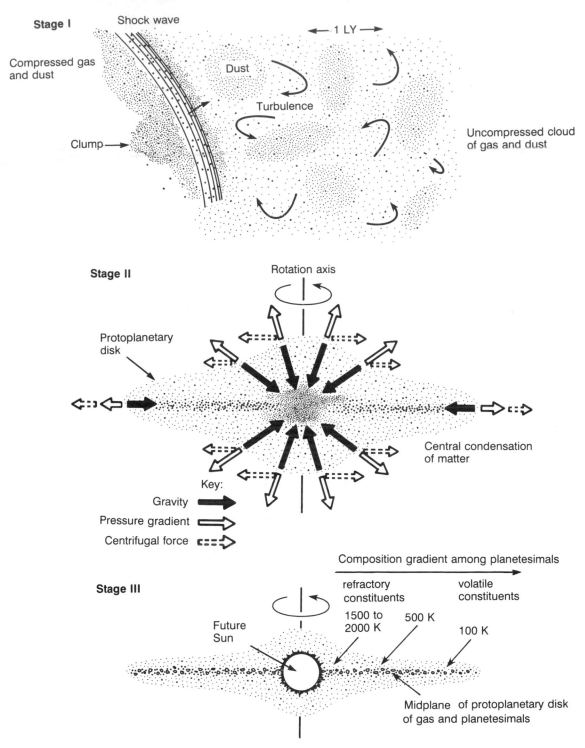

Figure 5.34. Summary of the stages in the birth of the Solar System (see also next page).
Details are discussed in the accompanying text and in chapter 3, The Birth of Stars.

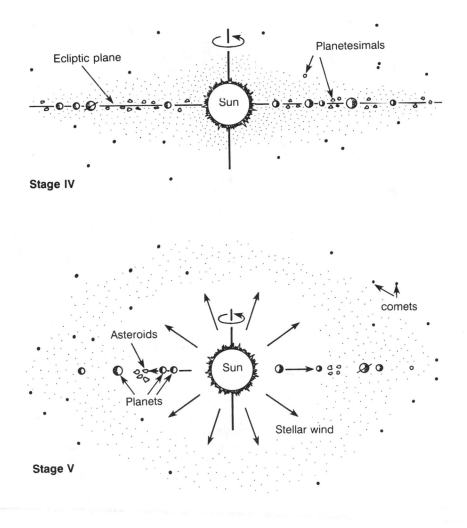

Figure 5.34
(continued)

Stage I

About 4.5 to 4.6 billion years ago, along one of our galaxy's spiral arms, a supernova exploded near an interstellar cloud of gas and dust. As its shock front swept through the cloud, it compressed many of its dense subregions to the point of gravitational collapse. One of the dense clumps thus formed (see chapter 3, The Birth of Stars) contained the raw materials—hydrogen (73.6% by mass), helium (24.8%), and heavier elements (1.6%)—for the birth of the Solar System. Before compression, the clump was roughly 0.1 to 1.0 LY (approximately 10^3 to 10^5 AU) across and it had a mass between 1.5 and 2 solar masses. After compression, its size was reduced by roughly a factor of ten, though there is at present little certainty about these values. The clump rotated slowly; its temperature was about 5–10 K; and most of its material was condensed into molecules and dust grains. Some of the molecules and dust grains had arrived with the shock wave from the supernova and differed in composition from the bulk material of the clump. (For the sake of balance, be aware that some scientists are not convinced that a supernova was responsible for the compression of the cloud to the point of gravitational collapse. They think that this point was reached by the gradual and natural evolution of the cloud, without any external, compressive forces. However, their theory does not readily account for the range of isotopic abundances found in the Solar System today.)

Stage II

As gravity pulled the clump together, it began to rotate faster. Three kinds of forces acted on the clump: gravity, the pressure gradient, and the centrifugal force resulting from rotation. (A fourth possible force—the force from magnetic fields—is disregarded here.) Gravity pulled the matter radially toward the clump's center, the pressure gradient pushed it outward, and the centrifugal force pushed it in directions perpendicularly away from the clump's rotation axis, as shown in figure 5.34. The net result was that the clump slowly contracted into a disk—the *protoplanetary disk*—revolving around a growing central concentration, the future Sun.

Stage III

For some time, the protoplanetary disk and the central concentration probably looked quite similar to the disk and nucleus of galaxies like NGC 3115 (figure 1.9), though, of course, they were very much smaller. Like the disk of a spiral galaxy, the protoplanetary disk rotated differentially, faster close to the central condensation and more slowly further out. As a result, neighboring segments of material that differed slightly in distance from the rotation axis orbited at slightly different speeds and, therefore, rubbed against each other. This created friction (that is, shear forces) and led to local turbulence. Yet another source of turbulence probably was thermal convection due to steep temperature gradients in the protoplanetary disk. The effect of turbulence was that the orbiting matter was slowed down and gradually spiraled inward toward the central concentration. As that concentration of matter grew, it heated up due to the release of gravitational energy and evolved into the Sun.

The disk heated up also because gravitational energy was released in it as matter spiraled through. Furthermore, the disk was heated by the nascent Sun, first by infrared radiation and later by optical and ultraviolet radiation. These heat sources were far stronger close to the Sun than farther out. Consequently, a temperature gradient was established from about 1500–2000 K in the region where Mercury eventually formed to approximately 100 K or less in the disk's outer fringes.

In the course of the growth of the protoplanetary disk, the concentration of its matter increased enormously. This was particularly the case along the disk's midplane, where a number of important and irreversible changes began to take place. The orbiting dust particles were so densely packed there that they collided very frequently with each other and coalesced into growing bodies, first of pebble size and then of boulder size and larger. Quite rapidly, many of the bodies reached diameters of tens to hundreds of kilometers and became the planetesimals, or "little planets."

At first, when the planetesimals were still very small, they had a rather loose and fluffy structure, similar to that of the dust grains. However, as they grew in size, self-gravity compacted them into hard, solid objects. Simultaneously, heat released by gravitational contraction and the decay of unstable isotopes, such as ^{26}Al, caused melting and differentiation.

The growth of the planetesimals was strongly affected by the disk's temperature. In the inner parts of the disk, where temperatures reached 1500–2000 K, the coalescing dust particles lost all but their most refractory constituents. Consequently, the planetesimals that formed there consisted only of silicate rocks and metals, without any volatiles. Toward the middle regions of the disk, corresponding to where Mars and the asteroids eventually formed, the temperature rose to only about 500 K. Hence, the dust particles retained many of their volatiles (such as H_2O, CO, CO_2, H_2S [hydrogen sulfide], N_2 and some organic molecules) and carried them along in the assembly of the planetesimals. Finally, in the outer regions of the disk that were heated the least, virtually all volatiles stayed with the dust particles, including those most easily vaporized, such as CH_4 and NH_3, and became part of the planetesimals. Thus, the temperature gradient of the disk led to a composition gradient among the planetesimals, from those with only refractory constituents to those composed of rocks intermingled with progressively more volatiles at increasing distances from the nascent Sun.

Stage IV

When the growth of the planetesimals first began and for some time thereafter, the gaseous component of the disk (consisting mostly of hydrogen and helium) was still thick and dense, and gas continued to spiral through it toward the Sun. The planetesimals participated only very little in this spiraling motion because they were massive and compact. They plowed right through the gas without being affected much and traveled in nearly circular or elliptical orbits around the Sun. Most of their orbits lay close to the midplane — the future ecliptic plane — of the protoplanetary disk, but some were inclined to various degrees. Large eccentricities and inclinations in the orbits were the result of continuous and ever-varying gravitational pulls that the planetesimals exerted on each other as they orbited the Sun. Planetesimals that were in elliptical and inclined orbits intersected the orbits of other planetesimals and, sooner or later, had near encounters or collisions with them. Some of the collisions were sufficiently energetic to shatter the planetesimals and disperse part or all of the resulting debris. The heavily cratered surfaces of many of the small and intermediate-sized bodies in the Solar System today are evidence of this violence. However, most often the collisions were rather gentle because near-circular and little-inclined orbits predominated. Hence, most collisions led to merger and growth.

The planetesimals that grew most rapidly were those that by chance acquired above-average masses early on and thus were most successful in gravitationally attracting and assimilating other smaller planetesimals. As illustrated by a sequence of computer simulations made by George W. Wetherill of the Carnegie Institution of Washington (shown in figure 5.35), dominant planetesimals established themselves at regular orbital intervals. In the course of time, they swept up nearly all of the other smaller bodies in their orbital neighborhoods and evolved into the terrestrial planets we observe today.

Collisions and accretion of solid planetesimals, as described so far, were the principal mechanisms of planet formation only in the inner part of the Solar System. In the outer parts, where temperatures were lower, planet formation was greatly influenced by the gas component of the protoplanetary disk. In fact, hydrogen and helium became the major constituents of Jupiter and Saturn and, to a smaller degree, became incorporated in Uranus and Neptune as well. At present it is not known whether, in the case of these four planets, solid planetary cores formed first, followed by the gravitational acquisition of envelopes of hydrogen and helium; or gaseous spheres formed first, followed by differentiation and the settling of rocky components toward their centers. Whichever sequence was the actual one, all four giant planets probably were, early in their formations, surrounded by their own protosatellitic disks of gas and dust, with temperature and composition gradients, from which satellites formed.

Stage V

While the planetesimals and planets were forming, the young Sun was also evolving. Within several hundred thousand to a million years of the initial compression of the primordial clump, the Sun became a T Tauri star and experienced the surface eruptions, rapid variations in light output, and mass loss in the form of a strong wind that are typical of that evolutionary stage. The wind blew for millions of years and swept away all of the gas and dust that had not yet coalesced into planets, satellites, or planetesimals, exposing these bodies to interstellar space.

Orbits, Rotations, and Composition Gradients

Some of the specific features of the Solar System can be explained in terms of the formation theory just described, but some cannot. Because the planets were the result of accretion of many small bodies and, in the case of the Jovian planets, of initially widely distributed gaseous material, their final orbits turned out to be averages of the individual orbits of the many small bodies and the gas. These averages added up to the near-circular and little-inclined orbits we observe today. The planets' directions of revolution around the Sun were also the same as the rotational direction of the protoplanetary disk, namely from west to east. The only exceptions were Mercury and Pluto, both of which acquired above-average orbital inclinations and eccentricities. We do not know what caused these results. However, the case of Pluto is not surprising, for it is very small and resembles more an asteroid than a genuine planet.

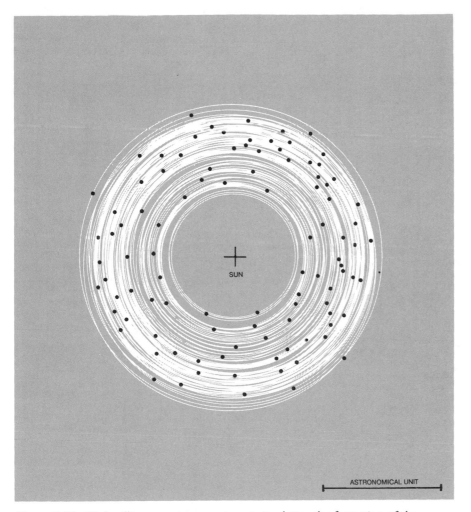

DIAGRAM 1

Figure 5.35. Wetherill's computer experiment simulating the formation of the terrestrial planets of the Solar System (continued on pages 246–247).

Diagram 1. The experiment begins with 100 planetesimals, each containing 2/100 earth masses and revolving around the Sun in random elliptical orbits as shown. Their total mass equals approximately that of Mercury, Venus, Earth, the Moon, and Mars.

Diagram 2. 30 million years later, near encounters and collisions leave only 22 bodies of various masses, as indicated by the sizes of the dots. Their orbits have become more elliptical and the range of their distances from the Sun has broadened.

Diagram 3. After 440 million years, the simulation has produced four fully-formed planets in isolated, nearly circular, and stable orbits. The largest of the planets is the outermost one. It has been built up from 34 of the original planetesimals.

The simulation was carried out in three dimensions, but only the projections of the orbits onto the midplane of the protoplanetary disk are shown. No gas and dust component or the perturbing effects of more distant giant planets were included in the experiment. (From Wetherill 1981, 162–174.)

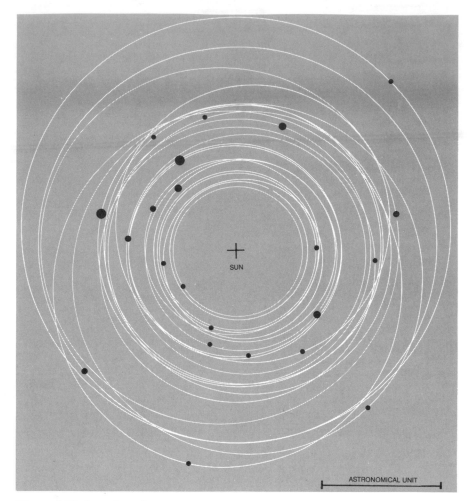

DIAGRAM 2

Figure 5.35 (continued)

Most of the collisions between planetesimals were probably not head-on, but off-center or at glancing angles. The same was most likely true of the coalescence of gases in the formation of the outer planets. Thus, all planets ended up with net rotations that, with the exceptions of Venus and Uranus, were also in the direction from west to east. However, it is not clear at present why the planets have the particular rotation rates we observe, or why two rotate from east to west and the rest from west to east.

The development of a composition gradient among the planetesimals combined with the substantial amounts of hydrogen and helium the outer planets managed to acquire, explains the different compositions in the planets today. The Earth, for example, formed in a part of the protoplanetary disk in which heating was so severe that most of the volatiles that originally were present in the dust grains were vaporized and driven off. Consequently, the Earth acquired

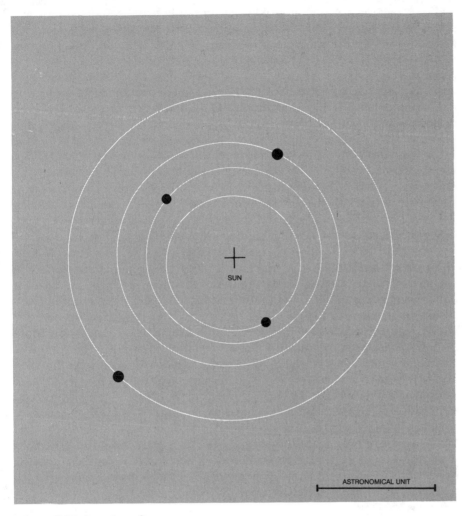

SUN

ASTRONOMICAL UNIT

DIAGRAM 3

Figure 5.35 (continued)

very few volatiles, compared to its bulk constituents of silicate rocks and iron–nickel metals. Most of the volatiles it possesses today—such as those constituting its ocean and atmosphere, and the CO_2 that became buried as dolomite and limestone in its crust—were probably brought to it by planetesimals from the vicinities of Mars and the Asteroid Belt. Perturbations of the planetesimals' orbits by near encounters with other planetesimals and nearby Jupiter deflected them inward toward Earth, which then captured them. As Wetherill's computer diagrams show, this kind of scrambling of the planetesimals' orbits was common in the protoplanetary disk. Thus, each planet had a much wider "feeding zone," from which it drew its raw materials, than today's spacing between planetary orbits.

The planet that had the greatest effect on the scrambling and broadening of planetesimal orbits was Jupiter with its large mass. It perturbed the planetesimals in its immediate vicinity so strongly that, for example, at 2.8 AU from the

Sun no planetary mass accumulated where, according to the Bode–Titius law, a planet should revolve. Most of the planetesimals of that region were dispersed to other parts of the protoplanetary disk. The few that survived there are today's asteroids. Likewise, Jupiter's perturbing effect removed many of the planetesimals from the neighborhood of the future planet Mars (at 1.5 AU), which explains why this planet acquired only about one-tenth the mass of Earth. Furthermore, many of the comets in the Oort Cloud, which extends 100,000 AU outward from the Sun, are believed to have been deflected there by Jupiter and the other giant planets.

The deflection of planetesimals from one part of the Solar System to another was responsible for the heavy bombardment that the planets experienced for hundreds of millions of years after they were formed. Only as the supply of planetesimals ran low, about 3.8 billion years ago, did the bombardment diminish. However, even today, the supply—the asteroids and comets—is not fully gone and the bombardment continues, though at a very low rate.

Despite the many features of the Solar System that are explained, at least qualitatively, by the theory just discussed, there remain numerous important and unresolved questions. For instance, how did it happen that a few, relatively tiny, planets swept up nearly all of the material between them? Or, put differently, why are there not more planets, each with a smaller mass? Why are the masses of Jupiter and Saturn roughly one order of magnitude greater than those of Uranus and Neptune, and two orders of magnitude greater than those of Earth and Venus? What was the time scale of planet formation? Was the main assembly phase completed in a hundred million years? Or did it take much more or much less time? Are planetary systems inevitable by-products of star formations?* Are they found only around single stars, or can binary stars have planets, too (perhaps in large orbits around close binaries and in relatively small orbits around widely spaced binaries)? Only by continuing investigations of the Solar and other stellar systems—both observationally and theoretically (in particular, through computer modeling)—can answers be found to such questions.

The origin of the Solar System dates back some 4.5 billion years, when the protosolar nebula contracted under the influence of its own gravitational forces. The greatest condensation of matter occurred at the center of the nebula and gradually evolved into the Sun. Orbiting the central condensation was a disk of gas, dust, and rapidly growing planetesimals. Collisions and coalescence among the planetesimals led to the formation of the small and intermediate-sized

---

*The first disk of rocky rubble surrounding another star was discovered in 1983 around Vega, with the U.S.–Dutch–British Infrared Astronomy Satellite (IRAS). The disk has a radius of about 80 AU, roughly twice the size of Pluto's orbit. Two more examples of stars circled by grainy or rocky rubble are Fomalhaut and Beta Pictoris, both visible from the Southern Hemisphere. Still other nearby stars are now being examined by the IRAS team for such circumstellar disks. In the case of Beta Pictoris the evidence even suggests, though not conclusively, that some of the rocky rubble may have coalesced into planetesimals or planets. Additional evidence for the existence of a planet or planets beyond the Solar System comes from a wobble in the motion of the nearby star Van Biesbroeck 8. (Similar wobbles reported during the past few decades in the case of several other nearby stars, and also interpreted as being due to planetary companions, have not been confirmed by recent, more sensitive observations.)

planetary bodies, while coalescence of the gas (at distances of Jupiter's orbit and beyond) produced the Jovian planets. All remaining uncondensed gas and dust were driven off into interstellar space by the strong wind that blew from the young Sun as it evolved through its T Tauri stage.

After the bodies of the Solar System were formed, they continued to evolve. In the case of most of them this evolution was strictly physical. For example, the Sun kept burning hydrogen and pouring fourth radiant energy. Venus, Earth, Mars, and most other bodies of intermediate size experienced major geologic changes and some acquired atmospheres. Many of the asteroids collided, some of them with the planets. The member bodies of the Oort Cloud, in the outer fringes of the Solar System, were periodically perturbed by gravitational forces of uncertain origin, and some of them were sent, as comets, in elongated orbits toward the inner part of the system. There, after numerous circuits around the Sun, they either crashed on one of the planets or the Sun, or they broke up and became part of the interplanetary debris.

However, the most remarkable evolution in the Solar System was not physical; it was biological. Within a relatively short time after the system's origin, perhaps even before the major accretion phase of the terrestrial planets had been completed, life emerged on Earth. The initially simple, microscopic organisms gradually became more complex and in the course of some 4 billion years evolved into an enormous variety of species, differing in size, habitats, and adaptive strategies. Most astonishingly, one of the species—*Homo sapiens*— evolved language and culture, the ability to think and to reason, and the urge to embark on a systematic investigation of the Universe near and far. The study of biological evolution on Earth will be the topic of the second part of this book.

One generation passeth away, and another generation cometh; but the earth abideth forever. The sun also ariseth, and the sun goeth down, and hasteth to the place where he arose. The wind goeth toward the south, and turneth about unto the north; it whirleth about continually, and the wind returneth again according to his circuits. All the rivers run into the sea; yet the sea is not full; unto the place from whence the rivers come, thither they return again.

Ecclesiastes—1:4–7

Exercises

1 Traditionally, the Sun's planetary system has been described as consisting of nine planets, the planets' satellites, asteroids, comets, and interplanetary matter. In the text, I did not follow this tradition, but defined just three families of bodies. Why? Discuss the most important characteristics of each of these families.

2 (*a*) Why are chondrites (ordinary and carbonaceous) said to be more primitive than the achondrites and the stony-iron and iron meteorites? (*b*) How do astronomers explain the origin of these different types of meteorites?

3 Distinguish between the following objects:
(a) meteors, meteorites, and meteoroids.
(b) asteroids and comets.
(c) stars and planets.

4 The radioactive isotope ^{14}C has a halflife of 5730 years. Analysis of skeletal remains unearthed by anthropologists at an ancient North American Indian village indicates that the bones contain only one-third of the ^{14}C originally present (assumed to have been the same as in our bones today). Using figure A, box 5.1, estimate the age of the bones.

5 (a) Calculate the average densities of Mercury, Venus, Earth, and Mars, using the planets' masses and radii listed in table 4.1. Compare your answers with the actual densities of these planets given in the table on page 211.
(b) Why are the densities you calculated in (a) larger than the planets' uncompressed densities? (Note: the average density of a spherical body $= M/\left(\frac{4\pi}{3}R^3\right)$, where M and R are the body's mass and radius, and $\pi = 3.14$.)

6 Describe the physical and orbital characteristics of the Galilean satellites, and discuss in which ways they and Jupiter resemble the Solar System. In which important ways do they differ from the Solar System?

7 Name the planetary bodies of intermediate size that have the most impact craters, the greatest geologic activities, the most water (liquid or ice), and the thickest atmospheres.

8 Why is Earth a geologically very active planetary body, but Mars is not?

9 Why does liquid water cover roughly two-thirds of the Earth's surface, but today is absent on the surfaces of its sister planets, Venus and Mars?

10 What mechanisms contribute to the formation and retention of planetary atmospheres?

11 Why is there no planet at 2.8 AU from the Sun, as predicted by the Bode-Titius law?

12 Water is thought to have been a relatively abundant molecular component of the protosolar disk. That is apparently why Ganymede, Callisto, the comets, and many of the other bodies in the other part of the Solar System contain so much of it. Why did the bodies in the inner part of the Solar System, such as the terrestrial planets, end up with much less water?

13 How does the presence of life affect the Earth's physical environment? (Think, for example, of the composition of the Earth's atmosphere; the existence of coal and oil deposits; the impact of plants, animals, and microorganisms on our climate, makeup of lakes and other bodies of water, and land forms.)

For **Suggestions for Further Reading** and **Additional References** see chapter 4.

PART TWO

BIOLOGICAL EVOLUTION

Introduction to Biological Evolution

Nature has no goal in view, and final causes are only human imaginings.
Baruch de Spinoza (1632–1677)
Ethica

. . . Owing to this struggle, variations, however slight and from whatever cause proceeding, if they be in any degree profitable to the individuals of a species, in their infinitely complex relations to other organic beings and to their physical conditions of life, will tend to the preservation of such individuals, and will generally be inherited by the offspring. The offspring, also, will thus have a better chance of surviving, for, of the many individuals of any species which are periodically born, but a small number can survive. I have called this principle, by which each slight variation, if useful, is preserved, by the term *Natural Selection*, in order to mark its relation to man's power of selection . . .
Charles Robert Darwin (1809–1882)
On the Origin of Species

Figure II.1. Drawing showing the simultaneous transcription of genetic information from genes of bacterial DNA (the long axial fibers) onto successive strands of RNA (× 26,000). The RNA molecules are at different stages of completion, and hence, of different lengths; thus a fern-like pattern is produced.

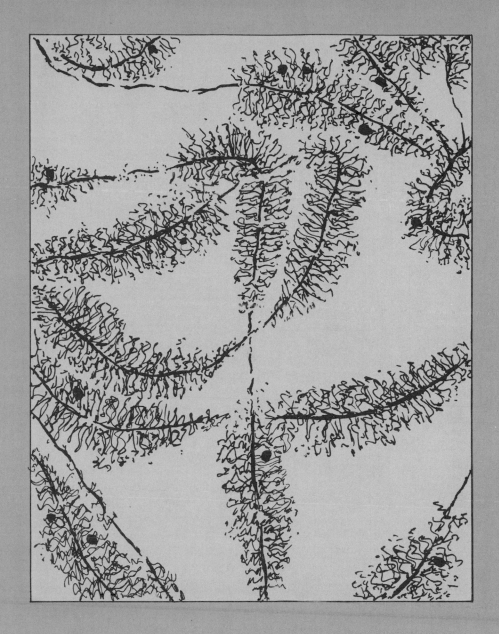

 **Introduction to Biological Evolution**

PART ONE OF THIS BOOK described the evolution of the Universe from the Big Bang to its present state, with particular attention to those events that were most crucial for the eventual genesis of life on Earth. Following the Big Bang, the primordial hydrogen and helium condensed into stars, galaxies, and galaxy clusters. Within the cores of stars the hydrogen was fused into helium, carbon, nitrogen, oxygen, and other heavier elements. At the end of their lives, the most massive stars exploded as supernovae and threw the newly synthesized elements into space, thus altering the composition of the interstellar gas and dust within galaxies. Finally, some 4.5 billion years ago, our Solar System was created in the disk of the Milky Way.

The third of the Sun's planets, Earth, was rather special. It satisfied all the conditions required for life to begin and to evolve. It possessed ample chemical raw materials of the right kind, a solid surface, and the correct mass to retain an atmosphere of moderate depth. Its distance from the Sun was such as to produce a climate that was and still is quite mild. Over large regions of the Earth the temperature was slightly above the freezing point of water, allowing the development of conditions suitable for the occurrence of a great variety of chemical reactions. Equally important, the Sun was born as a star of modest mass, thereby offering for billions of years a steady source of radiant energy, characterized by moderate intensity and a spectral distribution that peaks in the visual range.

Up to the time of the formation of the Earth, the evolution of the Universe was strictly physical. (Disregard for now the possibility that life-producing planets formed earlier elsewhere in our or in other galaxies, a topic discussed in the Epilogue.) This evolution was based on the existence of five elementary particles (electrons, protons, neutrons, neutrinos, and photons) and their interactions by four fundamental forces (gravitational, weak, electromagnetic, and nuclear forces). Furthermore, it was based on the presence of plenty of energy, which came into existence in various forms along with the particles during the Big Bang. In principle, physical evolution was and continues to be relatively simple and, in a statistical sense, predictable: Given the same starting conditions, physical evolution will follow a similar path every time. For example, should the Universe start over again, with the same initial conditions, the end result would again be stars, galaxies, and galaxy clusters, although they and their distribution would not necessarily be identical in every detail to those in our Universe.

With the origin of life on Earth an entirely new kind of evolution began— biological evolution. Biological evolution is consistent with all the laws of physics, but it is far more complex than physical evolution and much less predictable. This is an important statement and clearly requires justification. I shall present my justification in the next two sections. In the remainder of this introduction I shall discuss some of the characteristics of organisms, introduce concepts from organic chemistry needed in the chapters to come, describe the genetic code, and conclude with a brief history of the theory of biological evolution.

THE PHYSICAL BASIS OF BIOLOGICAL EVOLUTION

Biological evolution is consistent with the laws of physics because living organisms are made of ordinary matter. This matter consists of organic molecules, water, ions, and trace elements (see table II.1). The organic molecules are composed mainly of atoms of hydrogen, carbon, nitrogen, oxygen, phosphorus, and sulfur. Water molecules are composed of atoms of hydrogen and oxygen. The ions and trace elements include calcium, potassium, sodium, chlorine, iron, zinc, and others. The atoms of these materials are no different from those found in rocks, in stars, or in interstellar space; consequently, they are subject to the laws of physics, just as atoms are anywhere else in the Universe.

Even though organisms are subject to the laws of physics, they do distinguish themselves from nonliving objects. They possess unique structural organizations that give rise to a range of controlled functional processes geared toward their survival and reproduction. These functional processes give organisms the quality we call living.

In most organisms the functional processes are enormously complex as, for example, the metabolic activities in our cells, our sensory awareness of external and internal stimuli, the neuronal signals in our brains and the rest of our nervous systems, our bodily movements, and so forth. However, as discussed in

Table II.1. Major Chemical Elements Present in Living Organisms.

Major constituents of organic molecules (99.5% of all atoms)		Ions (0.5% of all atoms)		Trace elements (less than 0.01% of all atoms)	
H	Hydrogen	Ca^{++}	Calcium	Fe	Iron
O	Oxygen	K^{+}	Potassium	I	Iodine
C	Carbon	Cl^{-}	Chlorine	Cu	Copper
N	Nitrogen	Na^{+}	Sodium	Zn	Zinc
P	Phosphorus	Mg^{++}	Magnesium	Mn	Manganese
S	Sulfur			Co	Cobalt
				Cr	Chromium
				Se	Selenium
				Mo	Molybdenum
				F	Fluorine
				Sn	Tin
				Si	Silicon
				V	Vanadium

Note: Percentages refer to relative abundances (by number) of elements in the human body (they differ somewhat from organism to organism). For further details, see table 6.2.

chapters 6 and 7, such extreme functional complexity is not a prerequisite for being alive. Compared to the processes in our bodies and cells, those of some unicellular organisms appear quite simple, though they are still far more complex than most interactions among nonliving things. In addition, all life is part of an ecosystem that contains many components, interacting with each other in a multitude of ways.

The fundamental unit of biological organization is the cell. Cells are small, membrane-enclosed bodies, filled with a concentrated aqueous solution of chemicals, and containing all the biochemical machineries required for their metabolism and reproduction. Some organisms — the bacteria and protists — exist mainly as single cells, though a few form transient multicellular assemblages. Other organisms — the fungi, plants, and animals — consist of large numbers of cells precisely organized into specialized, hierarchical structures. In the case of the animals, these structures include tissues, organs, and organ systems. For instance, in our bodies, the circulatory system is composed of such organs as the heart, the blood vessels, and the lungs; the organs consist of various kinds of tissues, including muscles, membranes, and nerves; and the tissues are constructed from individual cells and their extracellular products.

The range of functional processes in our bodies includes the pumping of the heart and flow of blood through arteries and veins, exchange of oxygen and carbon dioxide across cell membranes and across the delicate linings of the alveoli of our lungs, ingestion and digestion of food, propagation of nerve impulses, movement of arms and legs, conduction of heat, and many more. Identical or nearly identical processes occur in other animals and similar ones are present in plants, fungi, and single-celled organisms. Ultimately, these processes have their origin in chemical reactions, which are based on interactions between atoms via the electromagnetic force. Again, this means that these processes are in principle no different from those found in the inanimate world and, like them, are subject to the laws of physics. It does not matter that they happen to take place in living things.

THE GENETIC BASIS OF BIOLOGICAL EVOLUTION

The unique structural organization of organisms and the accompanying functional processes are the result of biological evolution. This evolution distinguishes itself from strictly physical evolution in that it is based on the existence of genetic information and on the changes this information undergoes with time. Genetic information is the blueprint by which organisms reproduce, grow, and maintain themselves. It is present in nearly all cells, the only exceptions being certain nonreproducing cells such as the red blood cells of mammals.* With the exception of red blood cells, every one of the 10^{14} cells in our bodies carries a complete set of genetic information.

---

*The red cells of our blood do not contain genetic information and they are not capable of self-reproduction. They are produced by special cells in our bone marrow. When the red blood cells are formed, they do contain genetic information; but as they mature and acquire their hemoglobin (by which they carry oxygen from the lungs to the body cells), they gradually lose it.

In today's organisms, genetic information is encoded in long molecular strands called *deoxyribonucleic acids* or *DNA*. Molecules of DNA are constructed from atoms of carbon, hydrogen, nitrogen, oxygen, and phosphorus, and like the rest of the materials of organisms, this construction is based on physical laws. As discussed in the last section of this introduction, it is the way DNA is put together that allows it to carry information.

During each generation, DNA is replicated and passed on to the succeeding generation. Although this happens with great fidelity, DNA is subject to mutations and other changes that alter its informational content. In the course of many generations, these changes accumulate, giving species the remarkable ability to adapt to changes in their environment and to evolve. In contrast, objects created without the benefit of genetic information, such as raindrops, rocks, stars, and galaxies, possess no capacity for adaptation to their environments and they do not evolve in a biological sense.

Chance and Necessity

How did genetic information originate and how does it change with time? These two questions are central to any discussion of the origin and evolution of life. Both of them will be examined in the following chapters. Here I shall give just a brief answer to the second one.

The changes in genetic information come about in a two-step process of chance and necessity. In the first step, *random* mutations and their recombinations (if reproduction is sexual) alter the genetic information of a species from generation to generation,* thereby contributing to and maintaining a large degree of genetic variation among the individual members of a population. This variation is expressed by differences both in the outward appearances of the members of a species and in their internal chemistries. We can readily observe this in ourselves, for each of us is a unique individual different from everybody else. We all look different and our bodies react in their own particular ways to heat, cold, food, drugs, disease, and other stimuli.

To be more specific, the number of different ways genetic information can be expressed in human eggs and sperms, for example, has been estimated to be approximately 10^{2000}. This is a staggeringly large number and quite beyond our comprehension. It exceeds manyfold the total number of protons present in the Universe, which is a mere 10^{80}. Of course, only a minute fraction of those 10^{2000} possible sperms and eggs will ever be produced. Therefore, it is most improbable that any two will be identical. This means that every one of the 5 billion people inhabiting the Earth today is genetically distinct and different from everyone else, with the exception of multiple births from the same fertilized egg. Thus, the genetic variations of our species, like that of all other species, are enormously large.

---

*In the case of sexual organisms, *species* refers to a population of individuals capable of breeding among themselves. For instance, humans, the domestic cat, bald eagle, western hemlock, and common sunflower are species. In the case of asexual organisms, the concept of species is, strictly speaking, not applicable. Still, the word *species* is used for asexual organisms and refers to populations of organisms with common ancestry and characteristics that are judged to be relatively homogeneous.

The second step in biological evolution results from the fact that some members of every species are, because of their particular genetic makeup, more competent than others in the acquisition of food and shelter, defense against predators, resistance to disease, and adaptation to changes in climate and other physical conditions. Hence, some members are more likely than others to survive, reproduce, and pass their genetic information on to the next generation. This inequality in the likelihood of passing on genetic information constitutes a selection process. It continually pushes the genetic makeup of a species in directions dictated by the *necessities* of adapting to the biological and physical environments. The result is biological evolution.

It must be emphasized that the first step in biological evolution is strictly *random*. Mutations and other changes in genetic information occur according to chance, without regard to any benefit or handicap for the individual or the species. In contrast, the second step is imposed by the requirements of the environment and, hence, is *specific* and nonrandom. It acts like a filter on the reservoir of genetic variation created by the first step, mercilessly rejecting individuals with inadequate competence in the struggle for survival and passing on those best adapted to the current conditions. If the variations in a species do not meet the demands of the environment, its members are filtered out and the species becomes extinct. If they do meet the demands, the species survives and evolves.

The two-step theory of biological evolution was developed during the first half of the 19th century by two British naturalists, Charles R. Darwin (1809-1882) and Alfred R. Wallace (1823-1913). Today we often refer to it as "Darwin's theory of evolution." During the century and a half since its inception, the theory has been extended, refined, and thoroughly tested by laboratory experiments. In its broad outlines, it is accepted today by virtually all scientists. However, there continues to be heated opposition, mainly from adherents of fundamentalist religions, just as there has been in the past.

Ecosystems

The full complexity of life and its evolution can be appreciated only if we keep in mind that each species is part of the environment in which it lives. This environment is the totality of its biological and physical components. Each species interacts with and influences its environment and, in turn, is influenced by it. Hence, biological evolution does not occur in isolation. It is a multidimensional process containing countless feedback loops, with all the species evolving together and affecting each other's environment as they do so. The environment — including its biological and physical components — is commonly referred to as an *ecosystem*.

An ecosystem is never static, but is always changing or evolving. The rate of this evolution may vary greatly with time. In general, the evolution of an ecosystem depends on many factors, such as changes in climate or geology, the extinction of certain species, or the development of new adaptations in still other species. Sometimes these changes are sudden and severe, and new ecological niches arise rapidly. As discussed in later chapters, biological evolution progresses then in an explosive fashion. Species possessing sufficient genetic variation radiate into the newly created ecological niches, like air rushing into a vacuum, and evolution leads to new and distinct species within a relatively short span of time.

The combined effects of production of genetic variation and of selection according to the demands of the environment have created the multitude and diversity of life all around us today and of which we are part. This evolution began more than 4 billion years ago with random chemical reactions in the ocean and atmosphere of the primitive Earth. Countless generations of organisms have since come and gone. Each represented a temporary link in time. Species evolved and adapted to the ever-changing challenges of the environment. Many succumbed and died out, others prospered and gave rise to new species. In the course of time, biological competition and complexity increased. Life invaded the land and the air. We and the rest of present-day life are the products of this age-old struggle for survival, and in our DNA we carry the information accumulated by trial and error and by the persistent pressures of natural selection during all those years. This information is our most precious possession.

Unpredictability in Biological Evolution

Biological evolution is not only highly complex, but to a large degree it is also unpredictable. For instance, we cannot be certain what the environment will be like in the future. Hence, we cannot predict which selection pressures will be at work. Furthermore, we do not understand biological adaptation well enough to know which of several alternatives will do best in a given environment. Once a particular adaptive strategy is adopted, it will be refined by natural selection, but we cannot tell how this will be done.

Even if we had perfect knowledge of the evolution of the physical conditions of our planet and thoroughly understood biological adaptation, we still could not predict precisely how life would evolve. The reason is that the number of different ways genetic information can be written on DNA molecules is so large — 10^{2000} different ways in the case of human eggs and sperm — that the possibilities can never all be tried out by living organisms, even in a time span many times the present age of the Universe. Because mutation and the shuffling of genetic information by sexual reproduction are random processes, we can never know which of the many possible genetic variations will be present in a species. Hence, we cannot predict what the result of selection will be.

Of course, the constraints imposed by the physical environment force any evolving life into certain specific channels. For instance, there is little doubt that life on a planet like Earth would always be based on the chemistry of carbon, hydrogen, nitrogen, and oxygen; that it would start in the sea and eventually spread onto land and into the air; and that photosynthesis and respiration would become the major energy-producing mechanisms. But there is no way of predicting that insects would become the most populous class of animals, that after 3 or 4 billion years of evolution placental mammals would roam on land, and that a species of primates would evolve with the ability to think, to reason, and to develop culture.

TWO KINDS OF CELLS: EUKARYOTES AND PROKARYOTES

Traditionally, the various life forms that evolved on Earth have always been grouped into two kingdoms — plants and animals. Organisms that are rooted in the ground and obtain their food by photosynthesis were called plants, while those that move about and obtain their food by eating were called animals. However, after the discovery of microorganisms early in the eighteenth century, it became clear that some of them do not readily fit into this simple classification scheme. For example, euglenas are microscopic aquatic organisms that swim and swallow solid food. Yet they are also capable of photosynthesis. Are they animals or are they plants?

In order to deal with such problematic organisms, one of Darwin's disciples, the German zoologist Ernst H. Haeckel (1834–1919), introduced a third kingdom, the protists. It included protozoa, algae, fungi, and bacteria. But this still did not solve all of the taxonomic difficulties. Numerous other, more complex classification systems were, therefore, proposed. The one most widely accepted today is the five-kingdom system introduced in 1959 by the Cornell biologist Robert H. Whittaker (1920–1980), which consists of the *monera, protists, fungi, plants*, and *animals*. Examples of these five kingdoms, along with summaries of their most important characteristics, are given in figures II.2 and II.3.

In addition and complementary to the five kingdoms, biologists recognize today two superkingdoms — the *eukaryotes* and *prokaryotes* (terms that mean "true nuclei" and "prenuclear," respectively). The prokaryotes comprise the members of the monera, namely the bacteria. Their DNA exists freely in the cell interior and is not surrounded by a nuclear membrane (see below). The eukaryotes comprise the members of the other four kingdoms. Their DNA is usually contained within a membrane-enclosed body — the *nucleus*.

The eukaryotes and prokaryotes represent two distinct kinds of cells whose existence has been known only since the early 1950s, when the development of the electron microscope permitted detailed study of cellular architecture. It became apparent that differences between the two kinds of cells in reproduction, conversion of genetic information into protein structure, and metabolism are related to differences in their internal organization. These differences were recognized as constituting the most profound demarcation in the contemporary biological world. They are more profound than the differences between us and trees, fishes and mushrooms, or insects and algae.

Eukaryotes

Most eukaryotic cells have sizes in the range between 5 and 60 μm,* but some highly specialized cells, such as our nerve cells, may run the length of an entire arm; the largest single eukaryotic cells, such as the giant amoeba *Chaos chaos*,

---

*The symbol μ (pronounced "mu") is the Greek letter m. The combination μm is used as an abbreviation for the micrometer, which is a measure of length and particularly convenient for expressing the sizes of cells and objects of similar dimensions (see box I.2). One micrometer equals 10^{-6} meters (*micro* is Greek and means "one millionth") or 10^{-4} centimeters. If we stack 1000 objects, each one micrometer thick, on top of each other, we obtain the thickness of a dime.

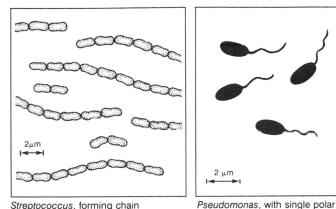

Streptococcus, forming chain

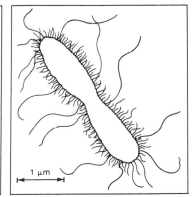

Salmonella typhi, surrounded by pili (short) and flagella (long appendages)

Pseudomonas, with single polar flagellum

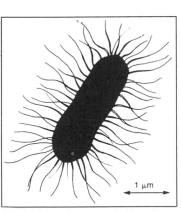

Escherichia coli, surrounded by pili

Fischerella, blue-green bacteria growing as branched filaments

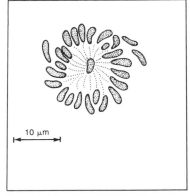

Caulobacter, forming rosette by the adherence of cells to one another

Figure II.2. Examples from the kingdom *Monera*. All its members are *prokaryotes* (also known as *bacteria*). They lack nuclear membranes and organelles, but possess ribosomes. Their body organization is single-celled, filamentary, or colonial. They have diverse nutritions, including absorption, photosynthesis, and chemosynthesis. They reproduce primarily asexually, but sexual recombinations occur in a number of species. About 3000 to 4000 species are identified.

(see figure 8.2) reach diameters of up to a few centimeters. All these cells are bounded by thin membranes through which they absorb nutrients from the environment and expel waste as well as products they manufacture, such as proteins, fats, and hormones. Their interiors are filled with a dynamic, aqueous solution called *cytoplasm*, which contains the various raw materials required for growth, self-maintenance, and reproduction: amino acids, fatty acids, sugars, minerals, and many more. The cytoplasm also contains a variety of filaments, fibers, and tubules—known as the *cytoskeleton*—which accounts for the cells' shape and movement. Furthermore, the cytoplasm contains numerous membrane-bound bodies called *organelles* ("little organs") that are the cells' living machinery. The most conspicuous organelles visible through the electron microscope are the *nucleus, endoplasmic reticulum, Golgi apparatus,*

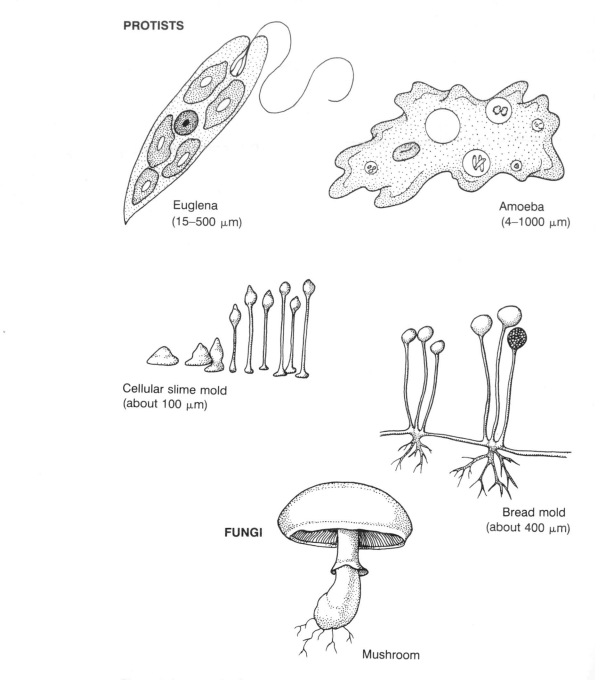

Figure II.3. Examples from the four kingdoms of eukaryotes:

Kingdom Protista: primarily single-celled or colonies of single-celled organisms; diverse nutrition, including photosynthesis, absorption, and ingestion; motile by flagella or other means; reproduction mostly sexual; about 20,000 known species.

Kingdom Fungi: primarily multinucleate; excrete digestive enzymes into surrounding environment and absorb dissolved nutrients; mostly nonmotile and living embedded in food supply; reproduction sexual and asexual; about 80,000 known species.

PLANTS

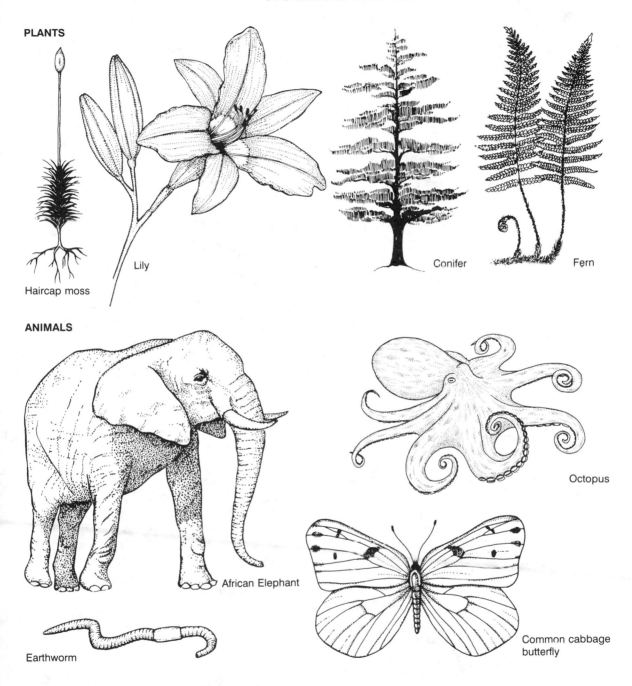

Haircap moss

Lily

Conifer

Fern

ANIMALS

African Elephant

Octopus

Earthworm

Common cabbage
butterfly

Kingdom Plantae: multicellular; cells are walled and frequently vacuolate; nutrition mostly photosynthetic; primarily nonmotile and living anchored to substratum; reproduction primarily sexual; about 300,000 known species.

Kingdom Animalia: multicellular; lacking the rigid cell walls characteristic of plant cells; high degree of structural differentiation; motile; nutrition by ingestion; reproduction primarily sexual; 1–2 million known species.

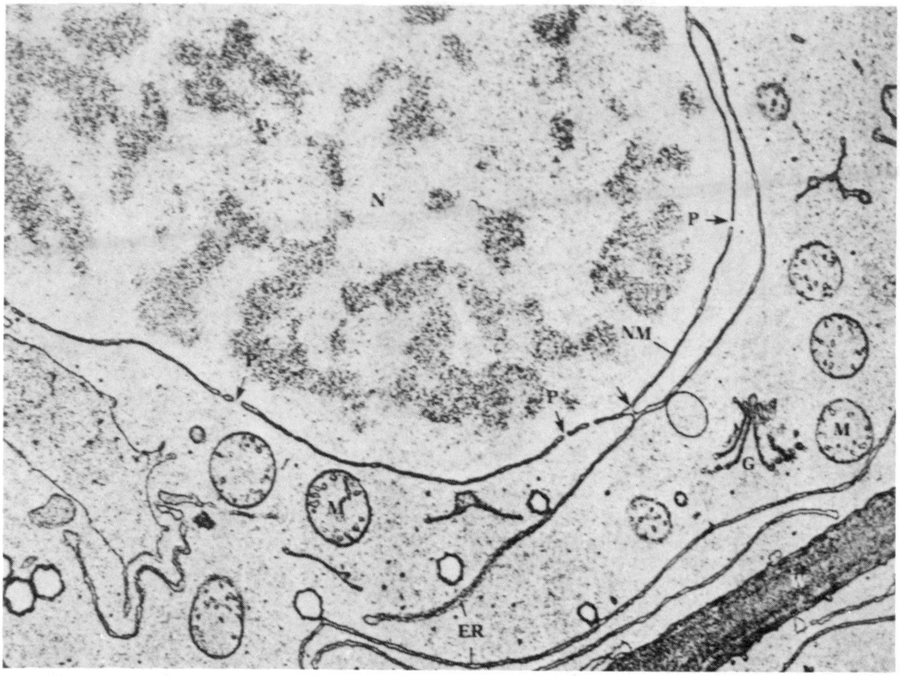

Figure II.4. A cell from a corn root (a typical eukaryotic cell) viewed through an electron microscope (× 16,700). Many details of the cell's structural components are visible: the dominant nucleus (N), the double-walled nuclear membrane (NM), pores in the nuclear membrane (P), sections of the endoplasmic reticulum (ER), a point where the endoplasmic reticulum and nuclear membrane interconnect (arrow), mitochondria (M), a Golgi apparatus (G), and the outer cell wall (W).

mitochondria, chloroplasts, and *lysosomes* (see figures II.4, II.5). The chloroplasts are present only in cells that carry out photosynthesis, such as the cells of green plants and algae. Each of the organelles is designed to perform specific tasks, much like the tools and machines in a workshop are designed for specific tasks. Other bodies present in the cytoplasm are the *ribosomes*, which are complexes of protein and nucleic acid and constitute the factories for protein synthesis (see below). Only eukaryotic cells possess the internal organization and biochemical sophistication needed to assemble into large and complex multicellular organisms, such as animals and plants.

Let us investigate the tasks performed by the organelles and ribosomes of eukaryotic cells. We will discover some of the ways by which the cells function and maintain themselves and will learn how genetic information governs biological processes. We will be able to compare eukaryotic and prokaryotic cells and gain some appreciation of the enormous differences evolution has created between them.

In chapter 1 we journeyed to the far reaches of the Cosmos; here we shall take an imaginary journey into the microscopic world inside one of the cells of the human body. The cell is filled with a watery fluid, so we will pretend to explore in a miniature submarine. The sub is about one-tenth of a micrometer in size, which makes it roughly a billion times smaller than a real sub. Despite

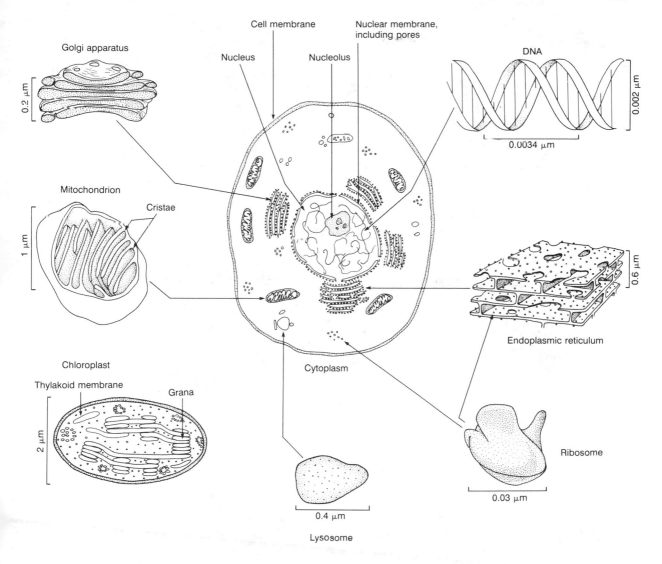

Figure II.5. The eukaryotic cell. The organelles shown are:

Nucleus, contains the cell's genetic information (written on molecules of DNA).

Endoplasmic reticulum, network of folded membranes; dots are ribosomes. The ribosomes are the sites of protein synthesis, according to instructions delivered by mRNA.

Golgi apparatus, packages and exports proteins and other products manufactured by the cell.

Mitochondrion, possesses folded internal membranes—the cristae—on which ATP is produced.

Chloroplast, present only in the cells of green plants and algae. Chloroplasts possess internal membrane systems—the thylakoids, stacked up into grana—on which energy of light is converted into energy of ATP.

Lysosome, filled with potent enzymes for breaking down foreign objects.

its smallness, we assume that our sub is sturdy enough to withstand the dangers ahead and that it has sufficient engine power to permit us to steer through molecular jams as well as against currents and out of whirlpools we might encounter. These assumptions are necessary because the cytoplasm of the cell contains dense concentrations of molecules, membranes, and organelles, resembling a miniature version of the Sargasso Sea with its great masses of free-floating seaweeds.

Enzymes. We start our journey in the nucleus of the cell. The first thing we notice in our microscopic environment is the incessant activity going on all around us. The watery medium is crowded with a great variety of molecules of all sizes and shapes. These molecules are in constant motion, like the materials and machinery in a busy factory. Simple molecules are being assembled into more complicated ones, and they, in turn, are being bonded together into still larger structures. The work is largely done by roundish looking proteins, called *enzymes*. They possess vise-like sections of different shapes, with which they firmly grab the molecular building blocks and force them into place, one against the other. Upon interlocking two molecules, they release their grip and go on repeating the task. In this way, the enzymes assemble long molecular chains, rings of various shapes, and intricate lattice structures. Some of these molecules are so small that, despite our own miniature size, we can barely see them. Others have sizes and shapes comparable to our own sub. Still others are strands of up to a hundred million times the length of the sub, but only a tiny fraction of its width. All of these molecules are very stable. When we bump into them, they merely bend and stretch a bit, but they do not break. Only certain enzymes with the correct grip seem to be able to undo the strong bonds holding the molecules together and tear them apart into simpler constituents. Generally this buildup or destruction of molecules is referred to as a *chemical reaction* and the enzymes that facilitate the reactions are *catalysts* (see footnote in box 3.4).

DNA. The largest molecular structures we find inside the nucleus are DNAs. They are the long, thin molecules mentioned above. They wind back and forth through the nucleus, and it is difficult to understand why they do not become inseparably tangled with each other. Altogether we count 46 of them.* Upon closer inspection, we find that each DNA consists of two parallel strands, connected by single- and double-ringed molecules called *bases*. The two strands wind around each other like those of a rope, forming a double-helical structure with the connecting bases in the middle.

As we examine the structure of a DNA molecule, we notice that the bases, which connect the two strands of the double helix, come in four different versions: *adenine* (A), *guanine* (G), *cytosine* (C), and *thymine* (T) (their

---

*We assume here that we are investigating a cell of the human body. The cells of every eukaryotic species have their own characteristic numbers of DNA molecules. For instance, the cells of the fruitfly *Drosophila* have 8 DNA molecules, those of the rattlesnake 36, and those of the potato 48. The DNA molecules are only about 0.002 μm thick, which is so thin that normally they can be seen only through the electron microscope. However, during cell division, when they coil up into the relatively thick, stubby bodies known as *chromosomes*, they become visible through the light microscope.

molecular structures will be described in the following section). The specific sequence of these bases constitutes genetic information, much as the sequences of letters on this page, taken from an alphabet of 26 letters, constitute information. Even though the sequence of bases is continuous, its information content consists of separate units — called *genes* — arranged in a linear sequence one after the other along the DNA. This is comparable to the subdivision of the continuous sequence of letters in an encyclopedia into separate articles. A DNA molecule may contain thousands of genes. A gene carries one of three different kinds of instructions: (1) assembly instructions for the manufacture of proteins, (2) control instructions for when to make certain proteins, and (3) instructions for making the machinery that assembles proteins (transfer and ribosomal RNA, see below).* One of the primary goals of our journey through the cell is to discover how the assembly instructions of the genes are converted into actual proteins.

RNA. Because the genetic instructions are written on molecules of DNA, observing them should give us a start in tracking down the process of protein synthesis. At several places we can see enzymes wedging themselves between the helical strands of DNA and pulling apart short sections corresponding to genes. This exposes the bases and, hence, the genetic information, and allows other enzymes to copy it. They do that by using molecular raw material present in the nuclear environment and constructing molecules very similar to DNA, except that they consist of only single strands with bases attached to them. These copies are called *RNA*, which stands for *ribonucleic acid.*†

The DNA molecules are the "master copy" of genetic information and are kept safely in the nucleus of the cell. The information is transcribed onto RNA molecules for the purpose of translating the information into proteins. There are three different kinds of RNA. One kind, the *messenger RNA* (mRNA), carries information for the assembly of proteins to the protein factories. The other two kinds, the *transfer* and *ribosomal RNA* (tRNA and rRNA), are part of the machinery that assembles proteins.

Let us follow one of the mRNA in the hope that it will take us to a protein factory. For a while, the mRNA meanders through the nucleus. As we steer our sub after it, we pass several DNA and RNA molecules, all sorts of smaller molecules, and large globularly shaped enzymes. Nowhere do we find a protein factory. Eventually, our mRNA approaches the membrane enclosing the nucleus, which is dotted with tunnels (the technical name is *pore*) leading to the cytoplasm beyond. Through the pores molecular raw materials enter the nucleus and other kinds of molecules leave. Among the molecular products leaving is our mRNA. It seems the protein factories might lie out there in the cytoplasm. We force our way out through one of the pores.

Ribosomes. Outside the nucleus we try to catch up with our mRNA, which is heading toward one of the many globularly shaped bodies scattered through-

---

*Some biologists define genes as referring only to the first of the three kinds of instructions carried by DNA and listed above, namely the assembly instructions for the manufacture of proteins.

†RNA contains the same bases as DNA, except that thymine is replaced by the almost identical base *uracil*. Thus, the bases of RNA are A, G, C, and U (more about this in the following section).

out the cytoplasm. These bodies are roughly of the size of our sub. They are *ribosomes* and, judging from the activity going on near them, are the factories we have been looking for. They are surrounded by millions of *amino acids*, the molecular building material of proteins. We count 20 distinct kinds of amino acids, differing from each other in shape and size. Each amino acid is attached to a small RNA molecule, namely a tRNA. Several of the ribosomes have strands of mRNA slowly running through them, much as a computer tape runs through the head of a tape reader, except that the ends of the mRNAs are not neatly wound up on spools. Simultaneously, the internal machineries of the ribosomes, which consist of enzymes and rRNAs, seize the tRNAs (each with an amino acid attached), line them up, and link the amino acids into growing chains as requested by the instructions. In the process, the tRNAs (now without amino acids) are released. As amino acid after amino acid gets added and the chains increase in length, they coil into helical structures, which, in turn, fold and twist into very specific configurations. The result is functional protein molecules.

It is fascinating to watch the ribosomes at work. They are busy and efficient assembly lines, manufacturing proteins according to instructions delivered by mRNAs. Usually a single strand of mRNA passes through several ribosomes at once, allowing the simultaneous construction of more than one protein from the same set of instructions. Depending on the number and on the particular sequence of amino acids in the assembled chain, different kinds of proteins result: globularly shaped enzymes for speeding up the chemical reactions in our bodies; proteins that act as chemical sensors or as transporters of certain molecules; and structural proteins for our bones, skin, connective tissues, and the linings of our organs, to name just a few.

Protein synthesis goes on at many places in cells. Some ribosomes exist freely in the cytoplasm. Others are attached to the exterior wall of the nuclear membrane. Still others are associated with vast networks of interconnected channels called the *endoplasmic reticulum* (figures II.4 and II.5). These channels are used to transport newly manufactured proteins to another set of channels (figure II.6; see also figure II.5) called the *Golgi apparatus* (named after the Italian physician and Nobel laureate Camillo Golgi [1843–1926], who first described this organelle in 1898). The Golgi apparatus packages the proteins and sends them through the cell membrane to the outside. Thus, the DNAs, RNAs, ribosomes, endoplasmic reticulum, and Golgi apparatus constitute the management, communication network, factories, and distribution system for protein synthesis in eukaryotic cells.

Energy Conversion. Any system as elaborate as that of protein synthesis requires ample energy. It is needed for each of the steps involved, from separating the two strands of the DNAs to assembling the proteins and distributing them. We have no difficulty tracking down where the energy comes from. Through its membrane the cell continually imports sugar molecules, which are the primary suppliers of energy. Most important among these sugars are molecules of *glucose*. The energy contained in glucose, and in the other sugar molecules, is liberated in a complicated sequence of chemical reactions during which the molecules are broken down into carbon dioxide (CO_2) and water (H_2O). The final part of the sequence requires oxygen (O_2) and occurs in special

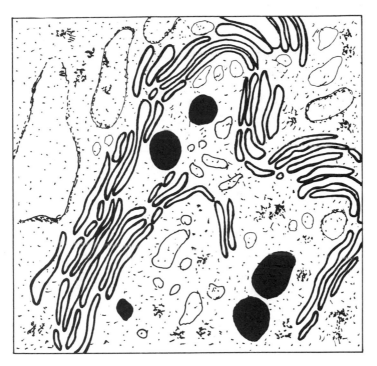

Figure II.6. Golgi apparatus from the green alga *Chlamydomonas* (× 78,000). Note the characteristic stacks of parallel, folded membranes of the apparatus.

rod-shaped organelles, the *mitochondria* (figure II.7; see also figure II.5). We find mitochondria everywhere in the cytoplasm. They are about ten times larger than our sub and we cannot miss them.

The energy released by the breakdown of sugar molecules becomes temporarily stored in high-energy bonds of molecules called *adenosine triphosphate* or *ATP*. The ATPs deliver their energy to wherever it is needed by the cell, such

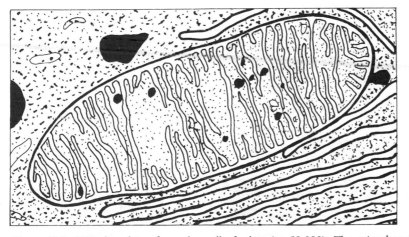

Figure II.7. Mitochondrion from the cell of a bat (× 50,000). The mitochondrion's double outer membrane and numerous folded inner membranes (the cristae) are clearly visible. Also visible, outside the mitochondrion, is a section of the cell's endoplasmic reticulum, with ribosomes attached to its membrane.

as in the various steps of protein synthesis or the transport of materials within cells and across cell membranes. (Contraction of muscle cells and conduction of nerve impulses are still other examples of processes utilizing energy stored in ATP.) The full sequence of chemical reactions by which cells convert the energy of sugars into high-energy bonds of ATP with the participation of O_2 is called *aerobic respiration*.

If we were exploring a cell from the green leaves of plants or from algae instead of one from our own body, we would find chloroplasts in addition to mitochondria and the other organelles. *Chloroplasts* ("green bodies") are disk-shaped bodies that harness energy by *photosynthesis* (figure II.8). They trap the energy of sunlight with the aid of molecules of chlorophyll and other similar pigments. Like the mitochondria, they temporarily store energy in high-energy bonds of ATP. However, instead of requiring O_2, they release it. They then use the ATPs to run their metabolic machinery. In particular, they use them to convert CO_2 and H_2O into energy-rich sugars and other carbohydrate molecules. Thus, as described more fully in chapter 6, photosynthesis and respiration are parts of the same cycle. In photosynthesis, energy from the Sun is used to elevate the carbon, oxygen, and hydrogen atoms of CO_2 and H_2O into energy-rich sugar molecules. Molecular oxygen is expelled in the process. In aerobic respiration, the sugars are broken down, with the participation of O_2, into the low-energy molecules CO_2 and H_2O, and the energy is put to use.

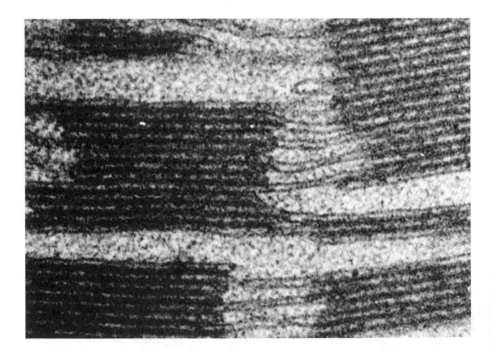

Figure II.8. Thin section of a typical chloroplast ($\times$ 60,000). The photograph shows the internal membranes of the chloroplast, called thylakoids, on which photosynthesis is carried out. It also shows the grana, regions where the thylakoid membranes form stacked aggregates.

We have now completed our tour through the eukaryotic cell and are ready to return to our normal macroscopic environment. We do this by disguising ourselves as a protein molecule manufactured by the cell and leaving via the Golgi apparatus. We must be careful, however. In the vicinity of the Golgi apparatus we are likely to encounter *lysosomes*, which are roundish saclike organelles filled with very potent enzymes. The lysosomes attack foreign objects, such as bacteria, viruses, or defective proteins, and engulf them. Within minutes, the enzymes break the objects down into simple molecular units, which are then recycled by the cell. Because we do not wish to test how well our sub would stand up to such an attack, we allow the Golgi apparatus to wrap us into a membrane coat, as it does with proteins ready for export. This makes us indistinguishable from proteins manufactured by the cell and we safely escape through the cell membrane to the outside.

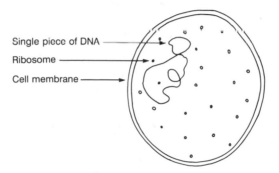

Single piece of DNA
Ribosome
Cell membrane

Figure II.9. The prokaryotic cell.

Prokaryotes

In contrast to the eukaryotic cells, most prokaryotic cells have sizes in the range from 1 to 5 μm. But there are exceptions. The smallest bacteria, the mycoplasmas, measure only about 0.3 μm across, while a few of the blue-green bacteria (also known as cyanobacteria, and formerly called blue-green algae, though they are not algae) may grow as large as small eukaryotic cells. Some prokaryotes aggregate into single-stranded and branched filaments or into globular colonies. However, none show evidence of cell specialization, nor do they generally depend on each other for survival.

Had we toured a prokaryotic cell, we would have found that, like the eukaryotic cell, it is filled with cytoplasm and bounded by a cellular membrane, through which it exchanges materials with the surroundings (see figure II.9). However, we would have found relatively little internal organization. There is no nuclear membrane, and instead of several molecules of DNA there is only one. This single DNA forms a circular loop and resides in the cytoplasm (figure II.10). There also are ribosomes. In fact, the entire process of protein synthesis, from the transcription of information from DNA onto mRNA to the translation of that information into protein structure is largely the same in both types of cells. The one rather interesting difference is that in prokaryotic cells the mRNAs begin passing through ribosomes and delivering their information while the other ends are still being copied from DNA—transcription of genetic information takes place simultaneously with translation of that information into protein molecules. We would have come across no organelles in the prokaryotic cell. In their absence, ATPs are produced on the cell's enclosing membrane and molecular

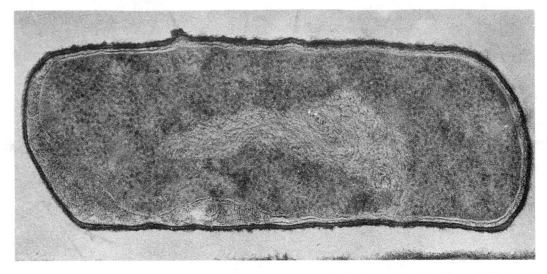

Figure II.10. Electron micrograph of a bacterium, *Bacillus subtilis* (a typical prokaryotic cell; × 75,000). No organelles are present and the DNA is not surrounded by a nuclear membrane, as is the case in eukaryotic cells.

materials are transported directly through the cytoplasm without the benefit of special channels.

These descriptions of eukaryotic and prokaryotic cells point to the profound differences existing between them. Eukaryotic cells are unquestionably more complex than prokaryotic cells. Nevertheless, important similarities also exist between them, which suggests that eukaryotic cells evolved from prokaryotic ancestors. That this is in fact the case becomes apparent when the biochemical processes of the two types of cells are compared. The structures of DNA and RNA are nearly identical in both cells and the processes of cell division in eukaryotic cells seem to be derived from those in prokaryotes. Both cells use similar steps in the synthesis of proteins. Furthermore, parts of the metabolic pathways of the eukaryotic cells are closely related to those of prokaryotes. The next two chapters will investigate these rather interesting relationships.

ORGANIC MOLECULES: THE FUNDAMENTAL BUILDING BLOCKS OF LIFE

This discussion of cell structure and function indicates that life's fundamental building blocks are organic molecules. The other constituents — water, ions, and trace elements — are mere support units. Organic molecules are so special for life because they possess a combination of rather unique properties:

1. They come in a number of small basic units that can be linked together into many different kinds of large and elaborate structures, such as chains, rings, lattices, long fibers, and globules.

2. Because of their diverse structures, organic molecules are able to participate in complex, self-regulating chemical reactions. These reactions are the source of all biological processes.

3. The chemical bonds holding organic molecules together are quite strong and not easily broken. That is why these molecules form and persist in such dissimilar environments as interstellar clouds, meteorites, the atmospheres of the Jovian planets, and organisms.

4. The primary constituents of organic molecules—carbon, hydrogen, oxygen, nitrogen, phosphorus, and sulfur—are among the most abundant elements on the surface of our planet.

Organic molecules are made by joining carbon with carbon, hydrogen, oxygen, nitrogen, and other atoms by chemical bonds.* Carbon is always connected by four bonds to other atoms, nitrogen by three, oxygen by two, and hydrogen by one. Examples of these bonds are found in methane (CH_4), ammonia (NH_3), and water (H_2O), which were described in the discussion of interstellar clouds (figure 3.4). The bonds between carbon and other atoms need not necessarily be single bonds, as in the case of methane. They may be double or triple bonds. For instance, in formaldehyde carbon forms a double bond with oxygen,

(figure 3.5); and in acetylene (a gas that burns brightly in the presence of oxygen and is used in welding and soldering) two carbon atoms are joined by a triple bond,

$$H - C \equiv C - H.$$

This characteristic of carbon—to form strong single, double, and triple bonds— gives organic molecules their great versatility and, together with the four properties discussed above, makes them the natural and, possibly, only workable choice for the construction of organisms.

Although organic molecules assume a virtually unlimited variety of structures, many of these structures exhibit certain similarities. This allows chemists to group organic molecules into distinct classes and greatly simplifies the way we think about them. The larger, so-called polymeric (meaning "many parts") organic molecules that are of biological significance may be conveniently placed into four major classes: the carbohydrates, lipids, proteins, and nucleic acids. The *carbohydrates* comprise the starches we eat and from which we derive much of our energy, the cellulose that gives wood its rigidity, and the glycogen

---

Organic chemistry deals with molecules composed of carbon, hydrogen and, quite often (but not necessarily), other elements. All chemistry not involving carbon *and* hydrogen is called *inorganic chemistry*. The words organ, organism, and organic are derived from the Greek word *organon*, meaning tool or instrument. They are also related to the Greek word *ergon*, which stands for work. Until the discovery during the past few decades of organic molecules in interstellar clouds, in meteorites, and in the atmospheres of the Jovian planets, it was thought that only living organisms could produce them.

by which we store energy in our bodies. The *lipids* include the fats of our bodies, the cholesterol in our blood, and many of the hormones, such as testosterone and estrogen (male and female sex hormones). The fats are primarily used for storage of energy and for the structural envelopes of cells and their constituent organelles. The *proteins* serve several functions. They may be structural proteins, giving our bones their strength, lining our organs, and holding our bodies together through their presence in skin, tendons, and connective tissues. Structural proteins are also associated with certain fats (see above) to form the membranes of cells and their organelles. Other proteins are chemical sensors and transporters of certain molecules. Still other proteins act as enzymes and are responsible for the smooth functioning of all chemical reactions going on in the cells, including reproduction, growth and self-maintenance. The *nucleic acids* are the DNAs and RNAs, which are carriers of genetic information.

Carbohydrates

The *carbohydrates* consist of carbon, hydrogen, and oxygen and are constructed by combining carbon with the atomic constituents of water. The sugars are the simplest carbohydrates, and the most important sugar molecules are *glucose, ribose*, and *deoxyribose*. The glucose molecule is one of the suppliers of energy in our cells (figure II.11). You may recognize ribose and deoxyribose, for the names ribo- and deoxyribonucleic acids indicate that these two sugars are present in RNA and DNA. Another sugar molecule is *fructose*. Bonded to glucose, it forms sucrose, ordinary table sugar (figure II.12). Glucose, fructose, ribose, and deoxyribose are ring-shaped molecules.

The chemical formula of glucose is $C_6H_{12}O_6$, meaning that it is constructed of six carbon, twelve hydrogen, and six oxygen atoms. It is manufactured during photosynthesis from CO_2 and H_2O, with the simultaneous release of O_2:

$$6\ CO_2\ +\ 6\ H_2O\ \rightarrow\ C_6H_{12}O_6\ +\ 6\ O_2.^*$$

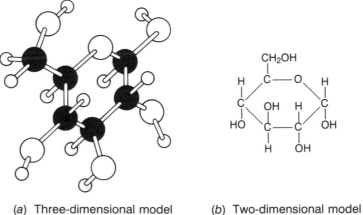

(a) Three-dimensional model   (b) Two-dimensional model

Figure II.11. Two ways of representing the glucose molecule. A third way is by the chemical formula $C_6H_{12}O_6$.

*The conversion of CO_2 and H_2O into glucose and O_2, as represented symbolically above, is an example of a chemical reaction. All of life's processes are based on chemical reactions.

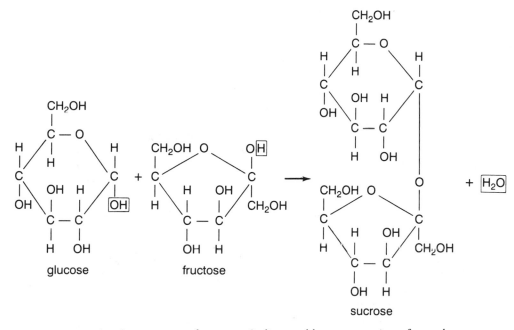

Figure II.12. Molecular structure of sucrose. Ordinary table sugar consists of crystals of sucrose molecules. Sucrose is formed by the bonding together of glucose and fructose, with the simultaneous removal of a water molecule (see rectangles). Glucose and fructose are ring-shaped carbohydrate molecules with the same chemical formula, $C_6H_{12}O_6$. However, the ring of glucose is hexagonal, that of fructose is pentagonal.

Glucose molecules are then assembled into long and often branched chains. Depending on the particular way the glucose molecules are bonded to each other, the result is either starch or cellulose (figure II.13). *Starch* is a major constituent of vegetables, fruit, and some roots (carrots, potatoes); while *cellulose* is a major structural component of plant tissues (found particularly in wood, plant stalks, and grass). We can digest starch, for our bodies produce the enzymes required to break the bonds between its glucose molecules. We cannot digest cellulose because we do not have the necessary enzymes, but insects, cows, and many other animals can feed on it. A part of their digestive tract is inhabited by microorganisms (bacteria and protists) that produce enzymes capable of breaking down cellulose.

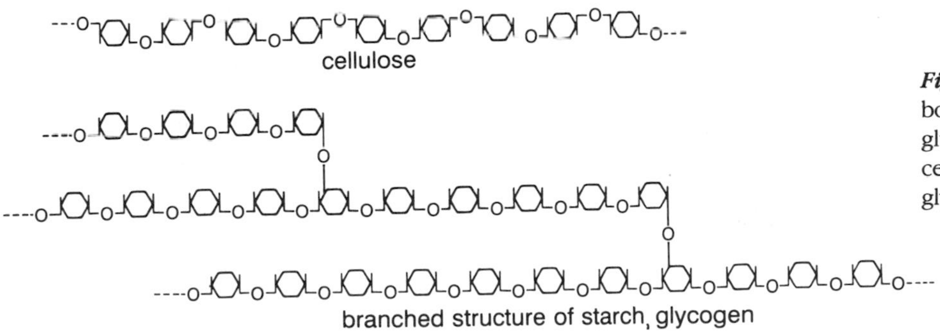

cellulose

branched structure of starch, glycogen

Figure II.13. The bonding between glucose molecules in cellulose, starch, and glycogen.

Upon digestion of starch, individual glucose molecules enter our blood stream and are carried to the cells to deliver their energy. That energy is converted into high-energy bonds of ATP, as discussed in the previous section. Excess glucose enters the liver and is stored for future use in long molecular chains identical to starch, except that they are more highly branched. These molecular chains are called *glycogen* (see box II.1). Whenever the glucose concentration in our blood—also known as blood sugar—runs low, such as with strenuous exercise, we feel tired. Glycogen is then broken down in the liver and glucose molecules are released into the bloodstream for transport to the active cells.

Lipids

The *lipids* consist of carbon, hydrogen, and oxygen (though some of them, the phospholipids for example, also contain phosphorus and nitrogen). In comparison to carbohydrates, the oxygen content of lipids is much lower, which makes all lipids insoluble in water. For instance, grease and oil (which are lipids) do not dissolve in water.

Of the several types of lipids present in organisms, I shall discuss only the *neutral fats*. They make up the major fat content of our bodies and are also present in milk, cheese, butter, and margarine. The neutral fats are composed of two different kinds of molecular units, *glycerol* and *fatty acids*. Fatty acids are chains of carbon with a carboxyl group* attached at one end:

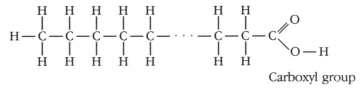

Carboxyl group

*The *carboxyl group*,

is very common in organic molecules. For instance, it constitutes the acid part of amino acids.

Sixteen- and eighteen-carbon fatty acids are the most common ones in our bodies. Glycerol is a three-carbon molecule:

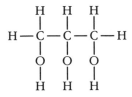

To make neutral fat molecules, three fatty acids are linked to a glycerol, with the simultaneous removal of three water molecules:

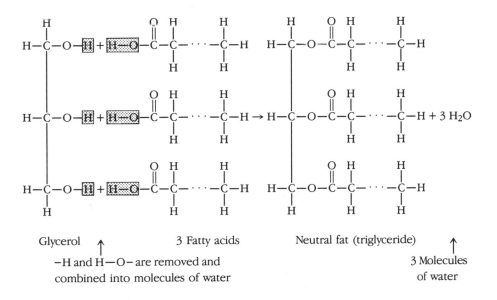

Glycerol   3 Fatty acids   Neutral fat (triglyceride)

−H and H—O− are removed and   3 Molecules
combined into molecules of water   of water

Neutral fats are also known as *triglycerides* because they consist of glycerol linked with three fatty acids. During medical checkups, it is now common to have one's blood examined for triglyceride concentration.

The atomic constituents of the water molecules liberated during the formation of neutral fats come from both the glycerol and the fatty acids, as shown by the shaded rectangles in the above chemical reaction. Such *dehydration* reactions are the usual way of assembling large organic molecules (polymers) from basic units (monomers). During the breakdown of polymers, the process is reversed and water must be supplied. The reverse reactions are called *hydrolysis*, meaning that they require the splitting of water into H and OH. (Hydrolysis is derived from the Greek *hydro-* and *-lysis*, meaning "water" and "loosening" or "splitting." The root *hydro-* is also present in the word *carbohydrate*.)

In some neutral fats, all the carbon atoms of the fatty acids are joined by single bonds, as in the examples above. These fatty acids contain the maximum possible number of hydrogen atoms and, therefore, are called *saturated*. In other fats, some of the carbon atoms of the fatty acids possess only one H and are linked to a neighboring carbon atom by a double bond:

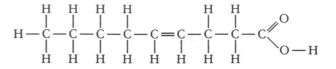

Such fats are said to be *unsaturated* or, if there are several such double bonds, *polyunsaturated*. Animal fats are mostly saturated, while those of vegetables, seeds, and nuts are usually polyunsaturated. There seems to exist a strong correlation between consumption of saturated fats with fatty deposits in the blood vessels and heart disease. Hence, nutritionists recommend that we eat only moderate amounts of meat and diary products.

Proteins

Proteins contain nitrogen and, in some instances, sulfur in addition to carbon, hydrogen, and oxygen. Of all the organic molecules found in organisms, the proteins have the most varied range of applications. Not only are they the major *structural materials* of organisms, they also act as *enzymes*, which catalyze all biochemical reactions from those of bacteria to those of man. Still other proteins serve as *hormones* or as *carrier molecules* in animals. An example of a protein hormone is insulin. It is produced in the pancreas (a gland) and assists in the transport of glucose across cell membranes, thereby speeding up the rate of energy use by cells. An example of a protein carrier molecule is hemoglobin (see below).

Proteins are polymers, consisting of large numbers of amino acids (the monomers) bonded together in long linear chains. Of the hundreds of conceivable amino acids, only twenty occur in the protein of organisms. The three simplest amino acids are glycine, alanine, and serine:

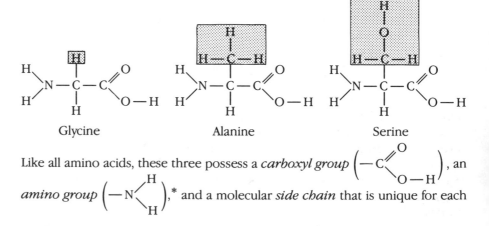

Glycine   Alanine   Serine

Like all amino acids, these three possess a *carboxyl group* $\left(-C\begin{smallmatrix}O\\\\O-H\end{smallmatrix}\right)$, an *amino group* $\left(-N\begin{smallmatrix}H\\\\H\end{smallmatrix}\right)$,* and a molecular *side chain* that is unique for each

---

*The name *amino* implies a relationship to ammonia (NH_3) and indicates the presence of nitrogen.

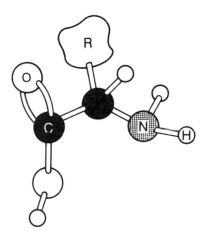

Figure II.14. Three-dimensional model of an amino acid. R designates the molecular side chain.

amino acid and distinguishes it from the rest. In glycine the side chain (marked by shaded rectangles in the above formulae) is a single hydrogen atom (H), in alanine it is CH_3, and in serine it is CH_2OH. If we label the side chains by the letter R, we obtain a generalized formula for amino acids (figure II.14):

Molecular side chain, characterizes the amino acid

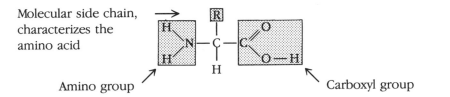

Amino group   Carboxyl group

Figure II.15 lists all twenty amino acids occurring in terrestrial life. In addition to carbon, hydrogen, oxygen, and nitrogen, two of the amino acids (cysteine and methionine) contain sulfur. In most amino acids, the side chains are linear or branched, but in the last five amino acids shown in figure II.15 they are ring-shaped. Unlike other life forms, animals have lost the ability to synthesize all of the twenty amino acids. For example, adult humans do not synthesize valine, leucine, isoleucine, threonine, lysine, methionine, phenylalanine, tryptophan, and histidine. We must obtain these nine "essential" amino acids from our diet.

The assembly of amino acids into linear chains occurs by dehydration reactions, like those involved in the making of cellulose, starch, glycogen and neutral fats. The OH from the carboxyl group of one amino acid and an H from the amino group of a second amino acid are removed (as H_2O) and the C and N thus exposed are bonded together:

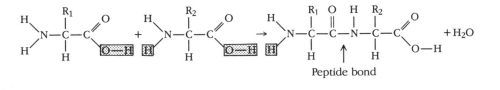

Peptide bond

First amino acid   Second amino acid   Peptide

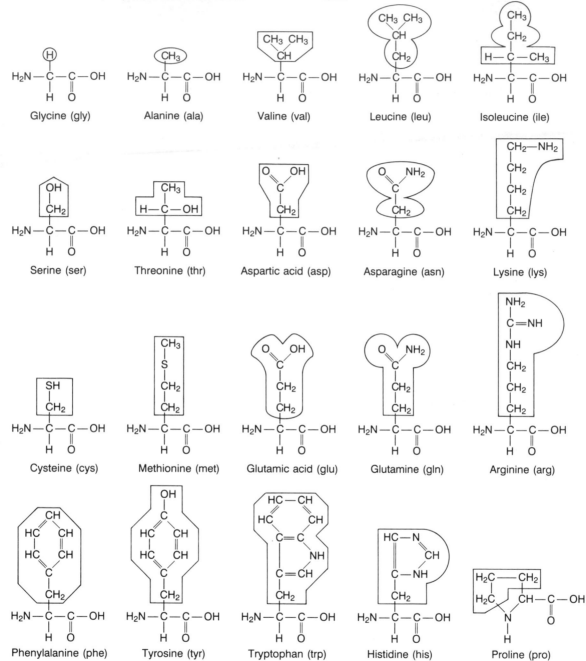

Figure II.15. The twenty amino acids present in protein.

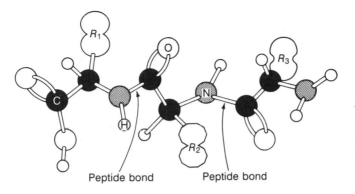

Figure II.16. A short peptide chain, consisting of three amino acids. The amino acids are linked by peptide bonds.

A third, fourth, and still other amino acids are then added, until chains containing hundreds and even thousands of amino acids are obtained. Such chains are called *peptides* (they are not yet functional proteins), and the bonds linking the amino acids are called *peptide bonds* (figure II.16).

In cells the assembly of peptides is far more complicated than indicated by the above description. Not only has a peptide bond to be formed, but it has to be formed between two particular amino acids. The instructions for the sequence in which particular amino acids are to be selected from the twenty existing in organisms and strung together into a peptide chain comes ultimately from the cell's DNA—the storage molecule of genetic information. The translation of these instructions from DNA into peptides involves RNAs and ribosomes and is a process that is central to what makes organisms alive. I shall discuss this translation process in chapter 6.

Linear chains of peptides represent the *primary* structure in the hierarchy of the structures that constitute a protein molecule. As a peptide chain grows, it forms a coiled strand that is the *secondary* structure in the hierarchy. Depending on the precise sequence of amino acids, the coiled peptide strand folds back and forth on itself, forms knot- or pretzellike configurations, or twists into a spiral. Such bending, folding, or spiraling represents the *tertiary* structure. Generally, tertiary peptide structures do not occur alone, but are intertwined with other tertiary peptide structures. The result is the *quaternary* structure in the hierarchy and constitutes a functional protein molecule (figure II.17).

The twisting and folding of peptide chains into functional proteins produce molecules that are either long and fibrous or compact and globular. Examples of these two types of proteins are collagen (fibrous) and hemoglobin (globular). *Collagen* is an important structural component of our bodies. Its synthesis begins with the assembly of amino acids, most of which are glycine and proline, into long peptide chains. Upon coiling into secondary structures, three of the peptides wind around each other into *tropocollagen* molecules. The tropocol-

(a)   (b)   (c)

Figure II.17. The structure of protein: (*a*) Diagram of a linear peptide chain—*primary structure*; (*b*) a coiled peptide chain—*secondary structure*; (*c*) three coiled peptide chains twisting around each other to form a fibrous protein, as, for example, a fiber of collagen. The twisted form of a single coiled peptide chain represents the *tertiary structure*. The combination of the three peptide chains twisting around each other represents the *quaternary structure* and constitutes a protein molecule.

lagen molecules are then joined in staggered fashion to form a collagen fibril (tertiary and quaternary structures), much like single strands of wool fibers are spun into yarn (figure II.18). From these fibrils still larger structures are assembled, analogous to the weaving or knitting of yarn into finished woolen products. If the collagen fibrils are combined into long parallel bundles, structures, like tendons and other connective tissues result, which have great strength but little stretch. If they are meshed into flat, interlacing networks, human skin or animal hide is formed, which are highly flexible. If the fibrils are interwoven with calcium-containing inorganic molecules (for example, calcium phosphate and calcium carbonate), bone is made, which is strong and hard without being brittle.

Hemoglobin is contained in red blood cells and is the carrier molecule of O_2 in blood. It consists of four coiled peptide chains, which are tightly folded together forming the *globin* part of hemoglobin (see figure II.19). In its midst, each of the peptide chains holds a relatively small, disk-shaped *heme* molecule (which is not a protein) that contains an atom of iron. The iron gives blood its deep red color. As blood flushes through the capillaries of the membranes lining the lungs, hemoglobin picks up O_2. The O_2 (bound to hemoglobin) is then carried by the bloodstream to the body tissues, where it is released. (The word *heme* is derived from the Greek and means "blood." *Globin* comes from the Latin *globus*, meaning "ball.")

The folding of hemoglobin is typical of all globular proteins. The nature of the folding and, hence, the uses to which the proteins are put—enzymes, hor-

Figure II.18. Collagen fibrils, carefully pulled away from human skin (× 42,000). The distinct dark bands visible on the fibrils are due to a partial overlapping of adjacent tropocollagen molecules (see text) and are about 0.07 μm apart.

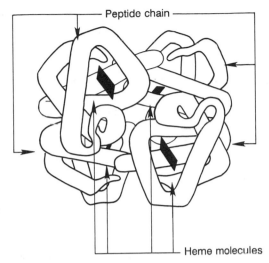

Figure II.19. Hemoglobin. The folding, twisting, and intertwining of the four coiled peptide chains of hemoglobin illustrate the *quaternary* structure of globular protein molecules.

mones, carriers—depends on the number and precise sequence of amino acids in the peptide chains. As discussed next, this number and sequence of amino acids is determined by the genetic information carried in coded form by DNA.

Nucleic Acids

The *nucleic acids* are made of carbon, hydrogen, oxygen, nitrogen, and phosphorus. Examples of nucleic acids are DNA and RNA. The major functions of DNA and RNA are to store genetic information in coded form, to participate in the conversion of this information into protein structure, and to pass on the information from generation to generation.

DNA generally consists of two molecular strands that spiral around each other like the two strands of a rope. That is why James D. Watson and Francis H. C. Crick, the discoverers of its molecular structure, referred to it as the "double helix" (see box II.2). *RNA* usually consists of a single molecular strand, whose structure differs only in minor ways from that of an individual strand of DNA.

Each of the strands of DNA or RNA is a polymer, whose monomers are called *nucleotides*. The nucleotides consist of three basic molecular units—a phosphate, a sugar, and a base (see figure II.20).

The *phosphate* is an inorganic molecule. The sugar molecule is *deoxyribose* in the case of DNA and *ribose* in the case of RNA (see above discussion of carbohydrates). The bases of DNA are *adenine* (A), *guanine* (G), *cytosine* (C), and *thymine* (T). The bases of RNA are the same as those of DNA except that thymine is replaced by a similar base called *uracil* (U). The bases A and G are double-ringed and called *purines*. The bases C and T (U) are single-ringed and called *pyrimidines*.

To make a strand of DNA or RNA, the nucleotides are strung together such that a phosphate is bonded to a sugar, the sugar is bonded to another phosphate, and so on, with the bases sticking off to the side (see figure II.21).

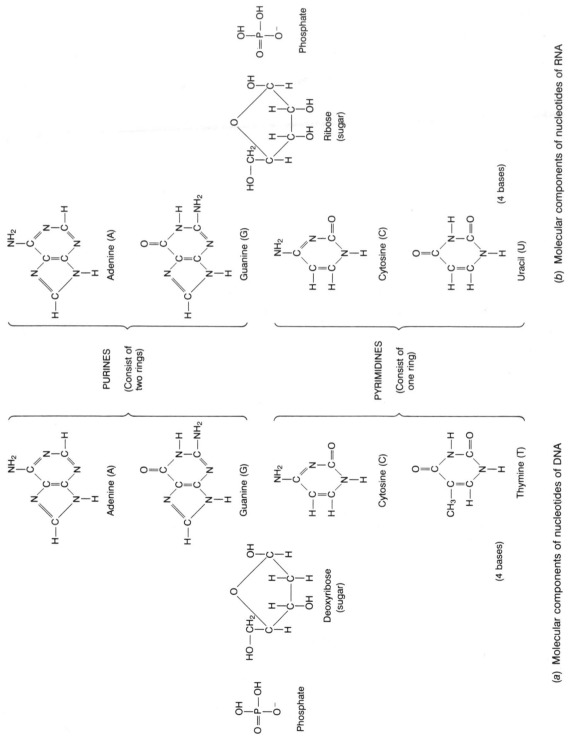

(a) Molecular components of nucleotides of DNA

(b) Molecular components of nucleotides of RNA

Figure II.20. Molecular components of nucleotides.

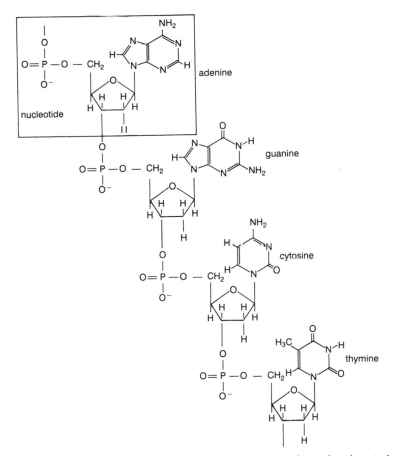

Figure II.21. A segment of a DNA molecule. For simplicity the chemical symbol for carbon (C) is deleted from the ring structures of this diagram. Both the assembly of nucleotides from their molecular constituents (phosphates, sugars, bases) and the assembly of DNA or RNA from nucleotides is accompanied by the release of water molecules. This can be seen by comparing the chemical structure of DNA given here with those of phosphate, deoxyribose, and the bases given in Figure II.20.

The sugar-phosphate sequence is often called the "backbone" of the strand. The sequence of bases, carried by the backbone, expresses genetic information (see the next section).

In the case of DNA, the two molecular strands run parallel to each other. The bases of the two strands point toward each other and are weakly bonded together by *hydrogen bonds* (see figure I.5), like the interlocking teeth of a closed zipper. Simultaneously, the two strands wind around each other in a double helix.

The bases of one strand of DNA are always bonded to the corresponding bases of the other strand according to the rule: A—T and G—C (or T—A and C—G). This ensures that a single-ringed base is always bonded with a double-ringed base, making the separation between the two sugar-phosphate chains the same all along the DNA molecule and providing for a neat, parallel structure

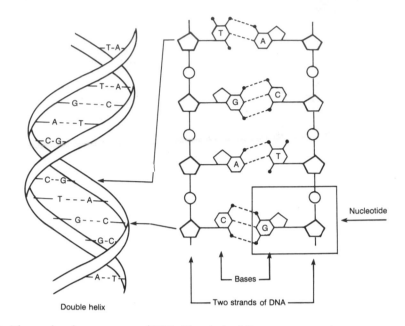

Double helix

Nucleotide

Bases

Two strands of DNA

Figure II.22. The molecular structure of DNA. The dashed lines represent the hydrogen bonds between bases A and T, G and C.

(figures II.22, II.23).* More important, this arrangement is the key to accurate replication and usage of genetic information, discussed further in chapter 7.

Before the nucleotides become bonded into strands of DNA or RNA, they exist in a form just slightly more complex than that described above. Instead of a single phosphate, they possess three phosphate molecules:

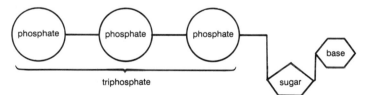

This arrangement of three linked phosphates is known as a *triphosphate*, and nucleotides possessing a triphosphate are said to be *activated*. For reasons connected with the bond structure linking the phosphates, a triphosphate is at an unusually high energy level. The energy is released when the bonds between the phosphates are broken. This is what happens when the nucleotides are assembled into strands of DNA and RNA. Two of the phosphates are broken off and the energy thus released is used to form the new bond between the single remaining phosphate and the sugar of the next nucleotide.

The principle for energy storage and energy use just discussed has been adopted by life for its most common energy-carrying molecule—*adenosine*

*Figure II.22 shows that A and T are bonded to each other by two hydrogen bonds, and G and C by three hydrogen bonds. No other bonding is likely to occur (such as A−C or T−G) because the number of hydrogen bonds would not match. The correct matching of bases is easy to remember: A and T are letters consisting of straight lines, while G and C are letters consisting of curved lines.

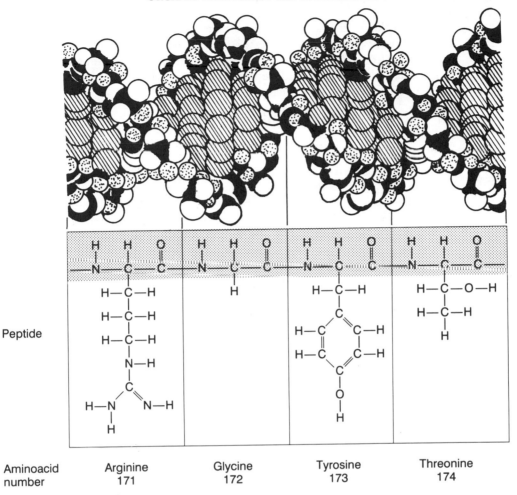

| Aminoacid number | Arginine 171 | Glycine 172 | Tyrosine 173 | Threonine 174 |

Figure II.23. Scale drawing of a section of a DNA molecule (top) and the corresponding peptide chain (bottom). The linear sequence of bases carried by DNA constitutes the instructions for the assembly of a linear sequence of amino acids, which constitutes a peptide chain. This direct correspondence between DNA and the peptide chain is emphasized in the diagram by the codon numbers and the amino acid numbers, 171–74. (See the genetic code section below for a definition of codon.) The black spheres of the DNA molecules represent the repeating units of deoxyribose sugar and phosphate, which form the helical backbones of the two-stranded molecule. The white spheres represent the base pairs, which connect the two sugar-phosphate strands.

triphosphate or *ATP*. ATP is nothing but a nucleotide with a triphosphate chain. Its sugar is ribose and the base is adenine:

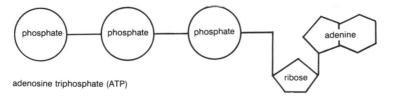

adenosine triphosphate (ATP)

By breaking off one of the phosphates (by hydrolysis), ATP is changed into *adenosine diphosphate* or *ADP* and energy is released for use by the cells. The change from ATP to ADP with the simultaneous release of energy is one step in a dual cycle. In the second step a phosphate is added to ADP, converting it back to ATP (a dehydration reaction):

Release of energy: $ATP + H_2O \rightarrow ADP + phosphate + energy$

Storage of energy: $ADP + phosphate + energy \rightarrow ATP + H_2O$

The second step of the cycle requires energy, which is obtained from sunlight (photosynthesis), glucose (respiration), or some other source.

The back and forth reactions between ATP and ADP plus phosphate, with the simultaneous release and trapping of energy, is an example of the economic use of molecular materials by cells. It is a recycling process in which the same molecules (ADP and phosphate) are used over and over again, while energy flows through the system and drives it. Similar recycling processes are found in many of the chemical pathways of living organisms.

On a more global scale, the flow of carbon, hydrogen, oxygen, nitrogen, and the other biologically important elements through the Earth's ecosystem also constitutes a recycling process. These elements become incorporated into biological organisms, only to be returned eventually to the ground, the water, or the air. As in the example of ADP and phosphate, this larger cycle requires the continuous input of energy, which ultimately comes from the Sun.

THE GENETIC CODE

In the previous section I introduced the molecular structures of DNA and RNA. But how does the sequence of bases in these molecules express genetic information? Focus on one of the strands of DNA and note that it carries its bases in a linear sequence, much as one side of a zipper carries its teeth in a linear sequence. There is one important difference, however, between the sequence of teeth of a zipper and the sequence of bases in a strand of DNA. The teeth of a zipper are all alike and hence contain no information. In contrast, there are four DNA bases—adenine, guanine, cytosine, and thymine, or A, G, C, and T (A, G, C, and U [uracil] in the case of RNA). Depending on how the four bases are arranged in sequence, a great many different kinds of messages can be written, just as many different messages can be written with the 26 letters of the alphabet.

Recall that DNA carries only three kinds of instructions: *assembly* instructions for how to make proteins (which involves mRNAs), *control* instructions for when to make certain proteins, and instructions for making *transfer RNAs* (tRNA) and *ribosomal RNAs* (rRNA), which are important constituents of the machinery that assembles proteins. DNA contains no other instructions. Of the three kinds of instructions, this discussion concerns only the first one.

Recall further that proteins (or their component peptide chains) consist of linear sequences of amino acids. The linearity of amino acids in proteins and of bases in DNA is no coincidence. They are directly related. The sequence of bases in DNA constitutes the instructions for the sequence in which amino acids are to be assembled into a given protein.

There are a number of conceivable ways by which the sequence of bases in DNA could express instructions for the assembly of amino acids. For instance, we could imagine that base A refers to one kind of amino acid, G to another, and so forth. A sequence of bases in DNA, such as CGACCATCAACT, would then have this meaning: assemble a peptide chain starting with the amino acid represented by C, add the amino acid represented by G, then the one represented by A, and so forth. Obviously, such a *singlet code* will not work because the four bases could give assembly instructions for only four kinds of amino acids.

Somewhat more complex and versatile than a singlet code is a *doublet code*, in which the pairs of letters AA, AG, AC, . . . , TT represent the instructions for the assembly of amino acids. In this code the above sequence, CGACCA . . . , would state: start with the amino acid represented by CG, add the amino acid represented by AC, and so forth. Such a doublet code consists of 16 (4 × 4) different letter combinations and, hence, is still insufficient for directing the assembly of the twenty different amino acids found in protein.

One further increase in the number of letters designating the amino acids constitutes a *triplet code*. Such a code consists of the combinations AAA, AAG, AAC, . . . , TTT and comprises 64 (4 × 4 × 4) different combinations. This is more than enough to represent all twenty amino acids and is, in fact, the code used in nature (see table II.2).

The triplets of letters of the genetic code, which specify the amino acids, are called *codons*. In table II.3 the codons of DNA are listed for all twenty amino acids. There are many synonyms among the codons, as would be expected because the number of codons far exceeds the number of amino acids. The code also includes two *initiator* and three *terminator codons* that signal the beginning and the end of a genetic message. Using the table as a dictionary, we can now translate the above genetic message—CGACCATCAACT—into the corresponding peptide chain:

CGA   CCA   TCA   ACT
 | | |
Alanine–glycine–serine

The final codon of the message, ACT, is the signal for terminating the assembly of the peptide chain.

The example just given points to an interesting correspondence between the genetic code and writing in western culture (see figure II.24). Both are linear in their representation of information. Both use letters and combine them into words. The letters of the genetic code are the bases and the words are the codons. There are four bases and all of the codons are three letters long. In contrast, writing in English requires 26 letters and English words are of variable lengths. If we think of the codons as words, we may compare a sequence of codons, which gives instructions for the assembly of one protein molecule and is called a gene, to an individual article in a reference book. A DNA molecule, which carries many genes or many articles, may then be thought of as a book. And all of the DNA molecules present in a cell, which together carry the totality

Table II.2. Combinations of Bases in Singlet, Doublet, and Triplet Codes.

Singlet code (4 words)		Doublet code (16 words)			Triplet code (64 words)			
A	AA	AG	AC	AT	AAA	AAG	AAC	AAT
G	GA	GG	GC	GT	AGA	AGG	AGC	AGT
C	CA	CG	CC	CT	ACA	ACG	ACC	ACT
T	TA	TG	TC	TT	ATA	ATG	ATC	ATT
					GAA	GAG	GAC	GAT
					GGA	GGG	GGC	GGT
					GCA	GCG	GCC	GCT
					GTA	GTG	GTC	GTT
					CAA	CAG	CAC	CAT
					CGA	CGG	CGC	CGT
					CCA	CCG	CCC	CCT
					CTA	CTG	CTC	CTT
					TAA	TAG	TAC	TAT
					TGA	TGG	TGC	TGT
					TCA	TCG	TCC	TCT
					TTA	TTG	TTC	TTT

of genetic information of an organism, correspond to the volumes making up an encyclopedia. All cells carry a full set of genetic information, at least when they first form, much as every library possesses an encyclopedia.

To conclude this section, consider how much genetic information is present in living cells. The DNA of a bacterium is typically 1 mm long and contains about 1 million codons. This is comparable to the number of words in five 500-page novels with 400 words per page. It also is approximately equal to the words contained in one volume of a large encyclopedia. In contrast, the 23 DNA molecules (chromosomes) of a human cell have a total length of about 1 meter and carry approximately 1 billion codons. This is roughly equal to the word content of 50 encyclopedias with 20 volumes each. This enormous difference in the quantity of genetic information in bacteria and in humans is yet another indication of the range in diversity and complexity that has resulted in the course of 4 billion years of biological evolution.

This Introduction covers a wide range of biological topics in preparation for the chapters to come. Physical and biological evolution were compared, with

Table II.3. DNA Codons for the Amino Acids.

AAA } phenylalanine	AGA }	ATA } tyrosine	ACA } cysteine
AAG	AGG	ATG	ACG
AAT } leucine	AGT } serine	ATT } "terminator"	ACT "terminator"
AAC	AGC	ATC	ACC tryptophan
GAA }	GGA }	GTA } histidine	GCA }
GAG } leucine	GGG } proline	GTG	GCG } arginine
GAT	GGT	GTT } glutamine	GCT
GAC	GGC	GTC	GCC
TAA }	TGA }	TTA } asparagine	TCA } serine
TAG } isoleucine	TGG } threonine	TTG	TCG
TAT	TGT	TTT } lysine	TCT } arginine
TAC methionine/"initiator"	TGC	TTC	TCC
CAA }	CGA }	CTA } aspartic acid	CCA }
CAG } valine	CGG } alanine	CTG	CCG } glycine
CAT	CGT	CTT } glutamic acid	CCT
CAC valine/"initiator"	CGC	CTC	CCC

Notes: The code used by nature for expressing genetic information is a triplet code. Each "word" or codon of the code consists of three bases and specifies an amino acid, an initiator signal, or a terminator signal. This table is a dictionary for translating between the codons of DNA and the amino acids.

emphasis on important similarities as well as differences between the two. The cell was introduced as the fundamental unit of biological organization. Prokaryotic and eukaryotic cells were distinguished and shown to represent the most profound demarcation in the biological world today. This demarcation is more profound than that existing between protists, fungi, plants, and animals, all of which are eukaryotes and evolved from prokaryotes, the most ancient living cells on Earth.

Biological evolution does not take place on the level of individual cells or organisms, but on the level of populations or species. This evolution proceeds by random genetic mutations together with a selection process that is the result of competition among the members of all species for available raw materials, energy, and living space. Because of this competition, species do not evolve in isolation, independently of each other. They evolve together, strongly affecting each other's success in the struggle for survival. The stage upon which this evolution takes place is the physical environment of land, sea, and air, all of which evolve as well. Together, these evolving biological and physical components make up the Earth's ecosystem. The energy that drives the ecosystem is the energy of sunlight (figure II.25).

The range of these topics reflects the fact that the biological world may be viewed and studied on many different levels — from the level of molecules to that of the ecosystem, from the microscopic to the global. Every level has its importance and validity, and to understand biological evolution requires an awareness and appreciation of each of them. In box II.2 and the next five chapters we will look at life from each of the different levels. Box II.2 describes briefly the history of biological evolution. Chapters 6 and 7 cover the origin and

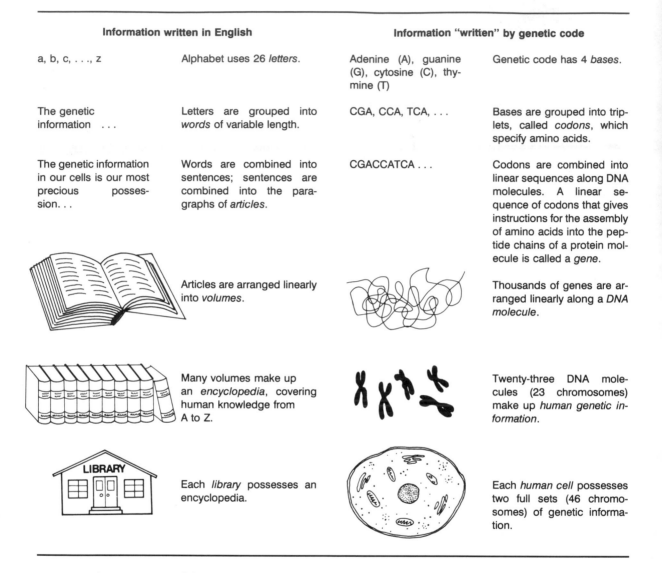

Information written in English		Information "written" by genetic code	
a, b, c, . . ., z	Alphabet uses 26 *letters*.	Adenine (A), guanine (G), cytosine (C), thymine (T)	Genetic code has 4 *bases*.
The genetic information . . .	Letters are grouped into *words* of variable length.	CGA, CCA, TCA, . . .	Bases are grouped into triplets, called *codons*, which specify amino acids.
The genetic information in our cells is our most precious possession. . .	Words are combined into sentences; sentences are combined into the paragraphs of *articles*.	CGACCATCA . . .	Codons are combined into linear sequences along DNA molecules. A linear sequence of codons that gives instructions for the assembly of amino acids into the peptide chains of a protein molecule is called a *gene*.
	Articles are arranged linearly into *volumes*.		Thousands of genes are arranged linearly along a *DNA molecule*.
	Many volumes make up an *encyclopedia*, covering human knowledge from A to Z.		Twenty-three DNA molecules (23 chromosomes) make up *human genetic information*.
	Each *library* possesses an encyclopedia.		Each *human cell* possesses two full sets (46 chromosomes) of genetic information.

Figure II.24. Comparison of the writing of information in English with the "writing" of information by the genetic code.

early evolution of life, focusing on developments that occurred mainly at the molecular level. Chapters 8, 9, and 10 concentrate on the origin and evolution of the major animal phyla, with particular emphasis on the vertebrates and primates. Here the emphasis will be mainly on the level of species and ecosystems. Throughout these discussions we shall find that biological evolution on any level — molecular, cellular, organismic, species, ecosystem — invariably affects the global environment, often in profound and far-reaching ways.

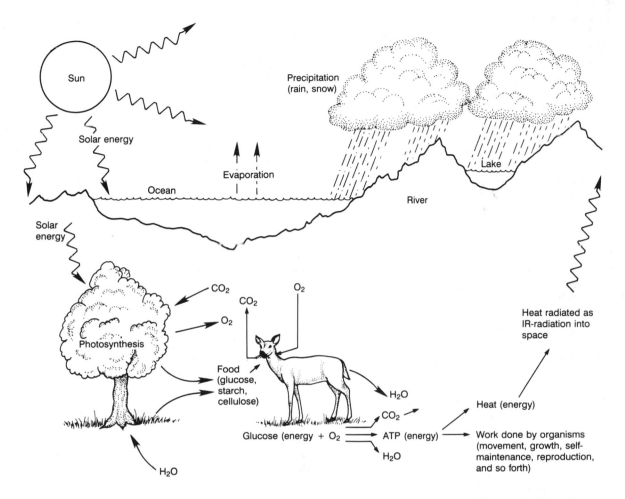

Figure II.25. Diagram of two of the many important cycles of materials that take place in nature and that are driven by the energy of sunlight: the cycle of water in the biosphere and the cycle of carbon dioxide and oxygen in the food chain. The food chain involves water also; thus the two cycles are interlinked.

BOX II.2.
Brief History of the Theory of Biological Evolution

"WHATEVER is must have been created." This thought has been accepted as self-evident throughout history and is the basis of legends, folklore, and religions. The ancient Sumerians, Egyptians, Greeks, and Hebrews believed that the act of creation happened only recently and that it was carried out by powerful or even omnipotent gods. They also believed that the world and all its content, including humans, were made right from the beginning much as they are today. For example, using the Old Testament as a basis, the Irish archbishop James Ussher (1581–1656) calculated with astounding certainty that God created the heavens and the Earth in the year 4004 B.C.

The concept of evolution—of changes occurring with time—is relatively new in the history of human thought. It arose only with the development of modern science, with its dual emphasis on careful observations and the construction of theories attempting to tie the observations together into a self-consistent intellectual framework. Along with the establishment of evolution as the modern paradigm of viewing the world, the time dimension became stretched as well, first to millions and hundreds of millions of years and, finally, during the present century, to billions of years.

The first suggestions that we might be living in an evolving Universe were voiced by the naturalists and philosophers of the eighteenth-century Enlightenment in western Europe. In 1721 the French philosopher Charles Louis de Secondat Montesquieu (1689–1755) wrote, in response to the discovery of flying lemurs in Java, "This would seem to corroborate my feeling that the differences between animal species can daily increase and similarly decrease; in the beginning there were very few species and they have multiplied since." Thus, breaking with tradition, Montesquieu proposed that species do not remain static, but evolve. In 1749, the French naturalist Georges Buffon estimated that the Earth is at least 70,000 years old, and six years later the German philosopher Immanuel Kant extended that estimate to millions or hundreds of millions of years.

In 1809, the French naturalist Jean Baptiste Lamarck (1744–1829) drew up an evolutionary tree of life, from microscopic organisms to humans, and offered the first systematic theory for the origin and evolution of the species. His theory may be summarized as follows:

1. All organisms possess a built-in drive to strive toward perfection.

2. Organisms have the capacity to adapt to circumstances.

3. Simple organisms are created spontaneously even today and then evolve toward greater complexity, according to principles 1 and 2.

4. Offspring inherit the characteristics or traits acquired by their parents.

The last of these principles, for which Lamarck is mainly remembered (and chastised), means, for example, that giraffes develop long necks by eating the leaves on the upper branches of trees and that they pass this trait (the long necks) acquired by their particular way of life on to their offspring. All but the second of Lamarck's principles turned out to be wrong. Still, through them he established the idea that a mechanism of evolution exists and paved the way for the progress made toward understanding this mechanism by succeeding generations of scientists.

Darwin and Wallace. Foremost among Lamarck's successors were the two British naturalists Charles Darwin and Alfred Wallace. They formulated the theory of biological evolution as we basically still understand it today. The major components of their theory may be stated in four points:

1. The biological environment is not static, but evolving.

2. The process of evolution is gradual and continuous.

3. Similar organisms are related and descendant from a common ancestor, and, very likely, all life forms can be traced back to a single origin.

4. Evolutionary change in biology is the result of natural selection, which consists of two steps (see pages 257–258)—the random generation of variation and the selection of those individuals with variations best adapted to their environment.

The first three of these principles were not original with Darwin and Wallace, but the fourth one was. In fact, it was so revolutionary that it completely changed our way of looking at the biological world,

BOX II.2 continued

including ourselves, stirring controversy and debate and stimulating a great deal of research up to the present time.

Initially, Darwin was a traditionalist and accepted the idea of the fixed nature of the species. He abandoned this belief while still a young man when he served as naturalist aboard H. M. S. *Beagle* during a five-year voyage around the world (1831–1836). His change of mind came about through the repeated observation that species on adjacent areas of continents or on adjacent islands (such as on the Galápagos Islands) are often different yet undoubtedly related and that there exists much similarity of structure between fossils and living forms in the same area. These facts, Darwin felt, could only be explained by a theory based on the concept of evolution. He worked out the main features of his theory in 1838, but did not publish it until 1859, under the title *On the Origin of Species by Means of Natural Selection*.

Wallace arrived at the theory independently during fieldwork in the East Indies. When he sent a draft of his ideas to Darwin, the latter was completely taken by surprise. He wrote to his friend Sir Charles Lyell (1797–1875, a noted British geologist), "I never saw a more striking coincidence. If Wallace had my manuscript sketch written out in 1842, he could not have made a better short abstract! Even his terms now stand as heads of my chapters." Darwin's and Wallace's theory was read as a joint paper at a meeting of the Linnean Society of London in 1858.

Darwin never tired of pointing out that evolution by natural selection depends on the continual production of variations among the species. He believed (erroneously, see below) that without this production whatever variations presently exist would, within only a few generations, become blended and obliterated through the mating process and evolution would come to a halt. He also believed (correctly) that mutations (known as "sports" in his time) are the source of new variations. However, neither he nor Wallace knew how phenotypic characteristics (the visible manifestation of an organism's genetic traits) are passed on from one generation to the next, nor how mutations affect this process.

Mendelian Genetics. Some of the gaps in Darwin's and Wallace's understanding of evolution were filled by the Augustinian monk Gregor J. Mendel

(1822–1884), who from 1856 to 1868 carried out genetic experiments in the gardens of his monastery in Brünn, Moravia (a province in today's Czechoslovakia). He bred and crossbred garden peas and flowers, focusing on specific characteristics such as tallness and dwarfness or presence and absence of color, to see how they are inherited. For example, in one of his experiments he started with purebred yellow-seeded and green-seeded pea plants (figures A, B). When he bred these plants with their own kind, the offspring were again either yellow-seeded or green-seeded (that is the meaning of "purebred"). However, when he crossbred, or hybridized, the two kinds of peas, the first generation of offspring (referred to as F_1, from "filial" or *filius*, meaning "brother") was always yellow-seeded. Somehow, the green color seemed to have become lost. Actually, it did not get lost, but was merely suppressed; for when he bred the F_1 generation with itself, the green color reappeared. Surprisingly, it reappeared in a definite ratio: for every three yellow-seeded plants in the F_2 generation there was, on the average, one green-seeded plant.

The result just described, and others like it, were at first very puzzling to Mendel. However, instead of giving up as some of his contemporaries did — including Darwin, who also conducted such experiments — he managed to deduce a number of conclusions, which have become the basis of what today we call Mendelian genetics:

1. Phenotypic characteristics are due to discrete hereditary "factors" (today we call them *genes*) that are passed on from parents to offspring.

2. The factors come in pairs (called *alleles* today, see chapter 7) with one member of each pair derived from each parent. For example, if, in the case of pea plants, we label the factors for yellow and green peas Y and g, pea plants may possess the combinations YY, Yg, gY, and gg. Yg and gY produce the same phenotypic characteristics and, hence, are equivalent.

3. The factors never blend, as Darwin thought. For example, Yg does not produce yellowish green (or greenish yellow) peas, as might be expected if yellow and green colors were blended. Instead, factors are distinct and become segregated during the formation of reproductive cells. A pea reproductive cell possesses either Y or g.

BOX II.2 continued

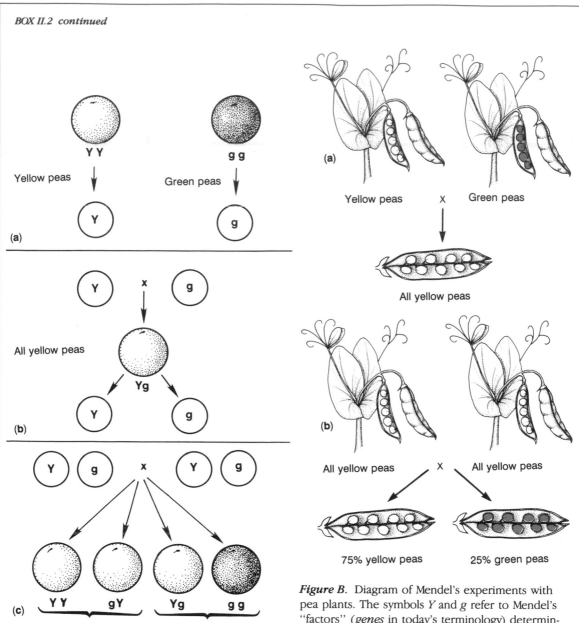

Figure A. Mendel's experiments with pea plants. (*a*) Crossing of purebred yellow-seeded and green-seeded pea plants yields an F_1 generation of plants with all yellow peas. (*b*) Crossing of F_1 generation pea plants yields an F_2 generation of plants, 75% of which have yellow peas and 25% of which have green peas.

Figure B. Diagram of Mendel's experiments with pea plants. The symbols Y and g refer to Mendel's "factors" (*genes* in today's terminology) determining pea color. (a) Purebred pea plants yield unique reproductive cells. (b) Crossing of Y and g reproductive cells yields an F_1 generation of plants whose peas are all yellow because Y is dominant and g is recessive. Reproductive cells from the F_1 generation possess either Y or g. (c) Crossing of reproductive cells from the F_1 generation yields an F_2 generation of pea plants, 75% of which have yellow peas and 25% of which have green peas.

BOX II.2 continued

4. During fertilization, factors become recombined (for example, as *YY, Yg = gY,* or *gg*). Because fertilization is a random process, the recombination of factors is random also.

5. Some factors are dominant and others are recessive. For example, *Y* is dominant and *g* is recessive (properties that are indicated by the use of upper and lower case letters, respectively). The reason for the terminology *dominant* and *recessive* is that pea plants possessing the combination *Yg* or *gY* are yellow-seeded, meaning that *Y* masks or dominates the presence of *g*. Plants possessing the combinations *YY* or *gg* are pure with regard to pea color, so their peas are yellow or green, respectively.

Unfortunately, Mendel's work never became widely known during his lifetime and his results had no immediate impact on the development of biology. Nearly forty years later, around the turn of the century, his results were rediscovered independently by three European botanists, Hugo de Vries (1848–1935) of Holland, Karl E. Correns (1864–1933) of Germany, and Erich Tschermak von Seysenegg (1871–1962) also of Germany. By then chromosomes had been discovered in plant and animal cells and it was recognized that chromosomes are the carriers of Mendel's factors or *genes,* in the modern terminology. Furthermore, the distinction was made between an organism's *genotype* and its *phenotype.* The genotype was defined as referring to an organism's full set of genes, the phenotype as the physical or visible expression of the genotype.

The Synthetic Theory of Evolution. Interestingly, the rediscoverers of Mendel's theory did not agree with all aspects of the theory of evolution as proposed by Darwin and Wallace. They suggested that biological evolution proceeds by sudden rather than by gradual changes (they called the sudden changes "saltations") and that its primary driving force is mutation. In fact, they thought that mutations occur ready-made for the production of specific and significant changes in the species, a notion that was reminiscent of Lamarck's erroneous idea of the inheritance of acquired characteristics.

These differences of opinion caused much tension in the biological community and led to intensive research with the aim of gaining a better understanding of evolution. Among the early participants of this work were Sergei S. Chetverikov (1880–1959) of Russia, Sir Ronald A. Fisher (1890–1962), and John B. S. Haldane (1892–1964) both of Great Britain, and Sewall Wright of the United States. They made use of statistical analysis, introduced the science of population genetics, and justified Darwin's idea that evolution is gradual. Whatever differences still remained were finally overcome during the 1930s and 1940s, mainly through the efforts of Theodosius Dobzhanski (American, 1900–1975), Julian S. Huxley (British, 1887–1975), Ernst Mayr (American), Bernhard Rensch (German), George Gaylord Simpson (American, 1902–1984), and G. Ledyard Stebbins (American).

In their so-called *synthetic theory* these men recognized that in the evolution of sexually reproducing species it is not the individual that matters, but the whole population of interbreeding individuals. Furthermore, they recognized that when mutations first occur they are not adaptive. In fact, most of them are detrimental to the organisms. Thus, the short-term effect of mutation on evolution is minimal or nil. The long-term accumulation of many mutations and the continual random shuffling of genes during sexual mating is the major source of variation in a population. Finally, the originators of the synthetic theory recognized that ultimately natural selection controls the direction, rate, and intensity of biological evolution.

Molecular Biology. While population genetics and the synthetic theory were being developed, progress was also made on another front. Chemists were trying to understand biology on the molecular level. They studied the structure of chromosomes and genes, the language of the genetic code, the changes in the genetic message caused by mutations, the translation mechanism from genotype to phenotype, the molecular processes involved in cell reproduction, and many other aspects of cell structure and function.

One of the early triumphs of this research was the discovery in 1953 of the molecular structure of *deoxyribonucleic acid,* or *DNA,* which is the molecular carrier of genetic information. The discoverers were two young scientists, James D. Watson of the United States and Francis H. C. Crick of Great Britain, who, together with Maurice H. F. Wilkins, also of Great Britain, were awarded a Nobel Prize in 1962 for their accomplishment. Their success stimulated research in laboratories around the world and led to the literal

BOX II.2 continued

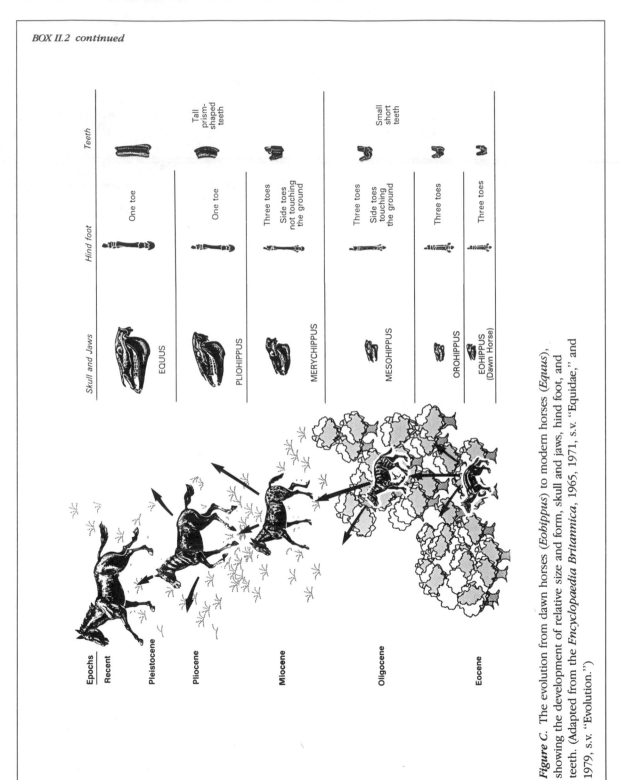

Figure C. The evolution from dawn horses (*Eohippus*) to modern horses (*Equus*), showing the development of relative size and form, skull and jaws, hind foot, and teeth. (Adapted from the *Encyclopaedia Britannica*, 1965, 1971, s.v. "Equidae," and 1979, s.v. "Evolution.")

BOX II.2 continued

explosion of *molecular biology* into one of the most active and best funded fields of science. Consequently, today we have a much more profound understanding of the process of biological evolution, as discussed in the chapters to come.

The Punctuated Theory of Evolution. Despite the enormous advances made during the past 150 years in our understanding of biological evolution, much remains to be learned. For example, a recent re-examination of the fossil record suggests that evolution may not be as gradual a process as has been assumed. Instead, it seems to progress episodically, at least in some lineages. This means that long periods of stability of individual species are punctuated by bursts of relatively rapid evolutionary change.

This "punctuated" evolution, as the two American paleontologists Niles Eldredge and Stephen J. Gould named it, is well illustrated by the fossil record of the ancestors of the modern horse family. The earliest ancestors of this family, the dawn horses, roamed through the forests of North America approximately 50 million years ago, changing very little for 4 million years or longer (see figure C). They were about as tall as fox terriers, had four toes on their front legs and three on their hind legs, and their teeth were small and short, adapted to eating leaves and other soft vegetation. Today's members of this evolutionary line include the domestic horse, the ass, and the zebra. They are rather large animals, their legs are single-toed and hoofed, and their teeth are tall and prism-shaped with heavy crowns, well suited for chewing grass and other abrasive roughage. These modern animals, too, have changed very little for close to 2 million years. The fossil record documents that there have existed at least seven other species in the direct evolutionary line from the dawn horses to their modern descendants, and that each of them remained virtually unchanged for several million years — corresponding to about a million generations — before evolving in a relatively short time into the next species. In fact, the changes from one species to the next occurred so rapidly and apparently in such confined geographic areas that very few fossils manifesting the transition have survived.

The punctuated progression just described for the horse family may well be the rule in biological evolution rather than the exception, as indicated by the fossil records of other lineages of plants and animals, including the lineage that led to humans. For periods of up to millions of years and, in some instances, even longer, species often change very little. Then, within a relatively short span of time, but still in a series of small steps, they undergo major evolutionary adaptations and new species arise.

What causes these variations in the rate of evolution? Are they due to variations in the rate of change of the environment, such as changes in climate, geographic isolation, or competition between species? Are they due to constraints in the embryological development of organisms? Are they due to rare mutations that happen to open up new opportunities of adaptations? Or are they due to instabilities arising from the complex and multifaceted interactions within entire ecosystems? Perhaps all of these factors are involved to varying degrees in different instances. Clearly, in the case of the horse family, the evolution toward hoofed feet and thickly crowned teeth was profoundly affected by the development of grasslands roughly 30 million years ago. However, at present there are no conclusive answers concerning the dominant causes that in general determine evolutionary rates. The subject remains, as much as ever, one of the most exciting and challenging areas of scientific pursuit.

Exercises

1 Discuss some of the similarities as well as differences between physical and biological evolution.

2 The process of evolution as explained by Darwin and Wallace has been characterized as being opportunistic. Explain why this characterization is fitting.

3 The process of biological evolution has been explained by a number of different theories, of which those based on teleological principles (evolution directed toward a future end), an *élan vital* (evolution directed by a "life force"), and natural selection (evolution according to Darwin and Wallace) are the most prominent ones. Of these theories, the first two are based on metaphysical or theological reasoning, while the third one is based on scientific observations and is consistent with the laws of physics. Discuss why, as far as science is concerned, an acceptable theory of biological evolution must be consistent with physical laws and why, therefore, the first two theories are rejected by modern biologists.

4 Dogs are the oldest domesticated animals and probably descended from the wolf, *Canis lupus*, somewhere in Eurasia between 14,000 and 12,000 years ago. All dogs belong to the same species, *Canis familiaris*, though there are more than a hundred different breeds today.
(*a*) Using Darwin's theory of evolution, speculate how the domestication of the dog and, in particular, the many breeds might have come about.
(*b*) Why are the different breeds all classified under the same species?

5 What are the distinguishing characteristics of eukaryotic and prokaryotic cells?

6 (a) List the five kingdoms introduced by Robert Whittaker to classify organisms, including some representative members of each.
(*b*) In addition, and complementary, to the five-kingdom classification, biologists today recognize two superkingdoms—the prokaryotes and eukaryotes. Which kingdoms belong to which of the two superkingdoms?
(*c*) Why are the differences between prokaryotes and eukaryotes more profound than, for example, those between mammals and trees, fishes and mushrooms, or insects and algae?

7 (*a*) If an electron micrograph of a cell shows that mitochondria are grouped around a particular region in the cytoplasm, what is probably happening in that region?
(*b*) What is probably happening if the grouping consists of ribosomes?

8 Discuss the major differences between carbohydrates, lipids, proteins, and nucleic acids in terms of their atomic constituents, molecular structure, and general biological uses.

9 Distinguish between saturated and polyunsaturated fats. Name some of the major food sources of each.

10 What do such animal tissues as tendons, hide, and bone have in common? In which ways do they differ?

11 Distinguish between *monomer* and *polymer*. Give examples of each.

12 Distinguish between dehydration reactions and hydrolysis. Give examples of each.

13 Distinguish between:
(*a*) purines and pyrimidines.
(*b*) RNA and DNA.
(*c*) mRNA, tRNA, and rRNA.
(*d*) ADP and ATP.
(*e*) Doublet and triplet code.
(*f*) Peptide, protein, and enzyme.

14 Distinguish between the genotype and phenotype of an organism.

Suggestions for Further Reading

"The Biosphere." 1970. *Scientific American*, September.* Eight articles on the grand-scale cycles of energy, water, oxygen, carbon, nitrogen, and minerals in our biosphere; followed by three articles on human food, energy, and material productions and their impact on the biosphere.

*The asterisk denotes books and articles that are general references for Part Two.

Clarke, B. 1975. "The Causes of Biological Diversity." *Scientific American*, August, 50–60. The author argues that the great diversity in genetic traits among individuals within a species appears to be actively maintained by natural selection.

Cloud, P. 1978. *Cosmos, Earth, and Man: A Short History of the Universe*. New Haven and London: Yale

University Press.* (Described in Suggestions for Further Reading, Introduction to Part One.)

Curtis, H. 1979. *Biology*. 3d ed. New York: Worth Publishers.* A college-level biology text.

Dawkins, R. 1976. *The Selfish Gene*. New York and Oxford: Oxford University Press. The author argues that genes, not individual organisms, are the unit of selection. The organisms are mere carriers ("robot vehicles") of the genes. Excitingly and persuasively written.

Dickerson, R. E. 1983. "The DNA Helix and How It Is Read." *Scientific American*, December, 94–111. Recent research results of how blocks of genes on bacterial chromosomes are turned on and off, thus initiating and stopping the translation of their message into protein structure. The article is complex and requires patient reading.

Duve, C. de. 1963. "The Lysosome." *Scientific American*, May, 64–72.

Evolution, A Scientific American Book. 1978. San Francisco: W. H. Freeman and Co.* Nine chapters (originally articles in the September 1978 issue of *Scientific American*) covering the history of the theory of evolution, origin of life, and the evolution of life from the earliest cells to humans. Though some of the articles are not quite up-to-date, the book is an excellent resource for anyone wishing to become familiar with the topic of biological evolution.

The Fossil Record and Evolution: Readings from Scientific American. 1982. San Francisco: W. H. Freeman and Co.* A collection of articles published in *Scientific American* between 1961 and 1982.

Gore, R. 1976. "The New Biology: I. The Awesome Worlds Within a Cell." *National Geographic*, September, 354–95.* A superbly illustrated, rather extensive article about the structure of cells and their organelles, DNA and protein synthesis, cell division, defense against infection, and genetic engineering.

Gribbin, J. 1981. *Genesis: The Origins of Man and the Universe*. New York: Delacorte Press. (Described in Suggestions for Further Reading, Introduction to Part One.)

Gross, J. 1961. "Collagen." *Scientific American*, May, 120–30.

Guttman, B. S., and **J. W. Hopkins, III.** 1983. *Understanding Biology*. New York: Harcourt Brace Jovanovich.* A college-level biology text.

Halstead, L. B. 1982. *The Search for the Past: Fossils, Rocks, Tracks and Trails, the Search for the Origin of Life*. Garden City, N.Y.: Doubleday & Co.* A topical and very informative book about the fossil record.

Hanson, E. D. 1981. *Understanding Evolution*. New York and Oxford: Oxford University Press.* A text dealing with the questions "What are living things?", "Where did they come from?", "What trends are apparent among them?" Recommended for students who want a solid understanding of biological evolution.

Judson, H. F. 1979. *The Eighth Day of Creation: Makers of the Revolution in Biology*. New York: Simon and Schuster. This lengthy but excitingly written book describes the discoveries and upheavals of understanding in biology since the mid–1930s.

Keeton, W. T. 1980. *Biological Science*. 3d ed. New York: W. W. Norton & Co.* A college-level biology text.

Kormondy, E. J. 1984. *Concepts of Ecology*. 3d ed. Englewood Cliffs, N.J.: Prentice-Hall.* A concise and readable text on ecosystems, populations (including human), and the flows of energy and materials through the biosphere.

Lewin, R. 1986. "Punctuated Equilibrium Is Now Old Hat." *Science* 231: 672–73. A report about two new investigations based on models of population genetics, that confirm the theory of punctuated equilibrium.

Margulis, L., and **K. V. Schwartz.** 1982. *Five Kingdoms: An Illustrated Guide to the Phyla of Life on Earth*. San Francisco: W. H. Freeman and Co.* A comprehensive and authoritative reference work on the five-kingdom classification of organisms, with excellent illustrations and extensive bibliographies for each kingdom. A must for anyone wishing to gain an appreciation of the enormity of biological diversity on Earth.

Mirsky, A. E. 1968. "The Discovery of DNA." *Scientific American*, June, 78–88. The history of the discovery of DNA, which began in 1869 with the work of Johann Friedrich Miescher (1844–1895), a Swiss biochemist.

Pauling, L., and **R. Hayward.** 1964. *The Architecture of Molecules*. San Francisco and London: W. H. Freeman and Co. (Described in Suggestions for Further Reading, Introduction to Part One.)

Perutz, M. F. 1964. "The Hemoglobin Molecule." *Scientific American*, November, 64–76.

Porter, K. R., and **J. B. Tucker.** 1981. "The Ground Substance of the Living Cell." *Scientific American*, March, 56–67. About the elaborate system of delicate filaments that support and move the cell organelles, as deduced from high-voltage electron microscopy.

Raven, P. H., R. F. Evert, and H. Curtis. 1981. *Biology of Plants*. 3d ed. New York: Worth Publishers.* A college-level botany text.

Reeves, H. 1984. *Atoms of Silence: An Exploration of Cosmic Evolution*. Cambridge, Mass.: The MIT Press.* (Described in Suggestions for Further Reading, Introduction to Part One.)

Rothman, J. E. 1985. "The Compartmental Organization of the Golgi Apparatus." *Scientific American*, September, 74–89.

Schrödinger, E. 1955. *What Is Life? The Physical Aspect of the Living Cell*. Cambridge: Cambridge University Press. A classic on the subject, by one of the foremost physicists of the 20th century.

Scott, R. M. 1980. *Introduction to Organic and Biological Chemistry*. San Francisco: Harper & Row, Publishers.* A college-level text. Some background in general chemistry would be helpful in using this book.

Seidler, N., and R. Gore. 1976. "Seven Giants Who Led the Way." *National Geographic*, September, 400–407. Brief illustrated biographies of Leeuwenhoek, Darwin, Mendel, Pasteur, Morgan, Watson, and Crick.

Stanley, S. M. 1981. *The New Evolutionary Timetable: Fossils, Genes, and the Origin of Species*. New York: Basic Books.* A thoughtfully written book about biological evolution, in particular about evidence suggesting that evolution is not quite what experts thought it was a decade or two ago.

Stebbins, G. L., and F. J. Ayala. 1985. "The Evolution of Darwinism." *Scientific American*, July, 72–82. Recent developments in molecular biology and new interpretations of the fossil record, which are gradually altering and adding to the synthetic theory of evolution.

Unwin, N., and R. Henderson. 1984. "The Structure of Proteins in Biological Membranes." *Scientific American*, February, 78–94. Cells, their nuclei, and organelles are bounded by membranes—extremely thin but tenaciously stable films of lipid and protein molecules. The article discusses the structure of this film and its adaptation to the two environments it separates.

Watson, J. D. 1968. *The Double Helix*. New York: Atheneum Publishers. An uncommonly candid account of the discovery of the structure of DNA as remembered by one of the key participants.

Additional References

Darwin, C. R. Letter to C. Lyell, 18 June 1858. In F. Darwin (ed.), *The Life and Letters of Charles Darwin* (2 vols.). New York: D. Appleton and Co., 1897, Vol. 1, p. 473.

Montesquieu, C. L. de S. 1721. Quotation from *The New Encyclopaedia Britannica*, 15th ed. (Micropaedia, Vol. 7, p. 7). Chicago: Encyclopaedia Britannica, Inc., 1979.

Figure 6.1. Discharge of lightning bolts during one of the eruptions of Surtsey, a volcano in the Atlantic Ocean off the coast of Iceland. The eruptions began on November 14, 1963, and continued for about 3½ years, creating an island more than 2 km across.

 Origin of Life on Earth

"**L**IFE BEGAN ON EARTH. There were no biological Big Bangs, nor extraterrestrial cultivators, just evolution from plausible nonliving beginnings." That is how John Scott (1981, 153), a biochemist at the Manchester Medical School in England, stated the basic assumptions scientists generally make about the origin of terrestrial life.* The question is, How did evolution from nonliving beginnings proceed? This chapter will attempt to answer this question, following theoretical developments that have gained widespread support from the scientific community since the late 1970s.

The story begins with the physical and chemical events that are believed to have taken place on the surface of the young Earth roughly 4 to 4½ billion years ago. At that time violent and nearly incessant volcanic eruptions occurred at many places, while at the same time the final accretion phase of the planet's formation drew to a close and its surface layers began to reach some semblance of equilibrium. The eruptions spewed forth large quantities of water, carbon dioxide, molecular nitrogen, and many other molecules, from which the first atmosphere and the juvenile ocean formed. Driven by energy from sunlight, lightning, volcanic heat, and meteorite impacts, the inorganic molecules reacted chemically with each other and produced a great variety of organic molecules—amino acids, sugars, lipids, the bases of nucleic acids, and many more. Gradually these molecules accumulated in the waters of the Earth until, in the words of John Haldane, "the primitive oceans reached the consistency of hot dilute soup" (Dickerson, 1978, 31). Today scientists believe that the temperature of the primitive ocean was probably close to the freezing point of water. Furthermore, its content of organic molecules may not have been as concentrated as Haldane had envisioned it, at least not throughout most of its volume. Nevertheless, many people still refer to it as the "primordial soup."

At present we are far from understanding all of the chemical reactions that took place on the young Earth. It seems that simple organic molecules assembled into larger and more complex molecules similar to those found in today's organisms, as suggested by laboratory experiments. For instance, amino acids probably assembled into peptide chains, and nucleotides assembled into short strands of RNA and other kinds of nucleic acids. Surfaces of clays, lava, rocks,

---

*A few prominent scientists (among them Fred Hoyle, Francis Crick, and Leslie E. Orgel) have questioned these assumptions and proposed that life originated somewhere else, possibly in one of the dark interstellar clouds or on a planet of some other stellar system. Their main reason for suggesting this is their belief that the 4.5 billion years are insufficient for life to come into existence and to evolve to its present-day complexity. A nonterrestrial origin of life could have happened 10 billion years or more ago, shortly after the Galaxy was formed and as soon as sufficient quantities of the necessary heavy elements had been synthesized by massive stars. Life could have come to Earth either by drifting through interstellar space in the form of bacteria-like spores, or, possibly, it could have been sent here in space probes by intelligent beings. This theory of the origin of life is known as *panspermia*.

I do not regard the theory of panspermia as a reasonable alternative, and it has found no common acceptance among biologists. Putting the origin of life 10 billion years into the past makes it somewhat easier to explain its present-day complexity. However, by removing the place of its origin to some distant place about which we know nothing, the theory of panspermia introduces unnecessary uncertainty and speculation. The theory cannot be tested experimentally, unless evidence for it is found in cosmic dust particles that happen to come our way or that have survived on such geologically inert bodies as the Moon or meteorites. Even if the theory turns out to be correct, it still does not solve the fundamental problem of how life actually originated.

sand, and other readily available substances may have served as catalysts facilitating these assembly reactions.

According to the theory of the origin of life that I am presenting here, short strands of RNA were the first molecules in the primordial soup that carried information, albeit very little, and they are regarded as the starting point of the evolution toward cellular life. (For alternative theories favored by some biochemists, see Cairns-Smith 1985 and Dyson 1985.) The short strands of RNA were capable of self-replication. In the process mistakes were made, so that the copied RNAs frequently differed from the original ones with regard to nucleotide sequence and length (recall that RNA nucleotides are of four types, with the bases A, G, C, and U). Some of the RNA molecules were more successful than others in surviving and replicating themselves under the prevailing conditions. Thus, the two components that form the basis of the Darwinian theory of evolution — random creation of variation and natural selection — may have been introduced very early among the chemical reactions in the primordial soup. As chemical evolution continued, different sets of RNA molecules coupled together into cooperative units. Some of the RNAs carried instructions for the assembly of primitive enzymes (peptide chains), while others acted as catalysts and contributed to the actual assembly of enzymes. Enzymes, in turn, helped in the replication of RNAs. Eventually, some of the coupled units of RNAs and their enzymes became enclosed by membranes and the first primitive cells — the protocells — were born. Life had emerged from among the random and spontaneous chemical reactions in the primordial soup.

This brief summary outlines the key components of the events that many biologists and chemists believe may have been central to the origin of life on Earth, and the remainder of this chapter will fill in some of the details. At present this theory is based on many assumptions and contains many gaps. It is not based on any direct evidence dating back to the primordial soup, for none has survived. Furthermore, laboratory experiments that attempt to simulate early Earth conditions and to reproduce the chemical reactions that took place then have been only partially successful in telling how the protocells arose. Scientists think they understand in general terms how the Earth's atmosphere and ocean came into existence. They believe they know something about the formation of organic monomers from inorganic raw materials and the assembly of polymers from monomers. However, they still have only a very limited understanding of the emergence of order and information among the chemical reactions in the primordial soup. And they know virtually nothing about the evolution from those initial chemical reactions to the formation of the first cells. The Nobel prize-winning German biochemist Manfred Eigen (1981, 114) characterized very aptly our current state of knowledge of how life began: "Anyone attempting [to re-create life] would be seriously underestimating the complexity of prebiotic molecular evolution. Investigators know only how to play simple melodies on one or two instruments out of the huge orchestra that plays the symphony of evolution."

We should not be discouraged by this lack of knowledge. Let us accept the problem of the origin of life as one of the great challenges facing science today. Let us also accept the fact that time inevitably diminishes and sometimes erases evidence of long-ago events. Hence, our first task is to discover the fragments of evidence that have survived. Our second task is to make good use of them. That is how Darwin and Hubble confronted their scientific challenges, which also dealt with events of long ago and for which much of the original evidence had been erased by time. Darwin deduced his theory of the origin and evolution of the species mainly from data gathered on a single trip around the globe, and Hubble based his proposal about the expansion of the Universe on measurements of recession velocities of about two dozen distant galaxies. Thus, there are precedents in the history of science that fragmentary information is no barrier to the development of feasible theories. Let us be optimistic that this will also be true of the current scientific attempts to reconstruct the events that led to life on Earth.

GEOLOGIC ACTIVITY ON THE YOUNG EARTH

When the Earth was formed roughly 4.5 billion years ago, it was a hot, partially molten mass without an ocean or much of an atmosphere. Most of the heat came from gravitational energy that was released when planetesimals collided and fell together to form the Earth, as discussed in chapter 4. Additional gravitational energy was released when the high-density iron and nickel of the proto-Earth sank toward the center to become the core of our planet, and lighter rocky material rose toward the surface to form the mantle and crust. Some of the energy also came from the decay of radioactive isotopes. These energies were liberated much more rapidly than they could be radiated away, and consequently they accumulated as heat. Only a fraction of this heat has been lost during the intervening eons. Even today, our planet's central temperature is still approximately 4300 K (see figure 5.11).

During the final accretion phase, Earth acquired a surface layer of low-density rocks rich in many kinds of volatiles including water, carbon dioxide, molecular nitrogen, and organic compounds. This rocky material was probably derived from carbonaceous chondrites, which bombard the Earth to this day, along with other types of meteorites, though at a much reduced rate. The outermost layers of the Earth radiated their heat into space and cooled to the point where they crystallized and hardened. They became the basaltic and granitic rock layers that form the crust of the Earth and float like rafts on the denser underlying mantle.

The crust is primarily responsible for maintaining the Earth's high internal temperature. It has a very low thermal conductivity, which slows the rate of heat flow from the Earth's interior to the surface. This can be seen in some of the desert caves in the western United States where the snow and ice that drift in during the winter stay throughout the hot summer months, even though they are separated from the surface by only a few meters of rock. Another feature that contributes to the maintenance of the Earth's high internal temperature is the presence of long-lived radioactive elements—uranium-235 and -238, thorium-232, and potassium-40—in the crust. The decay of these elements steadily releases heat and is the source of much of the geothermal energy that flows to the Earth's surface. Thus, the crust, enveloping the Earth, acts like an electric blanket: Its low thermal conductivity corresponds to the insulating qualities of the wool or polyester, and its radioactivity corresponds to the electric heat output of the blanket.

One consequence of the Earth's high interior temperature is that the outer part of the core and the mantle have never hardened into rigid structures, but have remained in molten or "pasty" states, resembling fluids of high viscosity.* Another consequence is that powerful convective currents are generated in the mantle, which relentlessly push and pull on the overlying layers (that is, on the *lithosphere*; see box 9.2) and prevent them from settling into a permanent configuration. That is why our planet's surface features continuously change over geologic time. Continents converge and break up, ocean basins come and go, and mountain ranges are lifted up and weathered down.

The pushing and pulling on the Earth's lithosphere by currents in the underlying mantle involve enormous forces and energies. Usually we are not aware of those geologic activities because they happen so slowly; but earthquakes, volcanic eruptions, geysers, and hot springs are reminders that we live on a restless and dynamic planet. This restlessness must have been much more severe when the Earth was first formed than it is today. The Earth's interior was hotter then and its temperature had not yet had time to adjust to a smooth gradient from the center to the surface. Consequently, the currents in the mantle must have been stronger. The lithosphere was still crystallizing and had not yet achieved its present thickness and rigidity. This crystallizing was slowed by the steady release of heat from the decay of the radioactive elements, which were initially much more abundant than they are today. Because it was thinner and less rigid than it is now, the lithosphere of the young Earth was more easily deformed by the mantle currents than it is today. As a result, earthquakes and volcanic eruptions, accompanied by huge lava flows (figure 6.1), must have occurred almost incessantly and with great intensity over large areas of the young planet's surface, much as they occurred on the Moon, Mercury, Mars, and, perhaps, on all planetary bodies of intermediate size.

In addition to earthquakes and volcanism, which are processes created by conditions within the Earth, there also was violence from outside. When our planet was formed and had reached approximately its final mass and size, there was still plenty of interplanetary debris—planetesimals, comets, rocks, and dust—left from the original protoplanetary disk, as discussed at some length in chapter 5. For hundreds of millions of years this debris kept falling onto the Earth at a high rate until most of it had been swept up. Even today some traces remain, as indicated by the roughly 30 tons of matter that fall onto Earth every day in the form of "shooting stars" and meteorites. The largest of the planetesimals that bombarded the young Earth probably weighed many billions of tons and were comparable in size to the asteroids that still orbit the Sun today. On impact, they shattered the crust, carved out huge impact craters, and threw molten and pulverized crustal material across the planet's surface.

Viscosity refers to how easily a fluid flows. For instance, honey is highly viscous, water is not. Viscosity results from interactions among the molecules making up the fluid and constitutes the fluid's internal resistance (friction) to distortions by external forces.

Table 6.1. Partial Listing of Molecules Referred to in Chapters 6 and 7.

CH_4	methane	H_2O	water
CO	carbon monoxide	H_2S	hydrogen sulfide
CO_2	carbon dioxide	NH_3	ammonia
$CO_3^=$	carbonate	NO_3^-	nitrate
C_2H_4	ethylene	N_2	molecular nitrogen
C_2H_6	ethane	O_2	molecular oxygen
HCN	hydrogen cyanide	$PO_4^\equiv$	phosphate
H_2	molecular hydrogen	$SO_4^=$	sulfate
H_2CO	formaldehyde		

ORIGIN OF THE EARTH'S ATMOSPHERE AND OCEAN

The earthquakes, volcanic outbursts, and bombardments by meteorites, which ravaged the young Earth so regularly, did not just churn the crust and produce large lava flows. As hot lava reached the surface and meteorites heated their impact areas to incandescence, gases that had been trapped in the rocks burst into the open in huge amounts—gases of water (H_2O), carbon dioxide (CO_2), molecular nitrogen (N_2) and, in lesser amounts, of molecular hydrogen (H_2), carbon monoxide (CO), methane (CH_4), ammonia (NH_3), hydrogen sulfide (H_2S), and many others (see tables 6.1, 6.2). Our planet was acquiring its first atmosphere.[*]

This kind of outgassing (releasing of gases) can still be observed today, although at a considerably diminished rate, in the hot springs and geysers of Yellowstone National Park, active volcanoes such as Mount St. Helens, and many other places of geothermal activity. For example, volcanic eruptions are usually accompanied by the emission of thick and often foul-smelling clouds of gases that billow for miles into the atmosphere.[†] The gases originate from within the

---

[*]Up to very recently, biologists who studied the problem of the origin of life assumed a different composition for the Earth's first atmosphere. They thought that, in addition to water, the major constituents had been H_2, CH_4, and NH_3. Such a hydrogen-rich composition is not consistent with the geological record. If these molecules had been present abundantly on the young Earth (rather than CO_2 and N_2), the hydrogen would have reacted with the oxygen of FeO (which the record shows was present) to form H_2O and Fe. There is no evidence that metallic iron (i.e., Fe free of any O) ever existed in the Earth's crust or upper mantle. Hence, H_2, CH_4, and NH_3 could not have been abundant molecular species in early outgassings.

[†]The gases discharged by present-day volcanoes, geysers, and hot springs include H_2O (steam), CO_2, N_2, H_2S, H_2, and others. The foul ("rotten egg") smell that is so noticeable near some volcanoes and hot springs is due to H_2S. Most outgassings today are recycled materials that have been on the surface of the Earth before. Water is usually ground water. The other materials became incorporated into sedimentary rocks during earlier epochs and are now once again released into the atmosphere. In contrast, the gases that poured forth from the primitive Earth were "juvenile" gases, released into the atmosphere for the first time.

Table 6.2. Relative Abundances (in Percent) by Mass of the Most Common Elements in the Sun, Entire Earth, Earth's Crust, and Human Body.

Element	Sun[a]	Entire Earth[b]	Earth's Crust[b]	Human body[c]	
Hydrogen	73.6	—	—	10.	(63.)
Helium	24.8	—	—	—	—
Carbon	0.29	—	—	18.	(9.5)
Nitrogen	0.093	—	—	3.1	(1.4)
Oxygen	0.77	30.	46.	65.	(25.)
Neon	0.12	—	—	—	—
Sodium	0.0029	—	2.1	0.1	(0.03)
Magnesium	0.046	13.	4.	0.04	(0.01)
Aluminum	0.0049	1.1	8.	—	—
Silicon	0.069	15.	28.	—	—
Phosphorus	0.0007	—	—	1.	(0.2)
Sulfur	0.038	1.9	—	0.3	(0.05)
Chlorine	0.0011	—	—	0.2	(0.03)
Argon	0.018	—	—	—	—
Potassium	0.0003	—	2.3	0.4	(0.06)
Calcium	0.0057	1.1	2.4	2.	(0.3)
Iron	0.16	35.	6.	—	—
Nickel	0.0084	2.4	—	—	—

Sources: [a]Noses 1982; [b]Press and Siever 1978, 14; [c]Keeton 1976, 28.

Note: The last column, given in parentheses, refers to abundance by number relative to a total of 100.

lava. They escape into the open when the hot lava reaches the surface of the Earth and is no longer subjected to the high pressures deep below the ground. Quite often the gases bubble forth in ways that give the resulting rocks a frothy appearance and make them lighter than water (figure 6.2). During some eruptions the gases burst forth so violently that they shatter the lava into fine-grained dust known as ash, which may be thrown hundreds or even thousands of kilometers into the surrounding areas.

Water

Despite the heat that accompanied outgassing on the early Earth, the average temperature of the atmosphere was probably not far above the freezing point of water, for the energy output of the Sun—our planet's major source of heat—

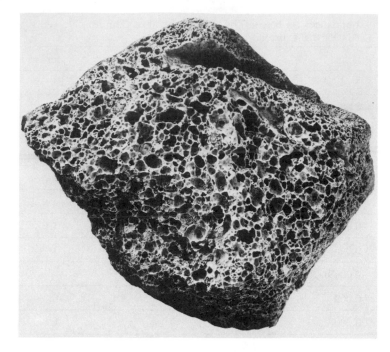

Figure 6.2. Basaltic volcanic rock filled with bubble holes.

was then only about two-thirds of what it is today. Hence, the water, which was the most abundant molecular species emitted by volcanism and meteorite impacts, did not remain in gaseous form for long. It condensed into rain or snow and fell to Earth. Very likely the early rains came down in torrents, accompanied by storms, lightning, and thunder. The first rivers began to flow, glaciers and ice caps accumulated in the polar regions, and low-lying basins filled with liquid water to become our planet's juvenile ocean.

Carbon Dioxide

Volcanic outgassings on the early Earth poured forth not only water but also many other gases. The most abundant gas after water was carbon dioxide. This presented a potential danger for the eventual origin and evolution of life on Earth because a planetary atmosphere that contains substantial amounts of CO_2 heats up to high temperatures by the *greenhouse effect* (see figure 6.3). This effect comes about as follows. Solar radiation, whose wavelengths correspond mainly to the colors blue through red (see Introduction to Part One), penetrates relatively easily through the atmosphere and is absorbed by the ground. The absorbed energy heats the ground, which reradiates it in the infrared part of the spectrum. Infrared (IR) radiation has long wavelengths and, unlike the original solar radiation, much of it becomes absorbed by the CO_2 in the atmosphere. The CO_2 molecules reradiate this absorbed radiation, with some of it being sent back toward the ground. The net result is that the IR radiation does not readily escape into space but becomes trapped in the atmosphere, heating it as well as the ground.

Fortunately, the greenhouse effect had only a small effect on the young Earth's surface temperature because the rains washed the CO_2 out of the

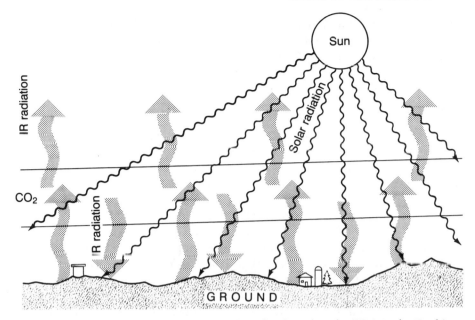

Figure 6.3. The greenhouse effect. Molecules of carbon dioxide (CO_2) in the Earth's atmosphere act somewhat like the glass roof of a greenhouse. They are transparent to the radiation from the Sun and let it pass to the ground. The ground heats up and reradiates energy as long-wavelength IR radiation. However, CO_2 molecules are much less transparent to IR radiation. They absorb most of the IR radiation, then reradiate some of it upward into space and the rest downward toward the ground, thus trapping much of the radiant energy and raising the temperature of both the atmosphere and the ground.

atmosphere before it could accumulate to dangerous levels.[*] This happened because CO_2 dissolves in liquid water (but not in gaseous water) and forms carbonic acid (H_2CO_3):

$$CO_2 + H_2O \text{ (liquid)} \rightarrow H_2CO_3.$$

The rains and rivers brought the carbonic acid in contact with the ground, where it leached positive ions of calcium (Ca^{++}) and magnesium (Mg^{++}) out of rocks and combined with them to form limestone ($CaCO_3$) and dolomite [$CaMg(CO_3)_2$]. The limestone and dolomite then became deposited as sediments on the ocean floor. Thus, most of the original CO_2 that had been released into the atmosphere became locked up in sedimentary rocks. (Note that after shale and sandstone, limestone constitutes the most abundant sedimentary rock type of the present-day terrestrial crust.)

The third most abundant gas released by volcanism and meteorite impacts on the early Earth was molecular nitrogen, N_2. It is chemically rather inert and

[*]Gaseous H_2O, CH_4, and NH_3 also produce greenhouse effects. However, neither of these gases accumulated in sufficient quantities in the atmosphere of the young Earth to have presented a problem. Gaseous H_2O condensed into rain or snow and fell to Earth, and CH_4 and NH_3 were never discharged in significant amounts. The little NH_3 that was discharges was rapidly destroyed by solar UV radiation or dissolved in water.

did not condense into liquid form at the temperatures that prevailed on Earth, nor did it combine with water. It stayed in the atmosphere and, in the course of time, accumulated to become its dominant molecular constituent.

Oxygen

Interestingly, molecular oxygen (O_2) — which today makes up 21% (by number) of our atmosphere and is essential for the survival of all life (except some bacteria) — was present in only very low concentrations in our planet's original atmosphere. Had it been abundant, it would have rusted the iron and other metals. The oldest known rocks show no consistent evidence of such rusting. Only rocks younger than about 2 billion years do and, thereby, attest to the abundant arrival of molecular oxygen in the atmosphere by that time. Molecular oxygen was released in appreciable quantities into the atmosphere only after some organisms acquired the ability to split water molecules as part of their photosynthetic activities (to be discussed in chapter 7).

The absence of O_2 from the early atmosphere was a fortuitous circumstance. Had it been present in large amounts, its high chemical reactivity would have made the development of life very difficult, if not impossible. Even today O_2 is inherently poison to all organisms. They can withstand its reactivity only because they possess special molecules that chemically bind with O_2 and let it react in slow, controlled steps.

Without O_2, the atmosphere of the young Earth did not have the ozone (O_3) layer that it has today. This layer is crucial for our survival, for it filters out most of the ultraviolet (UV) radiation from the Sun and prevents it from causing harmful chemical reactions in our cells, such as burns and damage to DNA and RNA (which may lead to cell death or cancer). In the absence of an ozone layer on the young Earth, UV radiation penetrated freely to the ground. But instead of being harmful, it provided the energy for triggering numerous chemical reactions that were important for prebiotic evolution, as discussed in the following sections.

The Next Step

Very likely, the first outgassings on the primitive Earth began long before the planet was fully formed and they continued with great intensity for hundreds of millions of years thereafter. Only when the initial burst of earthquake activity, volcanism, and meteorite impacts had abated, approximately 4.0 to 3.5 billion years ago, did the outgassing slow down. Geologic evidence indicates that by then the oceans covered large areas of our planet, and we may assume that the atmosphere had acquired a substantial fraction of its present content of N_2. Most of the outgassed CO_2 had been washed out of the atmosphere and locked up in sedimentary rocks, but not all of it. Some of the CO_2 together with other gases that had been discharged in small amounts, such as H_2, CH_4, NH_3, and H_2S, embarked upon a rather different and much more interesting evolutionary course. They reacted chemically with each other to form molecular structures of ever-increasing complexity.

ORGANIC CHEMISTRY ON THE EARLY EARTH, PHASE ONE: SYNTHESIS OF MONOMERS

There are good reasons to believe that conditions on the surface of the young Earth were quite suitable for the occurrence of a great variety of chemical reactions and the production of many kinds of molecules. Volcanic outgassings supplied ample amounts of atomic and molecular raw materials. The temperature was low, but not so low as to freeze water everywhere. Lightning, UV radiation from the Sun, geothermal heat from volcanoes and hot springs, and meteorite impacts supplied plenty of energy for driving the reactions. At the same time, the ocean and ground offered protection from too much UV radiation and heat that might have destroyed the molecular products.

Among the molecular raw materials on the surface of the young Earth, H_2O, N_2, and CO_2 were the most abundant. They offered a nearly inexhaustible supply of hydrogen, carbon, nitrogen, and oxygen. Hence, organic molecules must have been among the dominant products of the chemical reactions. This is the case in interstellar clouds, in which these elements are also abundant and in which organic molecules are manufactured copiously. The same was true of the Sun's protoplanetary disk, as manifested by the presence of organic molecules in meteorites.

Just how readily organic molecules form under conditions similar to those on the early Earth was first demonstrated experimentally by two American chemists, Stanley L. Miller and Harold C. Urey (1893–1981), in the early 1950s, using a glass apparatus as shown in figure 6.4. They filled the lower part of the apparatus, which included a small flask, with water to simulate the juvenile

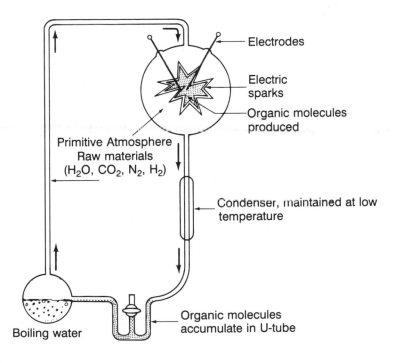

- Electrodes
- Electric sparks
- Organic molecules produced

Primitive Atmosphere Raw materials (H_2O, CO_2, N_2, H_2)

- Condenser, maintained at low temperature

Boiling water

- Organic molecules accumulate in U-tube

Figure 6.4. Apparatus used in Miller-Urey experiments. The lower part is filled with water, the upper part with gaseous CO_2, N_2, and H_2 (or with other gaseous raw materials containing C, N, O, and H). The water is heated to the boiling point, which circulates steam through the upper part of the apparatus. Electric sparks in the upper part induce chemical reactions that produce a great variety of organic molecules. In the condenser, which is maintained at a low temperature, water condenses into liquid form. As the liquid water returns to the small flask, organic molecules accumulate in the U-tube.

ocean. They pumped a gaseous mixture of CH_4, NH_3, and H_2 into the upper part to represent the primitive atmosphere. Then they boiled the water in the small flask to produce steam and to drive the gases in a closed circuit through the apparatus.* At the same time, they generated electric sparks in the larger upper flask to simulate lightning in the primitive atmosphere and to provide a source of energy for chemical reactions. Below the large flask, they cooled the circulating gases so that the water condensed and returned as droplets to the lower part of the apparatus, thus completing the circuit. After they had run the experiment for a day or two, the water in the lower part of the apparatus turned pink and later it became red. This indicated, and careful analysis confirmed, that organic molecules (such as hydrogen cyanide, formaldehyde, sugars, amino acids, bases) had been formed by the electric sparks and were trapped in the liquid water.

When Miller and Urey first carried out their experiment, it was commonly thought that hydrogen-rich molecules had been the major constituents of the Earth's primitive atmosphere (see first footnote, p. 308). Hence, they used the gases CH_4 and NH_3 instead of CO_2 and N_2. Since then, the experiment has been repeated many times with various combinations of CH_4, C_2H_4 (ethylene), C_2H_6 (ethane), CO_2, CO, NH_3, N_2, H_2S, H_2, and H_2O and with different sources of energy, from intense heat (from about 1100 to 1600 K) to UV radiation and electric sparks. Some experimenters added sand, clay, or lava rocks to simulate the catalytic effects the ground of the young Earth might have had on the chemical reactions. All of the experiments yielded rich mixtures of organic molecules.

Apparently, the main requirement for synthesizing organic molecules is the presence of hydrogen, carbon, nitrogen, and oxygen in some molecular form or other. The details — whether the atomic raw materials are supplied as H_2O, CO_2, N_2, and H_2, as H_2O, CH_4, NH_3, and H_2, or in some other form — matter little. Nor does it matter much what kind of energy source is used, as long as it is sufficiently intense to drive the chemical reactions. The chief limitations are that oxygen not be present as O_2 and that the newly formed molecules do not remain exposed to the energy sources for too long. If they are, the same energy source that creates them also destroys them. Protection from too much energy is achieved by dissolving the molecular products in liquid water (that is, in the juvenile ocean) or by allowing them to attach themselves to the surfaces of grains of sand, clay, mud, or lava rocks.

Note, however, that the Miller-Urey experiments do not yield many organic molecules of prebiotic importance if the only raw materials are H_2O, CO_2, and N_2. To enhance the production of organic molecules, H_2, CH_4, NH_3, or other hydrogen-rich molecules need to be present among the starting materials, at least in small amounts (without them, the chemical reactions are dominated by the reactivity of oxygen). This is particularly true of the production of hydrogen

---

*The steam in this experiment simulated evaporation from the juvenile ocean. Evaporation, by which at ordinary temperatures water changes from the liquid to the gaseous states, is too slow a process to be used in a laboratory experiment.

cyanide (HCN), which is a precursor molecule of many of the more complex organic molecules that are believed to have played significant roles in prebiotic chemistry.

Synthesis of Amino Acids, Sugars, and Bases

How are the simple starting materials converted into organic molecules in experiments simulating primitive Earth environments? Much attention has been paid recently to answering this question. Here are a few examples of these chemical reactions.

When molecules such as H_2O, CO_2, N_2, and H_2 are exposed to electric sparks, intense heat, or UV radiation, their bonds are broken and, temporarily, free atoms (H, C, N, O) and molecular fragments (for example, OH and CO) result. Almost instantaneously, new bonds reform between the atoms and the molecular fragments, but often in combinations that are different from the original ones. This breaking and reforming of chemical bonds constitutes the mechanism of chemical reactions by which new kinds of molecules are created. Such chemical reactions take place in statistically predictable ways and according to well-known physical laws. Given identical starting conditions, the molecules produced will, on the average, always turn out the same.

Two of the most common molecules that result during the initial stages of a Miller-Urey experiment are hydrogen cyanide (HCN) and formaldehyde (H_2CO). Both of them are important intermediates in the formation of still more complex organic molecules, in particular of amino acids. For instance, reactions among HCN, H_2CO, and water yield the amino acid glycine. The reactions involve several steps, but they may be summarized by this chemical equation:

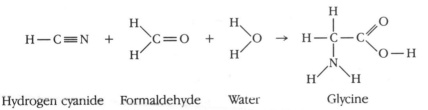

Hydrogen cyanide   Formaldehyde   Water   Glycine

Reactions involving more complex aldehydes than formaldehyde yield more complex amino acids. The nature of the end product depends somewhat on the sources of energy used and on the presence of certain atoms among the starting materials. If electric sparks are used, glycine, alanine, leucine, serine, threonine, asparagine, and other relatively simple amino acids result. If the energy source is heat of about 1600 K, some of the products are amino acids with ring structures, such as phenylalanine and tyrosine. If UV radiation is used and H_2S is present in the gas mixture, small amounts of sulfur-containing amino acids are formed.

In addition to being precursor molecules for the making of amino acids, formaldehyde and hydrogen cyanide may also be the starting materials for the synthesis of other organic molecules. For instance, reactions among five molecules of formaldehyde, in the presence of calcium carbonate, produce a complex mixture of end products, including the five-carbon sugar *ribose*:

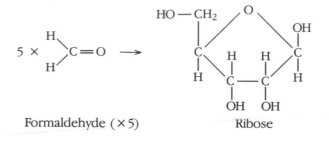

Formaldehyde (×5)   Ribose

As before, this formula merely summarizes the input material and the final product of the reaction. The actual reaction sequence is much more complex, involving several steps. In the first step, two formaldehydes react to form glycolaldehyde*:

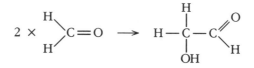

Formaldehyde (×2)   Glycolaldehyde

Glycolaldehyde may be either the precursor of ribose, if three more formaldehydes are added (as in the reaction shown above), or the precursor of the amino acid serine, if HCN and H_2O are added. In either case, the reactions are *autocatalytic*.[†] At first very little happens. Then suddenly, after several hours, ribose and serine are produced. Apparently, molecules of glycolaldehyde slowly form during the induction period. Once they exist, they act as catalysts for the production of more glycolaldehyde and then the reactions to ribose and serine run their course rather rapidly. Six-carbon sugars, such as glucose and fructose, are formed by similar reaction pathways.

Not all organic molecules of biological interest are as readily produced as the amino acids and sugars. For example, the syntheses of the bases of DNA and RNA call for rather high concentrations of starting molecules, which on the early Earth probably occurred only under unusual circumstances. *Adenine*, which is the easiest base to make (perhaps that is why it is the most common base found in organisms, being present in RNA, DNA, and ATP), requires the reaction of five molecules of hydrogen cyanide (top of page 317):

*Aldehydes are organic molecules of the type $R-\overset{\overset{\displaystyle O}{\|}}{C}-H$, where R stands for any number of possible molecular side chains. In the case of formaldehyde, R stands for H; in the case of glycolaldehyde, it stands for H_2OHC.

[†]As noted in box 3.4, catalysts are substances that speed up nuclear or chemical reactions without themselves undergoing any permanent changes. Chemical reactions in which the molecular products themselves act as catalysts are said to be *auto*catalytic reactions (*auto* means "same" or "self"). All of life's processes depend on chemical reactions. These reactions depend on the participation of catalysts, usually enzymes. In some reactions, vitamins act as catalysts as well. The problem of the origin of life may be regarded as a problem of discovering how enzymes and the genetic information that determines their construction came into existence.

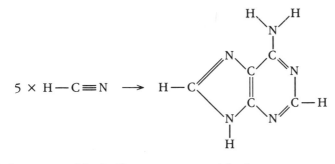

Hydrogen cyanide (×5)   Adenine

Again, this formula is merely a summary. In reality, several sequential reactions must occur, and the presence of UV radiation and ammonia are helpful. *Guanine*, the other double-ringed base, is very similar to adenine except that it possesses an oxygen atom and its side groups are slightly different. Its synthesis starts out identically to that of adenine, but in the final steps of the reaction sequence, water and cyanogen (N≡C−C≡N) or urea (H₂N−C−NH₂) need to be present.   O

L- and D-Amino Acids

These have been just a few examples of the numerous chemical reactions that occur in Miller-Urey type and other kinds of experiments simulating prebiotic terrestrial conditions. Many of the molecules produced are identical to those found commonly in present-day life. Others are rather novel and have unusual structures. For instance, many of the amino acids have two configurations, one being the mirror image of the other. In one configuration, the side chain points to the *left* as one looks from the carboxyl group to the amino group (for details, see the Proteins section in the Introduction to Part Two). In the other, the side chain points to the *right*. The two configurations are illustrated in Figure 6.5 for alanine.

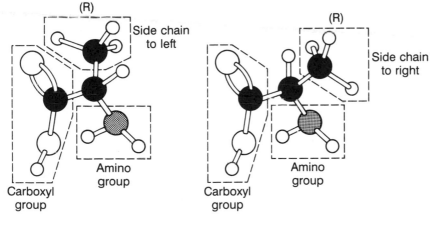

Figure 6.5. The two configurations of alanine.

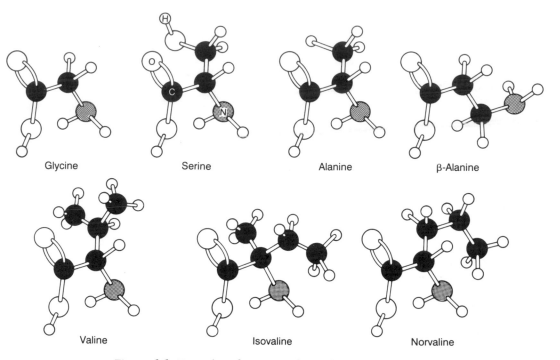

Figure 6.6. Examples of amino acids produced on the early Earth. Many of the amino acids existed in several isomeric forms and, with the exception of glycine, they occurred as both L- and D-types.

The two kinds of amino acids are known as L- and D-amino acids, where L and D stand for *levo* and *dextro* meaning "left" and "right." Both kinds are found in meteorites, and it is reasonable to assume that they were manufactured by prebiotic terrestrial chemistry as well. In contrast, with few exceptions today's organisms produce only L-amino acids. Somehow, during the origin of life, L-amino acids were selected over D-amino acids. Once that had happened, enzyme-aided biochemical reactions kept manufacturing chiefly the L-types. Was this selection an accident? Was it due to certain conditions on the early Earth, such as the particular molecular structures of clays and rocks, that may have acted as the first catalysts in prebiotic chemical reactions? Or do L-amino acids have some inherent advantage in the chemistry of life, perhaps due to some intrinsic bias in physical laws that is not yet understood? A number of answers have been suggested, but none of them is convincing. We don't really know.

In addition to L- and D-amino acids, other molecular variations that do not occur in present-day life show up in Miller-Urey experiments and, presumably, were also synthesized on the prebiotic Earth. An example is valine, which is produced in three different versions or *isomers* — valine, norvaline, and isova-line — as illustrated in figure 6.6. All three of these isomers have the same chemical formula, $C_5H_{11}O_2N$, and they are amino acids, but their side chains are put together differently.

We have seen that varying the conditions — the starting materials, energy sources, and catalysts — of the Miller-Urey experiment and others simulating early-Earth conditions produces a great variety of organic molecules. Some of the molecules form readily under almost any kind of condition as, for example, many of the amino acids. Others, like the bases, require rather special conditions and are more difficult to make (though some of them are found in meteorites).

On the young Earth, environmental conditions must have spanned an enormous range. In the rivers, lakes, and oceans water was amply available, while land parched by the Sun was characterized by dryness. The raw materials varied from molecules without hydrogen, like CO_2 and N_2, to local concentrations of H_2, CH_4, C_2H_6, NH_3, H_2S, and other hydrogen-rich molecules. In the northern and southern latitudes the temperature was well below freezing, while near thermal pools it was in the hundreds of degrees, and near lava flows it was well above 1000 K. Interplanetary debris ranging in size from microscopic dust to bodies kilometers across bombarded the Earth, bringing in new materials and locally heating the atmosphere and ground. The Sun's UV radiation penetrated freely to the ground. Primordial storms produced lightning and released concentrated forms of heat. Rocks and mineral grains of various sizes, textures, and compositions offered additional raw materials, besides those discharged by outgassing, and were available to serve as catalysts for many of the reactions.

From day to night, season to season, and year to year, the early terrestrial conditions fluctuated. We may expect that in the course of thousands to many millions of years a broad range of chemical reactions occurred and that organic molecules were produced in great abundance and in many different forms. Even molecules that are very difficult to make were produced occasionally. The American biochemist, George Wald (1954, 48), put it as follows: "Given so much time, the 'impossible' becomes possible, the possible probable, and the probable virtually certain. . . . Time is in fact the hero of the plot."

Once the molecules were formed, they were not easily destroyed as they would be today by bacteria and molecular oxygen. The oceans, rivers, and lakes offered protection from UV radiation and excessive heat as did sand, clay, and mud. Thus, organic molecules accumulated until, as Haldane suggested, the primordial ocean reached the consistency of dilute soup. We do not know exactly how concentrated the soup became. That depended on the rate of synthesis of organic molecules as well as on the rate of their destruction. It also depended on the volume of the ancient ocean. However, we may be certain that at least locally, such as in shallow marine basins, tide pools, and lakes, its concentration was sufficiently great for the next phase in chemical evolution to get started– the assembly of simple organic molecules (the monomers) into polymers (see box II.1).

ORGANIC CHEMISTRY ON THE EARLY EARTH, PHASE TWO: SYNTHESIS OF POLYMERS

The prebiotic synthesis of simple organic molecules was only the beginning of a complex and multifaceted evolution of chemical reactions. As amino acids, sugars, bases, and other molecules accumulated and their concentrations in the primordial soup and in the clay, sand, and mud of the young Earth increased,

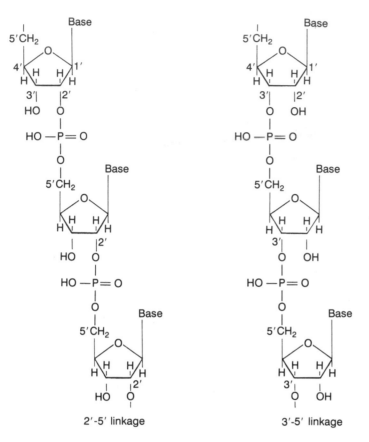

Figure 6.7. Chemically possible RNA "backbone" structures. The backbone structure of RNA, consisting of ribose sugars linked by phosphates, can exist in two forms: 2'–5' linkages and 3'–5' linkages. This notation means that the second carbon (as labeled in the diagram) of one ribose is linked by a phosphate to the fifth carbon of the next ribose, or the third and the fifth carbon atoms are thus linked. Very likely, both kinds of backbone structures formed in the primordial soup. During the development of life, the 3'–5' linkage was selected. The same linkage also became established in DNA.

further chemical reactions linked them together into larger molecular structures, namely polymers. Amino acids were joined into peptide chains; ribose, bases, and phosphates were combined into nucleotides; and nucleotides were connected into strands of RNA and, possibly, other kinds of nucleic acids. These are just a few examples of polymers that probably were constructed from monomers by prebiotic chemistry. Many early polymers were identical or similar to those found in nature today, but others were quite different, just as some of the monomers were different. For example, it is very likely that nucleotides were bonded together into short RNA and DNA polymers by both the so-called 3'–5' linkage, which is found in contemporary cells, and by the 2'–5' linkage (see figure 6.7).

The chemical reactions that produced polymers from monomers on the early Earth were probably the same dehydration reactions that take place in today's organisms (though initially they were not catalyzed by enzymes): A hydrogen atom was removed from one monomer and an OH fragment from another, and then a chemical bond was formed between the two monomers. The H and OH were combined into a molecule of H_2O, which was released into the environment (see Introduction to Part Two).

In today's organisms the assembly of monomers into polymers takes place in the interior of cells. There the monomers are well concentrated, so that collisions and, hence, reactions among them are frequent; the energy required for tearing the Hs and OHs from the monomers is amply supplied by ATPs; and enzymes are present to help remove the water molecules and speed up the formation of the bonds.

On the prebiotic Earth conditions for making polymers were considerably less favorable. Because cells did not yet exist, reactions took place in the environment at large. There the concentrations of monomers were generally low compared to the interiors of present-day cells; energy was available, but most of it not in forms best suited for driving the dehydration reactions; and enzymes were lacking. In the oceans and lakes the absence of enzymes presented a particularly serious problem. Without enzymes it was difficult to form bonds by removing water molecules from monomers and expelling them into surroundings that already consisted largely of water. The reactions were much more likely to run in the opposite direction, namely in the direction of taking water molecules from the surroundings, adding them to polymers, and splitting their bonds (a process called *hydrolysis*, see Lipids section, Introduction to Part Two).

Concentration Mechanisms

With all of these obstacles, how were the first polymers assembled? We may assume that in local areas conditions were occasionally suitable for polymerization. For instance, evaporation of water in shallow lakes and ponds during dry spells might have satisfied the requirement of concentrating the monomers by leaving most of the organic molecules behind on the muddy bottoms. The heat near hot springs or from freshly expelled lava flowing into bodies of water would have had similar effects. The periodic rising and falling of the water level in tide pools would have regularly concentrated the molecules, especially in hot, dry climates. Still another concentration mechanism would have been the freezing of lakes and ponds during winter. As water froze to ice, organic molecules would have become concentrated in the remaining liquid water. (The same method was used by American pioneers in the making of applejack. A barrel of cider was put outside during the freezing weather and left standing until most of the water in the cider was solid ice, leaving the liquid alcohol concentrated as applejack in a small volume near the center of the barrel. The colder the temperature, the higher was the proof and, hence, the alcohol concentration of the applejack.)

The concentration mechanisms just discussed were all a result of changes in the environment. Concentrations may also have been accomplished by the organic molecules themselves. For instance, in the laboratory when amino acids are put in water that is then heated to above the boiling point, peptides form and cluster spontaneously into spherical, membrane-enclosed structures called *proteinoid microspheres*. These structures are about 1 μm across and, during their self-assembly, concentrate in their interiors many of the organic molecules present in their vicinity. Another concentration mechanism occurs when certain combinations of organic polymers, such as peptides, carbohydrates, and nucleic acids, are mixed together in water. They spontaneously organize themselves into clusters that are marked off from their environment by structured layers of water molecules. These clusters are called *coacervates*. They have sizes of up to 500 μm and, like the proteinoid microspheres, concentrate organic molecules

---

THE clustering of peptides into proteinoid microspheres has been investigated by the American biochemist Sidney W. Fox and his collaborators since the 1950s. The name associated most frequently with coacervates is that of the Russian biologist A. I. Oparin, who during the early 1920s introduced the idea that the origin of life could be investigated in the laboratory. Oparin has experimented extensively with coacervates in order to understand prebiotic cellular organization. (The word *coacervate* comes from the Latin *coacervare*, meaning "to heap up.")

Until only a few years ago, proteinoid microspheres and coacervates were thought to have been direct precursors of ancient prokaryotic cells. Fox, Oparin, and other researchers spent much effort on demonstrating such a connection. However, neither proteinoid microspheres nor coacervates contain genetic information and they lack the internal organization of living cells. Today, the hypothesis that they may have been precursors of cellular life has been largely abandoned in favor of two theories: 1. The theory presented here, namely that biochemical evolution began with the self-replication of short strands of RNA in the open environment of the primordial soup. 2. The theory that prebiotic evolution began with crystals of clay able to evolve by natural selection and later switched to organic molecules as its basic building material (see Cairns-Smith 1985).

---

in their interiors. Both proteinoid microspheres and coacervates may well have played significant roles in concentrating organic molecules in the primordial soup of our planet (see box 6.1).

Energy Sources

The second requirement for assembling polymers — the availability of energy — was probably at first satisfied by lightning, heat, and UV radiation, even though these sources of energy are in general not very efficient in linking monomers together. As the number and diversity of monomers increased, the chemical energy stored in the bonds of some of them became available for dehydration reactions as well. Hydrogen cyanide, cyanogen, and other nitrogen-containing molecules similar to cyanogen were probably most effective in driving the reactions. Very likely, chains of phosphate molecules, the *polyphosphates*, were also important energy sources. In fact, polyphosphates may have been the precursors of ATP, which consists of a triphosphate chain attached to an adenine–ribose trunk. Polyphosphates must have been quite abundant on the surface of the early Earth. They would have formed readily by mild heating of minerals containing phosphate, particularly during dry periods when the minerals became concentrated (along with other materials, such as organic molecules) on the bottoms of lakes and ponds.

Catalysts

The third requirement for making polymers is the presence of catalysts for removing water molecules from the monomers and speeding up the reactions. Until enzymes evolved, the most obvious catalysts were the surfaces of mineral grains. Many clays consist of very thin sheets of silicates, that are separated from each other by molecules of water. The water layers would have given the organic monomers easy access to the silicate surfaces, where they could have become attached and been ready to undergo dehydration reactions with newly arriving monomers. Polyphosphates would have been amply available in the minerals as sources of energy. Experiments indicate that peptide chains as long as one hundred amino acids form in the presence of certain clays. The assumption that the surfaces of mineral grains acted as catalysts on the prebiotic Earth is supported by evidence from astrophysics. It appears that many of the molecules (including organic molecules) that are observed in the dark gas and dust clouds of spiral and irregular galaxies are also assembled on the surfaces of dust grains.

Experiments carried out with proteinoid microspheres suggest that they and other accidentally produced protein-like structures might have acted as catalysts as well. However, until the development of genetic information, the sequences in which amino acids were assembled into peptides and protein-like structures remained largely a matter of chance. Therefore, whatever catalytic properties such protein-like structures may have possessed lacked the specificity and efficiency that characterize enzymes in contemporary cells.

This section described some of the processes by which organic monomers are thought to have been assembled into polymers on the early Earth. However, at present scientists are far from understanding the full range of chemical reactions that took place then. No doubt, processes other than those described played important roles in the assembly of polymers as well. Our knowledge is limited because it is very difficult to duplicate in the laboratory the early terrestrial environment. As noted in the previous section, there are a great many variables to consider: an enormous variety of chemically possible precursor molecules, a broad range of likely concentration mechanisms, fluctuations in temperature of the environment from below freezing to well above 1000 K, numerous sources of energy, and the effects of many different kinds of inorganic materials from clays to rocks, sand, mud, lava, and sediments. Furthermore, experiments of prebiotic chemical evolution must be conducted in closed apparatuses in order to avoid contamination by microorganisms. In contrast, chemical evolution on the early Earth occurred in an open and ever-changing environment consisting of the atmosphere, water, and land. Above all, experiments are limited by the factor of time. Prebiotic chemical evolution took place over thousands to many millions of years, a time span that cannot be duplicated in the laboratory.

ORIGIN OF THE FIRST CELL

The last two sections described the enormous variety of organic molecules that are believed to have been produced during our planet's early evolution. Some of these molecules were probably similar or identical to those found in life today, while others were quite different. Many survived for long times in sheltered niches, while others were soon broken down into simpler components by the same sources of energy that created them. The simple components were then reassembled into new, more complex molecules. Initially, these chemical reactions occurred in a random, helter-skelter fashion and depended only on the prevailing physical conditions: the concentration of atomic and molecular raw

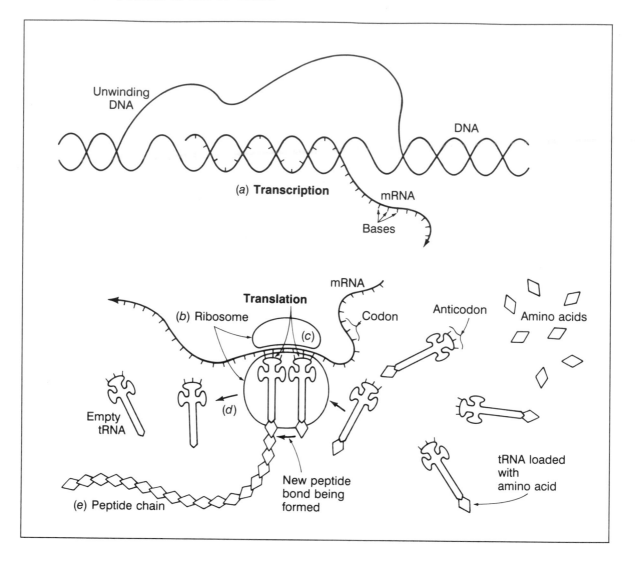

Figure 6.8. Protein synthesis. The information for the synthesis of proteins in modern cells is carried by molecules of DNA. The information consists of a sequence of bases, three of which constitute a codon, the basic unit of genetic information. Protein synthesis begins with the unwinding of a double-stranded DNA molecule and the *transcription* of information onto a molecule of messenger RNA [mRNA, step (*a*)]. The mRNA travels to a ribosome (*b*), the protein manufacturing plant. As the mRNA threads its way through the ribosome, each of its codons is matched up with the anticodon of a transfer RNA (tRNA) loaded with an amino acid (*c*). The amino acids are linked together by peptide bonds and the empty tRNAs are released (*d*). The sequence of codons of the mRNA (that is, the genetic information) has thus become *translated* into a sequence of amino acids, which is called a peptide (*e*). Upon folding into a three-dimensional structure, the peptide becomes a functional protein. Each of the steps of protein synthesis is catalyzed by enzymes. (For the sake of clarity, the bases of the DNA molecule are not shown.)

materials, the available sources of energy for driving the reactions, and, possibly, the presence of catalysts such as clays, rocks, certain ions, and accidentally produced small protein-like molecules.

Eventually, an *organizing mechanism* or, as Manfred Eigen calls it, an "organizing principle" emerged among the reactions. This organizing mechanism included information and the means of translating that information into chemical function. The emergence of this organizing mechanism was the beginning of a chemical evolution that, in the course of time, led to life. Today, such an organizing mechanism is present in the cells of all organisms, including those of our bodies. It consists of *genetic information* and a *biochemical machinery* that translates the information into *enzymes*, which, in turn, make the entire biochemical machinery run.

Before going on to the theory of how the organizing mechanism of life might have arisen, let us review how this mechanism works in today's cells (figure 6.8; for additional details, see the Introduction to Part Two). The components of the organizing mechanism are the following:

1. *DNA molecules* store *genetic information* by a code consisting of triplets of bases, called *codons*. The codons are arranged sequentially and partitioned into specific units, the *genes*, each of which codes for a particular protein molecule.

2. *Messenger RNAs (mRNA)* carry the information of a given gene from DNA to the protein assembly plants, the *ribosomes*.

3. *Transfer RNAs (tRNA)* carry *amino acids* to the ribosomes. For every one of the twenty amino acids found in nature, there exists one or more specific tRNA to which that amino acid becomes attached. Each of the tRNAs is characterized by a distinguishing triplet of bases, called an *anticodon*.

4. *Ribosomal RNAs (rRNA)* are essential components of ribosomes, the protein assembly machineries of the cells.

5. *Enzymes* are protein molecules that act as catalysts and are crucial for the smooth functioning of all biochemical reactions, including those of the organizing mechanism. (They function on the molecular level much as tools and machines function in a workshop.)

The organizing mechanism works as follows:

1. The information of a gene is *transcribed* (copied) from DNA onto mRNA.

2. The mRNA moves to a ribosome, enters and threads its way through it, thus *delivering* the genetic information.

3. As codon after codon of the mRNA pass through the ribosome, tRNAs with matching anticodons bind to the codons and deliver their amino acids, which become linked together by peptide bonds. The now free tRNAs are then released. The genetic information carried by the mRNAs—as a sequence of codons—is thus *translated into a peptide chain*.

4. As each peptide chain is assembled, it starts folding back and forth and assumes a unique shape, depending on its sequence of amino acids. The result is a functional protein—a structural protein, a regulatory or transport protein, or an enzyme protein.

In summary, the organizing mechanism of life directs the synthesis of proteins. The proteins—in particular, the enzymes—make the chemical reactions in an organism work, including the chemical reactions of the organizing mechanism itself.

Origin of the Organizing Mechanism

The next question is, "How did the organizing mechanism of life have its start among the disorder and randomness of the chemical reactions on the early Earth?" Note that the organizing mechanism in contemporary life is based on the presence of both *informational* and *functional* components. The informational components are DNA and RNA molecules. The functional components are RNA molecules and proteins (enzymes). Interestingly, the RNAs have dual tasks: they carry information and they perform functions. The DNAs and proteins have single tasks each: the DNAs carry information and the proteins perform functions.

There are a number of reasons why in today's cells the RNAs, DNAs, and proteins have these particular features. RNA molecules are usually single-stranded, consisting of a linear sequence of monomers (the nucleotides, which contain the bases A, G, C, and U). The sequential arrangement of monomers allows RNA molecules to carry information, similar to the way DNA molecules carry information. Furthermore, the single-strandedness allows RNAs to fold into three-dimensional shapes suitable for carrying out functions such as bringing amino acids to the ribosomes. Some RNAs are also capable of catalyzing certain biochemical reactions and thus act like enzymes, as was recently discovered by a number of researchers (see Cech 1986). Such catalytic RNAs act either alone or as part of RNA-protein complexes. Finally, short strands of RNA possess autocatalytic properties (see below). The three-dimensional folding of RNA has the additional advantage in that some of the shapes make the molecule resistant to destruction or alteration by hydrolysis and certain other chemical reactions.

In contrast, DNA molecules and proteins, the other components of the organizing mechanism in contemporary cells, do not have the RNAs' dual quality. The DNAs are generally double-stranded and, hence, are much less able than RNAs to fold into specific three-dimensional shapes that would allow them to carry out functions. They are mainly used for storing and reproducing genetic information. We may assume, therefore, that DNA was initially not part of the organizing mechanism but was introduced later on. Proteins are made of peptide chains and resemble RNA in that they also fold into a great variety of shapes, which gives them their excellent functional qualities. However, peptides are not capable of making faithful replicas of themselves and, hence, are not suitable as information carriers.

It is unlikely that all of the components of life's organizing mechanism—RNAs, DNAs, and proteins (enzymes)—came into existence at once in the primordial soup. They must have evolved gradually from primitive and inefficient precursors that were barely distinguishable from accidentally formed molecules. Here it is tempting to ask the old chicken and egg question, as biologists have since the 1930s: "Which came first, the informational or the functional components?" Experimental evidence gathered by Manfred Eigen and his coworkers, the American biochemist Leslie E. Orgel, and others suggests that neither came first. The informational and functional components of the organizing mechanism originated and evolved together. The start of this evolution is presumably to be sought among molecules of RNA because, of all the components of the organizing mechanism in contemporary cells, they alone carry information and perform functions.

RNA Quasi-Species

How did RNA molecules become the first component of life's organizing mechanism? A crucial laboratory experiment for finding an answer was carried out by Leslie Orgel. He demonstrated that short strands of RNA, such as $U-U-U-\ldots-U$ or $C-C-C-\ldots-C$ (poly-U and poly-C strands) are capable of making complementary copies of themselves in solutions containing activated nucleotides (that is, nucleotides with triphosphate chains). For example, poly-C strands form poly-G strands (recall that G is the complementary base of C), and they do so with a fairly high degree of fidelity, meaning that very few errors (the substitution of A, U, or C for G, in this example) are introduced. An important feature of this experiment is that the RNA strands have autocatalytic qualities and their replication proceeds without enzymes. This adds realism because, initially, enzymes would either not have been available in the primordial soup or, if present, would have been of poor and nonspecific quality. If zinc ions are added to the solution, much longer strands (up to 40 nucleotides in length) are copied and with greater fidelity than without those ions. The zinc ions act as inorganic catalysts, which, very probably, were available in the soup. Interestingly, today's RNA polymerases (enzymes that catalyze the formation of RNA) all contain zinc ions, which led Eigen (1981, 101) to ask, "Has nature perhaps 'remembered' how replication started?"

Orgel's experiment demonstrated that strands of RNA are capable of self-replication: They provide both the information and the function for making complementary copies of themselves. This result, as well as the dual roles RNA plays in contemporary cells, suggests that the first components in the evolution of life's organizing mechanism were indeed molecules of RNA or, at least, molecules similar to RNA.

We may imagine that life's organizing mechanism got its start when short strands of RNA—perhaps only two or three nucleotides long—were randomly assembled in the primordial soup. The original RNAs then began to replicate themselves by processes such as those in Orgel's experiment. In the absence of well-developed enzymes, errors were commonly made during replication, despite the RNAs' short lengths. Hence, the replicated copies differed frequently in minor and major ways from the original ones with regard to sequences of bases and lengths. *Variation*—the first requirement of Darwinian evolution—was thus introduced right from the beginning into the pool of RNAs in the primordial soup.

Along with the creation of variation among the RNAs, *natural selection*—the second requirement of Darwinian evolution—got its start as well. The different kinds of RNA were not all equally successful in replicating themselves, avoiding errors during replication, and surviving. For example, RNAs with an abundance of Cs and Gs probably tended to replicate themselves more faithfully than those with mostly Us and As; and RNAs with certain kinds of three-dimensional folding were more resistant to being broken up by water molecules and other chemically reactive compounds. It probably was also at this early stage that the particular bonding between nucleotides (the 3'–5' linkage, see figure 6.7) that we find universally in nature today won out over other, competing bonds.

In the course of time, the most successful RNA strands accumulated and gave rise to so-called *quasi-species* of RNA with variations in characteristics among their member RNAs, analogous to the variations observed today among the members of species of organisms. The quasi-species of RNA differed from one environment to another; and, as the environments changed, the quasi-species either changed also and adapted or they disappeared. Thus Darwinian evolution, based on variation and natural selection, appeared in the primordial soup long before life as we know it today had come into existence.

Very likely, the chemical reactions by which the earliest strands of RNA replicated themselves were aided by inorganic catalysts such as clays, lava, rocks, sand, and ions of zinc and other elements. Accidentally produced protein structures such as Fox's proteinoid microspheres might have placed a role as well.

In the absence of specific enzymes that catalyzed replication, Eigen (1981, 103) estimates that the early RNAs could at most have been about 50 to 100 nucleotides long, which is comparable to the lengths of today's tRNAs (see figures 6.9 and 6.10). Had they been much longer, the number of errors introduced during replication would have been so great that within a few generations their particular base sequences would have become changed beyond the range of viability.

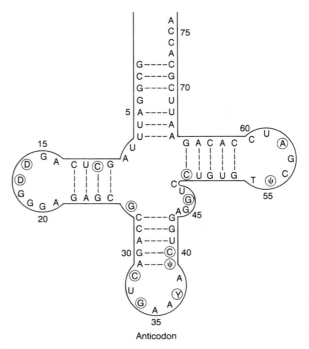

Figure 6.9. Two-dimensional representation of the nucleotide sequence (76 bases) of yeast tRNA (coded for phenylalanine). It illustrates the clover leaf shape that is typical of all tRNAs. The letters A, G, C, and U stand for the normal bases. The circled letters stand for unusual bases, all of which differ from the normal bases in only minor ways (for details see Watson et al, 1987, 384–385). Dashed lines indicate H-bonding between the bases (for clarity, not all H-bonds are shown).

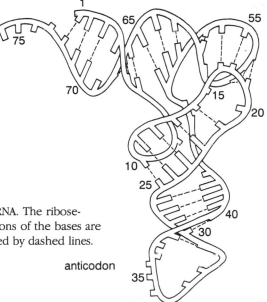

Figure 6.10. Model of the three-dimensional folding of yeast tRNA. The ribose-phosphate backbone is drawn as a continuous structure; positions of the bases are shown as bars; and the H-bonds between the bases are indicated by dashed lines. The numbers correspond to those of figure 6.9.

RNA Hypercycles

As long as the strands of RNA consisted of no more than 100 nucleotides, they could do little else than help in their own replication. They were not long enough to store information for the assembly of enzyme catalysts. However, enzymes were needed to allow the organizing mechanism to progress further. They were needed for the faithful replication of longer RNAs, and the longer RNAs were needed to store the information for the assembly of the enzymes. Furthermore, specialized functional RNAs were needed, analogous to today's tRNAs (and possibly rRNAs), to help in the translation of the information into enzyme structure.

According to theoretical studies by Eigen and his coworkers (1981, 104–107), the evolution toward longer strands of RNA and enzymes required the development of cooperative couplings between different RNA quasi-species. A simple example of such couplings (illustrated in figure 6.11) involves three quasi-species. The RNAs of one of the quasi-species carry information for the assembly of enzymes (albeit, very primitive ones) and the RNAs of the other two quasi-species function as tRNAs. With the aid of the tRNAs, the information of the first quasi-species is translated into primitive enzymes. The enzymes, in turn, aid in the replication of the RNAs of all three quasi-species. Such cooperative couplings are called *hypercycles*.

In reality, hypercycles in the primordial soup probably were never as simple as the one just described. We may assume that many more quasi-species of RNAs and kinds of enzymes than those indicated in the example were required, even for inefficient hypercycles. Hypercycles probably evolved from single RNA quasi-species that consisted of large numbers of RNAs with the usual variations with regard to length and base sequence. At first the participating tRNAs, helping in the assembly of peptides according to information carried by the longer RNAs, were quite inefficient. Likewise, the peptide chains, which functioned as enzymes, were quite inefficient also. They were probably not much more effective

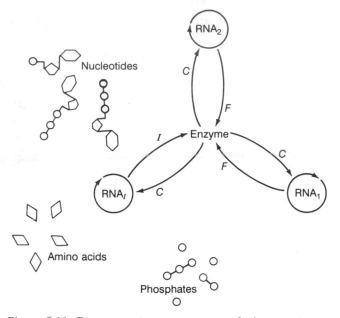

Figure 6.11. Diagrammatic representation of a hypercycle, consisting of three distinct RNA quasi-species (labeled RNA_I, RNA_1, and RNA_2) and enzymes. The RNA_I quasi-species provide information (arrow labeled I) for the assembly of the enzymes, while the RNA_1 and RNA_2 quasi-species assist in the actual assembly of the enzymes (that is, their RNAs carry out functions comparable to those of today's tRNAs, arrows labeled F). The enzymes, in turn, catalyze the replication of the three kinds of RNAs (arrows labeled C and circular arrows). The chemical reactions of the hypercycle take place in a solution containing an adequate supply of raw materials: nucleotides, amino acids, phosphates (suppliers of energy).

in this task than many of the accidentally produced peptide chains and other inorganic catalysts. In the course of time, however, the lengths of some of the participating RNAs increased, and their information content—coding for the assembly of enzymes—grew more distinct. Other RNAs remained short (50 to 100 nucleotides) and took on the role of today's tRNAs. The enzymes thus produced became more specific and efficient in their tasks. Slowly the original RNA quasi-species evolved into a number of distinct RNA quasi-species, with each quasi-species specializing in a particular task: to store information for the assembly of specific enzymes or to function as tRNAs. Thus the different RNA quasi-species, along with their enzymes, became firmly coupled and dependent on each other for survival. They evolved into hypercycles.

The longest RNAs of hypercyclic quasi-species were conceivably up to several thousand nucleotide units in length, as suggested by the gene lengths of certain contemporary viruses. That was roughly a ten-to-fiftyfold increase over the lengths of the RNA strands of the earlier, independent RNA quasi-species. This increase was possible because the RNAs of the hypercycles were replicated with the aid of specific enzymes, which greatly improved copying fidelity and efficiency.

We may assume that at this stage of prebiotic evolution the modern triplet code for storing genetic information came into existence as well. The code was needed to translate accurately the information carried by RNAs into enzyme structure. However, at present it is not clear whether the triplet code started out that way or was preceded by singlet or doublet codes. Singlet or doublet codes would have made the initial stages of the development of the organizing mechanism easier, but it is difficult to see how the switch from a singlet or a doublet code to a triplet code could have come about, because such a switch would have invalidated much of the previously accumulated information. Nevertheless, some researchers believe that the first code was singlet or doublet and that somehow, at a later stage during the origin of life, it became converted into the triplet code that is in universal use today.

Prebiotic evolution progressed from independent RNA quasi-species to hypercycles not only because the latter produced their own enzymes. It also progressed in this direction because, through the enzymes, the participating RNA quasi-species became dependent on each other and were forced to cooperate for the common good. They could not afford to outcompete and to eliminate each other. To survive, they had to cooperate. This cooperation greatly increased their reproduction rate and made them the dominant quasi-species in the soup.

The advantages of hypercyclic cooperation may be illustrated by an example from economics. Let us imagine a primitive human society in which every family is self-sufficient and competes with other families for available raw materials. Each family grows its own food, makes its own tools, builds its own shelters, and defends itself against competitors. Such an economy may be compared with the earlier, noncooperating RNA quasi-species, which competed with each other for the available raw materials. A much more efficient and productive economy results when the families begin to join into cooperative units, with each family specializing in specific tasks. Some families are farmers and grow food. Others are blacksmiths, butchers, bakers, or carpenters and specialize in the making of tools, slaughtering, baking, and building. Such a cooperative economy will easily outproduce and outcompete the simpler one in which each family fends for itself. The same was true of the hypercycles of coupled and cooperating RNA quasi-species, in which the different kinds of participating RNAs had specialized informational and functional tasks and in which they cooperated for the common good.

Another example of hypercycles is our present-day ecosystem, with its many interdependent and cooperative linkages among animals, plants, and microorganisms. Without the support from microorganisms, animals and plants could not exist. Likewise, many microorganisms depend on plants and animals for their nutrients. Furthermore, all animals depend (directly or indirectly) on plants for food and oxygen. In analogy, we may regard the RNA quasi-species that were linked into cooperating hypercycles in the primordial soup as our planet's first ecosystem.

Protocells

The arrival of hypercycles of coupled and cooperating RNA quasi-species constituted an important step forward in the development of life's organizing mechanism. However, there still remained a serious obstacle. There still existed no decisive feedback from the enzymes to the information content of the RNAs. Natural selection did not favor RNAs because they carried information for superior enzymes; it favored RNAs that were the most stable and reproduced most rapidly under the prevailing conditions, regardless of whether they also carried useful genetic information or not. Even RNAs that carried no information at all and performed no functional tasks benefited from the enzymes produced by other RNAs.

This feedback problem resulted from the fact that up to now all chemical reactions had taken place in the open environment of the primordial soup. As long as this remained the case, the obstacle could not be overcome. Only by enclosing the hypercycles with membranes and forming protocells did evolution progress further. Competition could then occur between the protocells, which forced natural selection to focus on the fitness of the protocells' entire enzyme-driven biochemical machineries. Cells that survived possessed, by definition, the most advantageous enzymes, and (because the cells survived) their genes were passed on to the next generation. Thus, the feedback from enzyme to gene—or from phenotype to genotype—that is characteristic of all contemporary life became established.

Besides solving the feedback problem, membrane enclosure had other advantages. It kept the enzymes produced by the hypercycles close to the RNAs and prevented their diffusion into the environment. It also created the possibility of concentrating biochemical raw materials—such as amino acids, nucleotides, and phosphates—by the development of selective transport mechanisms into the cells' interior.

At present no one knows just how the evolution from hypercycles to protocells came about. It was a step that probably required the parallel evolution of numerous major and interconnected biochemical capabilities: the evolution of suitable membranes; selective transport of the right kinds of raw materials through the membranes into the cell interiors; elimination of waste products; extraction of energy from energy-rich molecules; replication of genetic information with high fidelity; and efficient translation of the genetic information into enzymes and other kinds of proteins.

Though we do not know how these capabilities arose, it is clear that to function reasonably efficiently protocells required many more enzymes than were produced by the earlier hypercycles. Furthermore, these enzymes needed to be constructed with an ever higher specificity for catalyzing particular chemical reactions. Eventually, the information required for assembling the enzymes became so great that it could not have been stored and passed on from generation to generation if strands of RNA had remained the only information carriers. Too many errors would have been made during the replication of RNA, despite the assistance by enzymes, and any one of the errors might have been fatal to the cells. The only way out was the development of an error suppression mechanism.

The error suppression mechanism that evolved depended on the introduction of a double-stranded information carrier. That information carrier was DNA. As noted before, the two strands of DNA are complementary—the base adenine (A) of one strand is always bonded to thymine (T) on the other strand, and cytosine (C) is always bonded to guanine (G). Hence, the two strands carry identical information. The presence of the same information twice permitted the development of a "proofreading" mechanism during replication. While a copy is made of one of the strands of DNA, it is checked by proofreading enzymes against the other, complementary strand. Any errors detected are corrected.

This *error suppression* mechanism operates in all contemporary cells. It must also have evolved, at least in primitive form, in the protocells. It allowed their DNAs to store genetic information that extended over hundreds of thousands to millions of nucleotides, which probably was enough for the protocells' genetic requirements. The DNAs were replicated sufficiently faithfully that errors — that is, mutations — did not, in general, accumulate to harmful levels. The occasional errors that did occur contributed to the genetic variations of the protocell species and, hence, to their evolution.

Very likely, species of protocells arose separately on numerous occasions in the primordial soup. Nevertheless, the universality today of the structures of proteins and nucleic acids, the triplet code, the use of ATP, and the workings of the organizing mechanism suggest that competition and natural selection eliminated all but one of the early species of cells. The surviving species became the ancestors of all subsequent terrestrial life — from bacteria to protists, fungi, plants, and animals.

As discussed in the following chapter, there are good reasons to believe that the earliest cells were prokaryotes and that they reproduced by binary cell division, similar to the reproduction of contemporary bacteria. Furthermore, they probably derived their energy by fermenting energy-rich molecules, such as sugars, that they found in the soup.

No one knows how much time elapsed before chemical evolution made the transition to the first species of cells. It may have been a million years or less. Or it may have been many hundreds of millions of years. All we know, from the fossil record, is that by about 3.5 billion years ago single-celled life inhabited shallow areas of the sea. There are also indications that these cells were capable of photosynthesis and, hence, had already evolved considerably beyond the state of the earliest fermenters. For lack of more specific information, let us assume then, somewhat arbitrarily, that protocells arose approximately 4 billion years ago. This is a round figure that will suffice for discussions in the next chapter, even though it may be off by a few hundred million years one way or the other.

With the arrival of protocells in the primordial soup, the fundamental components of life as we know them today had come into existence, at least in rudimentary form: the organizing mechanism, consisting of DNAs, RNAs, and enzymes; the triplet code for writing genetic information; the means of selective absorption of raw materials and of expulsion of waste products across the cells' enclosing membranes; the extraction of energy from energy-rich molecules; and the capacity for cell reproduction and passing genetic information on to the next generation (see box 6.2).

The driving mechanism in the development of the protocells was Darwinian evolution. According to the theory presented here, this evolution began with the self-replication of short strands of RNA among the otherwise random chemical reactions in the primordial soup and progressed step-by-step to RNA quasi-species, hypercycles, and membrane enclosure of some of the hypercycles. Remember that the theory is based on many assumptions and contains many gaps. Some of the assumptions may turn out to be wrong. In fact, researchers in a number of laboratories today are developing and testing alternative theories, some of which rely on RNA and RNA-like molecules as the first replicator (see Trachtman 1984) and other that don't (see Cairns-Smith 1985 and Dyson 1985). Clearly, investigation of the origin of life is at present characterized by great uncertainties and enormous intellectual challenges. Nevertheless it is an exceptionally exciting field of scientific research, dealing as it does with one of nature's most fundamental secrets.

BOX 6.2.
Entropy: The Measure of Disorder in the Universe

THE evolution from initially random chemical reactions in the primordial soup toward an organizing mechanism and life was an evolution from disorder toward order. There is a branch of physics, called *thermodynamics*, that, among other things, deals with the increase and decrease of disorder. In that theory the concept of disorder is mathematically defined and referred to as *entropy*. The entropy of a system of atoms, molecules, or other kinds of particles (including electromagnetic radiation) is a quantity, like energy, that obeys definite physical laws. For example, the second law of thermodynamics states that *the entropy (that is, the disorder) of a closed system either remains constant or increases. It never decreases.*

A simple example will illustrate the meaning of the second law. Let us take a closed jar containing a few hundred peas, half of them yellow and the other half green. The peas are initially neatly separated by color into two layers, the yellow peas at the bottom and the green peas at the top. Let us now vigorously shake the jar, with the lid on tight. Soon the different-colored peas become thoroughly mixed and no amount of further shaking will separate them again by color. The shaking turned the initially ordered state of the system into a disordered state, as predicted by the second law.

The question now arises, "Does the second law not contradict our theory of the origin of life, in which order emerged out of disorder?" The answer is an unequivocal "No." The second law refers specifically to *closed* systems, that is, systems that are enclosed by rigid and fully insulating walls, through which pass neither energy nor matter. The primordial soup was not a closed system. It was *open*, and in open systems entropy can decrease, although it does not have to. For example, if we unscrew the lid of the jar of peas, the system becomes open. We can reach in, rearrange the peas by color, and return them to their original ordered state. Apart from patience, this rearranging takes energy.* Therefore, we conclude that *under suitable conditions and with the input of energy, order may arise from disorder in an open system.* Countless examples throughout nature confirm this conclusion, such as the formation of snowflakes in moist winter air, the crystallization of cooling lava into

---

*One could argue that shaking the jar adds energy and, hence, makes the system open. That is true. I used this example, despite its flaw, because it is easy to visualize. The problem can be remedied by extending the experiment to include, besides the jar, the person who does the shaking and enclosing the entire system within rigid and fully insulating walls. Another example of the increase of entropy, in which the question of being closed does not arise, is a bouquet of roses placed in a closed room. The fragrance of the roses, which is carried by *odorant* molecules, will soon permeate the entire room. No matter how long we wait, the odorants will not all return simultaneously to the roses. The initial ordered state (when all the odorants were neatly consolidated at the roses) evolved into a disordered state.

BOX 6.2 continued

rocks, the synthesis of helium from hydrogen in the hot interior of the Sun, and the collapse of interstellar gas and dust clouds into stars. In each of these examples, the final state is more ordered than the initial one. Furthermore, the ordering occurs at the expense of energy, which is released as heat into the environment.

The creation of order and the emergence of life in the primordial soup came about by the same principle. The soup was an open system because energy arrived from the Sun, the Earth's interior, and other sources. It passed through the system, driving chemical reactions and locally ordering raw materials—H_2O, CO_2, N_2, and so forth—into complex organic molecules. Eventually, the energy was emitted into space, mostly in the form of infrared radiation. The second law of thermodynamics was not violated.

The same ordering or organizing process goes on today in all of biology. With the aid of sunlight, green plants and other photosynthetic organisms combine raw materials into complex organic molecules. Animals take up the organic molecules and break them down into simpler molecules, thus satisfying their needs for energy and raw materials. The broken-down end products are recycled, serving as raw materials for further photosynthetic organizing processes. At every step in the cycle energy is given off as heat to the environment and, ultimately, emitted as infrared radiation into space.

The Earth's ecosystem is able to maintain itself because it is open and energy keeps flowing through it. Should the system become enclosed so that energy can flow neither in nor out, the second law of thermodynamics would quickly come into action. Photosynthesis would stop, animals would run out of food, and death and disorder would follow. There are indications that a scenario approximate to this was responsible for the mass extinction of dinosaurs, about

65 million years ago. According to one theory, an asteroid or comet collided with Earth, whirling enormous quantities of dust into the atmosphere and blocking out sunlight for months (see box 9.4). Photosynthesis came to a halt and plants, upon which the dinosaurs depended for nourishment, stopped growing. The dinosaurs began starving and finally they died.

Now consider the entropy of the entire Universe. In a thermodynamic sense, we may regard the Universe as a closed system. That is true even if the Universe is infinite and expands forever. Hence, according to the second law, the entropy of the Universe either remains constant or increases. In fact, its entropy increases slowly but steadily.

Virtually all of the increase of entropy in the Universe today is due to the radiation emitted by stars and other matter, including our ecosystem. For example, as stars evolve and acquire cores of heavy elements, their entropy tends to diminish. (This means that, averaged over time, the entropy of stars becomes less; for instance, a white dwarf has much less entropy than a main sequence star of comparable mass.) However, while stars evolve, they emit electromagnetic radiation, and that radiation contains more entropy than the decrease of entropy in the stars. Hence, the net effect is an increase in the total amount of entropy in the Universe. In the case of the Earth, the entropy of the radiation emitted into space exceeds the entropy of the radiation received from the Sun. Furthermore, that excess is greater than the decrease in entropy within the ecosystem. Hence, here too the total entropy increases. And so it is with all other systems in the Universe. Even though entropy may diminish locally, overall it inescapably becomes greater. The Universe's fate is evolution toward ever-increasing disorder.

Exercises

1 The concept of a "primordial soup" was first introduced in 1929 by the British biochemist John Haldane. Describe what is meant by this concept and how our current understanding of it differs from that of Haldane.

2 In 1871 Charles Darwin wrote to a friend about what he thought would happen to protein compounds if they were produced abiotically today (Dickerson 1978):

It has often been said that all the conditions for the first production of a living organism are now present which could ever have been present. But if (and oh! what a big if!) we could conceive in some warm little pond, with all sorts of ammonia and phosphoric salts, light, heat, electricity, etc., present, that a protein compound was chemically formed ready to undergo still more complex changes, at the present day such matter . . .

(a) Complete the passage in your own words, expressing what you think would happen to such protein compounds.
(b) Based on your answer, discuss the probability of life originating *de novo* on Earth today.

3 If you open an unrefrigerated bottle of soda pop, beer, or champagne, bubbles of CO_2 form, causing the bottle's content to overflow.
(a) Why is this so?
(b) Relate the formation of these bubbles and the overflowing to the escape of volatiles from hot, molten lava during volcanic eruptions.

4 According to geophysical evidence, what are thought to have been the major molecular constituents in the outgassings of the early Earth? What are thought to have been some of the secondary constituents?

5 Why are earthquakes, volcanism, and outgassings thought to have been much more prevalent on the early Earth than today?

6 (a) Describe the apparatus used in Miller–Urey experiments.
(b) Which early Earth conditions are simulated by Miller–Urey experiments?
(c) Which are not?

7 Using one of the molecular model kits listed in the bibliography of this chapter, build three-dimensional models of the molecules represented in figure 6.6. In the cases of serine and alanine build both the L- and D-types.

8 In contemporary cells, the synthesis of polymers takes place efficiently because the requirements of concentration of monomers, ready availability of energy in suitable form (ATPs), and presence of catalysts (enzymes) are well satisfied. How were these requirements met in the primordial soup, before the origin of cellular life?

9 Since the 1930s, biologists have been concerned with the problem of which came first, the informational or the functional components of life's organizing mechanism. How has this problem been resolved in recent years?

10 Why are RNAs, despite their information-carrying qualities, unsuitable as the chief storage molecules of genetic information in cellular life?

11 What evidence supports the theory that all contemporary life—bacteria, protists, fungi, plants, and animals—has descended from a common ancestor?

12 Summarize how protein synthesis is carried out in contemporary cells.

13 Define the concept of entropy. State whether entropy increases or decreases in the following situations.
(a) A mixing bowl in which flour, milk, eggs, and other ingredients are being mixed.
(b) A clump of interstellar gas and dust that is collapsing into a star.
(c) The Sun as it evolves.
(d) A woodstove while a fire is burning.
(e) Water when it freezes to ice.
(f) An ecosystem during springtime.
(g) Leaves after they have fallen to the ground.
(h) A large sample of the Universe, extending over several superclusters of galaxies, in the course of time.

Suggestions for Further Reading

Cairns-Smith, A. G. 1985. "The First Organisms." *Scientific American*, June, 90–100. An alternative theory of the origin of life to the one presented in this text. The author argues that clay provided the fundamental materials.

Cech, T. R. 1986. "RNA as an Enzyme." *Scientific American*, November, 64–75. The recent discovery that some RNAs can cut, splice, and assemble themselves overturns the theory that biochemical reactions are all catalyzed by protein enzymes.

Day, W. 1984. *Genesis on Planet Earth: The Search for Life's Beginning.* 2d ed. New Haven and London: Yale University Press. A clearly written text, with more advanced biochemistry and more detailed history of the subject than this text.

Decker, R., and **B. Decker.** 1981. *Volcanoes.* San Francisco: W. H. Freeman & Co. A readable, up-to-date account of all aspects of volcanoes, from their geographic distribution and important historical eruptions to their structures, origins, impact on climate, potential sources of power, and forecasting of eruptions. The many excellent photographs help the reader imagine what conditions may have been like on the early Earth, when volcanism was much more frequent than today.

Dyson, F. 1985. *Origins of Life.* Cambridge and New York: Cambridge University Press. This delightful little book begins with an historical development of the subject from Schrödinger and Von Neumann to Cairns-Smith, Eigen, Margulis, and Orgel. It then describes a model of the origin of life based on a metabolic machinery without replication. It ends by addressing some of the open questions raised by the model, such as "How did nucleic acids originate?" and "How much genetic information can be carried by a population of molecules without exact replication?" It also addresses such general questions as "What is life?" and "Why is life so complicated?"

Eigen, M., et al. 1981. "The Origin of Genetic Information." *Scientific American*, April, 88–118. The key article upon which the theory of the origin of life presented in chapter 6 is based.

Ferris, J. P., and **D. A. Usher.** 1983. "Origins of Life." Chapter 32 in *Biochemistry*, by G. Zubay, 1190–1241. Reading, Mass.: Addison-Wesley Publishing Co. A thorough, up-to-date article about the chemical processes that may have been involved in the origin of life on Earth. For the student with a strong background in organic chemistry.

Field, R. J. 1985. "Chemical Organization in Time and Space." *American Scientist*, March–April, 142–50. An interesting article about how chemical systems can spontaneously evolve to organized states. Such processes may have been associated with the beginning of life.

Fox, S. W. 1980. "Metabolic Microspheres: Origins and Evolution." *Naturwissenschaften* 67: 378–83. Fox summarizes his theory on the roles proteinoid microspheres may have played in the origin of life on Earth. However, the student is cautioned that in this theory chemical function is introduced first and there is no convincing explanation of how genetic information (based on RNA or DNA) arose. For this reason, the theory is not widely accepted today.

Lake, J. A. 1981. "The Ribosome." *Scientific American*, August, 84–97.

Layzer, D. 1975. "The Arrow of Time." *Scientific American*, December, 56–69. The author attempts to answer the question, "Why does time never go backward?" In doing so he introduces the concept of entropy and explains it with examples from physics, chemistry, and information theory. (Concerning entropy, also see the article "What is Heat" by F. J. Dyson referenced in the Suggestions for Further Reading, chapter 1.)

Miller, S. L. 1953. "A Production of Amino Acids Under Possible Primitive Earth Conditions." *Science* 117 (May 15): 528–29. The result of the original Miller-Urey experiment.

Miller, S. L., and **L. E. Orgel.** 1974. *The Origins of Life on the Earth.* Englewood Cliffs, N.J.: Prentice Hall. A somewhat dated but still useful book for the reader who is interested in the organic chemistry details of the subject.

Nomura, M. 1984. "The Control of Ribosome Synthesis." *Scientific American*, January, 102–14. Ribosomes consist of three RNAs and 52 proteins. The article focuses on how the assembly of ribosomes from these molecules is adapted to the needs of the cell.

Oparin, A. I. 1978. "Modern Concepts of the Origin of Life on the Earth." *Scientia*, (Bologna) 113: 7–25. The "father" of the modern study of life's origin offers his views on the general conditions and physicochemical processes that must have prevailed on the early Earth for life to get started. He also writes about the roles coacervates might have played.

Trachtman, P. 1984. "The Search for Life's Origin — and a First 'Synthetic Cell'." *Smithsonian*, June, 42–51. A description of the work and views of several experimenters in the United States and abroad, who are investigating life's origin on Earth. The article discusses some of the alternative possibilities indicated in the conclusion of chapter 6.

Wald, G. 1954. "The Origin of Life." *Scientific American*, August, 44-53. Though written more than three decades ago, much of what George Wald says is worth reading, particularly his discussions of the history of the subject, the probability of the origin of life, the self-organization occurring in chemical reactions, and his philosophical views.

Walker, J. C. G. 1977. *Evolution of the Atmosphere*. New York: Macmillan Publishing Co. A broad, interdisciplinary treatment of the structure and evolution of the Earth's atmosphere, including its interaction with organisms, rocks, oceans, and the interplanetary medium. Though mathematics is used freely in this text, much can be learned from it even when skipping over the quantitative details.

Molecular Model Kits

Molecular Models (Laboratory Kit). n.d. Sargent-Welch Scientific Co., distributed by Klinger Scientific Apparatus Corp., Jamaica, N.Y.

Molecular Models: Organic Structure Set. 1980. Suring, Wis.: Diversified Woodcrafts.

Organic Chemistry Set (Benjamin/Maruzen HGS Molecular Structure Models). 1969. Menlo Park, CA.: W. A. Benjamin.

Additional References

Dickerson, R. E. 1978. "Chemical Evolution and the Origin of Life," in *Evolution, A Scientific American Book*, 31. San Francisco: W. H. Freeman and Co.

Keeton, W. T. 1980. *Biological Science*, 3rd ed. New York: W. W. Norton and Co.

Noyes, R.W. 1982. *The Sun, Our Star*. Cambridge, MA: The Harvard University Press, Table 2.1.

Press, F., and R. Siever. 1982. *Earth*, 3rd ed. San Francisco: W. H. Freeman and Co.

Scott, J. 1981. "Natural Selection in the Primordial Soup." *New Scientist*, 15 January, 153.

Watson, J. D. et al. 1987. *Molecular Biology of the Gene*, 4th ed. Menlo Park, CA.: The Benjamin/ Cummings Publishing Company.

Figure 7.1. Fossil stromatolite from 3.5 billion year old sedimentary deposits near North Pole, Australia. This dome of matlike layers was formed by ancient prokaryotes and represents the oldest known direct evidence of life on Earth.

7

Evolution of Early Life

WITH THE ORIGIN of life on Earth approximately 4 billion years ago, a new era began in the evolution of matter. The physical laws were no longer the only ones determining the course of events. The laws of biology—based on the random production of variation and on natural selection—exerted their influence now as well. Over the eons, early life became more complex and diversified. Competition in the struggle for survival intensified and new ways of utilizing available raw materials and energy evolved. Gradually, life spread to all parts of our planet's surface, interacting with the physical environment and changing its composition, structure, and appearance. A worldwide ecosystem became established and the greening of the Solar System's third planet got under way.

When we keep in mind the humble beginnings of life, these far-reaching effects are both surprising and impressive. The protocells were primitive, unicellular prokaryotes that lived off organic molecules they found in the primordial soup. They metabolized their food by fermentation,* much as some bacteria and yeast cells do today, and they reproduced by simple binary cell division. Photosynthesis, the direct use of the energy of sunlight to manufacture organic molecules, had not yet evolved. Nor had respiration, the major metabolic pathway of the cells of all animals, plants, and other complex multicelled organisms. In fact, as discussed in chapter 6, the primary prerequisite for our modern kind of respiration, molecular oxygen (O_2), did not exist initially in the Earth's atmosphere. Had it been present, the early cells would have been killed by its high chemical reactivity. Finally, the eukaryotic cell and reproduction by sexual mating had not yet developed either.

A SURVEY OF THE EVOLUTION OF EARLY LIFE

The evolution of photosynthesis, respiration, eukaryotic cell structure, and eukaryotic sexual reproduction was slow in coming. The first step in this development was taken when the early fermenting prokaryotes began to spread across the primordial soup and multiplied to the point where the organic food supply—which initially was produced by abiotic, random chemical reactions—ran short. The competition for new sources of raw materials and energy intensified.

Photosynthesis

Some bacteria adapted by evolving ways to trap the energy of sunlight directly and use it to synthesize their own food from inorganic raw materials. They obtained the necessary carbon and oxygen from carbon dioxide (CO_2) and the hydrogen from such hydrogen-rich molecules as hydrogen sulfide (H_2S) and molecular hydrogen (H_2). *Photosynthesis* thus evolved.

However, unlike the photosynthesis carried out by modern blue-green bacteria (more about them shortly), algae, and green plants, the photosynthesis of

---

*Organisms that obtain their carbon from organic compounds are called *heterotrophs*. Besides organisms that live by fermentation, humans and all other animals are heterotrophs also, for we obtain our carbon by eating plants or other animals. The term is derived from the Greek roots *hetero-* and *-trophy*, meaning "other" and "nourishment," respectively.

those ancient bacteria did not release O_2. It was an *anaerobic* kind of photosynthesis, probably similar to the one still found today in the purple and green sulfur bacteria and the purple nonsulfur bacteria. There are indications in ancient rock records that anaerobic photosynthetic bacteria lived in the primordial soup 3.2 to 3.5 billion years ago.

In contrast to the fermenting heterotrophs, the photosynthetic bacteria were *autotrophs*, meaning they could "nourish themselves" starting with inorganic raw materials. They did not depend on already synthesized organic nutrients. In the course of time, as the early photosynthesizers became more efficient, they produced more and more of the organic matter in the soup. Eventually, they became the primary producers in the food chain of ancient life.

The development of photosynthesis required the simultaneous invention of two new biochemical tools—*chlorophyll* (or some other molecules with similar light-absorbing properties) and the *electron-transport chain*. Just as it is in contemporary photosynthesizers, chlorophyll was the actual light-gathering molecule by which ancient photosynthetic bacteria trapped the energy of photons and boosted electrons to high-energy states. The electron-transport chain was the tool by which the energetic electrons were returned to the unexcited, low-energy state in a sequence of small steps. This happened via a series of carrier molecules which, in bucket brigade fashion, passed the energetic electrons from one to the other with an accompanying release of energy. Some of the released energy was stored in molecules of ATP and became available for useful work, such as the synthesis of organic molecules. The rest of the released energy was lost as heat. As discussed below, both chlorophyll and the electron-transport chain were crucial for the development of still more advanced metabolic pathways during the succeeding evolution of early life (also see sections, The Evolution of Photosynthesis and The Making of the Eukaryotic Cell).

Despite their independence from the vanishing supply of organic nutrients in the soup, anaerobic photosynthetic bacteria still suffered from one serious shortcoming. They depended on molecules like H_2S and H_2 for their requisite supply of hydrogen atoms. Although H_2S and H_2 were available, as they still are today, in the swamps and in the vicinity of volcanoes and thermal pools of the early Earth, they were not nearly as universally abundant as was another hydrogen-rich molecule—water (H_2O). The problem with H_2O was that it is a particularly stable molecule and does not give up its hydrogen atoms as readily as do H_2S and H_2. A new and more powerful kind of photosynthetic apparatus was required to split H_2O into its atomic constituents. In the course of time, some bacteria evolved such an apparatus. Like their anaerobic photosynthetic predecessors, they used the hydrogen atoms in the synthesis of carbohydrates and other organic products. They released the oxygen, derived from the water, as O_2 into the atmosphere, where it gradually accumulated. This kind of metabolism is *aerobic photosynthesis*.

Aerobically photosynthesizing bacteria began to proliferate about 2.4 billion years ago and subsequently expanded into virtually all ecological niches*

Ecological niche refers to the range of conditions within an ecosystem in which a given organism or species lives and reproduces. *Ecosystem* refers to the sum total of physical and biological components in a given area. The word *ecosystem* comes from the Greek *oikos*, meaning "house" or "habitation." (See also the section Ecosystems in the Introduction to Part Two.)

containing water and sunlight, replacing the anaerobic photosynthesizers as the primary food producers on our planet. Their modern descendents—the blue-green bacteria, algae, and green plants—have maintained this position to this day.

Respiration

At the time when photosynthesis was emerging, some bacteria evolved in a rather different direction. They acquired the ability to derive energy from reactions between various kinds of organic compounds and such inorganic molecules as sulfate ($SO_4^=$), nitrate (NO_3^-), and carbonate ($CO_3^=$). This new kind of energy generation is called *anaerobic respiration*, where "respiration" refers to the fact that inorganic molecules participated in releasing the energy stored in the organic compounds, and "anaerobic" means that no O_2 was involved.

The metabolic pathways of the anaerobic respirers differed greatly from one bacterial species to the next, depending on the organic and inorganic raw materials used by them. Some of these unusual types of anaerobically respiring bacteria have survived with few changes to this day as, for example, species of *Desulfovibrio* (sulfate respirers), *Pseudomonas* (nitrate respirers), and methane-producing bacteria (carbonate respirers).

The reactions of anaerobic respiratory metabolisms, which combined organic molecules with oxygen-rich inorganic compounds, released so much energy that they were potentially harmful. Cells overcame this danger by adopting the electron-transport chain, which had already evolved earlier in the photosynthetic bacteria. Through the electron-transport chain, bacteria were able to control the chemical reactions of anaerobic respiration and let energy come off in a series of small steps, rather than all at once. As in photosynthesis, some of the released energy was stored in molecules of ATP and became available for useful work.

The application of the electron-transport chain in the respiratory pathways of anaerobic bacteria was a crucial preparatory step for the next major evolutionary step—the development of *aerobic respiration*. Aerobic respiration also required organic foodstuffs, but instead of depending on such inorganic molecules as $SO_4^=$, NO_3^-, or $CO_3^=$, it utilized O_2, which by then had become abundant in the atmosphere and oceans of the Earth.* The oxygen molecule was split into

---

*Organisms that only respire aerobically and, hence, require atmospheric oxygen are called *strict or obligate aerobes*. We, all other eukaryotes, and many bacteria are obligate aerobes. In contrast, some prokaryotes cannot survive in the presence of O_2 and are called *obligate anaerobes*. Examples are the bacteria of the genera *Desulfovibrio* and *Clostridium*. (Many of the latter cause diseases in humans, such as tetanus, gangrene, and botulism.) There also exist microorganisms, known as *facultative anaerobes*, that can tolerate O_2 but get along without it. Some facultative anaerobes possess both fermentative and aerobic respiratory metabolisms and have the remarkable ability to switch from one to the other, depending on the absence or presence of O_2. Examples are bacteria of the genera *Escherichia* (*E.*) *coli* and *Salmonella*, as well as yeasts. *E. coli* inhabit the large intestine of humans and other mammals. (*Salmonella* are responsible for such intestinal infections as dysentery, typhoid fever, and bacterial food poisoning.) The yeasts are eukaryotes which, under anaerobic conditions, are fermenters and are used in the making of beer, wine, and other alcoholic products. In the presence of O_2, yeasts respire aerobically and release CO_2 and H_2O. The release of CO_2 by yeasts is what makes dough rise.

two oxygen atoms, which were then allowed to react with hydrogen atoms to form water. The advantage of this reaction was that it released far more energy than did anaerobic respiration or any of the other metabolic pathways developed earlier. As in anaerobic respiration, the reaction between hydrogen and oxygen atoms was carried out safely with the aid of the electron-transport chain. The development of aerobic respiration was completed between about 2.0 and 1.5 billion years ago.

Eukaryotic Cells

Aerobic respiration, with its ability to liberate large amounts of energy through reactions between hydrogen and oxygen, became the basis for further evolutionary progress that eventually led to modern *eukaryotic cells*. Eukaryotic cells distinguished themselves from their prokaryotic ancestors by having *nuclei* that enclosed the genetic material. That material was distributed over several pieces of DNA, namely the *chromosomes*. All of the eukaryotes' metabolic activities involving O_2 were carried out within special organelles—mitochondria (respiration) and chloroplasts (photosynthesis). In addition, eukaryotic cells evolved other organelles that were responsible for synthesizing proteins, packaging and transporting proteins, and many other tasks, as discussed in the Introduction to Part Two.

With the development of eukaryotic cells, the binary cell division of prokaryotes evolved into a more complex process called *mitosis*. Mitosis ensured that the replicated chromosomes were properly divided into two equal sets and passed on to the offspring cells. Eventually, mitosis evolved into yet another kind of cell division—*meiosis*. Meiosis produced reproductive cells and was the basis for *eukaryotic sexual reproduction*.

The arrival of eukaryotic cells was a pivotal event in the evolution of life. Early prokaryotic organisms were microscopic and unicellular, and with few exceptions they have remained so to this day. Eukaryotic organisms, in contrast, had the potential to evolve into more complex multicellular and macroscopic forms, and within a few hundred million years of their arrival they realized this potential. By approximately 700 million years ago, the first jellyfish, worms, and other soft-bodied animals began to populate the waters of the Earth.

If life came into being approximately 4 billion years ago, the evolution from the first fermenting prokaryotes to complex multicelled life took roughly 3.3 billion years: about 500 million years for the development of anaerobic photosynthesis, another 1.1 billion years until O_2-releasing photosynthetic bacteria began to proliferate, still another 900 million years until aerobic respiration and the eukaryotic cell had evolved, and, finally, 800 million years more to the arrival of the first complex multicellular organisms (figure 7.2).

Another way to think about this evolution is to imagine that the age of the Earth is compressed into one year and to assume that our planet was formed on January 1 (when, according to this time scale, the Universe was already three years old). During much of January, the Earth acquired its early anaerobic atmosphere and juvenile ocean. The first fermenting prokaryotes came into existence in early February. Photosynthesis was invented in March. Free oxygen

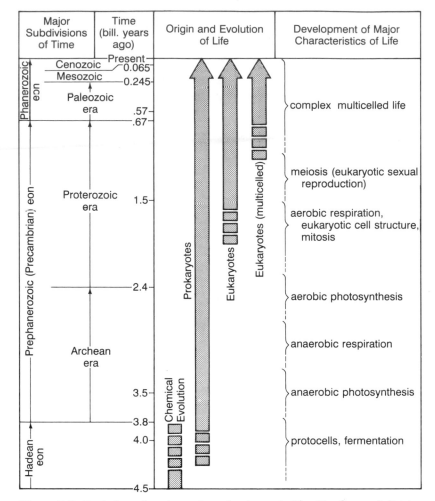

Figure 7.2. Evolution of prokaryotic and eukaryotic life. (For finer subdivisions of the Phanerozoic eon see table 8.1.)

started to become abundant by mid-July. Aerobic respiration and the eukaryotic cell structure developed during August. Sexual reproduction appeared some time in September. It was late October before complex multicellular organisms appeared in the sea. And the evolution of plants and animals into the forms that we see today took place during the final eight to nine weeks of this imaginary year.

THE EVIDENCE FOR EARLY LIFE

How have we learned about the evolution of early life? The most ancient cells lived more than 3 billion years ago in the primordial soup and we might expect that by now all evidence of them has vanished. Indeed, most of that evidence has been lost due to the continual reshaping of the Earth's surface features by geological activity. What little survived remained undetected until very recently.

For example, in the 1930s the fossil record could be traced back only about 570 million years. Ancient sedimentary rock deposits spanning that age show a clear and puzzling demarcation line. The rocks immediately above that line — therefore, younger than about 570 million years — contain fossils of a great variety of invertebrate marine animals. Some of them were encased in shells or possessed skeletons; others were soft-bodied and without any hard parts. Clearly, the ocean was then teeming with complex multicelled and macroscopic life.

In contrast, the rocks below the demarcation line appear to be devoid of any fossil remains. Did life suddenly come into existence, fully shaped in multicelled forms, approximately 570 million years ago? The evidence seemed to suggest this to early paleontologists. Hence, they called the time interval from the demarcation line to the present the *Phanerozoic eon*, meaning "eon of visible life." The preceding time span they called the *Precambrian eon*, because it was the eon before the Cambrian period, the earliest of the traditional eleven geological periods into which the Phanerozoic Eon has been divided (see box 8.2).

Charles Darwin was aware of and puzzled by the abrupt appearance of fossils of complex, macroscopic life at the beginning of the Phanerozoic eon, and in *On the Origin of Species* he wrote:

> To the question why we do not find rich fossiliferous deposits belonging to . . . periods prior to the Cambrian system, I can give no satisfactory answer. . . . The case at present must remain inexplicable; and may be truly urged as a valid argument against the views here entertained. (Darwin 1897, Vol. 2, 84–85)

Today we know that the Precambrian rocks are not devoid of fossil evidence, but much of that evidence stems from microscopic, unicellular organisms and so is not easily seen. The specialized laboratory techniques required to detect it were developed only during the 1950s. Since then microfossils have been found in dozens of Precambrian rock outcrops on all continents, from the Grand Canyon and the Rocky Mountains to many other sites in South Africa, Australia, Siberia, and elsewhere.

In addition to microfossils, evidence for the existence and evolution of Precambrian life comes from the effects it had on the physical environment. For example, some ancient rock deposits contain rusted iron and other metals, which implies that molecular oxygen was present, presumably released by early photosynthetic bacteria. Other ancient rocks contain unique isotopic abundances of carbon and other elements that can be explained only as being due to biological activity.

Yet a third line of evidence of the evolution of Precambrian single-celled life has been found in our own cells and those of other eukaryotes and prokaryotes, through a new understanding of their metabolic pathways and other biochemical processes. For example, when we put forth a sudden physical effort, like running a 100-meter dash or rapidly climbing a flight of stairs, our cells temporarily run out of O_2 (that is, we run out of breath) and are forced to halt or slow down their aerobic respiration. They then generate ATP (energy) anaerobically, much as fermenting bacteria do. Our cells can obtain energy in this manner for only short durations, but while they do they are reverting to the most ancient of all metabolisms — fermentation — from which all others have evolved.

None of the three sources of evidence — Precambrian microfossils, the ancient inorganic rock record, and biochemical processes in contemporary cells — would alone suffice to deduce the early history of life with any degree of confidence. Together, however, they point rather compellingly to the step-by-step evolutionary progression outlined in the previous section, from the first fermenting prokaryotes to the development of photosynthesis, respiration, eukaryotic cell structure, eukaryotic sexual reproduction, and, finally, the arrival of complex multicellular organisms. The remainder of this chapter will discuss these sources of evidence in more detail.

THE GEOLOGIC EVIDENCE

All continents include areas of low-lying rock outcrops that were laid down as sediments in the oceans during Precambrian times. Organisms that lived in the ancient oceans frequently became embedded in the sediments. Additional deposits in later eras often buried the older sediments to depths of thousands of meters, creating enormous pressures and temperatures and fusing them into compact, crystallized rocks (by a process called *metamorphism*). Simultaneously, whatever organismic remains the buried layers contained got crushed and destroyed. However, in some locations the burial of the ancient rocks never amounted to very much. Consequently, they did not experience extreme heating and compression, and some of the organisms trapped in them were preserved as recognizable fossils (see figure 7.3).

Microfossils

The most ancient fossils of early life discovered to date have been found in sedimentary rocks at North Pole in western Australia, with an estimated age of 3.5 billion years (as deduced from the decay of long-lived radioactive isotopes, see box 5.1). These fossils consist of mound-shaped structures resembling modern stromatolites (see p. 349), simple microspheroids, and complex filamentous microfossils (see figures 7.1 and 7.4). However, as the discoverers of these fossils (most of them are from the University of Western Australia and the Australian Bureau of Mineral Resources, Geology, and Geophysics) point out, they cannot be absolutely certain that the mound-shaped structures and microspheroids are of biological origin. And the rocks containing the filamentous microfossils appear to have been contaminated by carbonaceous matter some time after sedimentation took place, a process that may have introduced the microfossils. But most experts are beginning to agree that these objections are not valid and that the North Pole fossils are indeed 3.5 billion years old.

Fossils with a reported age only slightly less than that of the North Pole specimens come from the Onverwacht and Fig Tree cherts* of South Africa. The

---

*Chert is a hard, often dark-colored, fine-grained crystalline rock consisting of quartz. It forms as a result of the precipitation of silicon dioxide (SiO_2). Chert is particularly resistant to compression and contributes to the preservation of fossils embedded in it. Furthermore, the small grains make it easier to distinguish fossil imprints of microscopic size from random flaws in the rock's crystal structure.

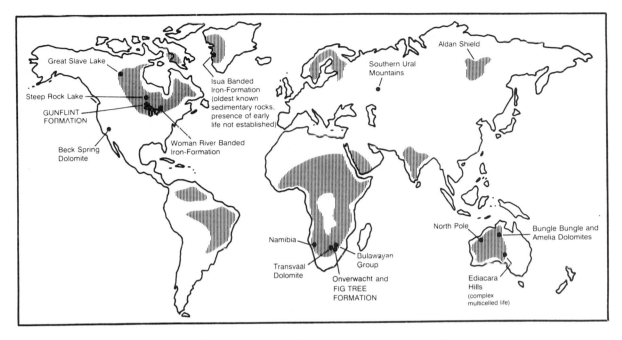

Figure 7.3. Selected sites of ancient rock formations that contain or are suspected to contain fossilized remains of early life. All continents include areas of low-lying rock outcrops that were laid down as sediments in the oceans billions of years ago and later became fused into rocks by additional deposits above them. These areas are referred to as *shields* and are indicated in the above map by shading. Most shields had formed and became permanent components of the Earth's crust by the end of the Archean era, about 2.4 billion years ago. Some of the ancient rocks have experienced relatively little geologic alterations since then and contain traces of fossilized remains of early life. (The concentration of many of the sites in English-speaking countries is probably an artifact of selection and does not reflect the actual global distribution of ancient microfossils.)

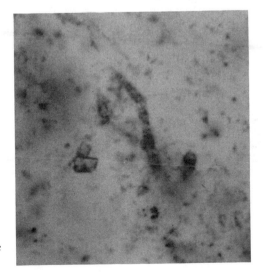

Figure 7.4. Microfossil in a rock from North Pole, Western Australia, with an estimated age of about 3.5 billion years. This filamentous microfossil is one of the most ancient remains of early life ever found, exhibiting a surprising degree of complexity.

ages of these rocks have been estimated at 3.4 billion years for the Onverwacht formation and 3.2 billion years for the Fig Tree formation. The Onverwacht microfossils, which are spheroidal or filamentous in shape, range in size from approximately 4 to 20 μm, which is similar to the sizes of modern bacteria. The fossils from the Fig Tree formation are spheroidal or rod-shaped. The spheroids have sizes like those from the Onverwacht site; the rods are somewhat smaller, with diameters of about 0.3 μm and lengths from 0.5 to 0.7 μm. A few of the Fig Tree fossils show double-layered membranes comparable in thickness to those of modern bacteria. The spheroidal fossils bear some resemblance to modern blue-green bacteria and may be the remains of aerobic photosynthesizers or of their precursors.

Despite these microfossils' similarity to contemporary bacteria, both in size and shape, a word of caution is in order here also. All of them show a very simple morphology and, in the words of William Schopf (1979, 643), "do not exhibit a degree of structural complexity that would make their biological interpretation wholly convincing." They may be pseudofossils that resulted from the deposit of carbonate material that was accidentally shaped into cell-like structures. Nevertheless, analysis of the relative abundance of the two stable isotopes of carbon, ^{12}C and ^{13}C, found in organic matter from the Onverwacht and Fig Tree formations suggests that it is of biological origin. For example, ^{12}C and ^{13}C are generally found in the ratio 99 to 1 in inorganic terrestrial material, such as atmospheric CO_2, but in biologically synthesized compounds the ratio is shifted slightly toward ^{12}C. Apparently, during photosynthesis plants fix somewhat more ^{12}C than ^{13}C. The ancient organic matter from the Onverwacht and Fig Tree sites show the same kind of ^{12}C enrichment as is found in more recent deposits of known biological origin and in contemporary biological matter. This strongly supports the view that the Onverwacht and Fig Tree fossils are indeed the remains of once-living organisms.

Evidence of Precambrian life somewhat younger than the Fig Tree fossils has been found in rocks of the Aldan Shield of Siberia and of the Woman River Banded Iron-formation of Canada, both of which have been dated at 2.8 to 3.1 billion years of age. Isotopic analysis of sulfur compounds in these rocks (similar to the $^{12}C/^{13}C$ analysis discussed above) suggests that they are the metabolic waste products of ancient bacteria that lived by anaerobic respiration. The bacteria used sulfate ($SO_4^=$) and converted it to hydrogen sulfide (H_2S) and water (H_2O), as members of the species *Desulfovibrio* still do today.

Desulfovibrio bacteria live in the muddy bottoms of ponds that are rich in organic matter and sulfate. The water and muddy environment protect them from atmospheric O_2 that would quickly kill them. In the water layers above the mud, purple and green sulfur bacteria are commonly found. They are anaerobic photosynthesizers: They use the energy of sunlight and split the hydrogen sulfide produced by the *Desulfovibrio*, converting it back to sulfate. Thus the *Desulfovibrio* and the purple and green sulfur bacteria constitute a simple ecosystem driven by sunlight, in which sulfate and sulfide are cycled back and forth. We may assume that the ancient anaerobically respiring bacteria of the Aldan Shield and the Woman River formations were similarly associated with anaerobic photosynthesizers that converted the sulfide back to sulfate. Otherwise, the sulfate would have rapidly become depleted.

Figure 7.5. Fossil stromatolites, exposed on the eroded surface of roughly 2-billion-year-old limestone near the Great Slave Lake, Canada. These concentrically laminated domes were formed in shallow water by ancient communities of blue-green bacteria.

However, be cautious, as with the interpretation of the North Pole, Onver wacht, and Fig Tree fossils, and keep in mind that the evidence for the existence of ancient sulfate respiring and anaerobic photosynthetic bacteria is not absolute. It is an indirect kind of evidence and requires further research to resolve whether those ancient sulfur compounds are indeed metabolic waste products or have some unknown abiotic origin.

Stromatolites

In contrast to the record of Precambrian life discussed so far, much of the evidence found in younger rocks clearly shows its biological origins. Good examples are the unusual dome-shaped fossil structures that were discovered around the turn of the century by the American paleontologist Charles D. Walcott (1850–1927) in the 2-billion-year-old Gunflint Iron-formations along the western shores of Lake Superior in Canada. These structures look like tall stacks of pancakes and reminded Walcott of modern stromatolites (see figures 7.5 and 7.6).* Stromatolites are matlike, sedimentary deposits in the form of domes or pillars, which today are the result of the life cycle of certain kinds of blue-green bacteria. The bacteria either trap sediment grains in a jellylike material they secrete or their metabolic activity causes sediments to precipitate. Because blue-green bacteria are aerobic photosynthesizers, Walcott suggested that the

*The word stromatolite is derived from the Greek *stroma*, meaning "bed," and *lithos*, meaning "stone."

Figure 7.6. Living stromatolites at Hamelin Pool, Shark Bay, Western Australia. The stromatolites are formed by certain kinds of blue-green bacteria, which, season after season, trap sediments in jellylike secretions and thus build up the layered domes. The unusually high salt content of the lagoon creates an environment unsuited for most organisms and allows the blue-green bacteria to flourish.

Gunflint fossil structures are also the product of aerobic photosynthetic bacteria. At first his suggestion was received with skepticism, largely because at the time there existed little other evidence that life dates back billions of years. Then in the 1950s, American paleobiologists Stanley A. Tyler (1906–1963) and Elso Barghoorn (1915–1984), discovered microscopic fossils in the Gunflint stromatolites that were undoubtedly the remains of living organisms and nearly identical to contemporary blue-green bacteria, and the doubt about the origin of the fossil stromatolites vanished (see figure 7.7). Today it is commonly accepted that they had indeed been built up in shallow water by aerobic photosynthetic bacteria, like their modern counterparts.

During the last fifteen years, fossilized stromatolites have been found in dozens of locations all over the globe. Many are of the age of those of the Gunflint formations or younger, but a few are older, such as those of the Bulawayan Group in Rhodesia (about 3.1 billion years in age), Steep Rock Lake in Canada (2.6 billion years), the Transvaal Dolomite of South Africa (2.3 billion years), and near the Great Slave Lake, Canada (2.0 billion years). These findings suggest that aerobic photosynthesis may have evolved very early, though it probably did not become widespread until about 2.4 to 2.0 billion years ago.

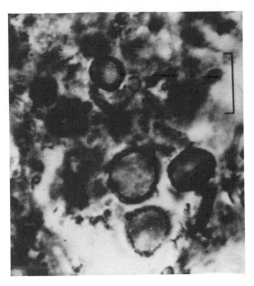

Figure 7.7. Spheroidal and filamentous microfossils in a stromatolitic rock from the Gunflint Iron formation, with an estimated age of roughly 2 billion years. The spheroidal microfossils have diameters from about 1 to more than 16 μm.

Two Precambrian Eras

The proliferation of aerobic photosynthetic bacteria about 2.4 billion years ago and the accompanying release of O_2 into the atmosphere, as manifested by ancient fossil and inorganic rock records, offers a convenient way of subdividing the long Precambrian eon into two eras: the *Archean* era, spanning the time from the oldest known sedimentary rocks, with an age of about 3.8 billion years

Figure 7.8. Rock from the Isua Banded Iron-formation of southwestern Greenland, the oldest known terrestrial rock formations, with an estimated age of 3.8 billion years. The dark bands of this rock are iron-rich, while the light bands are iron-poor.

(figure 7.8), to 2.4 billion years ago;* and the subsequent *Proterozoic* era, extending from 2.4 billion years ago to the beginning of the Phanerozoic eon, which is traditionally recognized as about 570 million years ago. (*Proterozoic* is derived from the Greek and means "before visible [macroscopic] life.")

In summary, the Archean era is the oldest time period for which there is evidence of the existence of life on Earth. Relatively few fossils have survived from this era, and the few that have are poorly preserved and show only very simple cell structures. Therefore, some doubts remain whether they are, in fact, remains of organisms or have some unexplained abiotic origin. In contrast, many of the microfossils of the Proterozoic era are well preserved and leave no question as to their biological origin. They attest to the progressive diversification, proliferation, and increase in structural complexity of early life. In particular, they indicate that the most dominant organisms of the Proterozoic era (at least for most of its duration) were the blue-green bacteria. Hence, this era may justly be called the "age of the blue-green bacteria."

Inorganic Rock Record

The arrival of aerobic photosynthesis and the widespread release of O_2 by a little over 2 billion years ago is further confirmed by the inorganic rock record, in particular by the *banded iron-formations* and *red beds*. The banded iron-formations are composed of alternating layers of iron-rich and iron-poor sediments, most of which have ages between 2.0 and 2.4 billion years. The iron-rich layers were laid down when iron that was dissolved in shallow ocean basins came in contact with O_2. This converted the dissolved iron into rusted, oxygen-rich iron, which then precipitated and settled to the ocean bottoms.[†] There it accumulated with other sediments to become the iron-rich layers found today in the banded iron-formations. Some of these iron-rich layers are only a few millimeters thick, but often can be traced for hundreds of kilometers, indicating that iron was once dissolved over wide expanses in the ancient oceans. The banded iron-formations are the source of many of the iron ore deposits that are mined today, as, for example, those of the Mesabi Range in Minnesota and the Marquette Range in Michigan.

The layering of the banded iron-formations suggests that periods when much oxygen-rich iron precipitated and accumulated on the floors of ocean basins alternated with periods when very little of it precipitated. This may have been

---

*The oldest known rocks on Earth are those of the Isua Banded Iron-formation of southwestern Greenland, with an estimated age of 3.8 billion years. However, minerals exist that are even older. They are zirconium-containing silicates ($ZrSiO_4$, known as *zircons*) from sandstone near Mt. Narryer and the Jack Hills area in Western Australia that have ages of approximately 4.15 and 4.3 billion years, respectively. Zircons are unusually hard and weathering-resistant minerals. These ancient zircons probably wound up in the sedimentary sandstones (which are much younger) after being eroded from the Earth's granite crust. The original crustal rocks have not been found, but the discovery of the zircons raises hope that remnants still exist.

[†]All iron originally present in the Earth's crust was in the doubly ionized, oxygen-deprived form Fe^{++}, called *ferrous* iron. When weathering washed ferrous iron out of ancient rocks, it dissolved in the water and was carried to the oceans by rivers. In the oceans it diffused over wide areas and became the dissolved iron referred to above. Only ferrous iron dissolves in water. The more highly oxidized, rusted iron, which is triply ionized (Fe^{+++}) and called *ferric* iron, is not soluble in water.

due to periods of rapid growth of aerobic photosynthetic bacteria followed by periods of less rapid growth, with a consequent periodic rise and fall of the O_2 level in the atmosphere and oceans. It may also have been due to fluctuations in the rate at which iron was washed out of rocks by rain and became exposed to oxygen.

Although most banded iron-formations have ages between 2 and 2.4 billion years, some are considerably older as, for example, the Woman River and Isua formations (figure 7.8). However, neither of these two formations can be unquestionably associated with early life. Hence, it is not clear whether the oxygen that precipitated the iron in those ancient rocks came from aerobic photosynthesizers or had a nonbiological origin, such as the splitting of water molecules by ultraviolet radiation from the sun. If the O_2 was released by photosynthetic bacteria, life may have come into existence even earlier than 4 billion years ago.

Very few of the banded iron formations are younger than 2 billion years. Apparently by then so much O_2 had accumulated in the atmosphere that dissolved iron could no longer spread out extensively over the oceans. The dissolved iron reacted with oxygen while it was still being carried by the rivers toward the oceans, and the resulting rusted iron was deposited (in the form of Fe_2O_3, called *hematite*) along with sand and gravel in relatively confined, onshore marine basins. Such deposits are known today as *red beds*.

The long history of banded iron-formations, followed by the relatively rapid transition to red bed deposits, indicates that the O_2 level in the Earth's ancient atmosphere rose exceedingly slowly for hundreds of millions of years. Only about 2 billion years ago did it reach levels at which it began to interact strongly and consistently with the environment. In part, it rose so slowly because the iron that was dissolved in the oceans acted as a buffer that absorbed and removed the oxygen as it was released by photosynthesis. This, in turn, gave early life a sufficient period of grace to adapt to oxygen's high reactivity and to evolve aerobic respiration.

The banded iron-formations and red beds are not the only geological evidence of the slow rise of O_2 in our planet's atmosphere 2 billion years and more ago. For instance, the gold–uranium deposits of the Dominion Reef and Witwatersrand System of South Africa, which were laid down between 2.0 and 2.8 billion years ago, offer independent confirmation. These deposits contain uranium and lead in the molecular forms UO_2 (called uraninite) and PbS (galena). In the presence of even small amounts of O_2 these molecules change rapidly into U_3O_8 (triuranium octoxide) and $PbSO_4$ (anglesite). The presence of UO_2 and PbS in the Dominion Reef and Witwatersrand rocks is a sure indication that the oxygen level up to about 2 billion years ago was indeed very low.

Eukaryotic Life

In response to the rise of O_2 in the Earth's atmosphere, aerobic respiration evolved and life made the transition from prokaryotic to eukaryotic forms. The oldest known evidence of this transition comes from a number of sites, such as the Bungle Bungle and Amelia dolomites of northern Australia and rocks in the southern Ural Mountains of Russia, with approximate ages of 1.5 billion years.

The fossil microbiotas preserved in these rocks are much more diverse than those found in older rocks. Those of the Bungle Bungle and Amelia deposits consist of associations between blue-green bacteria, growing in typical stromatolitic forms, and spherically shaped cells of various kinds. The spherical cells are between 1 and 5 μm in diameter and occur either alone or in clusters of up to a hundred. Most important, some of them contain what appear to be small, membrane-enclosed structures that may be remnants of organelles, suggesting that these cells were eukaryotes. Furthermore, a group of four tightly bound fossilized cells, found in the Amelia dolomite, closely resembles the spores produced by the mitotic cell division of modern green algae. Algae are eukaryotes and, hence, this ancient group of four cells is yet another indication that some of the organisms in these deposits may have been eukaryotes, though some researchers have questioned this interpretation. The microfossils found in the southern Ural Mountains have been identified by Russian and European paleontologists as originating from unicellular eukaryotic organisms that were buoyant and lived freely in the ancient sea as plankton.

As we go forward in time from 1.5 billion years ago, the fossil evidence for the existence of eukaryotic cells becomes progressively more common and convincing. The evidence comes from rock deposits of such widely dispersed areas as northern and southern Australia, southern India, arctic Canada, Montana, California, Utah, and Arizona. For instance, the California fossils are found in 1.2 to 1.4 billion-year-old strata of dolomite—the Beck Spring dolomite—of the Nopah Range, east of Death Valley. As in the Bungle Bungle and Amelia dolomites, the fossilized microbiota of the Beck Spring dolomite consist of host stromatolites that have been formed by several different species of blue-green bacteria and harbor a variety of other kinds of cells. Some of these cells are almost certainly of eukaryotic origin. They are up to 60 μm in diameter, which is much larger than any of the earlier prokaryotic cells and comparable to the sizes of many modern eukaryotic cells. Furthermore, they occur either singly or in clusters. Many contain minute dark spots and bands, suggestive of organelles. Some of them are smooth-surfaced, while others have surface ornamentations, such as granules and spines, similar to features found on contemporary green and yellow-brown algae.

In addition to having larger sizes and more complex internal and surface constructions than prokaryotes, many early eukaryotic organisms appear to have been plankton, as noted already in the case of the microfossils found in the southern Ural Mountains. The evidence comes from the fact that the fossilized remains of many of these early eukaryotic microorganisms are found uniformly distributed in such sedimentary rocks as shale, sandstone, and limestone, suggesting that they became trapped as the original materials of these rocks precipitated toward the bottom of the sea. That is what happens today to planktonic microorganisms. Some of these early eukaryotes may also have been motile, for the outer membranes of some of them contain small circular holes through which undulating flagella may have protruded. In contrast, most of the prokaryotes of that early time, such as those that formed the stromatolites, were members of benthic communities, meaning they were nonmotile and lived at the bottom of shallow seas.

Complex Multicelled Life

Life remained microscopic and mainly single-celled until about 670 million years ago. Then, within a relatively short span of time by geologic standards, macroscopic and multicelled marine forms appeared that bore a resemblance to modern jellyfish, worms, sea pens, soft corals, arthropods, and echinoderms. All of these marine creatures were soft-bodied; none possessed hard, mineralized shells or solid skeletons; and they inhabited shallow, near-shore marine environments. Some were free floating, others were anchored to the ocean floor, and still others crawled along in the mud.

It is interesting that with the arrival of complex multicelled life, stromatolites suffered a marked decline that has continued to the present. Apparently the blue-green bacteria that built the stromatolites could not stand up to competition with invertebrates. The age of the blue-green bacteria came to a close.[*]

The first convincing evidence of the existence of Precambrian animal life came in 1930 when the German paleontologist Ernst J. G. Gürich (1859–1938) found traces of multicellular organisms in roughly 600 million year old rocks in Namibia, Africa. Then, in the mid-1940s, the Australian geologist Reginald C. Sprigg discovered additional and more extensive evidence in the Ediacara Hills some 400 km north of Adelaide, Australia. There, a band of rock outcrops, exposed over a 140-km-long stretch and approximately 670 to 570 million years old, contains a variety of exceptionally well preserved fossils of multicelled, soft-bodied marine organisms (see figures 8.1, 8.18, 8.19). Martin F. Glaessner, of the University of Adelaide, has identified nearly two dozen species among the Ediacara fossils, as well as numerous surface tracks and shallow burrows made by the animals. Since the 1940s, other fossilized marine organisms of similar constitution and age have been found in rock deposits over much of the globe: in China, Siberia, the western U.S.S.R., Scandinavia, England, and Canada. They are collectively referred to as the *Ediacara fauna*, and the period of their existence — from about 670 to 570 million years ago — has become known as the *Ediacarian period*.

The discovery of the Ediacara fauna calls for a redefinition of the traditional geologic reckoning of time. The Phanerozoic eon should be extended back in time to 670 million years ago, to include the Ediacarian period (see table 8.1). The Ediacarian period becomes then the oldest of twelve geologic periods of the Phanerozoic eon, predating the Cambrian. In keeping with this change, the Precambrian eon needs to be renamed the Prephanerozoic eon and to end 670 million years ago, when visible life (complex macroscopic and multicellular life, visible with the unaided eye) began to appear. Preston Cloud (1982) has

[*]Even though blue-green bacteria have lost their dominant role as primary food producers, they still are abundant today. Thousands of contemporary kinds live in a great variety of habitats, from the ocean to lakes, shallow ponds, the soil, glacier ice, and thermal springs. However, few of them build stromatolites. Stromatolites are found today in only a few restricted environments, such as in the fresh water marshes of Andros Island in the Bahamas, in thermal springs of Yellowstone National Park, and along some marine habitats such as Shark Bay on the western coast of Australia. Special conditions are required for blue-green bacteria to build stromatolitic structures. For example, at Shark Bay, the stromatolites grow in a lagoon with an unusually high salt content. The salt keeps off invertebrates that graze on blue-green bacteria.

proposed such changes in the geologic time scale. It remains to be seen whether they will be adopted by the geologic community.

FROM FERMENTATION TO RESPIRATION

The previous section followed the evolution of early life as it is revealed by fossils and mineral deposits in ancient rocks. Additional insights into this evolution can be gained by studying the metabolic pathways and other biochemical processes in contemporary cells. Let us start by exploring the ways cells obtain energy.

During normal activity the cells of our bodies obtain energy by aerobic respiration. We inhale atmospheric oxygen; the blood stream carries it to our cells; and in the cells carbohydrates, in the form of glucose ($C_6H_{12}O_6$), are broken down and combined with the oxygen to form carbon dioxide and water, with the simultaneous release of energy. In contrast, during brief spurts of strenuous exercise that leave us breathless our cells gain energy quite differently. There is not sufficient time to bring enough O_2 to the cells, especially to the muscle cells, for carrying the respiratory reactions to completion. Our cells revert then to a primitive metabolism that does not use oxygen: glucose is converted to lactic acid. This *anaerobic* metabolism releases only a relatively small amount of energy compared to respiration and it corresponds in all important respects to bacterial fermentation. The only significant difference between them is that in our cells the end product is aways lactic acid, while in bacteria it may also be alcohol, vinegar, or some other simple organic compound.

It seems our cells "remember" their bacterial ancestry. Normally they meet their energy needs by aerobic respiration, but when there is a lack of oxygen, they fall back upon fermentation, the metabolism of the earliest prokaryotes. However, our cells can function in this mode for only a minute or two at a time. Fermentation does not free enough energy to sustain the needs of large, multicelled bodies like ours.

Fermentation

Let us take a closer look at fermentation. It consists of two separate reaction sequences that, like all biochemical reactions, are catalyzed by enzymes. In the first reaction sequence, called *glycolysis,** each glucose molecule is broken down into two pyruvate molecules and energy is released:

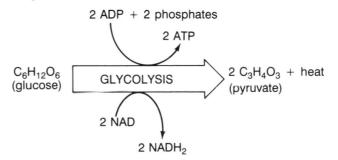

*The word "glycolysis" is derived from the Greek roots *glyc-* meaning "sweet" (or sugar) and *-lysis* meaning "loosening."

(See Introduction to Part Two, p. 274.) Some of the energy becomes stored in two molecules of ATP, and the rest is lost as heat. The four hydrogen atoms (H), which the two pyruvates are short in comparison to the original glucose, are picked up by two carrier molecules called NAD (*nicotinamide adenine dinucleotide*). Each of the NADs picks up two hydrogen atoms and is thereby converted to $NADH_2$.

The $NADH_2$ molecules keep their hydrogen atoms only temporarily. They enter the second reaction sequence of fermentation (which has no particular name) and react with the pyruvates. The $NADH_2$ molecules are stripped of their Hs and converted back to NADs, ready to participate in further reactions of glycolysis. The hydrogen atoms and the pyruvates combine to yield any number of different end products, depending on the species of bacteria involved: ethanol (drinking alcohol, C_2H_6O), acetic acid (vinegar, $C_2H_4O_2$), lactic acid ($C_3H_6O_3$), butyric acid ($C_4H_8O_2$), or some other organic compound of similar complexity. For instance, the reactions that yield drinking alcohol and lactic acid may be summarized as follows:

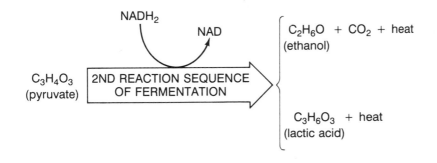

Molecular oxygen, O_2, does not participate in any of these reactions. That is why the French chemist Louis Pasteur (1822–1895) referred to fermentation as "the consequence of life without air."

Aerobic Respiration

Fermentation is a very wasteful process because its end products—lactic acid, ethanol, and so forth—still contain most of the energy originally present in glucose. All of these end products are rich in hydrogen. If the hydrogen atoms were combined with oxygen to form water, a great deal more energy could be extracted. That is exactly what happens during aerobic respiration and that is why this metabolic pathway evolved.

Aerobic respiration consists of three separate reaction sequences that are carried out in sequence and at two different locations in the cell. The first set of reactions is *glycolysis*, the same glycolysis as in bacterial fermentation. It takes place in the cytoplasm of the cell. The second and third reaction sequences are the *citric acid cycle* and the *respiratory chain*. Bacteria capable of aerobic respiration carry these reactions out on the inner surfaces of their enclosing membranes. Eukaryotic cells carry them out on the inner membranes (the cristae) of the mitochondria, the organelles described as the cell's "powerhouse" in the Introduction to Part Two.

During glycolysis each glucose molecule is broken down into two pyruvates and, simultaneously, two molecules of ATP and two of $NADH_2$ are produced. The ATPs store useful energy and are applied much as they are in fermenting cells. However, the pyruvates and $NADH_2$ follow rather different reaction paths.

The two pyruvates enter one of the mitochondria of the cell and start passing through the citric acid cycle.* With the participation of 3 H_2O molecules, each pyruvate is broken down into 3 CO_2 molecules and 10 Hs. The Hs are carried away by 5 carrier molecules, 4 of which are NADs and one is a similar kind of molecule called *flavin adenine dinucleotide* or FAD. In addition, one ATP molecule is formed:

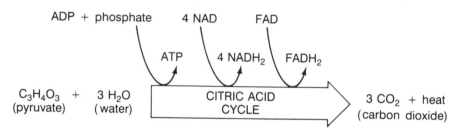

No oxygen is used during the citric acid cycle. Much of the energy that was present in the pyruvate is now stored in the $NADH_2$s and the rest has been converted to heat.

The 2 $NADH_2$s formed during glycolysis, the 8 formed during the citric acid cycle (4 from each of the 2 pyruvates), and the 2 $FADH_2$s, also formed during the citric acid cycle, now enter the respiratory chain. Only now does oxygen enter into the reactions. The reactions carried out by the respiratory chain are known as *oxidative phosphorylation* because they generate ATP by adding a phosphate to ADP and use oxygen. Each $NADH_2$ and $FADH_2$ gives up its hydrogen atoms and becomes once again an empty carrier molecule, NAD and FAD, ready for renewed use. The hydrogen atoms combine with oxygen to form water and a great deal of energy is released, some of which is stored in molecules of ATP:

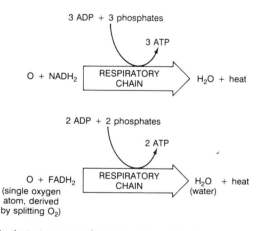

*The citric acid cycle derives its name from the citric acid (a 6-carbon molecule) that serves as a catalyst. The cycle is also known as the *tricarboxylic acid (TCA) cycle* or the *Krebs cycle*, after the British biochemist Sir Hans A. Krebs (1900–1981), who in the late 1930s worked out its chemical details.

The chemical reactions between hydrogen and oxygen atoms to form water molecules release enough energy to make them potentially explosive. To avoid this danger, the reactions of the respiratory chain must be carried out slowly and in a controlled fashion. This is done by the electron-transport chain, mentioned earlier, which is the central component of the respiratory chain. The electrons, which eventually bind the atoms together into H_2O molecules, are picked up by special carrier molecules, the *cytochromes*. Step-by-step the cytochromes lower the electrons into the final low-energy state that constitutes the bonds between the hydrogen and oxygen atoms. Because this occurs in a sequence of small steps rather than in one big jump (see figure 7.9), the energy comes off gradually and is usable for forming ATPs. No harm is done to the cell.

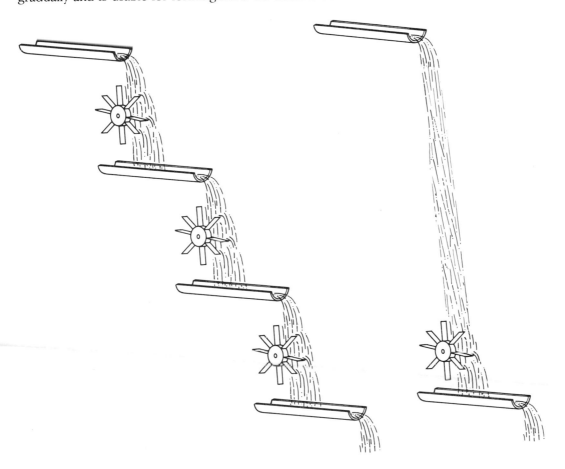

Figure 7.9. Using the energy of running water. If the water falls over a series of water wheels, as shown on the left, each wheel receives only a fraction of the total energy. Consequently, the wheels turn slowly and gently. If the water falls over the entire height at once, as shown on the right, the single wheel receives the full impact and turns rapidly and forcefully. The left example is analogous to the electron-transport chain, in which the energy of excited electrons is released in a controlled, step-by-step fashion and ATP is generated. The right example is analogous to the burning of wood, during which electrons fall in one single step from the high energy state of glucose to the low energy states of H_2O and CO_2.

The need for releasing the energy in a controlled, step-by-step manner becomes clear when we look at an example without such control, namely the burning of wood. Wood consists largely of glucose molecules bonded together into cellulose. When wood burns, the glucose molecules combine with O_2 from the air to produce CO_2 and H_2O. However, unlike aerobic respiration, no electron-transport chain is involved and the energy is released as rapidly as O_2 can get to the glucose. We feel that energy as the heat coming off a wood fire. If the same uncontrolled process occurred in our cells, they would be immediately destroyed.

Now let us reconsider the reactions of aerobic respiration, disregarding all details:

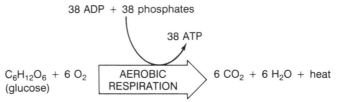

It turns out that roughly 40% of the total energy liberated is stored in ATP molecules and the rest is given up as heat. The heat energy is of no further use to the cell, except for helping to maintain body temperature, and is released to the environment. The energy stored in molecules of ATP, on the other hand, is useful energy. The ATPs deliver it to wherever it is needed by the cell, such as to the ribosomes for assembly of peptides, to the nucleus for replication of DNA, to the cell membrane for active transport of molecules in and out of the cell, and to hundreds of other places of cell activity. Incidentally, the 40% energy efficiency of aerobic respiration is higher than that of combustion engines, which typically have a rating of about 30%.

Anaerobic Respiration

Recall that when the oxygen supply is short or absent, our cells switch to a much simpler kind of metabolism that is anaerobic and similar to bacterial fermentation. It consists of glycolysis, followed by a second reaction sequence that converts pyruvate into lactic acid. The lactic acid is a waste product that our cells expel through the blood stream, to be turned back (with the expenditure of energy) into glucose in the liver. When our muscles work anaerobically beyond our level of physical conditioning, however, not all of the lactic acid is removed immediately. It accumulates and causes the sore and stiff muscles that we all have felt at one time or another.

The reactions of the anaerobic metabolism of eukaryotic cells may be summarized as follows:

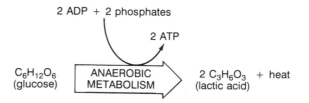

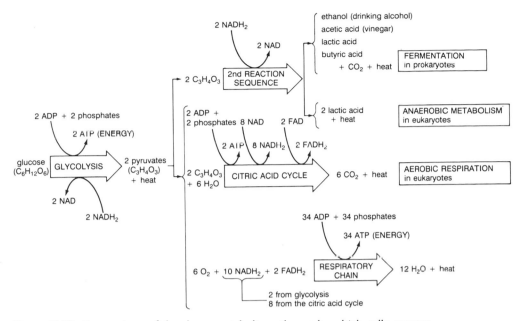

Figure 7.10. Comparison of the three metabolic pathways by which cells convert energy stored in glucose into energy of ATP. All three pathways—the fermentative pathways of prokaryotes and the anaerobic and aerobic pathways of eukaryotes—start with glycolysis, which breaks down glucose into pyruvate and produces two ATP per glucose molecule. In fermentation, the pyruvate is broken down by a second reaction sequence into alcohol, vinegar, and other molecules of similar complexity, without any additional production of ATP. The same is true of the anaerobic metabolism of eukaryotic cells, except that the end product is always lactic acid. In aerobic respiration, which requires the participation of O_2, pyruvate is broken down into CO_2 and H_2O and a total of 38 ATPs (2 by glycolysis, 2 by the citric acid cycle, and 34 by the respiratory chain) are produced per glucose molecule. Thus, aerobic respiration delivers far more useful energy to the cells than the other two metabolic pathways.

As in bacterial fermentation, this is a very inefficient way of harnessing energy. Only two molecules of ATP are produced per glucose molecule, which corresponds to an energy efficiency of a mere 2%, far less than the 40% efficiency of aerobic respiration.

All three metabolisms—aerobic respiration, the anaerobic metabolism of our cells, and fermentation—start with glycolysis (figure 7.10). This correspondence is too close to be coincidental and clearly suggests an evolutionary relationship. Fermentation was the first ATP-generating metabolism of life. Gradually, other pathways were added as the early prokaryotes kept adapting to the ever-changing environment and competed in the search for new sources of raw materials and energy. Eventually, as the O_2 level in the atmosphere of our planet began to rise, the citric acid cycle and the respiratory chain evolved and aerobic respiration came into existence. It allowed life to exploit the very substance that had been poison to it earlier, and to extract enormously more energy from glucose than by fermentation alone.

THE EVOLUTION OF PHOTOSYNTHESIS

As in the case of fermentation and aerobic respiration, the major stages in the evolution of photosynthesis are still apparent in the ways in which contemporary organisms harness the energy of sunlight. The purple and green sulfur bacteria and the purple nonsulfur bacteria harness it *anaerobically*; the blue-green bacteria, algae, and green plants harness it *aerobically*. Both kinds of photosynthesis take place in two phases, a light (photo) phase and a dark (synthesis) phase. During the light phase the energy of sunlight is captured and stored in molecules of ATP and, simultaneously, hydrogen atoms are picked up by special carrier molecules. During the dark phase the hydrogen atoms and energy stored in ATPs are used to synthesize organic molecules from CO_2. The dark phases of the two kinds of photosynthesis are very similar, but only the initial reactions of their light phases are. The final reactions of the light phases are quite different and give evidence of the evolution from anaerobic to aerobic photosynthesis.

Yet another indication of a relationship between anaerobic and aerobic photosynthesis is the fact that the reactions of both kinds of photosynthesis occur on special membranes, the *thylakoid membranes*. In the case of the prokaryotic photosynthesizers (the purple and green sulfur, purple nonsulfur, and blue-green bacteria), the thylakoids are either invaginated extensions of the membranes enclosing the cells or, in some species, they appear to be separate membranes that reside directly in the cells' cytoplasm (see figures 7.11 and

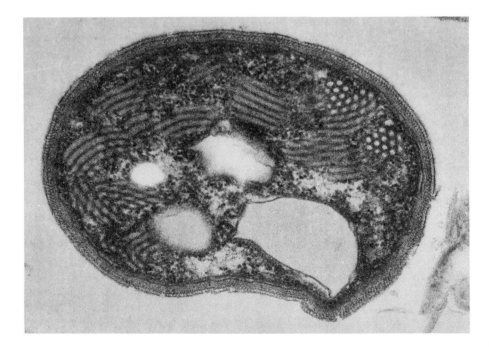

Figure 7.11. Electron micrograph of a contemporary purple sulfur bacterium (× 50,000). The thylakoid membranes can be seen as clusters of short parallel lines in the cytoplasm.

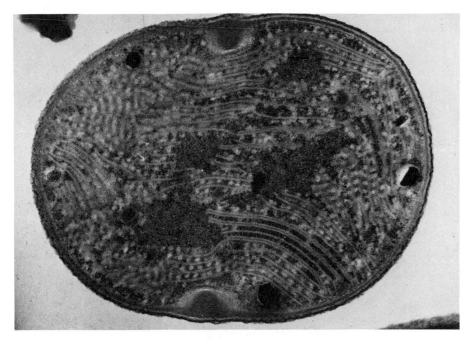

Figure 7.12. Electron micrograph of a contemporary blue-green bacterium, *Gleocapsa alpicola*. The thylakoid membranes are visible, in cross section, as long parallel lines in the cytoplasm.

7.12). In the case of the eukaryotic photosynthesizers (the algae and green plants), the thylakoids reside within special organelles, namely the *chloroplasts* (see figures II.5, II.8).

Anaerobic Photosynthesis

The light phase of anaerobic photosynthesis begins with the gathering of the energy of sunlight by molecules of chlorophyll (or by some other molecules with similar properties, such as the carotenoids) and the boosting of electrons to a high energy state (figure 7.13). Special carrier molecules (among them cytochromes) pick up the energetic electrons and via electron-transport chains return them to the low energy state by either a cyclic or a noncyclic pathway. In the process, some of their energy is used to generate ATPs, the rest is lost as heat.

If the electrons follow the cyclic pathway, they return to the chlorophyll and recycle through the system, generating more ATPs each time they complete a cycle. This pathway is called *cyclic photophosphorylation** because it is cyclic,

---

*Photophosphorylation may be compared with the operation of a flashlight (see figure I.8). The sunlight, which strikes chlorophyll and raises electrons to a high energy state, corresponds to the battery; the electron-transport chain corresponds to the electric circuit of the flashlight; and the generation of ATPs corresponds to the heating of the filament of the light bulb. Unlike cyclic photophosphorylation, the energy of the electrons in the flashlight is released in one single step, heating the filament of the bulb to incandescence.

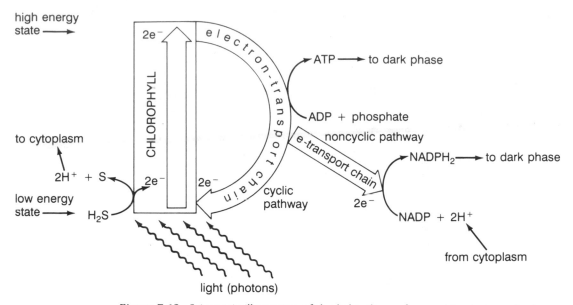

Figure 7.13. Schematic illustration of the light phase of anaerobic photosynthesis as carried out by green and purple sulfur bacteria.

uses the energy of photons, and generates ATPs by adding phosphate molecules to ADPs.

If the electrons follow the noncyclic pathway, they combine with positive hydrogen ions to form neutral hydrogen atoms, which are picked up by special carrier molecules called NADP (*nicotinamide adenine dinucleotide phosphate*). Each NADP picks up two Hs and is thereby converted to $NADPH_2$. The molecular structure of NADP is identical to that of the NAD of aerobic respiration except for an extra phosphate, indicating that the two kinds of molecules have a common ancestry.

The electrons and hydrogen ions used in anaerobic photosynthesis may come from various sources. The green and purple sulfur bacteria derive them from the hydrogen atoms of H_2S. The purple nonsulfur bacteria obtain them from the hydrogen atoms of H_2, ethanol, lactic acid, pyruvic acid, or some other hydrogen-rich molecule. With the energy of sunlight, the hydrogen-rich molecules are split and the hydrogen atoms are ionized, which means they are separated into electrons and hydrogen ions. The electrons are then shunted into the electron-transport pathways discussed above, while the hydrogen ions are released into the cytoplasm of the cell. Eventually, the electrons and hydrogen ions are recombined into neutral hydrogen atoms that are picked up by the NADPs, as noted above.

With the formation of molecules of ATP and $NADPH_2$, the light phase of anaerobic photosynthesis is completed. The ATPs and $NADPH_2$s are passed on to the dark phase, where their energy and the Hs are used together with CO_2s to synthesize glucose and other carbohydrates.

Aerobic Photosynthesis

The light phase of aerobic photosynthesis consists of two coupled photosystems labelled I and II (figure 7.14). *Photosystem II* splits water molecules into hydrogen and oxygen atoms. The oxygen atoms are combined into molecular oxygen (O_2) and expelled as waste to the environment. The hydrogen atoms are separated further into positive ions and electrons. The hydrogen ions are released into the cytoplasm of the cell and are eventually picked up by photosystem I. The electrons are boosted to a high energy state and passed on through an electron-transport chain to the unexcited state of the chlorophyll of photosystem I. Some of the energy liberated goes into forming ATPs, the rest is given off as heat.

Photosystem I possesses a cyclic and noncyclic pathway and resembles the single photosystem of anaerobic photosynthesis. Electrons are raised to a high energy state. Those that are shunted into the cyclic pathway return via an electron-transport chain to the unexcited state of the chlorophyll, generating ATPs in the process. Those that follow the noncyclic pathway are recombined with the hydrogen ions derived from the splitting of molecules of H_2O by photosystem II to form neutral hydrogen atoms. The hydrogen atoms are picked up by NADPs, and both the $NADPH_2$s and the ATPs are passed on to the dark phase for the synthesis of carbohydrates.

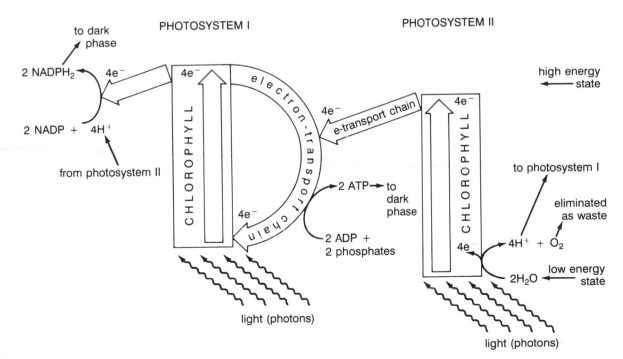

Figure 7.14. Schematic illustration of the light phase of aerobic photosynthesis. (This diagram is to be read from right to left, starting with photosystem II.)

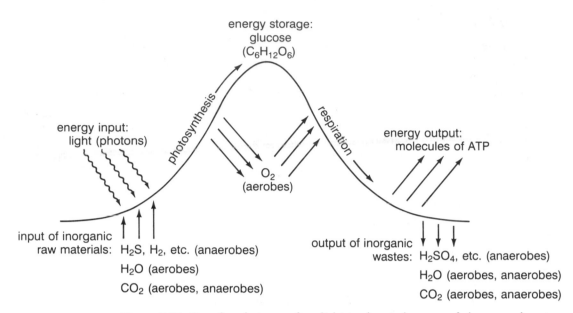

energy storage:
glucose
($C_6H_{12}O_6$)

energy input:
light (photons)

photosynthesis

respiration

energy output:
molecules of ATP

O_2
(aerobes)

input of inorganic
raw materials: H_2S, H_2, etc. (anaerobes)

H_2O (aerobes)

CO_2 (aerobes, anaerobes)

output of inorganic
wastes: H_2SO_4, etc. (anaerobes)

H_2O (aerobes, anaerobes)

CO_2 (aerobes, anaerobes)

Figure 7.15. Transfer of energy of sunlight to chemical energy of glucose and on to molecules of ATP. Starting with energy of sunlight, photosynthetic bacteria, algae, and plants assemble inorganic molecules into glucose by photosynthesis. Some of the energy of sunlight is thereby stored in the chemical bonds of glucose. Glucose that enters the food chain (for example, as food for animals) is broken down by respiration. The energy thus released is transferred to molecules of ATP, with the simultaneous release of inorganic wastes. The energy stored in ATPs is used for various cell activities such as DNA replication, protein synthesis, active transport of molecules in and out of cells, muscle contractions, and transmission of neuronal signals.

As in fermentation and aerobic respiration, the similarities between anaerobic and aerobic photosynthesis are too close to be accidental and thus manifest an evolutionary relationship. Anaerobic photosynthesis, with its single photosystem, developed first. In the course of time, a second photosystem emerged with a slightly modified chlorophyll that absorbed light of shorter wavelengths and could split water molecules. A virtually inexhaustible source of hydrogen atoms was thereby tapped and aerobic photosynthesis, with its two coupled photosystems, came into existence.

Cells capable of aerobic photosynthesis could produce far more food than their anaerobic predecessors because they utilized the energy of sunlight more efficiently and they could operate wherever water was available (figure 7.15). Thus, aerobically photosynthesizing bacteria (the blue-green bacteria) and their eukaryotic descendents (the algae and green plants) became the primary producers in the food chain of life. Gradually they spread to nearly all parts of the globe and brought about the greening of our planet.

THE MAKING OF THE EUKARYOTIC CELL

The biochemical reactions of aerobic respiration and photosynthesis are quite complex. They require many kinds of enzymes, hydrogen carrier molecules, and the electron-transport chain. Furthermore, the reactions of the citric acid cycle, the respiratory chain, and photosystems I and II do not take place in the cytoplasm of the cells, but on specially constructed membranes. In prokaryotes, these membranes are either inwardly-folded extensions of the membrane enclosing the cell or separate components in the cytoplasm. In eukaryotes, the membranes are part of the internal structure of special organelles, the mitochondria and chloroplasts.

Mitochondria and Chloroplasts

The *mitochondria* and *chloroplasts* are rather independent kinds of organelles and, in some ways, they appear to have a closer kinship with free-living bacteria than with the other organelles of eukaryotic cells. This was recognized in the 1840s in the case of mitochondria, when they were first studied through the light microscope. They looked so much like some species of bacteria, it was suggested that at one time mitochondria might have been free-living bacteria that became incorporated into larger cells. This symbiotic theory for the origin of mitochondria was revived during the early part of the present century by the Russian biologist Konstantin S. Merezhkovskii (1855–1921). More recently, Lynn Margulis (1982) of Boston University advocated the theory again, for both mitochondria and chloroplasts. She noted that both organelles possess their own DNA, mRNA, tRNA, and ribosomes, and that they manufacture some of their own proteins, much as bacteria do. Their reproduction goes on quite independently of the reproduction of the eukaryotic cells within which they reside, and it happens by a division process that is similar to the binary cell division of bacteria. (However, mitochondria and chloroplasts are absolutely dependent upon nuclear-coded proteins and cannot survive by themselves.) Finally, mitochondrial ribosomes are similar in size and sensitive to the same antibiotics as those of respiring bacteria.

The individualistic characteristics of mitochondria and chloroplasts prompted Margulis to propose the following sequence of events for their origin. At one time in the distant past, after the oxygen level of the Earth's atmosphere had begun to rise, prokaryotes capable of aerobic respiration began living in the cytoplasm of larger fermenting bacteria and established mutually beneficial partnerships (symbioses) with them. The cytoplasm of the fermenting bacteria provided a rich source of organic nutrients—the waste products of fermentation—while the respiring bacteria produced large amounts of ATP by aerobically breaking down the nutrients and extracting energy. At first such partnerships were probably quite casual, but with time they evolved into more stable unions. The guest cells were freed of the tasks of seeking food and fending for survival, for they lived now in a safe, stable, and nutrient-rich environment. They could specialize entirely on one task—namely that of producing ATPs—and they became very good at it. The extra energy supplied by the ATPs gave the hosts the opportunity to evolve new adaptations and to meet the demands of survival better than could bacteria that lived without the benefit of aerobically respiring guests. The partnerships proved to be so advantageous that eventually the guest cells resided permanently within the cytoplasm of the fermenting bacteria. The guest cells had become mitochondria, and the collection of host cells and mitochondria had become eukaryotes. (Why they are called eukaryotes is discussed below.)

Once eukaryotic cells possessing mitochondria had evolved, some of them took up partnerships with aerobically photosynthesizing bacteria. The eukaryotic cells offered shelter as well as the raw materials CO_2 and H_2O. The photosynthetic bacteria synthesized ATPs and organic molecules. Again, at first such unions were probably just occasional events. But in the course of many generations, host and guest cells evolved a mutual dependence that eventually became total. The photosynthetic bacteria had become chloroplasts and taken up permanent residence within their hosts. The modern, photosynthetic eukaryotic cell had evolved. (All eukaryotic photosynthetic cells — the algae and the cells of green plants — possess mitochondria in addition to chloroplasts, and they are able to carry out aerobic respiration.)

The symbiotic theory for the origin of mitochondria and chloroplasts is supported further by the fact that partnerships between different kinds of organisms are found commonly throughout the contemporary biological world. Quite often such partnerships lead to complete dependence between the participating organisms (see box 7.1). There is no reason to think that such partnerships did not also exist in the past and that they significantly affected the evolution of life.

However, a word of caution is in order. Although the symbiotic theory for the origin of mitochondria and chloroplasts is supported by much evidence and most biologists accept it as plausible, some do not. For example, Henry R. Mahler and Rudolf A. Raff of Indiana University assert that the genetic systems of mitochondria and chloroplasts are not as similar to those of bacteria as had been assumed earlier and they think that the symbiotic theory is unnecessary to explain the origin of these organelles. Mitochondria and chloroplasts could equally well have arisen through internal changes in eukaryotic protocells. Mahler and Raff (1975) caution that a "dogmatic adherence to the symbiotic theory is premature and may result in an unfortunate narrowing of experimental approach and interpretation in the study of organelles and their origins."

The Eukaryotic Nucleus

Cells do not become eukaryotes just by acquiring mitochondria and chloroplasts. They need to have their DNA enclosed by a nuclear membrane to be eukaryotes. It appears that such enclosures evolved along with the evolution of mitochondria and chloroplasts. In fact, this coevolution may have been unavoidable. The presence of mitochondria and chloroplasts within cells meant that O_2 was present in fairly large amounts. This was true regardless of whether they arose by symbiosis or through internal cellular changes. The O_2 created a potentially dangerous situation for the cells' DNA, which, as in all bacteria, initially resided freely in the cytoplasm and could have been easily damaged by reactions with O_2. The DNA needed to be protected. This problem was solved by the evolution of protective membranes around the DNA and the creation of nuclei. The cells became eukaryotes.

BOX 7.1.
Symbiosis

A PARTNERSHIP between two dissimilar organisms in any of various mutually beneficial relationships is called *symbiosis*, a word derived from the Greek and meaning "state of living together." Today symbiosis is commonly found throughout the biological world. For example, bees and flowers coexist in a symbiotic relationship. The bees fertilize the flowers and the flowers provide nectar. The alimentary tracts of cows, goats, and sheep are inhabited by a series of bacteria that help in the digestion of grass and other fodder. The roots of many plants are intimately associated with fungi or bacteria that assist in the absorption of nutrients. Still other bacteria living in the root systems of plants "fix" nitrogen by converting atmospheric nitrogen (N_2) into ammonia (NH_3), the only form in which plants can utilize nitrogen. Another example of symbiosis are lichens, those hearty organisms that live on the surfaces of rocks. Lichens are composed of a fungus, which supplies support and protection, and an alga, which photosynthesizes carbohydrates. Yet another example of symbiosis is the partnerships between prokaryotes that are thought to have led to the development of mitochondria and chloroplasts in eukaryotic cells.

Symbiosis may also be defined in a broader sense, namely one that distinguishes among three kinds of partnerships:

1. *Mutualism*, a partnership that is of mutual benefit to the participating organisms, as illustrated by the above examples.

2. *Commensalism*, a partnership in which one kind of organism obtains food, protection, or other benefits from another often larger organism without damaging or benefiting the second organism. An example is the remora, which attaches itself to whales, porpoises, sharks, and other large fish and is transported by them.

3. *Parasitism*, a partnership that is one-sidedly exploitative. For example, lice or fleas suck blood out of the animals on which they live but provide no benefit to their hosts (figure A).

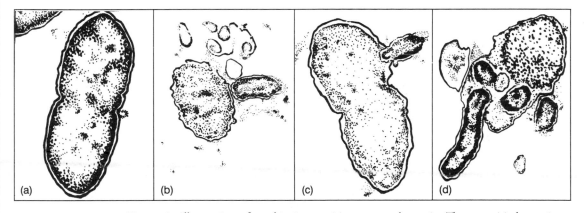

Figure A. Illustration of symbiotic parasitism among bacteria. The parasitic bacterium *Bdellovibrio bacteriovorus* penetrates through the cell wall of the host bacterium *Erwinia amylovora*, for the purpose of reproduction ($\times$ 28,300). (*a*) Uninfected host cell. (*b*) *Bdellovibrio* (smaller, elongated body) collides with and attaches itself to the outer cell wall of the host. (*c*) *Bdellovibrio* penetrates into a space between host's cell wall and the flexible membrane on its inner side. (Not shown: Once inside the host cell, *Bdellovibrio* elongates into a filament several times its original length and then segments into a proportional number of flagellated progeny, using nutrients obtained from the host's cytoplasm.) (*d*) A late stage of infection, showing liberation of *Bdellovibrio* progeny. The entire process of multiplication takes about four hours.

BOX 7.1 continued

These are but a few of the many examples of symbiotic partnerships found in nature. In fact, all life forms—from bacteria to plants and the largest animals—coexist in intricate, interdependent partnerships, and we may regard the entire ecosystem of our planet as being one giant, multifaceted symbiosis. The interdependence and cooperation are the result of a common evolutionary history and stand in contrast to competition, which is an equally important factor in evolution.

It seems that the stability that we usually find in nature is the result of a balance between these opposing forces—cooperation and competition. If one or several species gain too great a competitive advantage over others, the balance gets perturbed and a new ecological structure results. Similarly, if a cooperative partnership becomes particularly successful, the balance shifts also. This is apparently what happened when different bacteria joined into symbiotic partnerships that led to the origin of eukaryotic cells. The new cells were so successful that the balance of nature changed completely and life evolved in numerous, radically novel directions, as will be discussed in the next chapter.

We should not be surprised that cooperative and competitive forces have played such fundamental roles in biological evolution. According to the theory presented in chapter 6, these forces were already at work when life first arose in the primordial soup of our planet. Molecules of RNA and primitive proteins entered into cooperative partnerships and, simultaneously, different partnerships competed with each other for the available raw materials and dominance in the soup. The competitive struggle for survival continually selected those symbioses that were most adapted to the prevailing conditions. The cooperative interactions increased in complexity and, eventually, led to the origin of protocells. The same interplay between cooperation and competition can be observed today on an entirely different level—the level of human society. It has been and still is shaping the course of our history.

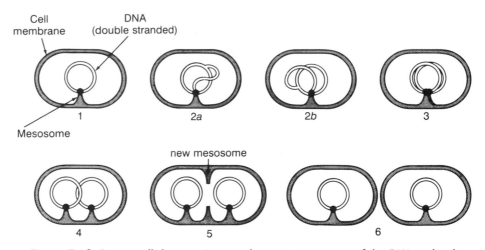

Figure 7.16. Binary cell division. In some bacteria, separation of the DNA molecules during binary cell division occurs with the aid of *mesosomes*, membranes that extend inward from the membrane enclosing the cell. 1. Diagram of the cell showing the mesosome, to which a single, circular DNA molecule is attached. (Helical structure of DNA is not shown.) 2*a, b*. DNA is being replicated. 3. Replication has produced two DNA molecules, each attached to a mesosome. 4. The mesosomes pull apart the two DNA molecules. 5. Mesosomes have pulled apart the two DNA molecules. Another mesosome grows from the cell's outer membrane inward, dividing the cell. 6. Cell division is completed. Each of the two offspring cells has inherited one of the DNA molecules.

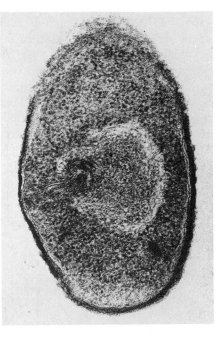

Figure 7.17. Electron micrograph of the bacterium *Bacillus subtilis*, showing the DNA in contact with a mesosome.

At present we don't know precisely how the evolution toward nuclear membranes proceeded. However, the way some contemporary bacteria reproduce by binary cell division gives clues of how it might have happened. As illustrated in figure 7.16, binary cell division begins with the replication of the circular DNA of the bacterium, temporarily giving the cell two identical pieces of DNA. Simultaneously, in some bacteria portions of the cell's enclosing membranes grow inward and attach themselves to the two pieces of DNA (see figure 7.17). The inward-growing membranes are called *mesosomes*, from *meso-* meaning "middle" and *-some* meaning "body." Gradually the cell membrane between the two mesosomes grows larger and the mesosomes separate in opposite directions, pulling the two pieces of DNA apart.* At the same time, an additional mesosome grows inward somewhere between the two pieces of DNA and divides the cell into two offspring cells, with each receiving one piece of the DNA. The cell division is thereby completed. It is easy to imagine how in some ancestral bacteria the need to protect the DNA from O_2 might have led to the evolution of larger mesosomes that eventually enveloped the DNA and thus became the nuclear membranes of eukaryotic cells.

EUKARYOTIC CELL DIVISIONS: MITOSIS AND MEIOSIS

However the nucleus of eukaryotes evolved and for whatever reason, one thing is certain: it changed the course of biological evolution. It changed it for two

---

*Not all bacteria grow mesosomes during binary cell division. In many bacteria the two DNA molecules become attached to the cell's enclosing membrane, which pulls them apart as the cell grows larger prior to division.

reasons. First, by enclosing the DNA in a membrane-bound structure and separating it from the chemical processes in the rest of the cell, the nucleus provided an environment in which the DNA could evolve to much greater lengths than was possible before. This meant that eukaryotic cells could carry considerably more genetic information and had the potential for evolving much more complex adaptations than did prokaryotes. Second, the nucleus permitted a new kind of organization and mechanical handling of the DNA, which led to the development of two new kinds of cell divisions—*mitosis* and *meiosis*. Mitosis is involved in the nonsexual reproduction of eukaryotic cells, while meiosis is the basis of their sexual reproduction.

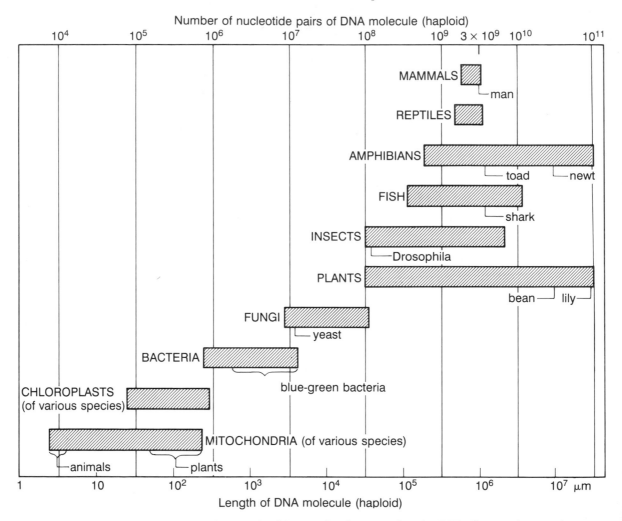

Figure 7.18. Length of DNA molecules. Note that the DNA of some plants and amphibians is longer than that of bacteria by a factor of 10^4 to 10^5. The total length of human DNA is approximately 10^6 μm or 1 m, which is equivalent to roughly 3 × 10^9 nucleotide pairs or 10^9 codons. Obviously, the length of DNA carried by an organism is not necessarily related to its phenotypic complexity. We may conclude that DNA does not carry useful information over its entire length. Much of its coding is "nonsense." (Adapted from Alberts, et al. 1983, 405, 530.)

Eukaryotic DNA

Eukaryotic DNA is typically ten to many hundreds of times longer than bacterial DNA. It varies from about 1 cm in very simple eukaryotes, such as yeasts and sponges, to about a meter or more in the case of humans, many other animals, and many plants (see figure 7.18).

Eukaryotic DNA is closely associated with a number of different proteins. Some of the proteins act as enzymes and help in the repair, replication, and transcription of the DNA. Others act as structural proteins and help keep the long strands of eukaryotic DNA organized into manageable units. The latter kinds of proteins are called *histones*. Histones are rich in positively charged amino acids and are arranged into roundish bodies known as *nucleosomes*. The positive charges of the histones cancel the negative charges carried by DNA on the phosphates of its backbone and greatly increase the degree to which DNA can be compacted in the cell nucleus. Each strand of DNA is associated with a sequence of tens to hundreds of thousands of nucleosomes, around which it is wrapped much like a long piece of thread might be wrapped around a sequence of spools.

Another feature unique to eukaryotic DNA is that it never exists in one piece, as does bacterial DNA, but always in two or more pieces. For example, there are 8 pieces of DNA in the cells of the fruit fly *Drosophila*, 48 in those of chimpanzees, gorillas, and orangutans, and 46 in those of humans. Apparently, only by dividing their DNA into several separate pieces are eukaryotic cells able to handle the DNA's great length.

During the initial stages of mitosis and meiosis (just after replication of DNA is completed, see below) the various pieces of DNA, together with their histones and other proteins, undergo several successive levels of folding and coiling (see figure 7.19). Finally they become the short, stubby bodies that are familiar from photographs made through the light microscope. In 1888, the German physiologist Wilhelm von Waldeyer-Hartz (1836–1921) named these bodies *chromosomes*, meaning "colored bodies," for their affinity for certain dyes.

Mitosis

Let us now turn to *mitosis*, the form of cell division by which eukaryotic cells reproduce without sex. Examples of mitosis are the repeated division of a fertilized egg and its development into an adult (see figure 7.20), the replacement of old or damaged cells in our bodies, the growth of plants, and the multiplication of yeast cells. Strictly speaking, mitosis refers only to the processes involved in the division of the cell nucleus: the replication of DNA and the equal distribution of the two sets of DNA among the offspring cells. The processes that affect the cytoplasm of the cell are called *cytokinesis*, which literally means "cell motion." Cytokinesis includes cleavage of the cell and the distribution of cytoplasmic constituents and organelles among the two offspring cells.

Mitosis and cytokinesis are dynamic and complicated processes, whose beauty is difficult to portray in words or static pictures. Both processes occur in innumerable variations and are never exactly alike from one cell division to the

The two strands of the DNA molecule wind around each other in a spiral or helical fashion.

During *interphase*, DNA is *replicated*. Two "sister" DNA molecules are created.

Each DNA double helix wraps around histone complexes like a thread around a sequence of spools.

During *prophase*, each DNA and its histones fold repeatedly, creating short, stubby *chromosome bodies*.

Each chromosome consists of two identical halves, the *chromatids*, held together at the *centromere*. (The two "sister" chromatids correspond to the two "sister" DNA molecules created by replication.)

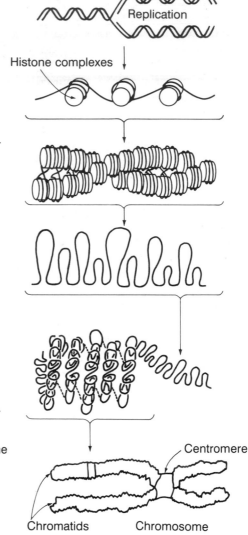
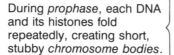

Figure 7.19. Schematic illustration of the different levels of packing of DNA into highly condensed chromosome bodies during eukaryotic cell division. (Adapted from Alberts, et al. 1983, 399.)

next. In particular, they are somewhat different in plant and animal cells, mainly because plant cells are surrounded by rigid cellulose walls and animal cells are not. Still, the processes of mitosis and cytokinesis of all eukaryotic cells share certain essential features that permit their description in a schematic and generally valid manner.

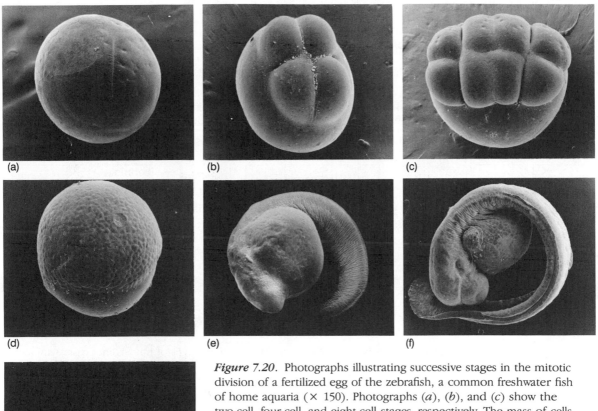

(a) (b) (c)

(d) (e) (f)

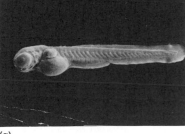

Figure 7.20. Photographs illustrating successive stages in the mitotic division of a fertilized egg of the zebrafish, a common freshwater fish of home aquaria (× 150). Photographs (*a*), (*b*), and (*c*) show the two-cell, four-cell, and eight-cell stages, respectively. The mass of cells shown in photograph (*d*), called a *blastula*, represents a later stage in development (also see figure 8.15). In photographs (*e*) and (*f*), the fish-like form becomes apparent, and the developing animal is considered to be an embryo. About four days after fertilization, the egg hatches and releases a newborn that is still incompletely developed, as shown in photograph (*g*).

(g)

Traditionally, the processes of mitosis and cytokinesis have been divided into four steps or phases, although in reality they are one continuous sequence of events. The four phases are *pro-, meta-, ana-*, and *telophase*. In addition there is a fifth phase —*interphase* — which refers to the time interval between successive cell divisions. The phases are described in figure 7.21 and below.

1. *Interphase* is the period between successive cell divisions. Except in very rapidly dividing cells, it is the longest of the five phases and is sometimes called the resting stage of the cell, even though during this phase the cell does everything but rest. It actively carries out its various metabolic processes needed for growth and self-maintenance as well as its specialized tasks, such as contraction if it is a muscle cell, production of hormones if it is an

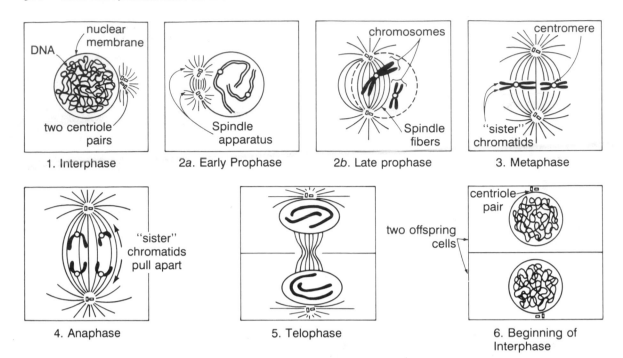

1. Interphase   2a. Early Prophase   2b. Late prophase   3. Metaphase

4. Anaphase   5. Telophase   6. Beginning of Interphase

Figure 7.21. Schematic illustration of the major stages of mitosis in a typical animal cell. 1. *Interphase*: The nucleus is compact and dark, with dispersed DNA. 2a. *Early prophase*: Patches of chromosomes emerge in the nucleus, and spindle apparatus begins to form around two pairs of centrioles. 2b. *Late prophase*: The nuclear membrane disintegrates and separate chromosomes become visible. The two centriole pairs, along with their radiating spindle fibers, separate. 3. *Metaphase*: The chromosomes are maximally condensed. Each consists of two chromatids, held together at the centromere. The chromosomes are aligned along a plane midway between the two poles. 4. *Anaphase*: The chromatids of each chromosome separate and are pulled by the spindle fibers toward opposite poles. 5. *Telophase*: The chromatids (now chromosomes themselves) are at the poles. They begin to uncoil and a new nuclear membrane forms around them. The spindle fibers disappear, though the centriole pairs remain. 6. The cell divides into two offspring cells by cytokinesis and a new interphase begins for each of the new cells.

endocrine cell, and ingestion and destruction of bacteria if it is a white blood cell. Interphase is also the period during which DNA is replicated.

2. During *prophase* the strands of DNA, which up to now have remained dispersed in the nucleus, coil into the short, stubby chromosome bodies described above. Because each DNA has been replicated during interphase, each of the chromosome bodies consists of two identical halves, called *chromatids*. The two chromatids of each chromosome are held together at a constricted region known as the *centromere*.

Also during prophase the nuclear membrane begins to break up into fragments and the *spindle apparatus* forms. The spindle apparatus consists of hundreds of thin fibers, some of which radiate outward from two clearly distinguishable *polar regions*, while others radiate away from the centromeres of the chromosomes. Each polar region is centered on a pair of tiny structures, the *centriole pairs*. Changes in the lengths of the spindle fibers are responsible for the dynamic processes of the next two phases.

3. With the onset of *metaphase*, the disintegration of the nuclear membrane is completed and the membrane fragments disperse into the cytoplasm. There is a flurry of activity as the chromosomes, guided by the spindle fibers, settle down in the middle of the cell with their centromeres aligned along a plane perpendicular to the polar axis.

4. During *anaphase*, the cell lengthens in the direction of the polar axis and the chromatids, which have been held together at their centromeres, break apart. One set of chromatids—which constitutes *one complete set of chromosomes*—moves toward the pole in one part of the cell and the other set moves toward the pole in the other part of the cell.

5. During *telophase* the cell becomes constricted in the middle and divides into two offspring cells. This happens such that each offspring cell receives one of the two sets of chromosomes that were separated during anaphase. Furthermore, the spindle apparatus now disappears (though a centriole pair remains in each offspring cell), the nuclear membranes reassemble around the chromosomes in each of the offspring cells, and the chromosomes begin to unwind and disperse throughout the nucleus. Cell division is completed and the two offspring cells enter interphase.

Meiosis

The sequence of nuclear changes that constitute the basis of sexual reproduction in eukaryotic organisms is called *meiosis*.* It is more complicated than mitosis because it is a *nuclear reduction* division that divides the set of chromosomes precisely in half. Meiosis leads to the formation of *reproductive cells* (also called *gametes*)—*sperm* in the case of males and *eggs* (or *ova*) in the case of females. For example, each human sperm and egg carries 23 chromosomes, not 46 as do fertilized eggs and our somatic (that is, body) cells.

The halving of the number of chromosomes during meiosis is necessary because when a sperm fuses with an egg during sexual mating, the chromosomes of the sperm and egg are combined. Without the halving, the fertilized egg would have twice as many chromosomes as each of the parental cells. Within just a few generations, this would lead to an intolerably large number of chromosomes.

In order to understand meiosis, it is important that you first become familiar with a few terms and concepts. Human fertilized eggs or somatic cells contain 46 chromosomes, which are called a *diploid set*. Half of the chromosomes of diploid sets—namely 23 in humans—come from one parent and the other half come from the other parent. Each half set of chromosomes is called a *haploid set* (*haplo-* and *diplo-* are derived from the Greek, meaning "single" and

---

*The terms meiosis and mitosis are derived from the Greek. *Meiosis* means "diminution." *Mitosis* has two roots, *mitos* and *-osis*, which mean "thread" and "process," respectively.

"double," respectively). In general, if the diploid set contains $2n$ chromosomes, then the haploid set contains n chromosomes ($n = 23$ in humans, 24 in apes, and 4 in *Drosophila*). During fertilization two haploid cells—an egg and a sperm—fuse to form a diploid fertilized egg (called a *zygote*).

Each of the 23 chromosomes of a human haploid set is uniquely identifiable, as illustrated in figure 7.22. Cytologists number them 1–22, plus X or Y for the twenty-third chromosome. Chromosomes with the same identification number are said to be *homologous*. For example, in human diploid cells the two

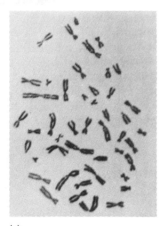

(a)

(b)

Figure 7.22. The chromosomes of a human somatic cell. Photograph (*a*) shows the 46 chromosomes—that is, a *diploid set*. Half of them—23 chromosomes, or a *haploid set*—are inherited from the mother, the other half are inherited from the father. In photograph (*b*) the 46 chromosomes are arranged in homologous pairs, lined up in order of size, and numbered 1–22, plus X and Y for the twenty-third pair. Because the twenty-third pair consists of the combination XY, the chromosomes shown here come from a male.

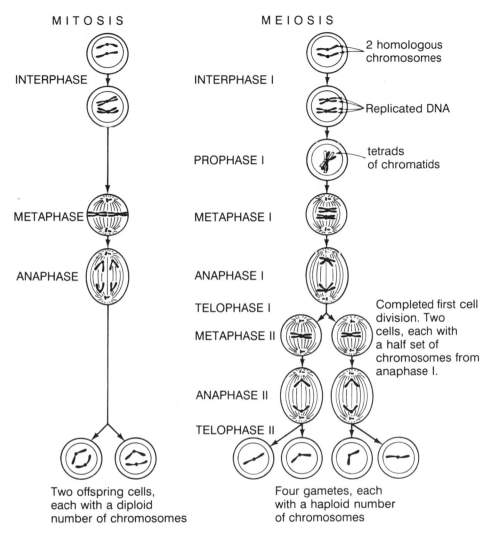

MITOSIS

MEIOSIS

INTERPHASE

INTERPHASE I
2 homologous
chromosomes

Replicated DNA

PROPHASE I
tetrads
of chromatids

METAPHASE

METAPHASE I

ANAPHASE

ANAPHASE I

TELOPHASE I
Completed first cell
division. Two
cells, each with
a half set of
chromosomes from
anaphase I.

METAPHASE II

ANAPHASE II

TELOPHASE II

Two offspring cells,
each with a diploid
number of chromosomes

Four gametes, each
with a haploid number
of chromosomes

Figure 7.23. Comparison of mitosis and meiosis. For clarity, the starting cell in each case contains only one set of homologous chromosomes. Note that *mitosis* consists of one cell division and produces two offspring cells, each with a *diploid* number (2 in this example) of chromosomes. *Meiosis* consists of two cell divisions and potentially produces four offspring cells (gametes), each with a *haploid* number (1 in this example) of chromosomes.

chromosomes numbered 1, which carry genetic information for the same phenotypic traits, are homologous chromosomes, or *homologs* for short. The same is true of the chromosomes numbered 2, 3, and up to 22. The only exceptions are the *X* and *Y* chromosomes. They code for some different traits and are thus only partially homologous. One important characteristic they code for is the gender of an offspring: Females possess the combination *XX*, males the combination *XY*.

Meiotic cell division takes place in special sex organs, for example the *testes* in male animals and the *ovaries* in female animals. The purpose of meiosis is to produce eggs and sperm containing haploid sets of chromosomes. This is accomplished by two cellular divisions carried out in sequence. The same phases and many of the mechanical processes that characterize mitosis can be identified in these divisions (labeled I and II, see figure 7.23).

1. *Interphase I* is characterized by the replication of DNA. The two copies of each chromosome remain joined as sister chromatids and behave as a unit throughout the first division.

2. During *prophase I* the strands of DNA coil into short stubby chromosome bodies (each with two sister chromatids), and homologous chromosomes pair up by a process called *synapsis* to form *tetrads of chromatids*. During this pairing, which persists until anaphase I, pieces of DNA may be randomly exchanged between homologous chromatids. This *crossing over*, illustrated in figures 7.24 and 7.25, has important consequences for evolution. It scrambles maternal and paternal genes and contributes greatly to genetic variation in sexually reproducing eukaryotic populations. This scrambling

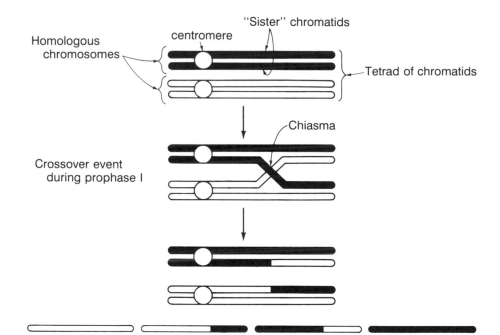

Chromosome distribution in reproductive cells, resulting from above kind of crossover event.

Figure 7.24. Chromosomal crossing over. This process occurs during prophase I of meiosis, when the four chromatids of homologous chromosomes — the tetrads — are tightly packed together. Segments of chromatids may break and become exchanged, or crossed over. The points of crossing over are often visible through the light microscope and are called *chiasmata* (singular *chiasma*, from the Greek capital letter Χ or "chi"). On the average, between two and three such crossover events occur on each pair of human chromosomes during meiosis. (See figure 7.25.)

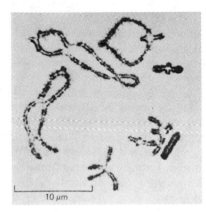

Figure 7.25. Light micrograph of tetrads of homologous chromosomes (from a grasshopper cell) undergoing meiosis. Multiple chiasmata can be observed on several of the tetrads (see also figure 7.24).

10 µm

occurs in addition to the random separation of maternally and paternally derived chromosomes during cell division of anaphase I (see below). Also, during prophase I, the nuclear membrane begins to break up and a spindle apparatus forms.

3, 4. During *metaphase I* and *anaphase I*, the tetrads line up along a plane and the spindle fibers pull the homologous chromosomes (each still consisting of two sister chromatids) apart. One of the resulting haploid sets of chromosomes migrates toward one part of the cell, the other set toward the opposite part. This is quite different from mitosis, which separates the sister chromatids of all the chromosomes and sends them to opposite sides of the cell.

The separation of homologous chromosomes during anaphase I of meiosis is completely random and is the major source of variation among sexually reproducing eukaryotes. It is impossible to predict for any of the tetrads toward which side of the cell the paternally and maternally derived chromosomes will travel, except that they will travel in opposite directions. Consequently, each of the separated haploid sets of chromosomes ends up with a random assortment of maternally and paternally derived chromosomes.

5. During *telophase I* the cell divides by cytokinesis into two offspring cells, with each receiving one of the half sets of chromosomes separated during anaphase I. Nuclear membranes form around the chromosomes and the first division is completed. The two offspring cells now enter the second division.

6, 7, 8, 9. A transient *interphase II*, during which there is no DNA replication, is quickly followed by *prophase II, metaphase II*, and *anaphase II*. Again, these phases closely resemble those of mitosis, except that there are only half as many chromosomes because their number was reduced during the first division. In particular, during anaphase II, the sister chromatids separate and are pulled toward opposite sides of the cells.

10. During *telophase II*, each of the two cells produced during the first division divides a second time to form four sex cells, each containing a haploid set

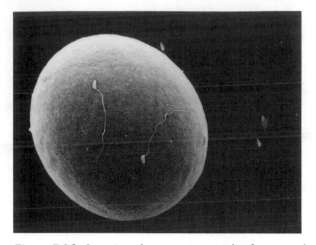

Figure 7.26. Scanning electron micrograph of a sea urchin egg with numerous sperm bound to its surface (× 400).

of chromosomes. Nuclear membranes reassemble and meiosis is completed. Due to the crossing over of pieces of DNA during prophase I and the random separation of homologous chromosomes during anaphase I, the four gametes produced by meiosis may inherit any combination of maternally and paternally derived genes from the original diploid cell.

Please note that meiosis is only the initial stage in the making of mature gametes. For example, cells developing into sperm need to grow *tails* (flagella), with which they swim, and *acrosomal vesicles*, which contain molecules for aid in penetrating an egg during fertilization (see figure 7.26). Eggs need to accumulate yolk for the nourishment of the embryo. Pollen grains (which contain sperm in the flowering plants) need to acquire durable covers and generate a pollen tube, through which the sperm pass to the egg for fertilization.

In human males, all four haploid cells produced by meiosis develop into viable sperm. The process of meiosis goes on at such a high rate that a healthy adult human male can make 100,000 sperm per minute—about 300 million per ejaculation—and he can keep that up throughout much of his life. In contrast, in human females only one mature egg results from each meiotic division. These divisions begin when the female is still an embryo, but they are for the time being arrested at the first division of meiosis. Only when she reaches sexual maturity, approximately at age thirteen, do the meiotic divisions resume and produce mature eggs—one per menstrual cycle (occasionally more). Altogether, about 450 eggs mature in the course of her reproductive life.

THE BENEFITS OF SEXUAL REPRODUCTION

At a four-day symposium entitled *The Origin and Evolution of Sex*, held during the summer of 1984 at the Marine Biological Laboratory in Woods Hole, Massachusetts, the participants agreed to define sexual reproduction as the process whereby a cell containing a new combination of genes is produced from two genetically different parent cells. (Morse 1984)

Clearly, this definition applies to eukaryotes, for in this case the genes of the zygote are derived from two genetically different parent cells, namely from a maternal and paternal gamete.

Prokaryotic Sex

Though prokaryotes reproduce by binary cell division and sometimes by budding or spore formation, in certain instances the above definition of sexual reproduction applies to them as well. The reason is that occasionally bacteria receive segments of DNA from other bacteria. In one mechanism, called *transduction*, a virus carries a bit of DNA from the cell it has grown in to a new cell, but this probably happens quite irregularly. A more important mechanism of gene transfer among bacteria is a process called *conjugation*. During conjugation, the membranes of two neighboring bacteria fuse at some point, a channel forms between the two cells, and a segment of DNA is transferred from one of the bacteria to the other (see figure 7.27).

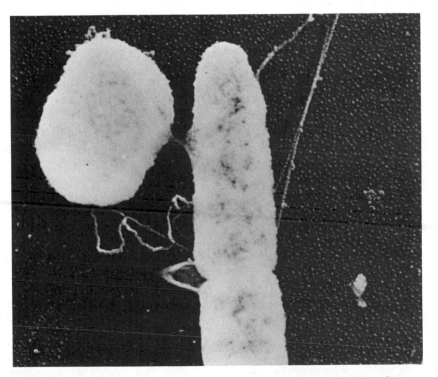

Figure 7.27. Gene transfer among bacteria by *conjugation*. The membranes of two neighboring bacteria (*Escherichia coli*) have fused and a channel has formed between them. A segment of DNA (not visible in this electron micrograph) is being passed through the channel from the oval-shaped donor (the "male") to the recipient (the "female").

Conjugation resembles eukaryotic sexual reproduction. Quite commonly, it is induced by genetic elements known as *plasmids*.* The plasmids bring about conjugation by forming proteins on the bacterial surface that make the cell sticky for other cells. This, in turn, induces the fusion of the cell to a neighboring cell and the formation of a channel between them. The plasmids of the first cell then make copies of themselves, which are transferred to the second cell. Quite often the plasmids promote the transfer of bacterial DNA as well, in addition to their own DNA. The plasmid-carrying cell (that is, the one that induced conjugation) thus becomes a "male" or gene donor, while the second cell becomes a "female" or gene recipient. After conjugation, the two bacteria break apart and go on to grow as before, though sometimes the donor cell dies because it has given away some of its genes.

After the transfer of a segment of DNA from one bacterium to another has taken place, part of the segment is sometimes spliced into the circular DNA of the recipient bacterium and thereby becomes part of its genetic makeup. This splicing is called *recombination* and requires a host of enzymes to orchestrate the breaking, inserting, and fusing of the DNA strands. Many of these enzymes are the same as those used in the repair of UV-damaged DNA, suggesting that prokaryotic sex was able to arise because part of the DNA repair machinery had already evolved earlier. Note that neither conjugation nor transduction entail cell reproduction, as does eukaryotic sex. Rather, these processes merely refer to transfer of genes from one bacterium to another.

Unlike eukaryotic sexual reproduction, recombination in bacteria is a rather imprecise process. Rarely does exactly half of the recombinant DNA come from one parent cell and the other half from the other parent cell. For example, in some cases the transferred segment of DNA carries just one or a few genes, and in others it carries nearly the entire set of the donor's genes. And in many instances, it involves only plasmids and not bacterial DNA.

Despite its imprecision, sexual reproduction has a profound effect on the evolution of prokaryotic populations. By shuffling and spreading genes, it creates an enormous degree of variation, upon which natural selection can act. Beneficial genes, which impart advantages on their carriers in the struggle for survival and hence have an above-average chance of being passed on to future generations, tend to spread and accumulate in a population; detrimental genes, which impart disadvantages, tend to diminish. The result is that bacterial populations are able to adapt remarkably rapidly to changing conditions in the environment.† For example, during just the past few decades, conjugation and

---

*Plasmids are small ring-shaped pieces of DNA, equivalent in length to only a few percent of the DNA molecules in a bacterium, but large enough to carry a number of genes. They are like little extra chromosomes that exist inside a bacterium and are replicated as the cell grows, just as the bacterium's own chromosome is. They are thus rather like parasitic pieces of DNA, because the bacterium perpetuates them even though it does not need them. However, the relationship quite often benefits the bacterium, for some of the plasmid genes may confer special properties on it, such as resistance to certain viruses or antibiotics. (The relationship is then an example of *mutualism*, see box 7.1.)

†Note that a cubic centimeter of sea or pond water today typically contains about 10^6 to 10^8 bacteria. This means that even when recombination occurs in only a small percentage of bacteria, the total number of recombinations may still be very large. Furthermore, because reproduction occurs

plasmid transfer have spread resistance to antibiotics widely among pathogenic bacterial species. Other bacterial species have acquired the ability to make bacteriocidal chemicals. And still others have evolved the ability to break down unusual organic compounds, such as those present in oil spills. Such rapid adaptations are not possible in the absence of sex; for in these cases adaptation depends on the occurrence of favorable mutations, which are relatively infrequent in a single organism.

Eukaryotic Sex

Given the adaptive advantages of sexual reproduction among prokaryotes, it is not surprising that it also evolved among eukaryotes. In fact, eukaryotes may have evolved from prokaryotes that already had sex, such as gene transfer by conjugation, though no one knows for certain. However, we do know that eukaryotic sex became much more complex and refined than prokaryotic sex. This probably happened in part because eukaryotic DNA is distributed over two or more separate chromosomes. Any process less precise in distributing chromosomes than meiosis followed by fertilization would mean that some offspring cells might end up with extra genes or chromosomes, while others might end up with too few. Such processes would rapidly degrade genetic information and lower the viability of eukaryotic organisms.

Despite the apparently plausible argument that sex evolved because of its adaptive advantages—namely the enormous degree of genetic variation it creates—some biologists have recently begun to question this argument. For example, Norton Zinder, a molecular geneticist at the Rockefeller University in New York, explains the problem this way: "How could an organism that only passed half of its genes to its offspring [through sexual reproduction] ever have competed with [an asexual] progenitor that passed all of them? It seems unlikely that the offspring produced sexually were 'fitter' than their asexually produced relatives" (Morse 1984, 155).

In the face of questions such as this one by respected scientists about the value of sex, the participants at the Woods Hole symposium on *The Origin and Evolution of Sex* agreed that at present no one really understands why sex persists. With this sentiment, they reflected the thoughts Charles Darwin expressed on this topic in 1862: "We do not even in the least know the final cause of sexuality. The whole subject is as yet hidden in darkness" (Morse 1984, 155). It is probably fair to say that today the subject is less hidden than it was during Darwin's time, before Mendelian genetics and the structure of DNA became known, but uncertainties do remain and the search to understand the full benefits of sexual reproduction continues.

In spite of these uncertainties, it is clear that with the origin of eukaryotic cells the pace of biological evolution began to quicken. The new cells were larger than their prokaryotic predecessors and contemporaries. They carried out their aerobic metabolisms with the aid of mitochondria and chloroplasts (if they were

roughly once every hour in many bacterial strains, new combinations of genes may spread rather rapidly through a population.

photosynthesizers), which increased their ability to generate energy quickly. They possessed more DNA and, hence, genetic information. And they were able to execute more complex functions, such as mitosis and meiosis. The combination of these abilities, including the ability to reproduce sexually through meiosis, was probably responsible for the increased pace of the evolution of eukaryotes.

In any case, only eukaryotes have evolved into complex macroscopic and multicellular organisms such as plants and animals. This evolution was already well under way by roughly 670 million years ago and it has continued to the present. It has created the enormous variety of organisms we see all around us today, some of which will be the focus in the remaining three chapters of this book.

Exercises

1 (a) Distinguish between heterotrophs and autotrophs; (b) obligate and facultative anaerobes.

2 Discuss the major stages in the evolution of early life described in the text, from the first fermenting prokaryotes to the arrival of eukaryotes. Make reference to the three sources of evidence that give us information of this evolution.

3 Figure 7.9 illustrates the differences in the mechanisms by which energy is released in the electron-transport chain and in the burning of wood. Describe additional examples illustrating these different ways of releasing energy.

4 (a) Discuss the biological and geophysical processes that gave rise to the banded iron-formations and the red beds.
(b) Why are few banded iron-formations younger than and few red beds older than 2 billion years?

5 For reasons of self-consistency, why does the discovery of the Ediacara fauna require that the beginning of the Phanerozoic eon be redefined and the traditional use of the term *Precambrian* eon be abandoned?

6 (a) Using an organic chemistry molecule set (some are listed at the end of chapter 6), construct a three-dimensional model of glucose.
(b) Simulate the chemical reactions of glycolysis by converting the glucose model into two models of pyruvate. Explain what is happening with the excess hydrogen atoms you will have.
(c) In an analogous manner, simulate the subsequent reaction sequences of fermentation, ending up with ethanol, acetic acid, or lactic acid.
(d) Repeat for the case of aerobic respiration, ending up with CO_2 and H_2O. Keep track of the molecules of ATP produced in fermentation and aerobic respiration. (*Hint*: Use figure 7.10 as a guide.)

7 Explain the purpose of (a) mitosis and (b) meiosis.

8 In the case of humans, explain:
(a) Which chromosomes determine the sex of offspring?
(b) Which combinations of these chromosomes are found in males and which in females?
(c) What percentage of newborns can be expected to be male and female, respectively?

9 Let (A, a), (B, b), (C, c), . . . refer to pairs of homologous chromosomes.
(a) Suppose gametes with the chromosomes (A, B) and (a, b) combine to make a diploid. When this diploid undergoes meiosis, what possible combinations of chromosomes can be present in the gametes?
(b) Repeat the exercise, starting with gametes having the chromosomes (A, B, C) and (a, b, c).
(c) From the pattern established in the previous two cases, derive a general expression for how many different combinations of chromosomes can be present in the gametes if the number of pairs of homologous chromosomes is n. Give numerical answers for the cases $n = 4$, 10, and 23.

10 The total length of a human haploid set of DNA is approximately 1 m, and the spacing between neighboring nucleotide pairs is approximately 0.00034 μm.

(a) Calculate the number of nucleotide pairs contained in a human haploid set of DNA. (Compare your answer with the value given in figure 7.18.)

(b) Calculate the number of codons that can be formed with these nucleotide pairs.

(c) If each codon corresponds to one word of written English, estimate how many encyclopedias the codons of a human haploid set of DNA correspond to. Assume an encyclopedia consists of 20 volumes, each volume has 1000 pages, and each page contains 1000 words. (Compare your answer with the estimate given on p. 290.)

11 Generally we think that only eukaryotic organisms are capable of reproducing sexually. However, if we accept the definition of sex as formulated by the participants of the 1984 Symposium *The Origin and Evolution of Sex* (see p. 383), bacteria are capable of sexual reproduction also. Accepting this definition discuss:

(a) Two mechanisms by which bacteria may reproduce sexually.

(b) The advantages that sexual reproduction imparts on bacteria, compared to asexually reproducing bacteria.

(c) The chief differences between prokaryotic and eukaryotic sexual reproduction.

12 Among the most primitive of all bacteria are the methane-producing bacteria or methanogens, for short. For instance, they do not possess cytochrome-containing electron-transport chains and their ribosomal RNA is different from that of most other bacteria. Methanogens are obligate anaerobes and live in swamps, lake sediments, and the digestive tracts of animals, where they metabolize H-rich compounds, utilizing CO_2 to produce methane (CH_4). Little is known at present about the details of their metabolic pathways. However, two of these pathways may be summarized as follows:

$$\underline{\quad} H_2 + CO_2 \rightarrow CH_4 + \underline{\quad} H_2O + \text{energy}$$
$$2\ CH_3CH_2OH + CO_2 \rightarrow CH_4 + 2\ \underline{\quad\quad\quad} + \text{energy}$$
$$\text{(ethanol)} \qquad\qquad (\underline{\quad\quad})$$

Complete the blanks in these equations.

Suggestions for Further Reading

Alberts, B., et al. 1983. *Molecular Biology of the Cell.* New York: Garland Publishing. An excellent, up-to-date text, written by six respected researchers in molecular biology. For advanced students.

Axelrod, R. M., and **W. D. Hamilton**. 1984. "The Evolution of Cooperation in Biological Systems." Chapter 5 in R. M. Axelrod, *The Evolution of Cooperation.* New York: Basic Books. The authors establish conditions under which cooperation in biological systems can evolve, based on reciprocity and without foresight by the participants.

Barghoorn, E. S. 1971. "The Oldest Fossils." *Scientific American*, May, 30–42. The article discusses the discovery of fossilized microorganisms at the Fig Tree, Gunflint, and Bitter Springs formations.

Chambon, P. 1981. "Split Genes." *Scientific American*, May, 60–71. Most genes in eukaryotic cells are discontinuous; that is, the DNA that codes for protein is interrupted by noncoding sequences. The article explains that after transcription those noncoding sequences are excised to make mature messenger RNA, which is translated into protein.

Darnell, J. E., Jr. 1983. "The Processing of RNA." *Scientific American*, October, 90–100. The problem of how two eukaryotic cells with the same set of genes can each make a different array of proteins. It appears that one source of this control of gene expression is due to what happens to the RNA between transcription and translation.

Day, W. 1984. *Genesis on Planet Earth: The Search for Life's Beginning.* 2d ed. New Haven and London: Yale University Press. (Described in Suggestions for Further Reading, chapter 6.)

Dickerson, R. E. 1972. "The Structure and History of an Ancient Protein." *Scientific American*, April, 58–72. The differences in the sequence of amino acids of cytochrome *c* taken from cells of different species. These differences provide a record of molecular evolution dating back 1.2 billion years, when the lineages leading to plants and animals appear to have diverged.

Eckholm E. 1986. "Is Sex Necessary? Evolutionists Are Perplexed." *The New York Times*, III, 1:1. The current scientific debate concerning the benefits and evolution of sexual reproduction.

Grivell, L. A. 1983. "Mitochondrial DNA." *Scientific American*, March, 78–89.

Groves, D. I., J. S. R. Dunlop, and R. Buick. 1981. "An Early Habitat of Life." *Scientific American*, October, 64–73. The tentative evidence of 3.5-billion-year-old microfossils found at North Pole, Australia.

Kornberg, R. D., and A. Klug. 1981. "The Nucleosome." *Scientific American*, February, 52–64.

Margulis, L. 1982. *Early Life*. Boston: Jones and Bartlett Publishers (Science Books International). A readable account of the early evolution of photosynthesis, respiration, eukaryotic cell structure, meiosis, and multicellularity by an eminent researcher in the field.

——. 1981. *Symbiosis in Cell Evolution: Life and Its Environment on the Early Earth*. San Francisco: W. H. Freeman and Co.

——. 1971. "Symbiosis and Evolution." *Scientific American*, August, 48–57. The theory of the origin of mitochondria and chloroplasts by the symbiosis of two kinds of cells.

Morse, G. 1984. "Why Is Sex?" *Science News* 126 (8 Sept.): 154–57. A report on the current controversy about the benefits of sexual reproduction, including the views expressed at the symposium "The Origin and Evolution of Sex," held in 1984 at the Marine Biological Laboratory, Woods Hole, Massachusetts.

Schopf, W. J., ed. 1983. *Earth's Earliest Biosphere: Its Origin and Evolution*. Princeton: Princeton University Press. Reports on current research by 21 members of the Precambrian Paleobiologic Research Group. The book contains much information not available in the more popular literature.

Stanier, R. Y., et al. 1979. *Introduction to the Microbial World*. Englewood Cliffs, N.J.: Prentice-Hall. An abridged version of the widely used text *The Microbial World* (by Stanier et al.).

Vidal, G. 1984. "The Oldest Eukaryotic Cells." *Scientific American*, February, 48–57.

Woese, C. R. 1981. "Archaebacteria." *Scientific American*, June, 98–122. The author recognizes a new type of bacteria, which in its biochemistry and in the structure of certain large molecules is as different from other prokaryotes as it is from eukaryotes. He calls them *archaebacteria* and suggests that along with true bacteria and eukaryotes they constitute three lines of descent from a univeral ancestor.

Additional References

Cloud, P., and M. F. Glaessner. 1982. "The Ediacarian Period and System: Metazoa Inherit the Earth." *Science* 217 (27 Aug.): 783–92.

Darwin, C. R. 1897. *On the Origin of Species by Means of Natural Selection* (with additions and corrections from the sixth and last English edition). New York: D. Appleton and Company (2 vols.).

Mahler, H. R., and R. A. Raff. 1975. "The Evolutionary Origin of the Mitochondrion: A Nonsymbiotic Model." *International Review of Cytology* 43: 1–124.

Schopf, W. 1979. "Precambrian Life" in R. W. Fairbridge and D. Jablonski, eds., *The Encyclopedia of Paleontology* (pp. 641–652). Stroudsburg, Pa.: Dowden, Hutchinson and Ross.

Figure 8.1. Fossil jellyfish from the Ediacara Fauna, with an estimated age of about 650 million years.

Origin and Evolution of the Major Animal Phyla

I IF WE PLACE a drop of pond water on a glass slide and examine it through a light microscope, we find a surprisingly varied and active world of life. Most of it is microscopic and single-celled. Some of the microorganisms are round or oval, others are surrounded by shells with interesting and sometimes very beautiful patterns, and still others are just blobs of cytoplasm enclosed by membranes with no particular shapes. They all are in search of food. Many have chloroplasts and obtain energy by photosynthesis. Others are predators that pursue their prey by thrashing about with long flagella, by beating mats of hairlike cilia in a coordinated fashion, or by extending and "walking" on temporary legs (called *amoeboid locomotion*, see figures 8.2–8.4). These movements are usually in response to information about the environment—sources of light, variations in temperature, or the presence of food and chemicals—that the organisms gather through special sense organs. However, because these organisms are just single cells, their sense organs do not consist of cells and tissues like ours. They are primitive, sensitive areas either on the organisms' outer membranes or in their flagella and cilia.

If we view the content of the drop of water through an electron microscope, with its greater magnification, we notice that the microorganisms are of two kinds, differing both in size and internal organization. The smaller organisms have the simpler organization and lack nuclei and organelles. They are the *bacteria* and *prokaryotes*. The larger ones have a nucleus, and their cytoplasm contains mitochondria, chloroplasts (if they carry out photosynthesis), endoplasmic reticula, and other organelles. They are microscopic and generally single-celled *eukaryotes* and belong to the kingdom *Protista* (see Introduction to Part Two). They go by the general names of *flagellates, ciliates*, and *amoebas* (see figure 8.5), though there are many species with a wide range of morphologies and ways of life. With the exceptions of the protists associated with shells, most contemporary microorganisms, including both bacteria and eukaryotes, are probably not much different from those that inhabited ponds, lakes, and the sea a billion years ago.

In our own macroscopic environment we see organisms that are quite different from those in the drop of water. They also are very different from the

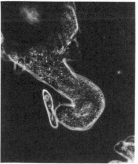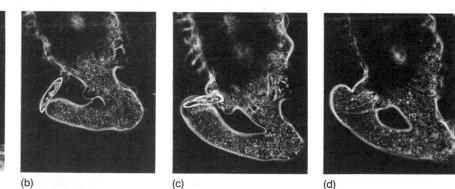

(a) (b) (c) (d)

Figure 8.2 . A giant amoeba *Chaos chaos* captures and engulfs a *Paramecium*. Though single-celled and seemingly disorganized, *Chaos chaos* is able to sense its prey, move toward it, and envelop it by sending out a pseudopod.

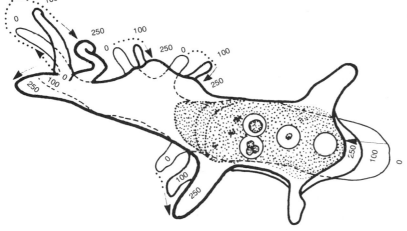

Figure 8.3. Amoeboid locomotion as illustrated by three superimposed sketches drawn from frames 0, 100, and 250 of a film of the movements of amoeba *Oscillosignum proboscidium*. An amoeba has no distinct head or tail, but at any place on its surface the enclosing membrane may become deformed into blunt projections that fill with cytoplasm and act as temporary legs or *pseudopods* ("false feet"). The pseudopods move so as to carry the amoeba a short distance forward. Then they withdraw and new pseudopods form at adjacent places. By repeating this process over and over, an amoeba manages to "walk" slowly about.

organisms that existed on Earth a billion years ago. For example, there are *fungi*, such as molds and mushrooms, which feed on dead organic matter or on other organisms. There are *plants*, such as ferns, conifers, and flowering plants, which find their nutrients in the ground and air (CO_2) and use the energy of sunlight. There are *animals*, such as worms, snails, insects, fishes, frogs, snakes, birds, sheep, and squirrels, which move around and survive by eating plants and other animals. And, of course, there we are — *Homo sapiens*. Unlike the protists, these organisms are complex multicelled eukaryotes and many are macroscopic (see box 8.1). (From here on, these organisms — that is, animals, plants, and fungi — will be referred to simply as complex multicelled organisms, even

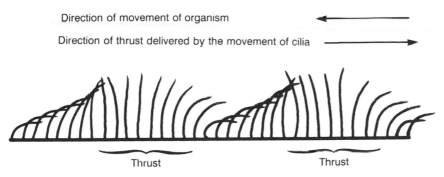

Direction of movement of organism

Direction of thrust delivered by the movement of cilia

Thrust   Thrust

Figure 8.4. Ciliated protists, small flatworms, and some larvae (e.g., those of sponges, marine annelids, and some cnidarians) are covered by thick mats of cilia. These organisms can move about by beating the cilia in rhythmic and wave-like fashions, as illustrated here with a single row of cilia. This *ciliary locomotion* may be compared with the propulsion of a racing shell by means of oars.

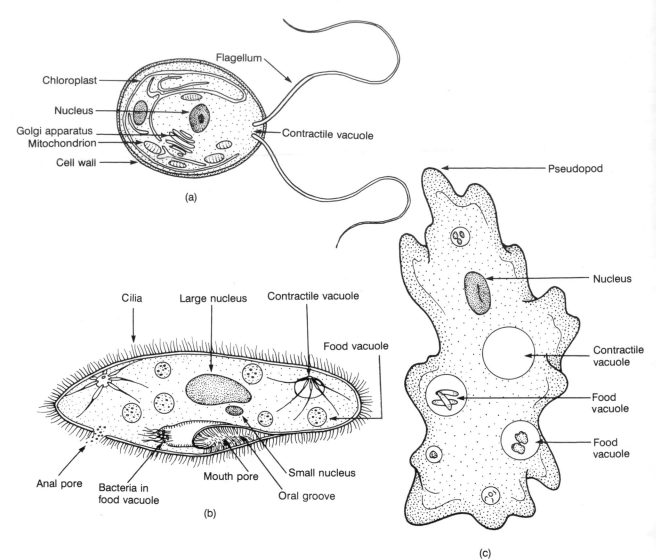

Figure 8.5. Three examples of single-celled eukaryotes, of the kingdom Protista: (*a*) *Chlamydomonas* (a flagellate), typically 30 μm long excluding the flagella. (*b*) A paramecium (ciliate), typically 100 μm long. (*c*) An amoeba, typically 4 μm to a few millimeters in size, though some amoebas may reach sizes of 1 to 2 centimeters.

though there also exist some multicelled, albeit much simpler, examples among the protists. Also note that not all animals are macroscopic. For example, rotifers, gastrotrichs, mites, and others are animals less than 1 mm in size and some can be seen only through the microscope.)

The main goal of the present chapter is to discuss some of the current theories concerning the transition from relatively simple single-celled to complex multicelled life and to examine in broad terms the origin and evolution of the major animal phyla. The discussion of animal evolution continues in chapters 9 and 10, which focus on the vertebrates and the primates. In general,

---

BOX 8.1.
Biological Taxonomy

THE terms *protozoa, metazoa, protophyta,* and *metaphyta* are common in biology. They are derived from the Latin roots *proto-, meta-, -zoa,* and *-phyta,* meaning "first in time," "occurring later," "animals," and "plants," respectively. These terms were coined when biologists placed all organisms into just two kingdoms—the animals and the plants. *Metazoa* referred to macroscopic (meaning "visible with the unaided eye"), multicelled animals; *metaphyta* to macroscopic, multicelled plants; *protozoa* to small, usually single-celled organisms that live like animals; and *protophyta* to small, usually single-celled organisms that live like plants.

Although these terms are well defined for such organisms as reptiles, ferns, amoebas, and blue-green bacteria, they become very confusing for many of the microorganisms that have both an animal-like and plant-like existence. Examples are the *Euglenas* (protists), which possess chloroplasts and are able to photosynthesize like plants, but also eat like animals by engulfing other microorganisms. The older terms also offer no clear classification choice for such organisms as the fungi, which, like plants, are attached to their source of food but do not photosynthesize, or the lichens, which are symbioses of an alga and a fungus (see box 7.1).

With the introduction of the five-kingdom classification scheme—*Monera, Protista, Fungi, Plantae,* and *Animalia* (see figures II.5 and II.6)—much of the confusion associated with the terms *protozoa, metazoa,* and so forth was eliminated, though their usage persists in the literature.

---

plants will not be discussed, except to refer to the major impacts that their evolution had on the animals.

In deducing the evolutionary history of animal life, three sources of information are chiefly relied upon: anatomical comparison between the adult forms of different kinds of present-day animals, the embryological development of animals, and the fossil record. Each of these three sources has its own limitations. The first two are limited because evolution is the product of a long sequence of adaptations by organisms to continually changing environmental conditions and to changing ways of exploiting those environments. Hence, it is not always possible to deduce accurately from modern animals and their embryology the particular sequences that gave rise to them. The third source of information, the fossil record, is in principle more informative and dependable. However, it is very fragmentary because much of that record has been destroyed by geological activities. Furthermore, with rare exceptions only the hard parts of animals—shells, bones, and teeth—have become fossilized and survived over the ages.

Despite these limitations in the available sources of information, comparative anatomists, biologists, and paleontologists have been remarkably successful in reconstructing the evolutionary past. Remember, however, that the limitations inherent in the data mean that we can only know the approximate routes by which present-day life evolved. We do not and probably never will know exactly how it happened.

THE SIMPLEST OF TODAY'S ANIMALS

Before discussing the current theories concerning the transition from single-celled to complex multicelled life, it is important to be familiar with the simplest

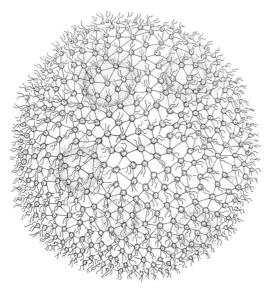

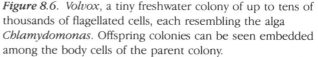

Figure 8.6. *Volvox*, a tiny freshwater colony of up to tens of thousands of flagellated cells, each resembling the alga *Chlamydomonas*. Offspring colonies can be seen embedded among the body cells of the parent colony.

of today's animals—*Volvox* (commonly regarded as a cell colony), sponges, cnidarians (jellyfish, hydrozoans, corals, sea anemones, sea pens), and flatworms—as well as with the early embryonic development of some of the more complex animals. The reason why you need to begin with these topics is that to date no fossil evidence of the origin of multicelled life has been found. Today's simplest animals and embryology constitute the only data upon which we can base our theories.

Volvox is one of the most primitive multicelled organisms (see figure 8.6). It can be found in freshwater ponds as a tiny, barely visible green ball that is composed of up to thousands of flagellated cells, each resembling the free-living alga *Chlamydomonas*. Like *Chlamydomonas*, each cell of *Volvox* possesses a red eye spot that is sensitive to light and it nourishes itself by photosynthesis.

The cells of *Volvox* are connected by thin cytoplasmic strands and are arranged in a single layer surrounding a spherical, jellylike mass. By exchanging signals through the strands, the cells coordinate the beating of their flagella so that *Volvox* moves about by rolling over and over (the Latin word for "rolling" is *volvere*, the origin of the organism's name). There are, however, no nerve cells. In fact, all of the *Volvox* cells are the same except that those of one of its hemispheres have larger eye spots and are primarily responsible for directing movement, while those of the other hemisphere have the task of reproduction. Because of its simple structure, with only a minimal degree of cell specialization, *Volvox* is generally regarded as a "cell colony" rather than as an animal.

More complex than *Volvox* are the *Porifera* or sponges. They live mostly in the ocean attached to rocks or other solid objects, though there are also a few freshwater species. Sponges grow in rounded lumps, sheets, or branched shapes and may reach sizes of up to a meter or two across. Their surfaces are covered by large numbers of tiny passages or pores that lead to a complex interior network of connecting canals and chambers (see figure 8.7.).

Sponges are constructed of large numbers of cells that are quite a bit more specialized than those of *Volvox*. Flat epithelial cells form the outer covering of

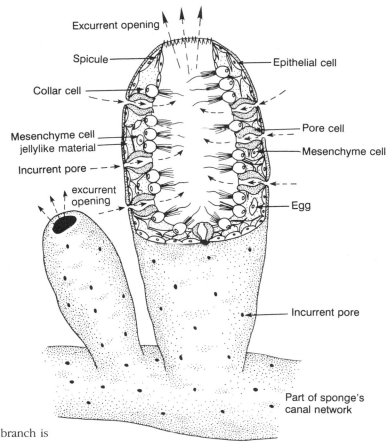

Excurrent opening

Spicule

Epithelial cell

Collar cell

Pore cell

Mesenchyme cell
jellylike material

Mesenchyme cell

Incurrent pore

excurrent
opening

Egg

Incurrent pore

Part of sponge's
canal network

Figure 8.7. Part of a sponge. One branch is
shown whole, the other partially sectioned.

the sponge. Embedded among them are pore cells with hollow tubes through
their middles that form the passages between the outside and the sponge's
inside. On the interior are collar cells with flagella, whose whipping action
draws water through the pores into the sponge and out through large openings
in the canal network. Contractile cells control the flow of water by constricting
or expanding the sizes of the pores. Finally, there are mesenchyme cells that
wander about in amoeboid fashion in the sponge's inside and help to digest and
distribute food. Some mesenchyme cells secrete calcium carbonate or silica to
form needleclike spicules and create a skeletal support system for the otherwise
limp cellular mass of the sponge. In certain warm-water species, the secretion is
a soft and flexible horny substance, which constitutes the natural sponges that
are used for washing and cleaning.

   Their porous construction and internal canal networks make the sponges
multicelled filter feeders of the simplest kind. The beating of flagella forces
water into the sponge through the pores, circulates it through the inner canal
system, and expels it through the large openings. As the water passes through
the pores, microorganisms are filtered out, engulfed and digested by the collar
cells, and the extracted nutrients are distributed by the mesenchyme cells to all
the other cells for growth and energy. There are no sense organs, nerve cells,

muscles, digestive cavity, circulatory system, or other kinds of organ systems. All the activities necessary to sustain the sponge are carried out on the cellular level and whatever coordination is required is achieved by the direct transmission of signals from one cell to another. Sponges are constructed according to the simplest possible plan for animals—the *cellular level* of organization.

The *cnidarians* (also called *coelenterates*) are mostly marine animals whose basic body plan is also quite simple, though it shows a distinct increase in complexity over that of the sponges. The general shape of these animals is a cylinder that is closed at one end and open at the other, with tentacles surrounding the open end. The single opening serves both as a mouth and an anus: It is an entrance for food as well as an exit for undigested food residues. The hollow cylindrical cavity also has a dual function. It serves as a stomach for digesting food and as a distribution system for circulating water and nutrients to the cells, and so it is called the *gastrovascular cavity* (*gastro-* refers to "stomach" and *vascular* to tubes or vessels for circulating body fluids, as in "cardiovascular"). There is no separate circulatory system.

The bodies of the cnidarians may be oriented in two different directions. Some resemble a bell or an umbrella, with the opening at the bottom and the tentacles hanging loosely downward (see figure 8.8). This kind of body is called the *medusa* form (from an imagined resemblance of the waving tentacles to the snake-entwined hair of Medusa, one of the three Gorgon monsters of Greek mythology, who lived in the underworld). Examples of medusae are the jellyfish, which are free-swimming and propel themselves forward by rhythmic pulsations of the bell (see figure 8.9). The bodies may also be oriented with the opening at the top and the tentacles pointing upward or to the sides, like the petals of flowers. This kind of body is called a *polyp* (meaning "many-footed"). Polyps are sessile (attached to the ocean bottom). Examples are the sea anemones (figure 8.10), corals, and hydras, many of which are brightly colored. They vary from white and yellow to pink, purple, and violet, and truly are a beautiful sight when an outgoing tide reveals them in the shallow pools of a rocky ocean shore.

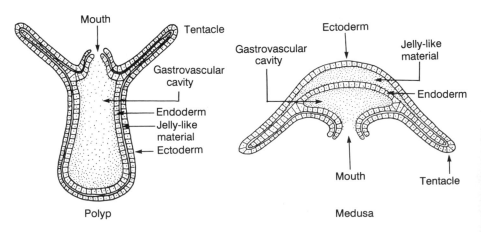

Figure 8.8. Cross sections of the body construction of cnidarians. Because the diagrams are two-dimensional, only two tentacles are shown in each case.

Figure 8.9. A jellyfish, a cnidarian of the medusa form.

Figure 8.10. Sea anemones. Sea anemones are cnidarians of the polyp form, whose often colorful tentacles give them a superficial resemblance to flowers. They are carnivores that prey on small animals that they catch with their tentacles and stinging cells.

The bodies of the cnidarians, both medusae and polyps, have a two-layered construction. There is an outer cell layer called the *ectoderm* and an inner cell layer called the *endoderm*, between which lies a surprisingly stiff jellylike material (hence, the name "jellyfish"). Because of the presence of two cell or tissue layers,* the cnidarians are said to be constructed on the *tissue level* of organization.

The two layers are constructed of cells specialized for numerous tasks (see figure 8.11). Many of the cells, of both the endoderm and ectoderm, possess contractile fibers that serve as primitive muscles and allow the animals to swim, reach for food with their tentacles, and move water and objects in and out of the gastrovascular cavity. Gland cells in the endoderm secrete digestive enzymes. Some ectoderm cells contain harpoonlike stinging capsules used to inject poison into passing prey. There are ovaries and testes that produce eggs and sperm for reproduction. There are also sensory cells and nerve cells, which form a network that pervades the entire body and permits the animal to respond adaptively to external stimuli. There is, however, no controlling group of nerves, or brain.

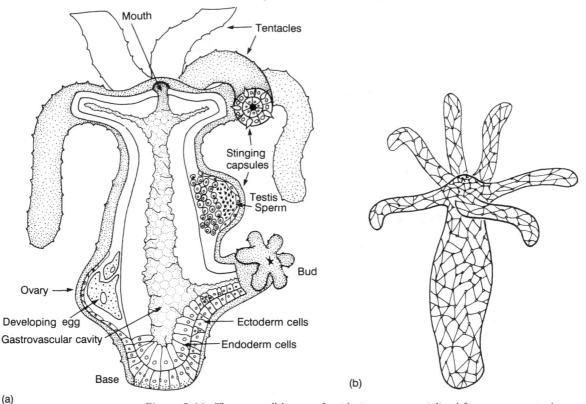

Figure 8.11. The two cell layers of cnidarians are specialized for numerous tasks, as illustrated here in the case of the hydra. (*a*) Cross section. (*b*) Nerve net.

*A *tissue* is an aggregate of cells, usually of similar kinds, that together with the intercellular fluid associated with them form one of the structural materials of most animals and plants. Examples of tissues in our bodies and those of many other animals are epithelial tissues (skin, linings of body cavities), connective tissues (bone, cartilage, ligaments, tendons), muscle tissues, and nerve tissues.

Figure 8.12. The body construction of the flatworms, typically about 2.5 cm long. (*a*) General body plan. The pharynx (suctorial eating organ) is shown extended. When not in use, it is drawn through the mouth into the body. (*b*) Cross section showing internal organs.

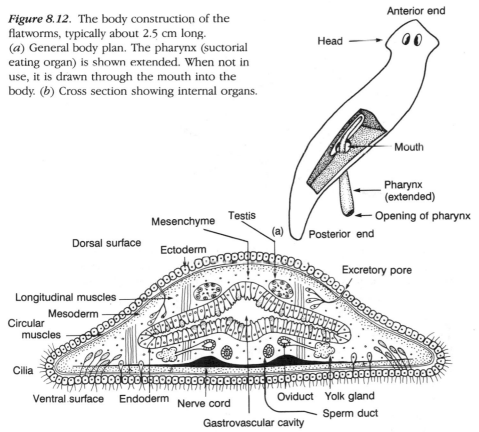

The next level of complexity in body construction beyond the cnidarians is found among the *Platyhelminthes* or *flatworms*, which include creeping fresh-water and marine species, swimmers, parasitic flukes, and tapeworms. The flatworms distinguish themselves by being the simplest animals having a head with eyes and other sense organs, anterior (front) and posterior (rear) ends, and ventral (belly) and dorsal (back) surfaces (see figure 8.12). Most important, like all more complex animals, they have a *mesoderm*, a third cell layer between the ectoderm and endoderm, which gives rise to organs and organ systems.*

In flatworms, the organ systems are quite primitive and few in number. However, there are muscle tissues that serve in locomotion, though beating cilia aid in the task (see figure 8.13). The muscles are not just contractile fibers, but genuine muscle cells as in human bodies, and they are arranged in circular and longitudinal tissue layers. There is an excretory system with tubes running throughout the body, ending in many tiny openings to the outside (see figure 8.14). Its main function is probably not excretion of wastes, but regulation of the water content of the body tissues. Many species possess an unusual reproductive system, with rows of ovaries and testes as well as a penis and a copulatory sac, all in the same animal. There is also a central nervous system that relays information picked up by the sensory cells to the rest of the body. It consists of

---

*As with most generalized statements in biology, there is an exception to this one. The acoel worms, which are discussed below and which are classified as flatworms, do not have a mesoderm.

Figure 8.13. An aquatic flatworm. As their name implies, flatworms are flat, elongated animals. They move along the sea floor and rocks by creating muscular waves in their bodies. Locomotion is aided further by beating cilia and the secretion of a sticky mucus.

a primitive brain located in the head and two nerve cords. The nerve cords run lengthwise through the animal and are connected by cross linkages resembling the rungs of a ladder. The flatworm's central nervous system is the simplest example of the kind possessed by all of the more complex animals. It does not, however, entirely replace the nerve net, which persists and is important to the animal. (Incidentally, localized nerve nets are present in most animals, as, for example, the nerve net in the walls of the human intestine.) Because of the

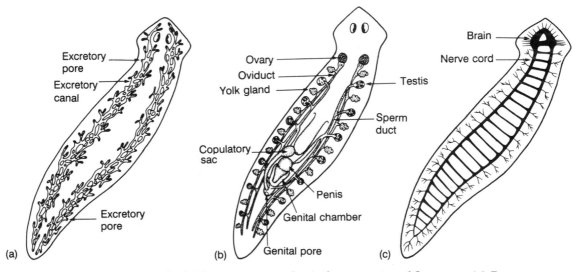

Figure 8.14. The organ system level of construction of flatworms. (*a*) Excretory system. (*b*) Reproductive system. (*c*) Nervous system.

presence of these organ systems, flatworms are said to be constructed on the *organ system level* of organization.

Despite the complexity found in flatworms, their gastrovascular cavity still has just one opening that serves as both a mouth and an anus. There are no such specialized digestive organs as an esophagus, stomach, liver, and intestine. There is no circulatory system. It is not needed, for the body of the flatworm is sufficiently flat and simple that nutrients can reach all cells by diffusion through the endoderm.

ORIGIN OF COMPLEX MULTICELLED LIFE: EARLY VIEWS

Volvox, sponges, cnidarians, and flatworms are organisms that have increasingly complex body construction, from the colonial level through the cellular and tissue levels to the organ system level. Furthermore, this sequence of increasing complexity is paralleled by a sequence of body forms from *spherical* through *radial* to *bilateral symmetries. Volvox* is approximately spherically symmetric; the sponges and cnidarians are radially symmetric; and the flatworms, like most more complex animals, are bilaterally symmetric.* Like most bilaterally symmetric animals, flatworms have, in addition to equal left and right sides, a definite anterior end (usually the head), a posterior end (often the tail), and dorsal and ventral surfaces.

Today biologists regard these increases in complexity as accidental and do not believe that they are evidence of any evolutionary sequence among these simple organisms. However, biologists of the past interpreted these increases in complexity differently. They saw in them an evolutionary progression from single-celled life to the higher animals, an idea that goes back to Aristotle (384–322 B.C.). The first serious modern proposal of such a connection was made by European biologists in the second half of the nineteenth century, in particular by Ernst H. Haeckel, the German zoologist who introduced Darwin's evolutionary theory to his country. According to him, *Volvox* represents the transition stage from single- to multicelled life and the sponges, cnidarians, and flatworms are successive stages in the evolution toward the higher animals.

Haeckel found compelling support for his theory in the embryological development of the higher animals, which shows a parallel increase in complexity. As illustrated in figure 8.15 for the frog, that growth begins with a single fertilized egg, or zygote, which Haeckel compared to the single-celled ancestors of the animals. By repeated cell division, the zygote develops into a *blastula*, a spherical cell layer that surrounds a hollow, fluid-filled cavity and resembles a *Volvox*. After further cell division, the blastula folds in on itself, or invaginates,

---

*Spherical symmetry means sameness in all directions outward from a central point, like a sphere. *Radial symmetry* means sameness in all directions perpendicularly away from a central axis, like a cylinder or cone. And *bilateral symmetry* means two equal sides, like the left and right sides of the human body. In describing the body shapes of animals by reference to geometric symmetries, it is understood that the symmetries may not be exact. For example, branched sponges are not exactly radially symmetric and the two sides of a flatworm may not be exactly equally developed. Still, reference to these symmetries is useful, for they characterize in simple yet meaningful ways the basic body shapes of most animals.

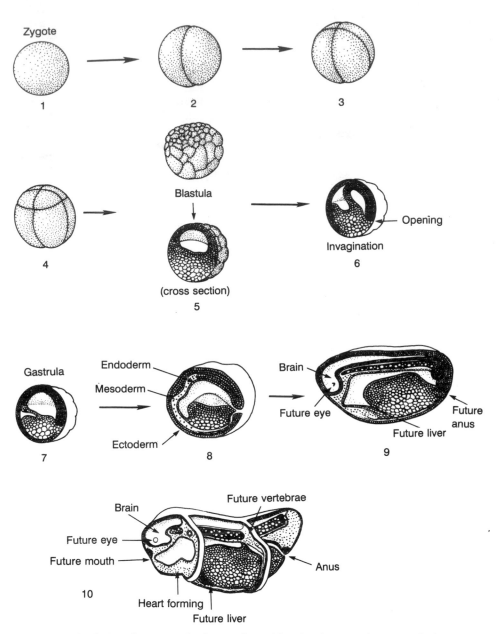

Figure 8.15. Development of a frog embryo. The development begins with the repeated division of a *zygote*, or fertilized egg (1–4), to form a hollow, fluid-filled sphere of cells, the *blastula* (5). Then one area of the blastula folds inward (6) to form the *gastrula* (7). The inner cell layer of the gastrula, the endoderm, develops into the lining of the digestive system, including the esophagus, stomach, liver, and intestine. Its outer cell layer, the ectoderm, develops into the skin, nervous system, and sense organs. Between ectoderm and endoderm a third cell layer, the mesoderm, arises. It develops into the internal organs and organ systems, connective tissues (for example, bone, muscles, and tendons), and the lining of the body cavities (8–10). (Also see figure 7.20.)

as if an imaginary hand were pushing the cell layer inward. This creates a *gastrula*, a structure with a single opening and an endoderm and ectoderm, resembling the body construction of the cnidarians. Finally, in the space between the ectoderm and endoderm, the mesoderm begins to grow. At this stage, the embryo is bilaterally symmetric and its construction corresponds approximately to that of flatworms.

Thus, Haeckel saw in the embryological growth of the higher animals a recapitulation of their evolutionary ancestry. The fact that mammalian embryos pass through stages in which they have gills, like fishes, and tails, like reptiles, he took as further evidence of his theory, and he stated quite generally that *"ontogeny recapitulates phylogeny."** Haeckel called his theory the "fundamental biogenetic law" and first published it in 1874 under the title *Natürliche Schöpfungsgeschichte* (English translation, *History of Creation*, 1876).

The biogenetic law was hailed by many of Haeckel's contemporaries as a major scientific breakthrough and it stimulated much discussion and research. However, some biologists were not so quick to accept it, for they found serious inconsistencies in the assumptions upon which it was based, particularly with regard to the origin of complex multicellular life. For instance, the blastulae of cnidarians do not develop into hollow gastrulas by invagination, but by cells wandering in from the ectoderm. This leads first to the formation of a solid gastrula. Only later, during the larva stage, do the gastrovascular cavity and mouth develop by a hollowing-out process from the inside. Thus, if ontogeny really recapitulates phylogeny, then the ancestors of the cnidarians do not lie on the main lineage to the higher animals. Nor do those of *Volvox* or the sponges; for the cells of modern *Volvox* possess chloroplasts that are certainly not present in animal embryos or adults, and the single-layered structure of sponges has no embryological analog among the higher animals. Furthermore, the main body opening of the sponges is excurrent (that is, water flows through it to the outside), which is unique among the animals. Because of these divergent features, it is commonly accepted today that Haeckel's biogenetic law, though useful if interpreted cautiously, is not generally valid. In particular, most biologists agree that *Volvox*, the sponges, and the cnidarians probably arose separately from single-celled ancestors and are distinct from the lineages that led to the flatworms and more complex animals.

ORIGIN OF COMPLEX MULTICELLED LIFE: CURRENT VIEWS

How then did the lineages that led to the more complex animals have their start? Two theories are currently discussed seriously by biologists. The first theory was introduced more than 75 years ago by the German zoologist Otto Bütschli (1848–1920), who proposed that the original ancestor of the more complex animals was a small, flat, two-layered creature that crept along the sea bottom in search of food. He reasoned that the flat, two-layered form would have maximized the animal's contact with the ground, where food (microor-

Ontogeny refers to the embryological growth of an individual and *phylogeny* to the evolution of the species.

ganisms) was located. At the same time, it would have minimized the strength required of its cells for support of the body structure. The cells of the animal's ventral layer became specialized for absorbing and digesting food, while those of its dorsal layer were for protection. Because the animal moved about by creeping, Bütschli thought its shape was probably bilaterally symmetric. He suggested further that there would have been an advantage if the animal could have raised its flat body during feeding to form a temporary gastrovascular cavity for retaining the food it caught and concentrating the digestive enzymes it secreted.

Interestingly, at about the time when Bütschli formulated his theory, a flat, two-layered animal, very similar to the hypothesized one, was discovered in seawater aquaria by Franz E. Schulze (1840–1921) of the University of Graz, Austria. Schulze named the animal *Trichoplax adhaerens*. However, *Trichoplax* was not accepted as evidence in support of Bütschli's theory, for it was thought to be the larva of a hydrozoan (a class of the cnidarians), and soon Bütschli's theory, as well as *Trichoplax*, were forgotten.

Only in 1969 was *Trichoplax* rediscovered. The German zoologist Karl G. Grell of the University of Tübingen detected it in some algae sent to him from the Red Sea. He found that this simple animal, consisting only of a ventral and a dorsal cell layer with some mesenchyme fiber cells between them, is not a larva but a fully grown animal capable of reproduction. He also discovered that the dorsal cells are all flagellated, while some of the ventral cells are flagellated and others are nonflagellated gland cells. Furthermore, he noted that quite often the animal raises itself up to form a temporary gastrovascular cavity, just as proposed by Bütschli for his hypothetical animal. The only significant difference between *Trichoplax* and Bütschli's animal is that *Trichoplax* is not bilaterally symmetric. Its shape is irregular, without any constant symmetry.

The close correspondence in structure and feeding habit between *Trichoplax* and Bütschli's hypothetical animal greatly bolsters the hypothesis that flat,

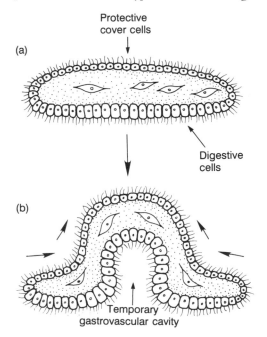

Protective cover cells

(a)

Digestive cells

(b)

Temporary gastrovascular cavity

Figure 8.16. Trichoplax adhaerens. (*a*) General cross section. (*b*) Shape while feeding.

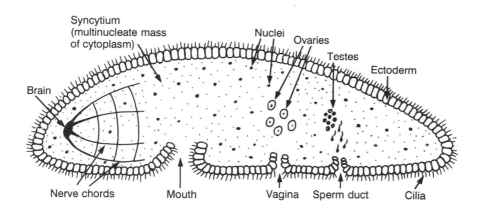

Figure 8.17. Longitudinal section of an acoel flatworm.

two-layered creeping organisms were the ancestors of the higher animals. These organisms may have arisen from flagellated single cells that failed to separate during cell division. During subsequent evolution, the temporary gastrovascular cavity of these organisms became permanent, the cells became further specialized, a mesoderm arose, and a primitive nervous system developed that coordinated the animals' body functions and movements. In body complexity and feeding habit, these animals came to resemble the modern flatworms, and, according to the theory, were the ancestors of all of today's higher animals.

The second theory of the origin of the higher animals begins with the recognition that the simplest of today's flatworms—the acoel flatworms—are small, ciliated marine organisms that lack a gastrovascular cavity (hence, the adjective *acoel*). Instead there are digestive cells that form a single syncytial mass of cytoplasm (see figure 8.17). *Syncytial* means that the cytoplasm has many nuclei with no cell membranes separating them from each other. This syncytial interior is bounded on the outside by an ectoderm consisting of ordinary, membrane-enclosed cells. The ectoderm has only a simple opening through which food is taken in. Furthermore, these worms crawl around by means of their cilia. In short, the acoel worms look and live very much like ciliated single-celled protists.

Because of the resemblance between the modern acoel worms and the ciliates, the Jugoslav biologist Jovan Hadži (1884–1972) proposed in 1953 that the evolutionary lineages that gave rise to the more complex animals started with ciliated protists that had developed syncytial interiors. In the course of time, membranes evolved around the nuclei, which provided strength and made larger body structures possible. Individual cells began to specialize for feeding, digestion, and locomotion. The opening of the mouth grew inward to form a gastrovascular cavity, similar to the formation of a gastrula (see figure 8.15). Gradually, such animals acquired body complexities and feeding habits resembling those of flatworms, at which point evolution continued as indicated above in the description of Bütschli's theory.

Figure 8.18. *Charniodiscus arboreus*, a fossilized polyp cnidarian from the Ediacarian period that may be an early ancestor of today's sea pens.

The two theories concerning the origin of the more complex animals just described — Bütschli's theory and Hadži's syncytial theory — are the ones that are currently favored by biologists. Remember, however, that these two theories are based only on indirect evidence from embryology and the adult forms of present-day animals, not on fossil remains. Hence, even though many biologists think that one or the other of them is on the right track of explaining and identifying the actual origin of the more complex animals, they cannot be sure.

Whatever the true origin of the animals was, it is known from the fossil record that by 670 million years ago, at the beginning of the Ediacarian period, the transitions from single- to complex multicelled life had been completed (see box 8.2). To date, more than two dozen species of soft-bodied marine animals, apparently belonging to several genera and phyla, as well as simple surface tracks resulting from creeping or burrowing activities, have been discovered in Ediacara rock strata from all over the world — Australia, Africa, China, Siberia, the western USSR, Scandinavia, England, and Canada. Those ancient marine animals resembled modern jellyfish, worms, sea pens, soft corals, arthropods, and echinoderms. They had no hard, mineralized shells or solid skeletons.

The transition from single- to complex multicelled life could not have come about suddenly, however, as the currently known fossil record suggests. Organisms such as jellyfish, worms, arthropods, and the rest of the Ediacara fauna are too complex and constitute too diverse a group to have arisen suddenly. At present no one knows just how long this transition took. Hundreds of millions of years may have been involved. But it may also have taken place in just 10 million years. Perhaps fossil traces of this evolution exist in some pre-Ediacara rock deposits and will eventually give us a more definite answer.

The most common fossil specimens from the Ediacarian period are cnidarians, both of the polyp and the medusa forms. An example of the polyp form is *Charniodiscus arboreus*, which may be related to today's sea pens. It lived in colonies anchored to the sea bottom and left beautiful, leaflike fossils up to one meter in length (see figure 8.18). Examples of the medusa type are jellyfish (see figure 8.1), which probably lived much like their modern counterparts. Other fossils found abundantly in the Ediacara rocks resemble today's annelids, the segmented worms. One of them, *Dickinsonia*, is an oval-shaped creature whose complex array of ridges and channels shows bilateral symmetry. Perhaps the most interesting Ediacara fossil is *Spriggina* (see figure 8.19). It is about five centimeters long, has a well-developed head covered by a prominent triangular shield, and its elongated segmented body suggests that it may be related to the ancestors of both the annelids and the arthropods.

There can be little doubt that complex multicelled life of the Ediacarian period had not only spread throughout much of the ancient sea, but it had also acquired a surprising degree of diversity and established complex ecosystems (see the section on ecosystems in the Introduction to Part Two). Besides animals, these ecosystems included many single-celled organisms and probably some multicelled marine plants. The single-celled algae and blue-green bacteria,

Figure 8.19. Fossilized annelid worm, *Spriggina*, from roughly 650 million year old sandstone of the Ediacara Hills of southern Australia.

BOX 8.2.
The Geologic Time Scale of the Phanerozoic Eon

IN the discussion of the history of the theory of biological evolution (see box II.2) I noted that by the beginning of the nineteenth century the idea that species are evolving was accepted by many philosophers and naturalists in Western Europe. Furthermore, they realized that the surface of the Earth is also changing and that the time scales involved are of the order of millions of years or longer.

These concepts of change stood in sharp contrast to the traditional western doctrine of creation and were not readily accepted by the population at large. The theories needed to be defended. To do so, and to strengthen and extend them, scientists kept searching for new evidence. The most rapid progress in this effort during the first half of the nineteenth century was made by geologists in England, in part because of

the way the bedrock tilts on this island, allowing easy access to most strata of the Phanerozoic eon (the eon of "visible life"). The tilt points downward toward the southeast and thereby exposes progressively older rocks toward the northwest. Each of the rock strata is distinct and obviously was formed under different conditions. Furthermore, each rock stratum contains its own characteristic fossil specimens. Thus, the geology of England resembles an open book that reveals a great deal about the past to those who can read it.

During the initial, somewhat crude attempts by British geologists to understand the geological and biological past of their country, they classified the much folded and very old rockbeds of Wales and the west as the *Primary stratum* (see figure A). They designated the next younger rocks, toward the central

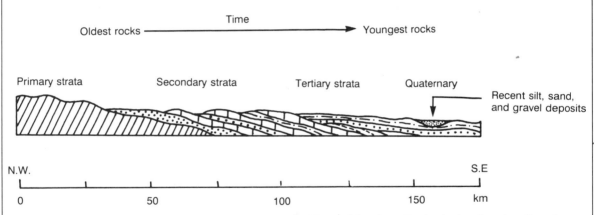

Figure A. Section across part of Wales and Southern England, showing the tilt and general relationship of the four major strata of sedimentary rocks.

part of England, the *Secondary stratum*. And they labeled the youngest rocks, in the southeast, the *Tertiary* and *Quaternary strata*. By the mid-1800s, when it was recognized that each rock stratum is characterized by its own particular set of fossil specimens and that the fossils constitute an evolutionary sequence, the terms *Paleozoic* (meaning "ancient life"), *Mesozoic* ("middle life"), and *Cenozoic* ("re-

cent life") were introduced to refer to the times when the Primary, Secondary, and Tertiary and Quaternary strata were laid down. These time segments, called *eras*, were then subdivided further into *periods* and, because much of the pioneering work was done in England, many of the names of the periods were derived from English locations and tribes.

BOX 8.2 continued

Era	Period	Origin of Period's Name
Paleozoic	Cambrian	*Cambria*, the Latin name for Wales
	Ordovician Silurian	The Ordovices and Silures, two old Welsh tribes
	Devonian	Devon, a county in southwestern England
	Carboniferous	Reference to extensive coal beds
	Mississippian Pennsylvanian	Two states in the United States; refer to the Lower (earlier) and Upper (later) Carboniferous
	Permian	Perm, a region west of the Ural Mountains in Russia
Mesozoic	Triassic	Rock strata in the Alps characterized by three (whence *Tri*-assic) distinct subdivisions—a thick marine stratum separating two continental strata
	Jurassic	Jura Mountains along the Swiss-French border
	Cretaceous	Latin *creta*, meaning "chalk"; refer to the prominent white chalk cliffs on both sides of the English channel
Cenozoic	Tertiary Quaternary	Original names given to the rock strata of southeast England

Since the initial naming of the periods, additional field data have allowed geologists to subdivide each period into shorter units, the *epochs*. The epochs of the Tertiary and Quaternary periods are as follows:

Period	Epoch	Meaning of Epoch's Name
Tertiary	Paleocene	Ancient epoch
	Eocene	Earliest epoch
	Oligocene	Few
	Miocene	Less — Of these three epochs, Oligocene rocks have the fewest, Miocene somewhat more, and Pliocene rocks still more fossils of currently living species.
	Pliocene	More
Quaternary	Pleistocene	The *most* fossils of currently living species
	Recent	Most recent, or present, epoch

Up to roughly the middle of the present century, no fossil evidence was known of complex multicelled organisms older than the Cambrian. Thus, the beginning of the Phanerozoic eon was originally defined to coincide with the beginning of the Cambrian. Then, during the mid-1940s, the Australian geologist Reginald C. Sprigg found fossils of large, multicelled organisms in the Precambrian rocks of the Ediacara Hills, some 400 kilometers north of Adelaide, Australia. This prompted the introduction of a new period, the Ediacarian period, preceding the Cambrian. More recently, Preston Cloud proposed that the Phanerozoic eon be extended backward to include the Ediacarian period, a proposal adopted in this book (see pp. 355–356 and table 8.1).

together with the plants, were the primary food producers. The animals, which moved about in search of food by creeping along the sea floor, burrowing in the mud, and swimming or drifting in the sea, were the consumers. Finally, various kinds of nonphotosynthetic bacteria were the decomposers.

DIVERSIFICATION OF THE ANIMAL KINGDOM: PHYSICAL CONSIDERATIONS

The ancient fossil record gives clear evidence that during the Ediacarian period multicellular organisms were already experimenting with many different ways of life and survival strategies, and that these organisms were evolving into a diverse range of species. To understand how this far-ranging diversification came about, recall that the environment in which complex multicellular life first arose was populated by single-celled organisms that offered relatively little competition to multicellular life. In fact, the bacteria and protists created the foundations of potential resource bases for the animals and plants, for they were both producers of food and decomposers of waste. Thus, to early multicelled life virtually the entire water-covered surface of the Earth—oceans, rivers, and lakes—constituted an untapped environment, ready-made for invasion and occupation and offering countless opportunities for biological experimentation.

Then, as today, evolution proceeded by a two-step process: first, by the creation of genetic variation by mutation and sexual reproduction, and second, by the preferential survival of those individuals that were most competent in the competition for available raw materials, living space, and participation in mating processes. In general terms, it may be said that in the animal kingdom success in this competition came most likely to those organisms that were able to move about efficiently and respond adaptively to their particular habitat.

The abilities that meant success in the animal kingdom were, of course, important characteristics among single-celled life also. However, by the transition from simple single-celled organisms to complex multicelled animals, many new opportunities for moving about and adapting to the environment were opened up. For example, early single-celled organisms, consisting of single fluid interiors surrounded by membranes, were severely limited in size, strength, locomotor skills, and responsiveness. Complex multicellular organisms, on the other hand, were limited much less so, mainly because multicellularity allowed them to evolve cell tissues, organs, and organ systems. It also allowed them to evolve bilateral symmetry, a head containing many of the sense organs, and a tail, which improved still further their locomotor skills and their abilities to respond adaptively. The flatworms are the simplest known examples in the animal kingdom possessing these advantages of tissue, organ, and organ system specializations and body shape.

Evolution of Digestive, Circulatory, and Excretory Systems

One of the most significant specializations that occurred beyond the flatworm stage of organization was the formation of a second opening in the gastrovascular cavity. This permitted food to travel in a one-way direction from one of the openings, the mouth, to the other, the anus. Furthermore, it permitted different sections of the *digestive tract*, as this one-way passageway is called, to carry out specialized functions, such as the digestion of food (in the stomach), regulation of glycogen in the body (via the liver), absorption of nutrients (in the intestine), and elimination of waste (via the rectum). This specialization meant that energy and nutrients could be extracted much more speedily and thoroughly from food than before, which, in turn, allowed the animals to evolve new, energy-utilizing adaptation, such as increased size and stronger muscles.

Increase in body size made new demands on the mechanisms responsible for the distribution of nutrients and elimination of metabolic wastes. No longer could nutrients released in the digestive tract reach all body cells by simple diffusion through the endoderm, nor could metabolic wastes (such as ammonia and urea, which result from the breakdown of amino acids) be eliminated directly through the ectoderm to the outside, as is the case in animals that are small and flat. Only the evolution of separate circulatory and excretory systems could accomplish this. In time, such systems did evolve. At first, they worked simply by the back-and-forth sloshing of fluid through the body cavity as the animal constricted and distended its various body parts during movement. In the process, the fluid absorbed nutrients at the intestine and carried them to the cells, while it picked up metabolic wastes at the cells and eliminated them through the ectoderm. In the course of time, primitive systems of this sort were replaced by special *closed circulatory systems* consisting of vast networks of passages that extended throughout the body, with their own fluids and pumps. In short, blood vessels (arteries and veins), blood (later including hemoglobin for carrying oxygen), and hearts evolved. *Excretory organs*, such as kidneys, capable of filtering the blood and increasing the efficiency of eliminating wastes were also added.

The evolution of efficient digestive, circulatory, and excretory systems ensured that ample energy was available to the animals that came to possess them. However, these systems would not have evolved if there had not been simultaneous evolutions of other physiological systems that could make use of the energy and give the animals advantages in the competition for survival. Of these, the most important ones were the *muscular, skeletal*, and *nervous systems*.

Evolution of Muscular, Skeletal, and Nervous Systems

Improvements in the function and distribution of muscle tissues allowed animals to carry out their movements with more force and agility, including the movements of avoiding or fending off adversaries, grabbing food, swallowing it, and pushing it through the digestive tract. However, body movements could not be carried out efficiently as long as animals had nothing but soft bodies. There had to be components with a certain degree of stiffness to which the muscles could be anchored and against which they could pull during contraction. Initially, this stiffness was provided by the confinement of fluid within the body cavity. It was a stiffness comparable to that of an inflated inner tube, which also is due to the confinement of fluid (that is, air). This kind of stiffening, known as a *hydrostatic skeleton*, was quite limiting and permitted only relatively slow and clumsy movements. Only the development of new skeletal systems based on hard, strong components connected by movable joints promised improvement. In time, such systems did evolve, as, for example, the chitinous skeletons of the insects and crustaceans, the shells of the mollusks, and the cartilaginous and bony skeletons of the chordates.

Finally, improvements in the functioning of the nervous system made possible adaptive responses by the other newly developed systems, from digestion to circulation, excretion, and muscle action. The improvements of the nervous system were due to a number of different factors. The nervous system became further centralized. Brains evolved. Sense organs were refined for improved sight, smell, taste, hearing, touch, and awareness of temperature. And nerve cells, including their connections (synapses), became more efficient in delivering sensory information to the brain, interpreting that information, and delivering the response information to the various organ systems.

DIVERSIFICATION OF THE ANIMAL KINGDOM: ORIGIN OF THE MAJOR PHYLA

The increase in organ system complexity just described took place in most animal lineages that evolved beyond the flatworm stage. However, it did not follow the same course in all of them. For example, the flatworms were bilaterally symmetric with a head and a tail, and their flat structures were well suited for creeping along on the bottom of the sea in search of food. But creeping along on the sea bottom was only one way among many of finding nutrients and fending for survival. Burrowing in the mud, moving along the sea floor with the aid of legs, and swimming in the water were other adaptations. Most important, these differences in locomotion contributed significantly to the evolution of structural differences among the species and to the diversification of the animals into numerous distinct phyla.

Many kinds of adaptations were tried out by the early descendants of the ancestral flatworms. Some proved successful, and the species experimenting with them survived and evolved. Other adaptations were failures and eventually faded from the scene. Clearly, this book cannot do justice to the full range of this early evolutionary experimentation. I will consider only those adaptations and phyla that turned out to be most successful with the regard to the number of species they produced, the range of habitats that were invaded as a result of them, and the general impact they had on subsequent evolution.

Burrowers: Nemertea, Nematoda, and Annelida

Burrowing is a technique that allows animals to consume the microorganisms and other food particles in mud. A roundish, elongated shape that tapers at both ends and has no appendages is particularly suited for this way of life. A tough, flexible outer covering, a hydrostatic skeleton, and muscles that permit waves of contractions to run the length of the body during locomotion are further advantages. Animals that adopted this burrowing kind of life style evolved into the phyla Nemertea (proboscis worms), Nematoda (round worms), and Annelida (segmented worms).

The *proboscis worms* are of particular interest for they are the simplest animals past the flatworm stage that have both mouths and anuses and, hence, genuine digestive tracts (see figure 8.20). They also have circulatory systems, but no hearts. Their most distinctive characteristic is a long muscular tube, the *proboscis*, located in the anterior part of the body, that can be extended for capturing prey and aiding in defense. From the simple nature of the organ

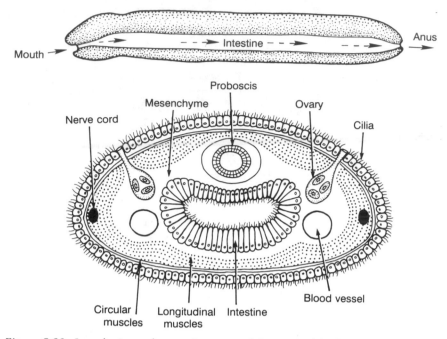

Figure 8.20. Lengthwise and crosswise views of the internal body construction of a *nemertean* (*proboscis worm*), the most primitive modern animal possessing an anus.

systems of these worms, we may conclude that they branched off very early from the trunk lineage that led to the more complex animals.

The *round worms* have become quite a successful phylum, judging from the sheer numbers of them. They are found in the sea, in fresh water, and in the soil, and they range from the tropics to the polar regions. A spadeful of garden soil may teem with millions of them. Most round worms are very small, but some attain lengths of up to one meter. Their bodies are approximately round, with a tough outer covering secreted by the underlying ectoderm. Unlike most other animals, including the flatworms and proboscis worms, the round worms possess only longitudinal muscles. This limits the type of movements they can execute. In water they thrash about in an apparently random and aimless manner, but in mud, which offers more resistance, the movements are undulatory and resemble the gliding locomotion of snakes. Interestingly, these worms have neither a circulatory nor an excretory system, although their digestive tract does have a mouth and an anus. Like the proboscis worms, the round worms are believed to date back to near the beginning stages of the animal kingdom.

Of all the burrowers, the *annelids* or *segmented worms* have the best-suited bodies and muscle constructions for life under ground. The most familiar present-day members of this phylum are the earthworms (see figure 8.21) and leeches, though there are also a great number of less well-known marine and freshwater species.

As the name implies, the bodies of the segmented worms are divided into segments, which are nearly all alike except for the head and tail segments. Each

Figure 8.21. The earthworm. (*a*) The segmented body construction of this animal. (*b*) Cross section through one of the earthworm's segments, showing, besides cell layers and organs, the *coelom*.

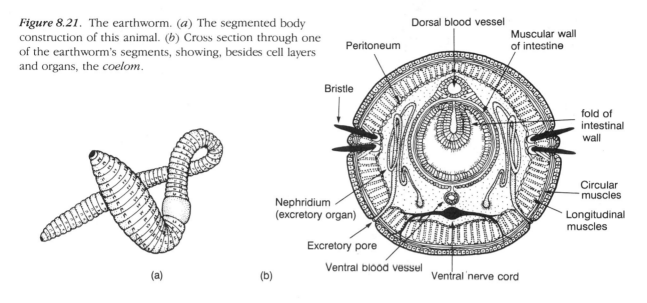

(a)   (b)

segment is separated from its neighbors by membranes called *septa*. Each segment has its own circular and longitudinal muscle tissues, nerve ganglia, blood vessels, and excretory organs. Only the digestive system is independent of the segmentation. It is differentiated, from front to rear, into a mouth, pharynx, esophagus, intestine, and anus.

Apart from segmentation or *metamerism*, as this feature is also called, the other novel characteristic of the annelids is a cell layer, the *peritoneum*, that completely lines the body cavity between the other body wall and the inner digestive tract. This lined body cavity is called a *coelom* (derived from the Greek *koilos*, meaning "hollow") and contains most of the organs and organ systems, as well as the body fluid. It is not present in any of the animals discussed so far, at least not in fully developed form. It is a characteristic of only the more complex animals, including the annelids and the other animals discussed below.

One great advantage of segmentation, and probably one of the reasons for its evolution, was its great usefulness in pushing the body through mud. Each segment, filled with coelomic fluid confined by the peritoneum and septa, functions as a hydrostatic skeleton that can be narrowed by contraction of the circular muscles and distended by contraction of the longitudinal muscles. Coordination of the muscle actions among the segments creates waves of body extensions and contractions—called *peristalsis*—that slowly and repeatedly work their way from the rear forward and push the animal's front end through the mud (see box 8.3). Without segmentation and local confinement of the coelomic fluid, muscle contractions would squeeze the fluid forward as well as backward along the length of the body, with little control and without providing a focused forward push at the head.

One of the least specialized and simplest of the annelid worms is *Nereis* (of the class Polychaeta), illustrated in figure 8.22. This sea-dwelling worm is an excellent example of metamerism. Each of its segments possesses, in addition to the features described above, a pair of short footlike appendages with horny bristles (called parapodia) that serve as paddles in swimming. Because of its

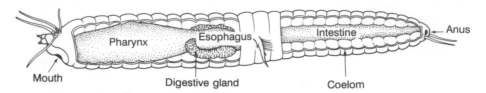

Figure 8.22. The primitive annelid worm *Nereis*, an excellent example of segmentation, or metamerism. The segments are nearly all alike, with the exception of those in the front and rear and the digestive system, which is differentiated and runs lengthwise through the animal.

simplicity, *Nereis* is thought to have a structure similar to that of the early ancestors of the annelid phylum.

Crawlers: Arthropoda

Crawling across the ancient sea floor on legs provided access to the food in the uppermost layers and on the surface of the mud, as well as in the water just above it. The earliest of the crawlers are believed to have been descendants of primitive annelid worms that had given up their burrowing existence. Like the annelids, these ancient crawlers were also segmented. Furthermore, each of their segments had a pair of legs much like the parapodia of *Nereis*, and they possessed external skeletons made of a tough, horny substance called *chitin*. These animals belonged to the phylum Arthropoda (see figure 8.23).

From their initially crawling marine mode of life, the arthropods evolved and diversified into the most successful of all phyla, with the largest numbers of species and individuals. They spread into virtually all geographic areas and habitats and today comprise such diverse classes as the crustaceans (lobsters, shrimps, prawns, crabs, barnacles), centipedes, millipedes, arachnids (spiders, scorpions), and insects. The phylum also includes the trilobites, an extinct class that flourished in the seas of the Cambrian and Ordovician periods and did not become extinct until the close of the Paleozoic era.

The chitinous external skeleton of the arthropods probably originated from the rather thin, tough outer covering of the ancestral annelids. In the course of further evolution and adaptation to new habitats, this exoskeleton became not only the support structure for the arthropod body, but it also found application as protective armor, biting jaws, piercing beaks, walking legs, pincers, wings, tactile sense organs, sound-producing and auditory organs, optical lenses, and numerous other specialized structures. In short, chitin became to the arthropods what steel and, more recently, plastic have become to industrialized man.

The fossil creature *Spriggina*, mentioned earlier, and the members of the presently living genus *Peripatus* (which live in tropical forests and belong to the small phylum Onychophora) have external and internal body constructions intermediate between those of the annelids and the arthropods, which supports the view that these two phyla are closely related (figure 8.19). For example, *Peripatus* has a segmented body with numerous short, unjointed legs and a thin flexible outer covering like the annelids. But it also has claws and an open

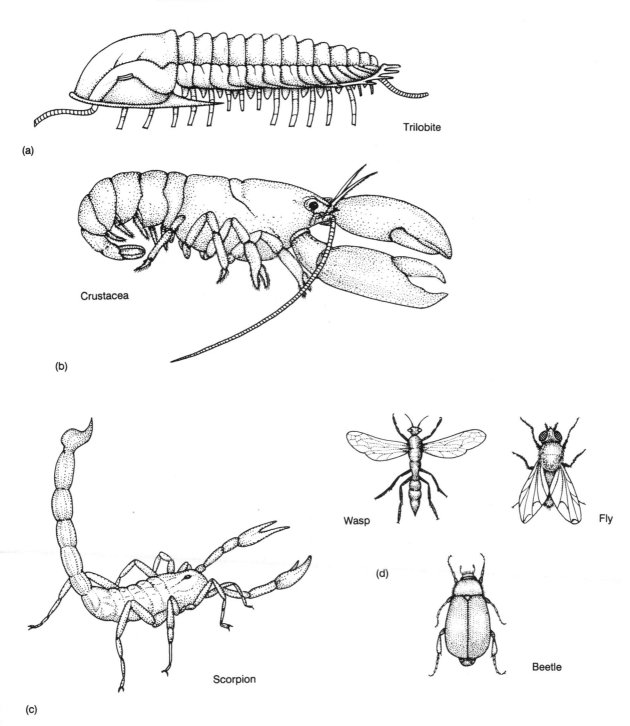

Figure 8.23. Representatives of the major classes of the phylum *Arthropoda*.
(*a*) Trilobite (extinct). (*b*) American lobster (class Crustacea). (*c*) Scorpion (class
Arachnida). (*d*) Insects (class Insecta).

BOX 8.3.
Animal Locomotion

THE ability to move about is a characteristic shared by most animals. It allows them to go after food, find a mate, escape from danger, seek shelter, and in general carry out the countless other activities that are part of daily life.

Since the origin of complex multicelled life more than 670 million years ago, many forms of animal locomotion have evolved. Some are rather specialized adaptations to a particular way of life. For example, hydras (polyp cnidarians) move about by somersaulting. Squids and octopuses propel themselves by forcefully ejecting water through an opening near their fronts, much like the way a jet engine propels an aircraft by ejecting heated gases. Insects, birds, and bats fly. And humans walk on two legs. Many other animals still employ the basic techniques of locomotion that date back to the earliest stages in the evolution of the animals and their diversification into distinct phyla—burrowing (peristalsis), creeping (pedal locomotion), crawling or walking, and swimming. Let us look at these four techniques in more detail.

Peristalsis (from the Greek, meaning "to constrict" or "to place around") is the form of locomotion used by segmented invertebrates, such as the earthworms and their relatives, to burrow through soil and mud. It consists of rhythmic wavelike body deformations that run from the front of the animal to the rear. The body deformations are due to alternating contractions of the circular and longitudinal muscles, in one body segment after the other. The contractions of the circular muscles make the segments thinner and longer, while the contractions of the longitudinal muscles make the segments thicker and shorter. The

resulting body waves slowly push the animal forward, as illustrated in figure A.

In addition to muscle contractions, a number of other factors contribute to making peristalsis an effective way of locomotion underground. The body segments that are compressed into the thick form are stationary and become firmly wedged by friction into the excavated burrow, thereby preventing the animal from sliding backward. The segments that are stretched into the thin form are the ones that move forward. They do so with little friction because they are thinner than the burrow. Finally, the animal's fluid-filled segments (that is, the hydrostatic skeleton) act like a hydraulic system that translates all muscle contractions into high pressure, distributed over a relatively small area, at the animal's tapered front end. This pressure allows the animal to force its way through even heavily compacted soil.

Pedal locomotion (also known as "muscle sheet movement") is the locomotion by which snails, slugs, some flatworms, and numerous other invertebrates move across the ground, rocks, and other objects. It is a form of creeping that results from waves of muscular contractions passing either forward or backward along the animal's ventral surface (the muscular foot in the case of snails and slugs).

Forward waves begin by raising the back end of the underside, compressing it, and putting it down slightly forward. The wave, or fold, that results works its way slowly forward until it reaches the animal's front end. In forward waves the muscles of the raised

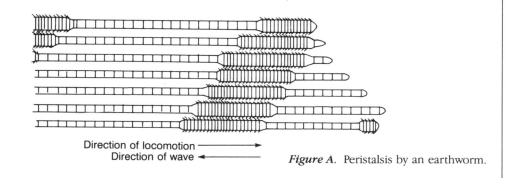

Direction of locomotion ⟶
Direction of wave ⟵

Figure A. Peristalsis by an earthworm.

BOX 8.3 *continued*

Figure B. Pedal locomotion by a snail, using forward waves.

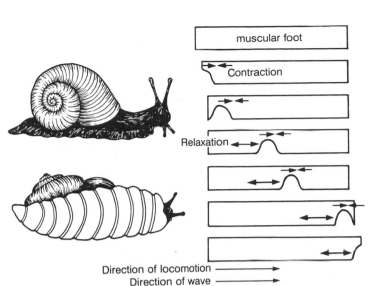

Direction of locomotion ⟶
Direction of wave ⟶

fold contract, while those in contact with the ground remain relaxed, as illustrated in figure B.

Backward waves begin by raising the front end of the foot, extending it, and putting it down slightly forward. This also creates a wave or fold, but this wave works its way slowly backward, as section after section of the foot is raised and extended forward. In backward waves the muscles of the foot in contact with the ground contract, while those of the raised fold are extended.

Crawling or walking is carried out with the aid of legs. Muscles pull or push the legs backwards, then raise the legs and bring them forward. During the backward motion, the legs are in contact with the ground and move the animal forward.

Compared with peristalsis and pedal locomotion, crawling and walking are very energy-efficient means of covering a given distance. The chief reason is that the bulk of the body is kept in steady forward motion, while only the legs, with their relatively small mass, go

through cycles of acceleration and deceleration as they move back and forth. In addition, legs act as levers that, depending on their length and structure, can translate relatively short and slow muscle contractions into great locomotory speeds. Furthermore, vertically directed legs hold the animal's body off the ground, which greatly reduces friction and energy expenditure during locomotion.

Over the ages, crawling and walking have evolved independently numerous times in different phyla, such as in the annelids (for example, *Nereis*), arthropods, echinoderms, and the vertebrates (for example, the land-living tetrapods). The fundamental principle of moving with the aid of legs is the same in all these examples, but the details vary greatly. They depend on the nature of the animal's skeleton, the size and structure of the legs, the number of legs, the strength and geometrical arrangement of the muscles, and many other factors. Figures C and D illustrate the

direction of locomotion ⟶
direction of push by parapodia ⟵

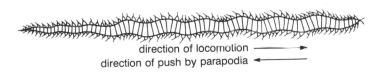

direction of locomotion ⟶
direction of push by legs ⟵

Figure C. Walking locomotion by *Nereis*.

Figure D. Walking locomotion by the centipede *Scutigera*.

BOX 8.3 continued

walking locomotions of two invertebrates — those of *Nereis* and of the centipede *Scutigera*.

Two distinct mechanisms of *swimming* have evolved among the animals. One is based on the ejection of water from a body cavity. This mechanism is used by invertebrates, such as the squids and octupuses mentioned above, the medusa type cnidarians (for example, jellyfish), and many of the bivalve mollusks (for example, scallops and clams). The other mechanism is based on the forward thrust obtained from undulatory body motions. Both invertebrates and vertebrates use this mechanism, such as the marine annelids, some of the mollusks, fishes, and mammals that have returned to the sea (for example, whales and dolphins). The undulatory mechanism of swimming is by far the most common way of moving through water among the animals.

Animals using the undulatory mechanism of swimming are usually elongated and many have the rear part of their body widened into a flat vertical flap as, for example, the caudal fins of fishes. In both invertebrates and fishes, the undulatory motions are created by the alternating contraction of the longitudinal musculature on the right and left sides of the body. This creates body waves that start in the front and pass backward along the body axis (see figure E), increasing both in amplitude and speed as they approach the tail section. These left and right sweeping motions displace water backward and thereby produce a forward thrust on the animal, much like the way the side-to-side motions of a single oar at the stern of a boat provides a forward thrust. In the case of whales and dolphins, the undulatory motions are not from side to side, but up and down and their caudal fins are oriented horizontally.

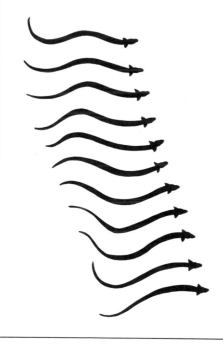

Direction of locomotion
Direction of undulatory body waves

Figure E. Swimming locomotion of an eel.

circulatory system similar to that of the arthropods. The adult anatomy of today's centipedes offers additional evidence of an ancient kinship between the arthropods and the annelids. The centipede body is composed of a head and an elongated wormlike trunk with many leg-bearing segments, similar to those of *Nereis* (see figure 8.24). Still further evidence for a common origin of the

Figure 8.24. A centipede (class *Chilopoda*). Centipede means "hundred-legged," though the number of pairs of legs of centipedes ranges from 15 to more than 170. Each trunk segment bears a single pair of legs, of which those on the first segment are modified as large poison claws.

arthropods and annelids comes from similarities in their embryological developments, as will be discussed shortly.

Pedal Locomotors: Molluska

Crawling on legs was not the only possible mechanism of locomotion across the ancient sea floor. For example, the flatworms had no legs and moved about by creeping. So did the members of another phylum, the mollusks. Their entire undersides consisted of a thick muscular "foot," with which they created shallow wavy protrusions that allowed them to creep over rocks and other hard surfaces. This way of getting about is called *pedal locomotion.*

The muscular foot and other structural features of the mollusks have proved remarkably adaptable to many different ways of life. As a result, these animals have become highly successful and today they constitute the second largest phylum—Molluska—after the arthropods (see figure 8.25). It includes such apparently dissimilar classes as the gastropods (slugs, snails), pelecypods (clams, oysters, mussels, scallops), and cephalopods (squids, octopuses, nautiluses); and its habitats range from the sea to the shore, freshwater, and moist land everywhere.

Despite the mollusks' great variation in body structure, they are all built according to the same basic plan. This plan is best illustrated by the simplest known member of this phylum, the chiton (figure 8.26). This sluggish animal, which lives on microscopic algae growing on rocks near the seashore, consists of four distinct parts (see figure 8.27): (1) a large ventral muscular *foot* used in

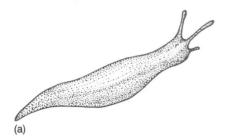
(a)

(b)

(c)

Figure 8.25. Representatives of the major classes of the phylum *Molluska*.
(*a*) Snails and a slug (Gastropoda).
(*b*) Mussels and oysters (Pelecypoda).
(*c*) Octopus (Cephalopoda).

Figure 8.26. A *chiton*, the simplest of living mollusks.

locomotion, (2) a *visceral mass* above the foot, which contains most of the organs, (3) a specialized tissue layer, called the *mantle*, that covers the visceral mass, and (4) a *shell* secreted by the mantle which, in most chitons, consists of eight overlapping plates. The shell is composed of a horny substance reinforced with calcium carbonate and attached to the body by a set of retractor muscles that permit the chiton to pull it down tightly over its body.

Very likely, the early mollusks were grazers at the bottom of the sea and resembled today's chitons. The evolution of this phylum into the diverse range of species characterizing it today was a consequence of the animals' invading different habitats and facing various environmental challenges and opportunities. For example, many species remained grazers, but their protective shells evolved beautiful spiral patterns and became large enough that the animals could completely retreat into them, as most snails do. Other species became diggers and filter-feeders. Their shells evolved into two hinged halves that offered protection from all sides, as in the clams and scallops. Other species, such as the nautiluses, took up swimming and hunting, and their shells became enlarged so as to include gas-containing floatation chambers for buoyancy control. And still other species carried the swimming mode of life even further and became active, intelligent hunters. Their shells became greatly reduced or lost all together, and the foot evolved into a crown of large and rather versatile tentacles surrounding the head, as in squids and octopuses.

Embryological studies suggest that the mollusks are related to the annelids and arthropods. For example, the single opening of the gastrulas of all three phyla becomes the mouth, while the anus develops later as an entirely new opening. For this reason, mollusks, annelids, and arthropods are collectively called *protostomes*, meaning "mouth first." Furthermore, the early cleavage of the zygotes of the mollusks and annelids follows a spiral pattern (see figure 8.28), and the cells resulting from cleavage show deterministic differences.

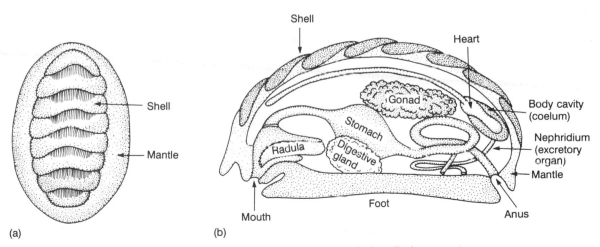

Figure 8.27. The body construction of the chiton, representative of all mollusks. (*a*) Dorsal view. (*b*) Section showing the animal's foot, the visceral mass containing most of the organs, the mantle, and the shell.

Deterministic means that if the two cells resulting from the first cleavage of a zygote are separated, each will develop into a fixed half of the gastrula, but neither is capable of developing further without the other. Put differently, identical twins do not occur. Still another common embryological feature of the mollusks and annelids arises during the free-living larval stage, through which many animals pass. Their larvae, which are called *trochophores*, distinguish themselves by a wheellike band of cilia around the middle (the term *trochophore* is derived from the Greek roots *trochos* and *pherein*, which mean "wheel" and "to carry," respectively). Interestingly, arthropods have neither spiral cleavages nor do their larvae resemble those of the mollusks and annelids, though they are protostomes.

Despite the fact that mollusks, annelids, and arthropods share certain basic embryological characteristics, which point to a common phylogenetic origin, their relationship is not a close one. Very probably, the mollusks diverged from the other two phyla long before either typically molluskan or annelidan and arthropodan characteristics appeared in the ancestral lineages, for most mollusks show little sign of segmentation. However, a few zoologists think the relationship is closer. As evidence they point to the eight-sectioned shell of the chiton and the segmentation of the monoplacophoran mollusk *Neopilina*.

There is an interesting story that goes with *Neopilina*. For many years this creature, which resembles both the chitons and the gastropods, was thought to be extinct, for the only evidence of its existence were fossils from Cambrian to Devonian times. Then, in 1952, Danish scientists dredged up ten living specimens of this animal in the Pacific Ocean and, in the meantime, other specimens have been found in the South Atlantic and the Gulf of Aden. It is an animal adapted to life at great depths and the structures of its gills, retractor muscles, and excretory system are segmented.

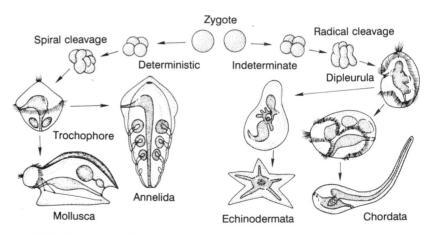

Figure 8.28. Embryological characteristics in mollusks, annelids, echinoderms, and chordates. In mollusks and annelids the early cleavage of cells follows a spiral pattern, their cells are deterministic (identical twinning does not occur), and their free-living larvae are typically trochophores (that is, they have radial symmetry). In echinoderms and chordates, the early cleavage of cells follows a radial pattern, the cells are indeterminate (identical twinning can occur), and their free-living larvae are typically dipleurulas (that is, they have bilateral symmetry).

Swimmers: Chordata

One form of locomotion and source of phylogenetic diversity remains to be discussed—swimming.* Swimming gave its practitioners access to all levels of the ancient sea. Most species that adopted this form of locomotion belonged to the phylum Chordata. In time, they evolved into three subphyla: the Urochordata (tunicates), Cephalochordata (lancelets), and Vertebrata (jawless, cartilaginous, and bony fishes, amphibians, reptiles, birds, mammals).

The distinguishing characteristics of the early chordates were

1. A flexible rod, called the *notochord* (whence the name Chordata), that ran lengthwise through the dorsal part of the body,

2. A hollow dorsal nerve cord that lay just above the notochord and extended from the anterior brain to the rear part of the body, and

3. *Gill slits* through which water flowed from the mouth and pharynx.

The notochord functioned as a support for the soft tissues and as an anchorage for the muscles used in swimming. The nerve cord served as a pathway for the transmission of information signals from the sense organs to the brain and response signals from the brain to the muscles, glands, and other organs. The gill slits were initially used for filtering small food particles from the water, but in time they evolved into an organ for extracting oxygen from the water.

These three characteristics—notochord, hollow nerve cord, and pharyngeal gill slits—are still the distinguishing marks of all chordates today (see figure 8.29), though they are not present in the adult forms of all species. For example, in humans the notochord and gill slits show up only temporarily during early embryological development.

One of the simplest of today's chordates, and an example of what the early ancestors of this phylum might have been like, is the *lancelet*, or *amphioxus*. This small animal (typically 5 cm long) is a semitransparent fishlike creature, found in shallow marine waters around the world. Although it can swim with

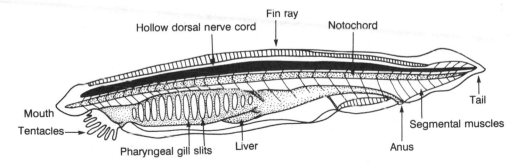

Figure 8.29. Longitudinal section of the body of the lancelet, or amphioxus, illustrating the notochord, hollow dorsal nerve chord, and pharyngeal gill slits found in all chordates.

*Note that swimming in the case of the nautiluses, squids, and octopuses is a secondary adaptation that evolved from the initial creeping mode of life of the ancestral mollusks.

typical fishlike undulations of its body, most of the time it lies half-buried in the sand or mud and feeds by straining microorganisms out of the water it draws into its mouth. Its musculature is V-shaped, segmented into bundles,* and strongly developed, and its elongated body ends in a tail that extends well beyond the anus. The circulatory system is closed, but the heart is much simpler than that in adult vertebrates. It is a simple pulsating tube, just as in the embryos of all vertebrates, that pumps blood through the arteries, capillaries, and veins. After the blood passes through the capillaries surrounding the intestine, it passes through the capillaries of the liver before returning to general circulation. All of these features are adaptations to the swimming way of life that are not generally found in other phyla, and they point to a common origin of amphioxus and the vertebrates.

Before concluding this section, consider a few interesting embryological characteristics that distinguish the chordates from the mollusks, annelids, and arthropods. For example, the single opening of the gastrulas of chordates becomes the anus, and the mouth develops later. Hence, chordates are called *deuterostomes*, meaning "mouth second." The early cleavage of the zygotes of chordates follows a *radial* pattern, and cells resulting from cleavage are *indeterminate* (see figure 8.28). If these cells (after the first few divisions) are separated, each will develop into a normal adult. Identical twinning, quadrupling, and so on, are therefore possible. Furthermore, the free-living larvae of the chordates are typically *dipleurulas*, meaning they have bilateral symmetry.

The chordates share many embryological characteristics with another common phylum, the *echinoderms*, which also are classified as deuterostomes. The echinoderms are marine animals with basically a sessile way of life, though most are capable of slow movements across the sea floor. Most of them are characterized by radial symmetry in body construction, which is a secondary adaptation to sessility and probably evolved from bodies that initially were adapted to a swimming way of life. The echinoderms include such well-known members of the intertidal zone as the sea stars, sea lilies, sea urchins, and sea cucumbers (see figure 8.30). The embryological similarities between the chordates and echinoderms suggest that these two phyla are related, though clearly they diverged very early. Likewise, the differences in embryological development between the protostomes and deuterostomes suggests that they diverged even earlier.

THE FOSSIL RECORD

The evolution of multicelled life into distinct phyla was already well under way 670 million years ago. During the Ediacarian period this diversification accelerated and by about 570 million years ago, at the beginning of the Cambrian period, most of the 35 phyla that are known to have evolved during the Phanerozoic eon (of which 26 survive today) had become established. The

---

*Segmentation in the vertebrates is evident not only in their muscle development, but also in the vertebral column and rib cage construction. There is no reason to believe that segmentation in amphioxus and the vertebrates is related to segmentation in the annelids and arthropods. It is an adaptation that proved useful and probably arose independently in the chordates.

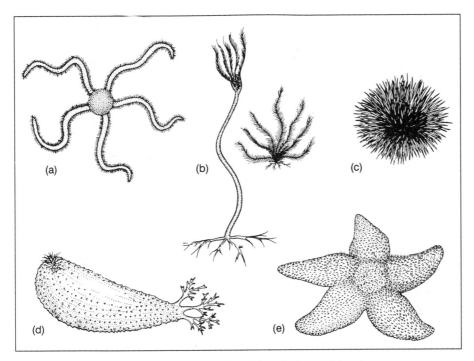

Figure 8.30. Representatives of the classes of the phylum *Echinodermata*. (*a*) Serpent star (Ophiuroidea). (*b*) Sea lilies (Crinoidea). (*c*) Sand dollar and sea urchin (Echinoidea). (*d*) Sea cucumber (Holothuroidea). (*e*) Sea star (Asteroidea).

evidence of this explosive diversification comes from the fossils that are abundant in Cambrian rocks all over the world.

However, nowhere are the Cambrian fossils better preserved than in a thin band of shale on the slopes of Mt. Wapta, just north of Burgess Pass, high in the Canadian Rocky Mountains (about 10 km, and a 1000-meter elevation gain, by trail north of Field, British Columbia). The Burgess shale, as it has become known, was laid down at the bottom of a shallow sea some 550 million years ago by a series of mud slides from overlying submarine shelves. The slides swept along all kinds of creatures that lived on those shelves and buried them alive. No oxygen reached them afterwards and, hence, no scavengers or bacteria fed on them. That is how their fossil remains became so exceptionally well preserved, showing the finest details, including delicate appendages, internal organs, and the contents of the guts in some of them.

These fossils were discovered accidentally in 1909 by Charles D. Walcott while he was traveling through the Canadian Rocky Mountains and examining the extensive Cambrian rock formations there. (He is the same person who discovered the fossil stromatolites in the Gunflint Iron-formations; see chapter 7.) In the area north of Burgess Pass, Walcott stumbled over a slab of shale with a perfectly preserved trilobite. That aroused his interest and he began to search for more fossils. During the following summers he returned to the task and by 1917 had collected nearly 50,000 specimens belonging to some 130 species and

Figure 8.31. A group of trilobite specimens from the Jinetz Shales of the Middle Cambrian, Jince, Czechoslovakia. These early marine arthropods were abundant during the Lower Paleozoic. They survived until the Permian, roughly 250 million years ago.

70 genera. There are jellyfish and other cnidarians, various worms, and arthropods, as would be expected from the presence of similar animals among the fossils of the earlier Ediacarian period. However, whereas the earlier animals were all soft-bodied, a great many of the Cambrian animals possessed shells or hard skeletons.

The most common fossils in the Burgess shale and other Cambrian deposits are the *trilobites* and their relatives. The bodies of these now-extinct arthropods were oval in shape with three characteristic lobes (whence their name) running longitudinally down their backs (see figures 8.31 and 8.23). In addition, a head can be distinguished, as well as a trunk with a variable number of segments. Each segment had a pair of jointed legs and feathery gills on a branch of the legs. At the front of the head were two feelers. Furthermore, there was a gut running lengthwise through the body, muscle fibers stretching along the back, and a body armor constructed of calcium carbonate and chitin. Judging from their abundance in the Cambrian fossil record, the trilobites must have been dominant species over wide stretches of the ancient sea floor. They crawled and some of them probably paddled about with their jointed limbs, scanning their environment with large compound eyes and feeding on algae and small animals.

A second abundant group among Cambrian fossils are the worms, in particular annelids. An example is *Canadia* of the class Polychaeta (see figure 8.32). Its general body shape and short parapodia with tufts of bristles indicate that it may have been an early relative of *Nereis*. Other common Cambrian worms were the *priapulids* (meaning "little phallus"), which are very rare today. They seem to have been segmented, yet sufficiently different from the annelids to constitute a separate phylum. They were predators that captured prey with a hook-bearing proboscis and chewed it with chitinous teeth that lined their pharynx. One species of priapulid fed on small shellfish and worms, including others of its own kind. These eaten animals were found undigested and fossilized in the priapulids' guts. Another interesting Cambrian worm was *Aysheaia* (see figure 8.33). It resembled today's *Peripatus* and, like it and *Spriggina* of the Ediacarian period, showed similarities to both the arthropods and annelids.

A third very common group of animals of Cambrian times were the *brachiopods** or lampshells (so named because of a vague resemblance to early

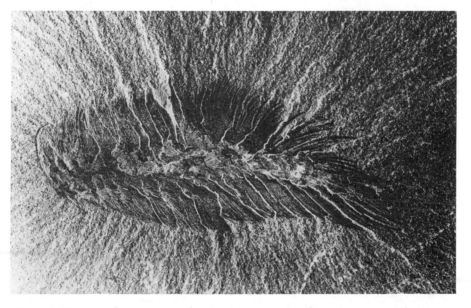

Figure 8.32. *Canadia spinosa*, a fossil polychaete worm from the Burgess shale and, possibly, an early relative of *Nereis*.

*Biological terms are often derived from Latin and Greek sources, and it is both instructive and interesting to know their literal meanings. For example, the names of the phyla introduced in this chapter mean the following:

Annelida — adorned with rings
Arthropoda — jointed feet
Brachiopoda — arm-foot
Chordata — chord or string
Cnidaria — like a nettle
Echinodermata — prickly skin
Mollusca — soft
Nematoda — thread

Nemertea — one of the Nereids (in Greek mythology, the Nereids are sea nymphs who attend Poseidon, the god of the sea)
Onychophora — claw carrier
Platyhelminthes — flat worm
Porifera — pore bearer
Priapulida — little phallus

Figure 8.33. *Aysheaia pedunculata*, a fossil segmented worm with parapodia from the Burgess shale. Like today's *Peripatus* and extinct *Spriggina*, this fossil animal shows similarities to both the annelids and the arthropods.

Roman oil lamps). Brachiopods are not numerous today. They have two horny shells and look superficially like small clams. However, while clams carry their shells vertically, brachiopods carry theirs horizontally, with ventral and dorsal halves. The brachiopod body plan is rather unique, the intestine is bent into a U-shape, and two coiled arms, or brachia, are attached on either side of the mouth. Each of the arms bears a row of tentacles with which the animal creates a current of water that passes through the mouth and carries along food particles as well as oxygen. Many fossil brachiopods are found entangled in the

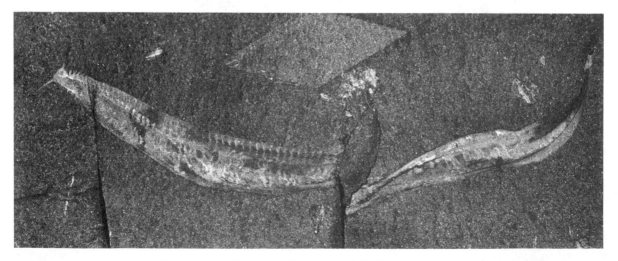

Figure 8.34. *Pikaia gracilens*, a small fishlike fossil animal from the Burgess shale.

siliceous spikes of sponges, suggesting that the animals attached themselves to those spikes for protection and feeding.

A rare, but from our human perspective very important, Cambrian fossil is *Pikaia* (see figure 8.34). It is important because its shape and internal construction resembled today's lancelets. It had an elongated streamlined body, a notochord, gill slits, and segmented muscle groups arranged in the V-shaped pattern typical of fishes. About thirty fossil specimens of *Pikaia* have been found in the Burgess shale. They are the earliest known members of the chordate lineage.

Besides the Cambrian fossils just described, there are several bizarre ones, with no resemblance whatever to any living species. One, *Opabinia*, was a creature with five compound eyes, a trunk in front of its head with grasping spikes at its tip, and overlapping plates along the sides of its segmented body. The plates in the front look as if they were used as paddles for swimming, while those in the back may have served as rudders. Another eccentric animal, *Hallucigenia*, seems to have crawled on seven pairs of stiff spines. From its back protruded seven pairs of waving tentacles, each with what appears to be a mouth as its end. These animals were experiments that proved unsuccessful and rather rapidly fell to the wayside.

The present chapter traced the evolution of complex multicelled life from its beginnings in the ancient sea some 670 or more million years ago, paying particular attention to the origin of the major animal phyla. The transition from microscopic, mainly single-celled to complex multicelled life was probably made more than once, giving rise to such diverse results as the plants, sponges, cnidarians, and flatworms.

The ancient flatworms are thought to be the ancestors of all of the more complex animals, from the annelids and most other worms to the arthropods, mollusks, echinoderms, and chordates. The phylogenetic differences among animals arose very early, probably as a result of differences in the selection pressures that were exerted on the early descendants of the ancient flatworms as they colonized the sea. Animals that sought their nourishment by burrowing in mud evolved into the worm phyla, those using pedal locomotion to move across the sea bottom evolved into mollusks, the crawlers evolved into arthropods, those that specialized in a sessile way of life evolved into echinoderms, and swimmers evolved into chordates.

In the course of time, the competition for survival increased steadily as the various phyla improved their particular adaptations and invaded each other's territories. The invasions brought animals with different adaptive specializations face to face and greatly increased the complexity and interdependency of the evolutionary process. All species, including microorganisms and plants, which also competed and evolved, affected each other in a multitude of minor and major ways, leading to the enormous diversification among the species that has characterized the Earth's ecosystem for the past 500 million years. At the same time, the evolution of the physical environment—such as the accumulation of oxygen in the atmosphere, the rising and receding of the oceans, the collision and breakup of continents, the building and eroding of mountain chains, the periodic warming and cooling of the climate, and, possibly, occasional bombardment by large meteorites — had its impact on biological evolution as well.

Many species were unable to keep up in this struggle. They were but fleeting experiments in the flow of evolutionary change and succumbed. Other species were temporarily successful. They managed to adapt and to evolve into new species as the physical and biological environments changed. Some of these lineages flourished for hundreds of millions of years and then either became extinct or were forced to retreat into restricted and specialized ecological niches. Still other lineages played rather insignificant roles early on, but eventually hit upon the right combinations of body structure and function that allowed their descendants to prosper and become dominant. Figure 8.35 summarizes this evolutionary pattern in the case of the most important animal and plant phyla. The following two chapters single out the phylum that eventually gave rise to the mammals and humans — the chordates — and discuss its evolution in more detail.

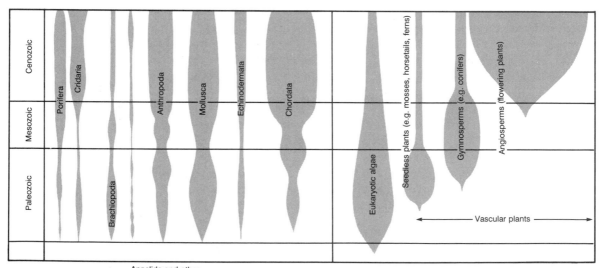

Figure 8.35. The fossil history of major animal and plant phyla. The widths of the columns correspond approximately to the relative numbers of fossil representatives from a given age.

Table 8.1. The Geologic Time Scale of the Phanerozoic Eon

Era	Period	Epoch	Duration of epochs (10^6 years)	Duration of periods (10^6 years)	Millions of years ago (from start of period)	
Cenozoic (65 million years duration)	Quaternary	Recent	0.01	2	2	Rapid evolution of hominids (human beings), rise of civilizations
		Pleistocene	2.0			
	Tertiary	Pliocene	3	63	65	Widespread diversification of mammals (beginning of Age of Mammals), birds, insects, and flowering plants
		Miocene	20			
		Oligocene	13			
		Eocene	17			
		Paleocene	10			

Major mass extinction, including the disappearance of the dinosaurs

Era	Period	Epoch	Duration of epochs (10^6 years)	Duration of periods (10^6 years)	Millions of years ago (from start of period)	
Mesozoic (160 million years duration)	Cretaceous			80	145	Major diversification of modern bony fishes; first placental mammals; end of the Age of Reptiles; second great diversification of insects; first flowering plants
	Jurassic			65	210	Flourishing of dinosaurs; first birds
	Triassic			35	245	First dinosaurs and mammals; beginnings of Age of Reptiles; first coniferlike plants

The most severe mass extinction in Earth history

Era	Period	Epoch	Duration of epochs (10^6 years)	Duration of periods (10^6 years)	Millions of years ago (from start of period)	
Paleozoic (345 million years duration)	Permian			40	285	Major diversification of reptiles; decline of amphibians; last of the trilobites and placoderms
	Carboniferous[a]			75	360	Major diversification of amphibians (Age of Amphibians); first reptiles; first great diversification of insects; great coal-forming forests of fernlike plants
	Devonian			50	410	Major diversification of fishes (Age of Fishes); invasion of land by vertebrates; first amphibians; first insects and seed plants
	Silurian			30	440	First fishes with jaws (acanthodians); invasion of land by plants and by arthropods
	Ordovician			65	505	Origin and expansion of many invertebrate classes

Note: All ages are approximate.

[a]In North America, the Carboniferous period is often divided into two separate periods: the Mississippian period (40 million years duration), corresponding to the Lower (earlier) Carboniferous, and the Pennsylvanian period (35 million years duration), corresponding to the Upper Carboniferous.

continued

Table 8.1. continued

Era	Period	Epoch	Duration of epochs (10^6 years)	Duration of periods (10^6 years)	Millions of years ago (from start of period)	
	Cambrian			65	570	First widespread appearance of organisms with hard skeletons; final stages of the major diversification begun during the Ediacarian period; first vertebrates (jawless fishes)
			Transition from soft-bodied organisms to organisms with hard skeletons			
——	Ediacarian			100	670	Earliest known fossil evidence of soft-bodied, multicelled marine organisms; rapid evolution of the major body plans of the animal phyla

Exercises

1 Collect a small sample of water from a stagnant, unpolluted pond, swamp, or other similar source (salt water will do, too). Place a droplet of the water on a microscope slide and cover with a coverslip. Examine the droplet through a microscope with 100–400 × magnification.

(*a*) Describe the different kinds of organisms you see, paying particular attention to their shapes. For example, look for ciliates (such as paramecia), flagellates (such as euglenas), volvoxes, and algae (many of which are filamentous). You may also see a background of tiny rod-shaped or spherical organisms, which are bacteria.

(*b*) Observe which of the organisms move about and describe their movements. (Some may be moving so rapidly that you will find it difficult to follow them.)

(*c*) View a clear plastic ruler with a mm scale through the microscope and determine the size of the field you are seeing. Express it in μm. (*Note:* 1 mm = 1000 μm.)

(*d*) Estimate the sizes of the organisms you described in (*a*).

(*e*) Estimate the speed with which some of the organisms move about by measuring the times it takes them to cover a given distance. Express your estimates in μm/sec or mm/sec.

2 Describe (*a*) flagellar, (*b*) ciliary, and (*c*) ameboid locomotion. Give examples of organisms specializing in each of them.

3 Name the four principal tissue types present in the more complex animals and give examples of each.

4 Compare the cellular, tissue, and organ system levels of organization of animals, and list some examples of animals of each type.

5 What is meant by "ontogeny recapitulates phylogeny?" Discuss why this so-called biogenetic law is not a generally valid law and needs to be applied with caution.

6 Describe and compare the two theories discussed in the text concerning the origin of flatworms and more complex animals.

7 (*a*) Describe the major components of the digestive tract as found in the more complex animals of today.

(*b*) Briefly describe the major developments in the origin and evolution of the digestive tract, as well as the advantages that each of these developments gave to the animals.

(*c*) Give examples of living animals that illustrate the major stages in the evolution of the digestive tract.

8 (*a*) Describe the four major types of animal locomotions discussed in the text: burrowing, creeping, crawling or walking, and swimming.

(*b*) What advantages does each of these types of locomotion offer?

(*c*) Give examples of animals specializing in them.

9 (a) Describe the main functions of chitin, shells, cartilage, and bone.
   (b) Give examples of animals in which these organic building materials are used.

10 Biologists explain such "inventions" in animal phylogeny as the anus, coelom, stiff skeletal systems, segmentation, and walking legs, as being the result of natural selection. Why then do virtually all ecosystems contain organisms that lack these advantageous features?

11 (a) Name the three defining characteristics of the chordates and discuss the functional significance of each.
   (b) One or more of these characteristics are absent in the adult forms of some living chordates. Give examples of this fact and explain why such species are still classified as chordates.

12 Name the six periods of the Paleozoic era, list their durations, and describe the major evolutionary events that took place during each of them.

Suggestions for Further Reading

Attenborough, D. 1979. *Life on Earth: A Natural History*. Boston: Little, Brown and Co. A popular exposition of animal and plant life, both of today and of the past, with excellent color photographs.

Berg, H. C. 1975. "How Bacteria Swim." *Scientific American*, August, 36–44. The author explains that bacteria do not swim by beating or waving their flagella, but by rotating them like propellers. The article begins with an historical summary of observations of bacteria through the microscope, starting with the work of Anton van Leeuwenhoek (1632–1723) of Holland.

Buchsbaum, R. 1976. *Animals Without Backbones: An Introduction to the Invertebrates.* 2d ed. Chicago: The University of Chicago Press. This classic text on invertebrates is surprisingly readable considering its detailed coverage.

Case, G. R. 1982. *A Pictorial Guide to Fossils*. New York: Van Nostrand Reinhold Co. An extensive collection of photographs of fossils, selected to be useful to students as well as to the professional paleontologist.

Fortey, R. A. 1982. *Fossils: The Key to the Past*. New York: Van Nostrand Reinhold Co. A broad overview of paleontology, recommended as a first introduction to the subject. The book is concisely and clearly written and includes many superb photographs of fossils, most of them from the British Museum (Natural History).

Garcia-Bellido, A., P. A. Lawrence, and **G. Morata.** 1979. "Compartments in Animal Development." *Scientific American*, July, 102–110. Animals seem to be composed of a number of compartmental units, within which key genes are active and determine the development to the adult form.

Glaessner, M. F. 1961. "Pre-Cambrian Animals." *Scientific American*, March, 72–78. The Precambrian fossil discoveries in the Ediacara Hills of South Australia, by one of the key participants.

Grant, P. 1978. *Biology of Developing Systems*. New York: Holt, Rinehart and Winston. A college-level text on cell biology, genetics, and biological development.

Levi-Setti, R. 1975. *Trilobites: A Photographic Atlas*. Chicago and London: University of Chicago Press.

Macurda, D. B., Jr., and **D. L. Meyer.** 1983. "Sea Lilies and Feather Stars." *American Scientist*, July–August, 354–65. Living crinoids provide a window on an ancient class of marine invertebrates.

Milne, D., et al., eds. 1985. *The Evolution of Complex and Higher Organisms*. Washington, D.C.: Scientific and Technical Information Branch, NASA (SP-478). A report prepared by the participants of workshops held in 1981 and 1982 at the NASA Ames Research Center (Moffett Field, Cal.) on the evolution of complex terrestrial life, the effects of extraterrestrial events on this evolution, and whether knowledge of such events could contribute to a better understanding of the nature and distribution of complex extraterrestrial life. Remarkably readable, considering the advanced nature of the report.

Morris, S. C., and **H. B. Whittengton.** 1979. "The Animals of the Burgess Shale." *Scientific American*, July, 122–33.

Richardson, J. R. 1986. "Brachiopods." *Scientific American*, September, 100–106. Though not numerous today, these clamlike bivalves are not as sedentary or evolutionarily stagnant as has been thought.

Satir, P. 1974. "How Cilia Move." *Scientific American*, October, 44–52.

Skinner, B. J., ed. 1981. *Paleontology and Paleoenvironments: Readings from American Scientist*. Los

Altos. Cal.: W. Kaufmann. A collection of 21 articles published in the *American Scientist* between 1953 and 1980 on topics from the origin and diversification of life to invertebrate and vertebrate paleontology and ancient geographies.

Stanley, S. M. 1984. "Mass Extinctions in the Ocean." *Scientific American*, June, 64–72. The article focuses on the brief intervals over the past 700 million years when many marine species died out and suggests that these mass extinctions may have been brought on by cooling of the sea.

Stent, G. S., and D. A. Weisblat. 1982. "The Development of a Simple Nervous System." *Scientific American*, January, 136–46. Two questions are addressed: "How do networks of neurons generate animal behavior?" and "How do the neurons and their specific connections arise during the development of an animal from a fertilized egg?" The authors offer preliminary answers by tracing the pedigree of neurons in the embryonic growth of dwarf and giant leeches.

Storer, T. I., et al. 1979. *General Zoology*. 6th ed. New York: McGraw-Hill Book Co. A college-level text.

Thurman, H. V., and H. H. Webber. 1984. *Marine Biology*. Columbus, Oh.: Charles E. Merrill Publishing Co. A well-illustrated introductory text to the subject of life in the oceans and marine ecology, including chapters on the physical ocean and plate tectonics.

Ward, P. 1983. "The Extinction of the Ammonites." *Scientific American*, October, 136–47. These animals, which resemble the nautilus, suddenly died out some 65 million years ago. More mobile, shell-crushing predators may have been the reason.

Webb, P. W. 1984. "Form and Function in Fish Swimming." *Scientific American*, July, 72–82. Interest in the subject of how fishes swim dates back 2500 years, but only during the past two decades has it been studied quantitatively. The article describes these latter studies, which correlate a fish's form with its habit of swimming.

Wells, M. 1968. *Lower Animals*. New York: McGraw-Hill Book Co. A small but worthwhile and well-illustrated book about invertebrates, focusing on how they behave, how they detect what is going on around them, and what they are able to do about it.

Yates, G. T. 1986. "How Microorganisms Move Through Water." *American Scientist*, July–August, 358–65. An article about the hydrodynamics of ciliary and flagellar propulsion. For the student with a physics background.

Figure 9.1. *Cephalaspis lyelli*, a heavily armored, jawless fish from the Lower Devonian, Glamis, Scotland.

 Evolution of the Vertebrates

W HEN THE CHORDATE lineage first arose, there were no immediate signs of the dominance its members would eventually exert in the waters, on land, and in the air. The earliest known chordates, the *Pikaias* from the Burgess shale, were rather small and inconspicuous animals (see figure 8.34). Like their modern descendants, the lancelets, they probably lived on the muddy bottom of the sea, filter-feeding on whatever food particles the currents happened to carry by. For many millions of years these animals remained on the sidelines, while the annelids, arthropods, mollusks, brachiopods, and echinoderms made impressive evolutionary advances and successfully invaded the various habitats of the ancient sea.

Despite their inconspicuous nature, the early chordates already possessed the three structural characteristics that formed the basis of their later success — a notochord, hollow dorsal nerve cord, and pharyngeal gill slits. The notochord became associated with the evolution of elaborate internal skeletons, including backbones and appendages (for example, fins in fishes; limbs in amphibians, reptiles, and mammals; and wings in birds), that lent stability to the bodies and greatly improved the speed, agility, and diversity of locomotion. The nerve cord, which ran the length of the body, evolved into complex and efficient central nervous systems that coordinated the transmission of nerve signals between the brain and the rest of the body. This increased the animals' responsiveness to their environments and contributed further to improving locomotion. Finally, part of the bony structures that supported the gills evolved into jaws with teeth and thus revolutionized the chordates' feeding habits and modes of life.

The present chapter will follow the major stages of this evolution, focusing entirely on the *vertebrates*, which constitute the major subphylum of the chordates (the other subphyla are the urochordates and cephalochordates). Beginning with the most ancient and primitive of the vertebrates, the jawless fishes, the story continues chronologically to the most recently evolved members, the mammals (see table 8.1 for references to geologic time periods).

THE EVOLUTION OF THE FISHES

The *Pisces*, or fishes, comprise five classes, shown in figure 9.2:

Agnathans, or "jawless fishes;" extinct except for lampreys and hagfishes; the extinct agnathans are called ostracoderms.

Acanthodians, or "spiny fishes;" extinct fishes with fins supported by stout spines; first vertebrates with jaws.

Placoderms, or "plated fishes;" extinct fishes with primitive jaws and heavily armored with bony plates.

Chondrichthyes, or "cartilaginous fishes;" extinct and surviving fishes, including sharks, skates, and rays.

Osteichthyes, or "bony fishes;" extinct and surviving fishes, including sturgeons, perch, salmon, cod, trout, and lungfishes.

All but the agnathans possess jaws and, hence, are *gnathostomes*, or vertebrates with "true jaws."

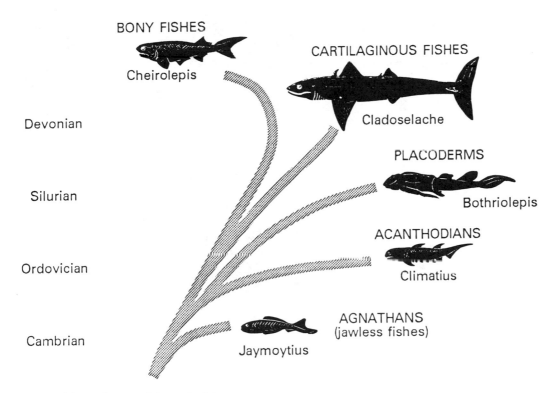

Figure 9.2. Evolution of the early fishes.

Agnathans

The first chordates that actively swam about in search of food were the *jawless fishes*, which, like all jawless vertebrates, belonged to the class Agnatha. The oldest known fossil evidence of the agnathans dates back to the late Cambrian period, roughly 520 million years ago. It consists of fossilized scale-bearing plates that were found in the late 1970s in northeastern Wyoming by John E. Repetski of the U.S. Geological Survey. The fish, named *Anatolepis*, was probably from 3 to 8 centimeters long and lived in the sea. Other fossil evidence of *Anatolepis*, also fragmentary in nature, comes from Ordovician rocks of Greenland, Spitzbergen, and additional sites in North America.

The oldest fossilized complete skeletons of agnathans date from the Silurian period and have been found in England. Two genera have been identified, one of which is *Jamoytius* (see figure 9.3). Its members resembled today's lampreys

Jamoytius

Lamprey

Figure 9.3. The Silurian jawless fish *Jamoytius* and one of its modern relatives, the lamprey.

Figure 9.4. Head and pharyngeal region of a lamprey, a modern jawless fish.

and hagfishes (see figure 9.4). Like them, these early fishes were unarmored and had small tubular bodies, with a suctorial mouth at one end and a propulsive tail fin at the other. The mouth seems to have been well suited for vacuuming up food particles from the sea bottom. Along the sides, behind the eyes, these early agnathans had a row of circular gill openings. There may have been lateral and dorsal fin folds that served as stabilizers in swimming, but the fossil evidence is not clear on this.

In contrast to *Jamoytius*, most agnathans of the Paleozoic era seem to have been covered by armors of thick bony plates or scales (see figure 9.1). A number of these fishes also had substantial deposits of bone in the head region. This left remarkably clear impressions of the brain, nerves, and blood vessels in some of the fossil remains. But none of the agnathans had a bony vertebral column. That evolved only much later, with the origin of the Osteichtyes (see below).

Beyond these common features, the early agnathans varied widely in shape and armor development, indicating that they had already adapted to a range of habitats. For example, some were flat-bottomed with a ventrally placed round mouth, a dorsal pair of eyes, and an upwardly bent, or heterocercal, tail. The ventrally placed mouth and dorsal eyes suggest that these agnathans wallowed in the bottom mud, sucking up detritus and filtering out food particles. Other early agnathans were narrow and deep-bodied; their tapered front ended in a transverse mouth; their eyes were laterally placed; and their tail was bent downward. They are believed to have swum mostly near the water's surface, feeding on plankton that grew there. Still other agnathans probably swam at varying depths and preyed upon soft-bodied animals. The evidence comes from rows of slender fossilized plates that surrounded the mouths of some of these early fishes. These may have functioned as primitive teeth and are the first tentative sign that jaws were developing in the vertebrates.

The fossil record indicates that most agnathans possessed dorsal or ventral fins. But probably none had paired pelvic fins and only a few had paired pectoral fins, which are common among modern fishes. The absence or only limited development of such paired fins restricted the agnathans' maneuverability in the water and indicates that when these fishes were the dominant vertebrate class, the technique of swimming was still in a rudimentary stage of development.

Acanthodians

When the ancient agnathans were at the height of their development, other classes of chordates began to appear and compete with them. Among the earliest of them were the *Acanthodii*, or "spiny sharks," which arose late in Silurian times, reached a peak of abundance and diversity during the early part of the Devonian, and then declined rapidly. They were small marine and freshwater fishes that resembled sharks in general body proportions, but were probably not closely related to them. Along their backs, bellies, and sides they possessed an unusually large number of median and paired fins, which consisted mainly of a stout and sometimes quite large spine with a small web of skin behind it (see figure 9.5). These oddly constructed fins, which served as stabilizers and rudders in swimming, attest to the continued evolutionary experimentation among the ancient fishes. Unlike most of their contemporaries (for example, the placoderms, discussed below), the acanthodians did not have an armor of heavy bony plates, but were covered by small diamond-shaped scales. Most important, they were the first known fishes with true jaws, a development that had far-reaching consequences in subsequent vertebrate evolution (see box 9.1).

The importance of the origin of vertebrate jaws cannot be emphasized enough. The existence of jaws profoundly affected the food chain and stiffened the competition for available resources among chordates and nonchordates alike. The fishes, which up to now had fed mainly on soft nutrients, were now able to prey on each other and on other large animals. The different survival strategies that arose among the fishes, as a consequence of this increase in

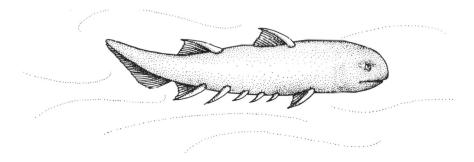

Figure 9.5. An acanthodian or "spiny shark" from Silurian and Devonian times.

BOX 9.1.
The Origin of Vertebrate Jaws

VERTEBRATE jaws originated from the gill arches of ancient fishes. These arches were (and in today's fishes still are) the bony or cartilaginous support structures of the gill pouches. They were shaped roughly like letter Vs laid on their sides (> > >) and situated in pairs on both sides of the trunk behind the mouth. Some of the early agnathans possessed ten or more pairs of gill pouches and gill arches, compared to seven in the modern shark and five in the bony fishes.

It appears that in the course of many millions of years of evolution the front two pairs of gill arches were lost (see figure A). The third pair gradually changed so that these two apexes of the Vs became hinged, allowing the sides of the Vs to be closed or opened. Simultaneously, this hinged pair of arches slowly advanced forward into the mouth, and the left and right ends of the Vs fused together to create an upper and a lower jaw. Muscles evolved that carried out the closing and opening motions. Finally, bony plates that initially were scales or part of the armor around the mouth migrated inward to the jaws and became teeth. The result was the first pair of true vertebrate biting jaws.

Evidence that vertebrate jaws originated this way comes chiefly from embryology. For example, all vertebrate embryos pass through a stage in which they possess *visceral arches*, which correspond to the gill arches of adult fishes. As the embryo develops, the front pair of visceral arches enlarges and gradually becomes transformed into jaws (if the animal is a gnathostome). Still further evidence comes from the fact that in fishes the nerves that run to the jaws lie in series with those that run to the gills.

The origin of vertebrate jaws is yet another example of nature's conservative ways. Structures that were originally used for filtering food from water and mud first became transformed into supports for gill pouches and then into jaws. Nature rarely discards things that work. When new needs arise in evolution, old ''inventions'' become altered or are developed further to meet the challenges of the new situations.

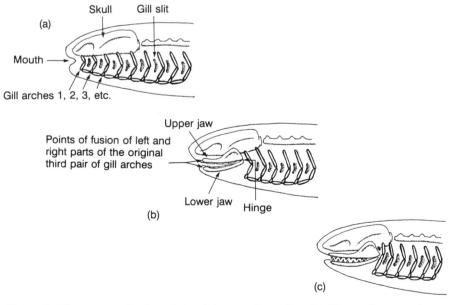

Figure A. Three stages in the origin of the vertebrate jaw. (*a*) The earliest vertebrates (the ostracoderms) had no jaws. (*b*) The front two pairs of gill arches were lost; the third pair evolved into hinged upper and lower jaws. (*c*) Bony plates near the mouth became teeth.

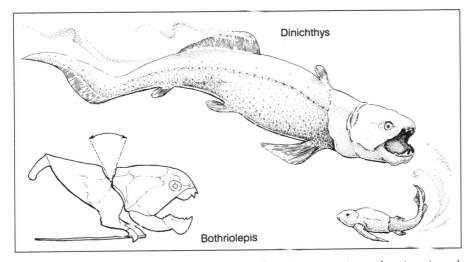

Figure 9.6. Two representatives of placoderms from Devonian times: the gigantic and fierce *Dinichthys*, or "terrible fish," that grew up to 3 meters in length, and the much smaller *Bothriolepis*, with its jointed appendages.

competitive pressure, included the evolution of larger size, heavier armor, and greater speed and maneuverability in swimming. We may assume that still other strategies involved an increase in the keenness of the senses, development of chemicals that discouraged predators by their smell or taste, refuge in dark and inaccessible places, camouflage, and production of prodigious numbers of offspring. Like their modern descendants, most of the ancient fishes relied on combinations of these strategies, though a few specialized in just one or two. The result was a remarkable diversification and the beginning of the "Age of Fishes."

Placoderms

The most spectacular class of early jawed fishes was the *Placodermii* or "plated" fishes, which lived mainly in the sea, though some were freshwater species. They arose and reached their dominance during the Devonian period and then declined, becoming extinct before the end of the Paleozoic.

As their name implies, the placoderms were covered by plated armor, which has been interpreted as protection against predators. This probably was the case with many of these early fishes. However, some of them were the largest animals in the sea and were themselves predators. What might have been eating them? No fully satisfactory answer to this question is known.

The plates of the placoderms consisted of bone and, from one species to the next, varied greatly in size, shape, and the area of the body covered by them. In some placoderms only the head and front part of the trunk were armored, while in others the armor extended over the entire body, including the fins. As a result, many of these fishes looked rather grotesque. For example, in *Dinichthys*, a gigantic placoderm from the Upper Devonian, the head and trunk shields were joined by a hinge, with a wedgelike gap between the shields (see figure 9.6). This arrangement allowed the fish to raise its head up high while also dropping the lower jaw, making possible an enormously large bite. Certainly, *Dinichthys* deserved its name, which means "terrible fish." Another bizarre but very small placoderm was *Bothriolepis* (figure 9.6). Its head and the front part of its boxlike trunk were armored, while the finned rear part of its body was naked. From the front of its trunk shield extended a pair of long, pointed appendages, which were also armored. The appendages were attached to the trunk by hinges and they had a hinged joint near their middle, which increased the range of movement. Furthermore, fossil imprints left by *Bothriolepis* show a well-developed pouch extending from the gut. This pouch is interpreted as having been a lung used for breathing and constitutes the first of several lines of evidence indicating that lungs, at least in a primitive form, arose during the Devonian period and subsequently became rather common in most vertebrate lineages.

As noted before, all of the placoderms possessed jaws. In many of them (as, for example, in *Dinichthys*), the jaws were powerful and formed awesome eating instruments. In others they were rather weakly developed and were probably used primarily for eating plants or soft-bodied animals. Clearly, these fishes had adapted to a broad range of habitats and modes of life, and they illustrate particularly well the enormous diversification that followed the invention of jaws.

Cartilaginous Fishes

Each of the classes of fishes discussed so far was, at one time, the most abundantly represented class in the waters of the Earth, before giving way to the next. However, with the exception of two modern orders of agnathans—the lampreys and hagfishes—none of them survived in the long run. They all became extinct by the end of the Paleozoic.

While these fishes flourished, there were others that did not rely on such specialized survival strategies as heavily plated armor or large spiny fins. Among them were the *sharks*, or cartilaginous fishes, and the *bony fishes* (see figure 9.7). Though it is true that over the ages many of them also adopted specialized modes of life, as the great variety of their outward appearances today demonstrates, their most characteristic adaptation was the perfection of the swimming mode of locomotion. This included increased speed and maneuverability. Their bodies gradually became more streamlined, allowing them to glide through the water with little friction. Their tails, or *caudal* fins (see figure 9.8), evolved toward larger size, and the back and forth motions of the caudal fins became coordinated with the sinuous undulations of the trunk, creating a powerful forward thrust. On their backs and bellies they evolved slender median fins that functioned as stabilizers to prevent rolling and side slip. Like the placoderms, they evolved two sets of paired fins along their sides, just behind the head and near the anus—the *pectoral* and *pelvic* fins (which evolved into fore- and hind limbs in land-living vertebrates). These paired fins controlled upward and downward motions and acted like rudders and brakes.

The *sharks*, or *Chondrichthyes*, first show up in the fossil record of mid to late Devonian times, mostly through fossils of their teeth and other small hard

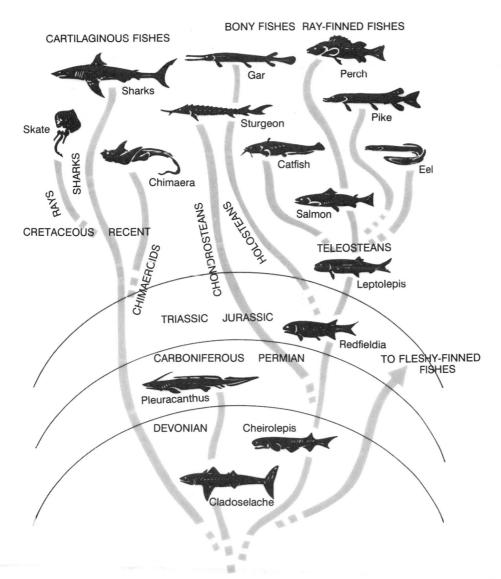

Figure 9.7. Evolution of cartilaginous and bony fishes.

parts (see figure 9.9). They possessed internal skeletons of cartilage, which probably had evolved from earlier bone. Because cartilage is lighter than bone, sharks were more buoyant than the acanthodians, placoderms, and bony fishes. However, cartilage and flesh are still denser than water and, because the sharks possessed neither lungs nor a swim bladder, they tended to sink. The ancient sharks overcame this difficulty, as their modern descendants still do, by spreading their pectoral fins horizontally and using them like the bow planes of a submarine for getting lift. Of course, this worked only during forward motion and meant that the sharks had to swim continuously to keep from sinking.

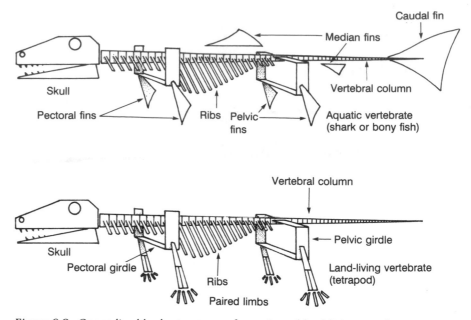

Figure 9.8. Generalized body structures of aquatic and land-living vertebrates.

All the ancient species of sharks died out toward the close of the Paleozoic and gave way to types with more efficiently constructed fins. These more advanced sharks developed throughout the Mesozoic and have changed little since, despite fluctuations in climate and competition from the bony fishes and aquatic reptiles and mammals. By numbers, sharks are not and never have been very abundant, compared with the bony fishes and today's land vertebrates. However, their long survival with relatively little change indicates that they are among the most highly adapted inhabitants of the sea.

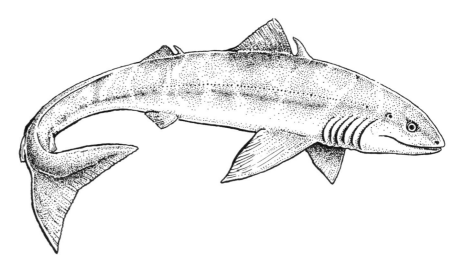

Figure 9.9. *Cladoselache*, a shark of late Devonian times.

Bony Fishes

The *bony fishes*, or *Osteichthyes*, first appeared in freshwater lakes and rivers of the Devonian period. In those ancient fishes, the internal skeleton was still largely cartilaginous. Their outsides were covered by heavy, diamond-shaped scales and their tails were bent upward, much as in sharks. Furthermore, they had lungs through which they could take up oxygen. This was of particular advantage in oxygen-poor ponds and swamps, in which many of them lived and where gills alone could not do the job.

During the Mesozoic and on into the Cenozoic eras, these and other features of the bony fishes gradually changed to those that we find in most of their present-day descendants. Their internal skeleton became increasingly bony, including the skull, vertebrae, ribs, gill arches, spines supporting the fins, and girdles to which the pectoral and pelvic fins were attached. The scales became thinner and rounded, and their outsides became lubricated with a slippery mucus, reducing the friction with the water and providing additional protection. The tail grew straight and took on a symmetrical double-lobed form.

During the early part of this evolution, probably already in mid-Devonian times, two major groups of bony fishes became established: the *Actinopterygii*, or "ray-finned" fishes, and the *Sarcopterygii*, or "fleshy-finned" fishes. In the ray-finned fishes the paired fins consisted mainly of bony rays supporting a web of skin. Most importantly, the lungs of these fishes became transformed into swim bladders, into and out of which they could diffuse air and other gases. This allowed them to control their bouyancy and hence their depths in the water without having to be in motion. This, together with their streamlined bodies and the other structural advances discussed above, allowed them to evolve into the most successful of all fish species. Today they can be found in all waters of the Earth, from mountain streams to rivers, lakes, and all levels of the sea. They range in size from a fraction of an inch to several meters and they have evolved a startling variety of body forms. Most of the well-known game and commercial fishes are ray-finned fishes: the sturgeon, herring, salmon, trout, eel, cod, perch, halibut, flounder, bass, tuna, and many more. Without question, the ray-finned fishes are presently near the height of their evolution and there are no signs that this will change in the foreseeable future.

The members of the second group of bony fishes, the fleshy-finned fishes, were initially rather similar to the ray-finned fishes, though there were important differences (see figure 9.10). Their paired fins did not consist merely of skin supported by bony rays, but of flesh supported by an internal skeleton. This skeleton was composed of median bones from which other smaller bones radiated out to the sides and the ends. Furthermore, the fleshy-finned fishes had two dorsal fins rather than a single one, as had the contemporary ray-finned fishes. Their lungs did not become transformed into swim bladders, but were retained and used for breathing. And, finally, unlike all other fishes, their nostrils had internal openings that allowed them to inhale air without opening their mouths.

The fleshy-finned fishes evolved in two main lineages: the *Dipnoi*, or "lung-fishes," and the *Crossopterygii*, or "lobe-finned" fishes. The lungfishes were abundant from the late Devonian to the early part of the Mesozoic era, but since

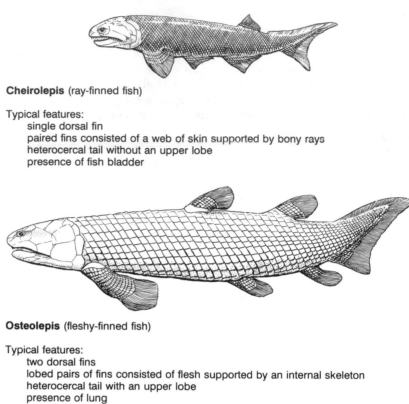

Cheirolepis (ray-finned fish)

Typical features:
    single dorsal fin
    paired fins consisted of a web of skin supported by bony rays
    heterocercal tail without an upper lobe
    presence of fish bladder

Osteolepis (fleshy-finned fish)

Typical features:
    two dorsal fins
    lobed pairs of fins consisted of flesh supported by an internal skeleton
    heterocercal tail with an upper lobe
    presence of lung
    nostrils with internal openings

Figure 9.10. Comparison of representatives of the two major groups of bony fishes of Devonian times: *Cheirolepis*, an actinopterygian, or ray-finned fish, and *Osteolepis*, a sarcopterygian, or fleshy-finned fish. The ray-finned fishes were members of the lineage that led to the modern bony fishes. The fleshy-finned fishes were members of the lineage that led to the amphibians and other land-living vertebrates.

then have largely become extinct. Only five species survive today: one in Australia, three in Africa, and one in South America. The Australian lungfish, *Neoceratodus*, lives in the rivers of Queensland and has the interesting habit of walking with its fleshy fins along the river bottoms. During the dry season, when the rivers become reduced to small stagnant pools, it survives by coming to the surface to breathe air. The African and South American lungfishes, *Protopterus* (see figure 9.11) and *Lepidosiren*, are able to live for months in the sun-baked mud of dried-up ponds by breathing air through narrow holes they dig in the mud. When the rains fall and the ponds fill with water, they wriggle themselves free of the softened mud and swim off.

    The second group of fleshy-finned fishes, the lobefins, flourished during late Devonian times, but then declined and, shortly after the beginning of the Permian period, most of them disappeared. In some of their habits they probably resembled the lungfishes, and we may assume that they too were

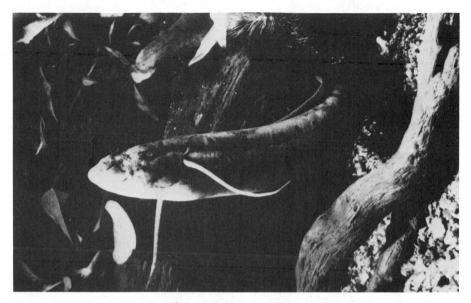

Figure 9.11. The modern lungfish *Protopterus dolloi*. This fish and its relatives live in the rivers and swamps of central and western Africa. During the dry season, they survive by burying themselves in the mud and breathing air through narrow holes they dig in the mud.

able to use their paired fins as primitive walking legs and their lungs as breathing organs.

Until 1938, lobe-finned fishes were thought to have been extinct. In that year a fishing trawler dredged up one of them in the Indian Ocean near East London, South Africa. It was described as steely blue, weighed some 60 kg, and was 1.5 m long. The specimen belonged to the family *Coelacanthidae* and was named *Latimeria*.* Since then numerous other specimens of this interesting fish have been caught in the waters near the Comoro Islands, between Madagascar and Africa (see figure 9.12). It is typically two meters long, has powerful jaws and thick bony scales, and looks much like its ancestors, as suggested by

---

*The fishermen took the fish to their home port of East London, where the curator of a small local museum, Miss Courtney-Latimer, examined it. She sent a sketch of the fish to Professor J. L. B. Smith of Rhodes University College, Grahamstown, who was an authority on African fish. He immediately realized the great importance of the specimen as a living fossil. Unfortunately, before he could get to East London, the fish's soft parts had begun to rot and were thrown away. On the basis of what remained, Smith identified the fish as a member of the coelacanths, which were thought to have been extinct for 70 million years. He named the fish *Latimeria* to show his appreciation to Miss Courtney-Latimer for having brought it to his attention.

Latimeria was hailed as a major scientific discovery. An intensive search was launched to catch another specimen and an offer of a large reward was posted up and down the coast from East London. In 1952 a second specimen was caught some 3000 kilometers away near Anjouan, one of the tiny Comoro Islands, midway between northern Madagascar and Africa. The natives of these islands were familiar with the fish and caught one or two every season at depths of about 200 to 300 meters. Though not particularly tasty, the natives consider the fish edible when dried and salted, and they use its rough scales to rub down inner tubes when mending punctures.

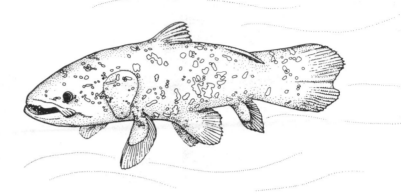

Figure 9.12. *Latimeria*, a living coelacanth, in the waters near the Comoro Islands. The coelacanths are lobe-finned fishes, most of which became extinct by the close of the Paleozoic. One group of Devonian lobe-finned fishes (and close relatives of the coelacanths), the rhipidistians (see figure 9.18), is believed to have been the ancestors of the amphibians and all other land-living quadrupeds.

the fossil record. Furthermore, its heart, kidney, and stomach are very primitive compared to those organs in most modern fishes, and its pectoral fins can be turned through an arc of 180 degrees.

Despite their only brief success roughly 400 million years ago, the lobe-finned fishes are of particular interest to us because some of them used their fins and lungs to move from water onto land. They were the direct ancestors of the amphibians (see figure 9.13) and the earliest pioneers of an entirely new phase in evolution—the vertebrate invasion of the land and, later, the air of our planet.

THE INVASION OF LAND AND AIR: PLANTS, AMPHIBIANS, REPTILES, AND BIRDS

Plants

Before turning to the vertebrate invasion of the land and air, let us consider another similar development that took place slightly earlier among the plants: they, too, were moving from water onto land. Up to the middle of the Silurian, the land of our planet was barren and, except for the presence of swamps, rivers, and lakes, probably looked much like the plains and some of the mountainous areas of Mars today. Then, some 425 million years ago, while the fishes evolved from primitive, jawless creatures to the masters of the sea and fresh waters of the Earth, the first land plants began to appear. (Some forms of algae may have preceded the plants in this movement onto land, especially in wet areas along rivers, lakes, and the seashore. But the timing and extent of algal spreading onto land is conjectural and at present a matter of controversy.)

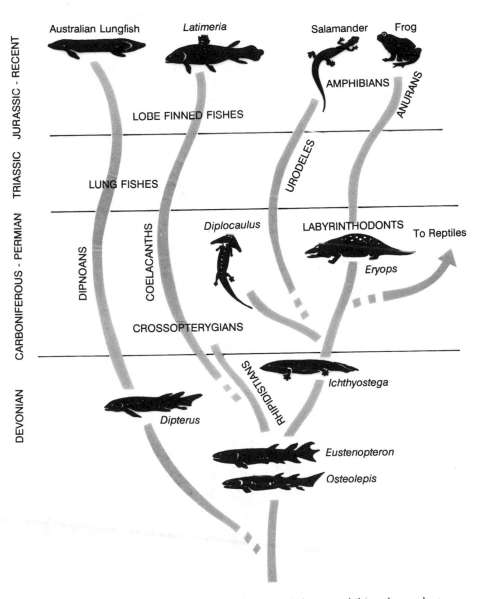

Figure 9.13. Evolution of the fleshy-finned fishes and their amphibian descendants.

The earliest land plants consisted of simple branched stems that had little strength and grew only a few centimeters tall (see figure 9.14). They lacked roots and leaves, and they carried out photosynthesis directly in their stems, though they had already acquired vascular tubes for conducting water up the stems. Their only modern survivors are the whisk ferns, which grow in the tropics and subtropics. The early land plants grew in warm, humid environments, where ample water was available. At first, they probably formed small, scattered patches of stems along the edges of estuaries, swamps, and other bodies of water in the ancient equatorial zone. But gradually they spread further and gained an ever-growing foothold on land.

In time, the plants developed roots for clinging and absorbing water more efficiently, stiffer stems of wood for greater strength, and spreading branches with leaves for increased absorption of sunlight. In addition, they evolved the ability to grow new cell layers around the already existing stem, which widened their girth and reinforced their structural strength. As in modern trees, this

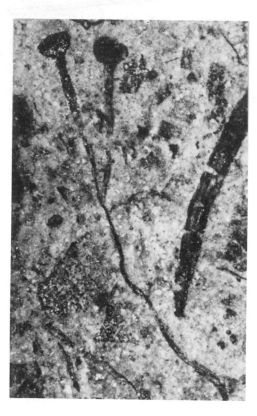

Figure 9.14. Cooksonia, fossil remains of the earliest known vascular plant, from Upper Silurian rocks of South Wales. The plant had simple branched stems and no leaves, but it already possessed vascular tubes for conducting water up the stem.

Figure 9.15. Modern club mosses. These plants are relatives of some of the most ancient land plants. They reproduce by spores borne on leaves specialized for reproduction, which give the stems a club-shaped appearance. Unlike today's species, many of the ancient club mosses were tall and, together with the horsetails and ferns, were major components of the forests of Carboniferous times.

Figure 9.16. A fossil horsetail from Carboniferous coal shales, Grundy County, Illinois. These simple plants possessed true roots, vascular stems, and leaves. The leaves grew in circles at each of the stems' nodes.

ancient seasonal growth shows up as rings in the fossil remains. These plants resembled (and probably were related to) today's club mosses (see figure 9.15), horsetails (see figures 9.16 and 9.17), and ferns, except that many were much taller. Some of them grew up to 30 meters tall and formed the Earth's first forests. By the Carboniferous, the period following the Silurian and Devonian, the forests had become enormous and were the source of most of our present-day coal deposits.

As the forests spread inland, invertebrates of various kinds were the first animals to follow them. The trees and shrubs provided ample food, protection, and, at least initially, they constituted a new, unoccupied and unexploited territory. The earliest invertebrates to settle the forests were the arthropods— centipedes, scorpions, spiders, and somewhat later, insects. Without doubt, other invertebrates, such as worms and mollusks, participated in the colonization of the early forests also, though no fossils have been found to confirm this. In any case, by the time the vertebrates were ready to make their first tentative forays onto land during Devonian times, a rich ecosystem of low-growing shrubs and tall trees, including a diverse invertebrate fauna, skirted many of the waters of the Earth.[*]

Amphibians

The early forests represented an hospitable new environment for all animals that managed to make the transition from water onto land and strongly affected their further evolution. However, in the case of the vertebrates, the forests were probably not the chief cause of this transition. It seems the vertebrates moved

*The recent discovery of fossil burrows in ancient dry-land soil near Potters Mills in central Pennsylvania suggests that animals and, presumably, plants appeared on land earlier than previously thought. The discovery was made by Gregory J. Retallack and Carolyn R. Feakes, both of the University of Oregon. The soil in which the burrows were found dates from the late Ordovician.

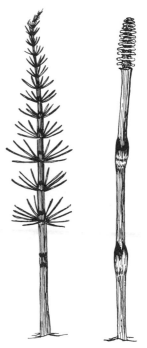

Figure 9.17. Modern horsetails. Like their ancestral relatives, today's horsetails grow circles of leaves at each node of their stems.

onto land mainly because of fluctuations in the climate, a theory suggested by the survival adaptations of today's lungfishes.

The Devonian period was generally characterized by warm, humid conditions, with ample rainfall at least near the major bodies of water in the tropics and temperate zones. However, as in modern times, the climate fluctuated. Both seasonally and, sometimes, for much longer stretches of time, the warm and wet conditions were interrupted by dry spells. During those periods, many of the ponds, lakes, and rivers became reduced to muddy, stagnant pools or they dried up entirely. As in the case of the modern lungfishes, lungs must then have been essential organs for survival. They allowed the fishes to obtain oxygen even when gills were of no further use. Paired fins that were supported by skeletons and that could be moved by muscles also proved advantageous. They could be used as primitive walking legs either for following the remaining water in pools that were drying up or for seeking out the first water-containing puddles after the rains started to fall again. No doubt, paired fins also proved advantageous in going after food, escaping from danger, and finding a mate.

The fishes in which these features—well-functioning lungs and reasonably strong and movable paired fins—were best developed were the *rhipidistians*. They were lobe-finned fishes of late Devonian times and close relatives of the coelacanths and *Latimeria*, introduced in the previous section. An early rhipidistian example was *Eusthenopteron*, whose fossilized pelvic fins contain bones corresponding to the pelvis, femur, tibia, fibula, and some of the other smaller bones of the hind limbs of ancient amphibians and other land-living vertebrates (see figures 9.18 and 9.19). These fossil bones suggest that *Eusthenopteron* was a fish that, in addition to swimming, carried out walking motions with its paired fins. Quite probably, this fish also possessed well-developed lungs that allowed it to survive for considerable stretches of time by breathing without having to use its gills. Furthermore, its skull, jaws, teeth, internal nasal opening,

Figure 9.18. *Eusthenopteron foordi*, a fossil rhipidistian, from Devonian deposits near Quebec, Canada.

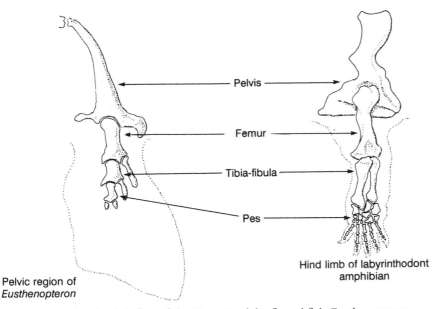

Pelvis

Femur

Tibia-fibula

Pes

Hind limb of labyrinthodont
amphibian

Pelvic region of
Eusthenopteron

Figure 9.19. Fossilized pelvic fins of the Devonian lobe-finned fish *Eusthenopteron*.

bone-enclosed notochord, and general body shape closely resembled those of the ancient amphibians.

Despite these features, *Eusthenopteron* is usually classified as a fish. When conditions demanded it, it may have made short forays onto land and managed to survive there for brief periods of time, but it seems to have been better adapted for life in water. The gills were its primary organ for oxygenating the blood, while the lungs were of secondary importance. Furthermore, it had a well-developed fishlike tail as well as median dorsal and ventral fins that allowed it to swim competently.

In contrast to *Eusthenopteron*, there were other vertebrates in late Devonian times that, though similar to it, were better adapted for life on land. The earliest known members of this group were the *ichthyostegids*. Their fossils, which are found in sediments of eastern Greenland, suggest that their primary organs for oxygenating the blood were the lungs, while the gills were of secondary importance. Their pectoral and pelvic fins were stronger than those of any of the fishes. Their paired bony appendages were well developed, ended in toelike digits, and almost certainly were walking limbs. Furthermore, the ichthyostegids' skulls were greatly elongated in front of the eyes, which was quite atypical of the fishes. These features indicate that these vertebrates were *amphibians* and like their modern relatives—the frogs, salamanders, and caecilians (a little-known, tropical group without legs)—led a double life. They lived both in the water and on land.

Life on land presented new obstacles that the fishes never had to face. For example, there existed the ever-present problem of gravity. In water, an animal's weight is largely canceled by buoyancy, but on land weight is felt fully. Amphibians adapted to this situation by evolving stronger backbones, heavier pectoral

Figure 9.20. Branchiosaurus, a small labyrinthodont amphibian from the Permian, found in red beds near Bad Kreuznach, West Germany; 18 cm.

and pelvic girdles, and more robust ribs. Their muscles, too, became more powerful, in particular those of the paired fins, which were now the main means of locomotion. And the tails, which in the earliest amphibians were still fishlike, gradually became modified into elongated, tapered forms.

Another problem of life on land was desiccation, or drying out. Returning frequently to water reduced this danger. In addition, ancient amphibians evolved tough outer skins, which limited the loss of body fluids. Today the skins of most amphibians are very thin and in many of them function as respiratory organs, assisting the lungs, though this may be a secondary adaptation that ancient amphibians did not possess.

Yet a third problem unique to life on land was reproduction. Like the fishes, most of today's amphibians lay their eggs in water, or in moist places, and fertilize them externally. The young that hatch from the eggs pass through various aquatic larval stages (for example, tadpoles), during which they breathe with gills and propel themselves by fin-bordered tails. Then, during metamorphosis into the adult form, limbs emerge and body proportions change. Furthermore, in many amphibians the tail is resorbed (for example, in frogs and toads) and the gills are replaced by lungs (the latter is not true in the case of certain aquatic salamanders). Fossil remains of Devonian amphibian larvae suggest that reproduction was then carried out similarly, which restricted the animals to swamps or the onshore environments of lakes and rivers.

Despite their continued dependency on the ancestral habitat of freshwater, the amphibians were a highly successful group for about 100 million years following their origin. During the early part of that time, they were, after all, the only vertebrates living on land and, hence, had little competition. They diversified into a great number of forms and specialized modes of life, and during the Carboniferous they became so abundant that this period is called the "Age of Amphibians."

Some of the amphibians of the Carboniferous period, the *labyrinthodonts*,* grew to lengths of 3–4 meters and, with their powerful jaws lined with sharp, pointed teeth, were the dominant animals in their habitats. Others remained small and assumed snakelike or froglike forms (figure 9.20). Still others returned to the water and lost many of their land adaptations. Then during the Permian period the amphibians' numbers declined along with the rise of a new group of vertebrates — the reptiles.

Reptiles

The emergence of the reptiles as the dominant land-vertebrates during the Permian, the last period of the Paleozoic, was a consequence of significant changes in the physical and biological environments of that time. The continents, that for the previous few hundred million years had been separate and distinct, began to converge and merge into a single, global landmass — called *Pangaea* (see box 9.2). Mountain ranges formed and the general topography of the land became more varied. As a result, the climate also became more varied, both geographically and seasonally. In particular, there was a general cooling trend and the Southern Hemisphere became widely glaciated. Widespread biological changes took place as well. By the end of the Paleozoic, up to 90% of all marine species became extinct, including the last of the trilobites, the placoderms, and most of the formerly common brachiopods. The ancient mollusks, crustaceans, echinoderms, and bony fishes disappeared and gave way to their modern successors. In short, the close of the Paleozoic was a time of major geologic restructuring, climatic changes, and biological extinction and renewal.

It was in this setting that reptiles gained the advantage over their amphibian competitors. However, the origin of the reptiles dates back further, probably to the early part of the Carboniferous, when some amphibians (perhaps certain species of labyrinthodonts) evolved a number of new adaptations. Foremost among these adaptations was a change in methods of reproduction. The details of the early stages of this change are not known, but very likely a variety of reproductive strategies were tried out, all characterized by reducing the need of external water habitats for developing eggs and larvae. The strategy that proved most successful was one that, in the course of many millions of years, evolved into the kind we observe today among the reptiles. The animals fertilized their eggs internally, by a physical mating between males and females. The females then laid the eggs on land and, after a period of incubation, fully formed baby reptiles hatched from them.

*When the fossilized teeth of the labyrinthodonts are cut crosswise and examined under the microscope, the enamel covering the teeth is seen to be folded into a complex, labyrinthine pattern — whence the name "labyrinthodont." The same pattern is already present in the teeth of the lobe-finned fishes, providing yet additional evidence that they represent the ancestral lineage of the amphibians and other land-living vertebrates.

BOX 9.2.
The Theory of Continental Drift

UNTIL the early 1960s, most earth scientists regarded our planet as rigid and unchanging. They held this view despite repeated suggestions for more than a hundred years that the continents were at one time joined together and subsequently drifted apart. The first to make this suggestion was Antonio Snider Pellegrini of Italy (1812–1885), who based his theory on the close fit between the coastlines on both sides of the Atlantic (a fit that was pointed out as early as 1620 by the English philosopher Francis Bacon). Toward the close of the nineteenth century, the Austrian geologist Eduard Suess (1831–1914) repeated Pellegrini's suggestion and postulated the former existence of a single supercontinent in the Southern Hemisphere, which he called *Gondwanaland* (after *Gondwana*, a region in central India).

During the early part of the present century the first comprehensive theory of continental drift was developed. The most influential contributor to the theory was Alfred L. Wegener (1880–1930), a German meteorologist. He based his belief in continental drift not only on the evidence from geography, but also on evidence from biology and paleontology. For example, the flora and fauna of today and of the past on both the west coast of Africa and the east coast of South America appear to be related.

Wegener suggested that the continents, made of lighter rocks, float like rafts on denser underlying material and for the past 200 million years have been drifting apart from a single, primordial supercontinent, which he called *Pangaea*, meaning "all land." At first, Wegener's proposal attracted worldwide attention, but during the succeeding decades fell into disrepute, particularly among North American geologists, because the movement of rocky continents through rocky ocean floors (regarded as analogous to the way ships plow through water) was thought to be impossible. No mechanism was then known that could explain how continents move over the Earth's surface.

This situation changed drastically after World War II, chiefly on account of two undertakings. A global network of seismic stations was established to pinpoint the locations of earthquakes, and extensive explorations of the world's seafloors were carried out.

It was discovered that the majority of earthquakes occur along narrow, well-defined belts, many of which are paralleled by deep, trough-shaped ocean trenches. Examples are the earthquake zones along the Pacific Rim, and the deep ocean trenches in the eastern Pacific and off the west coast of South America. A second, equally significant result of the post–World War II geologic explorations was the discovery of a 60,000-km-long suboceanic chain of mountains that runs from the Atlantic around Africa to the Indian and Pacific Oceans (see figure A). Most interestingly, along the axis of this ridge system stretches a crack-like rift valley, in which molten magma from the Earth's interior continually wells up. Furthermore, at many places, the ridge system is offset by sharp fractures in the ocean floor that cross the rift valley perpendicularly.

Based on these discoveries, a rigorous and self-consistent theory of continental drift was worked out during the 1960s. It was recognized that the Earth's surface is not one continuous, rocky layer, but consists of a number of separate tectonic plates that meet each other along distinct boundaries. The plates consist of the crust and the mantle material immediately below it, down to a depth between 60 and 100 km. They are extraordinarily rigid structures and move as units across the planet's surface, driven by convective currents deep within the Earth. As they move they carry the continents along and thus give rise to continental drift. Today there are seven major tectonic plates and several smaller platelets, whose names and geographic distribution are shown in figure A.

The boundaries between the tectonic plates are of three kinds—*subduction zones*, *suboceanic ridges*, and *transform faults*. Subduction zones are usually found along the earthquake belts and deep ocean trenches. Suboceanic ridges run along the 60,000-km-long rift valley. And transform faults are the fracture lines that cross the rift valley at right angles. The dynamic processes of continental drift are most apparent along these plate boundaries. Hence, in order to understand continental drift, we must concern ourselves with what happens along these boundaries. First let's review in somewhat more detail the outer structure of the Earth.

BOX 9.2 continued

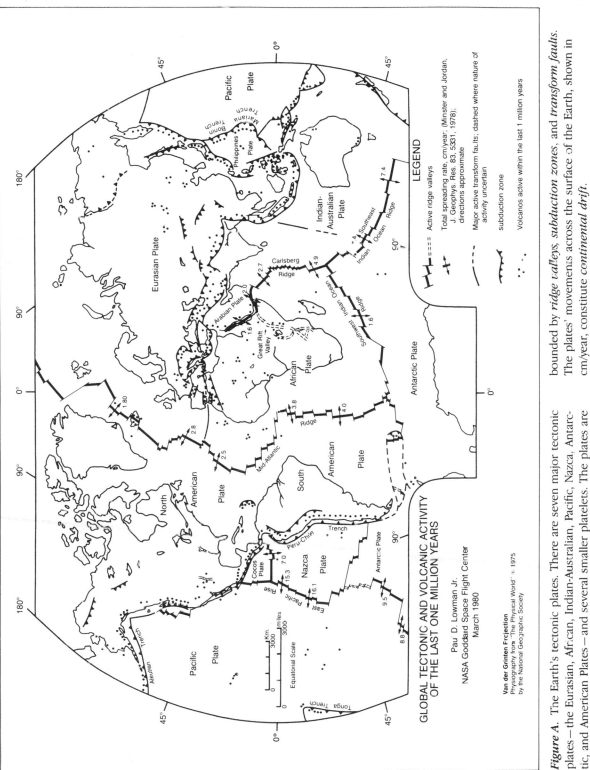

Figure A. The Earth's tectonic plates. There are seven major tectonic plates—the Eurasian, African, Indian-Australian, Pacific, Nazca, Antarctic, and American Plates—and several smaller platelets. The plates are bounded by *ridge valleys, subduction zones,* and *transform faults.* The plates' movemens across the surface of the Earth, shown in cm/year, constitute *continental drift.*

BOX 9.2 *continued*

Geologic Structure of the Earth

As noted in chapter 5 (see figure 5.11), the Earth consists of three major structural components—the *crust, mantle*, and *core*. The crust is the outermost rocky layer, with an average thickness of 35 km under the continents and 6 km under the oceans. These thicknesses vary with surface topography. For example, under low-lying continental plains the crust is thinner than 35 km, while under high-rising continental mountain ranges it may reach a thickness of up to 70 km, with the bulk of it extending deep into the mantle. This great thickness is needed to support (by the force of buoyancy) the weight of the mountains above, much as the bulk of an iceberg lies submerged and supports (by the force of buoyancy) the tip that extends above the ocean surface.

The continental and oceanic crusts float like rafts on the underlying mantle because their average densities—2.7 g/cm^3 for the continental crust and 2.9 g/cm^3 for the oceanic crust*—are less than that of the mantle (3.3 g/cm^3 for the upper mantle). The oceanic crust, with its greater density and compactness, floats at a lower level than the continental crust and thus forms the basins in which the oceans have accumulated. The lower boundary of the crust is called the *Mohorovičić Discontinuity* (or *Moho*, for short) after the Croatian geologist Andrija Mohorovičić (1857–1936), who discovered it in 1909.

The mantle of the Earth extends from the crust to a depth of approximately 2900 km, where it makes contact with the core. It consists of two parts, the *upper* and the *lower mantle*, whose common boundary lies at a depth of about 700 km and is characterized by a rapid increase in density to about 4.3 g/cm^3. Apart from an increase in density with depth, the mantle also exhibits structural variations. Its uppermost layer, down to between 60 and 100 km, is extremely rigid and, together with the crust, forms the *lithosphere*. The lithosphere consists of a number of distinct segments, which are the tectonic plates

---

*The upper layers of the continental crust consist mainly of *granite*, which is rich in aluminum and silicon oxides. The oceanic crust is largely *basalt*, which also contains aluminum and silicon oxides, but which in addition is rich in magnesium, calcium, and iron oxides. This difference in composition is responsible for the greater density of the oceanic crust compared to that of the continental crust.

introduced above. (The term *tectonic* is derived from the Greek *tekton*, meaning "builder," and is most appropriate for describing the Earth's lithospheric plates.)

Below 100 km, the mantle is much less rigid and has a pasty consistency. The layer between the lithosphere and a depth of 700 km (the lower part of the upper mantle) is known as the *asthenosphere* (from the Greek *asthenes*, meaning "weak"). Seismic data suggest that the asthenosphere is convectively unstable, with its material slowly circulating between its upper and lower boundaries. Similar convective circulations probably also take place in the lower mantle. The forces driving these convective currents are at present not well understood, but the rapid increase in temperature with depth is probably in part responsible (much as steep temperature gradients are responsible for convection in the Earth's atmosphere and in stellar interiors, see box 3.3).

Tectonic Plate Movement

To return to the discussion of tectonic plates and the dynamic processes that take place along their boundaries, let us look at *subduction zones* (see figures A and B). They are sites where an oceanic plate passes beneath another plate (either a continental or an oceanic plate) and descends into the asthenosphere. As it does so, it drags the rest of the plate after it, with far-reaching, global effects. It also touches increasingly hotter rock and slowly begins to heat up. Over a time span of 1 to several 100 million years, it reaches thermal equilibrium with the asthenosphere and is resorbed.

Oceanic plates. Closer to the surface, the subduction of an oceanic plate manifests itself in a number of different ways. Frequently, it drags the ocean floor down to great depths and creates the deep ocean trenches mentioned above. Examples are the Bonin and Mariana Trenches south of Japan (greatest depths 10,595 m and 11,020 m, respectively) and the Peru–Chile Trench off the west coast of South America (8055 m). Friction between the subducting and overriding plates is responsible for the earthquake belts and chains of volcanoes—volcanic *island arcs* (if the overriding plate is oceanic) and volcanoes of coastal mountain ranges (if the overriding plate is continental)—that are commonly found along subduction zones. For example, along the Pacific Rim these

BOX 9.2 continued

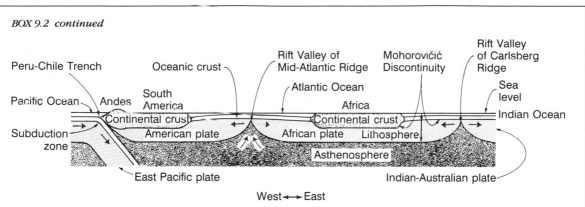

Figure B. Cross section through the lithosphere, showing the subduction of the Pacific Plate under the South American Plate and the seafloor spreading in the Atlantic and Indian Oceans.

chains of volcanoes are so densely spaced that they have become known as the "ring of fire."

Rift valleys are sites where two neighboring oceanic plates are being pulled apart. This produces a break between the plates, through which hot, viscous rock from the asthenosphere is able to rise to the surface. As the rock rises, it partially melts (because of the great reduction in pressure), spreads laterally, and then cools and welds itself to the trailing edges of the diverging plates. In this way, the plates on both sides of the rift valley grow at the same rate at which they are pulled apart.

The growth of oceanic plates produces long, narrow ridges, parallel to the rift valley. As the plates move away from each other (typically at a rate of a few centimeters per year), they carry the ridges along and thus form the ridge systems that are so characteristic of ocean floors. Not surprisingly, the ridges are older with increasing distance from the rift valley. However, nowhere are the ocean ridges older than 200 million years indicating that this is the maximum age before they reach a subduction zone and return to the mantle.

An excellent example of a rift valley and seafloor spreading is the Mid-Atlantic Ridge, which runs down the middle of the Atlantic and forms the boundary between the American Plate in the west and the Eurasian and African Plates in the east (see figure B and *National Geographic Atlas of the World*, 1981). Topographic maps of the floor of the Atlantic show ridge after ridge, at some places for more than a thousand miles, on both sides of the central rift valley.

They are a sure sign that the Atlantic has been growing, as the American Plate, along with North and South America, and the Eurasian and African Plates, along with Eurasia and Africa, moved apart. This movement began about 200 to 180 million years ago near the equator and then spread north and south, thus forming the Atlantic. Earlier than that, North and South America, Europe, and Africa had been firmly joined together.

Transform faults are boundaries along which tectonic plates slide past each other, without being either created or consumed. They are fractures in the lithosphere that always occur near rift valleys and cross them perpendicularly, with the rift valley and individual ridges on one side being offset relative to those on the other side. The sliding is frequently jerky and accompanied by powerful earthquakes, as one plate drags the other along by friction before fracturing. Transform faults are the result of shear forces that act on the plates as they move across the Earth's surface. (Tectonic plates are very rigid structures and tend to fracture rather than deform when lateral forces act upon them.)

Transform faults are found in great numbers all along the suboceanic ridge system. They also occur on continental plates as, for example, the San Andreas fault in California. There the Pacific Plate slides past the American Plate in a northwesterly direction. If the slippage continues at its present rate, San Francisco (which lies on the Pacific Plate) will move northwestward into the Pacific in about 40 million years, and neighboring Berkeley (which lies on the American

BOX 9.2 continued

North ↔ South

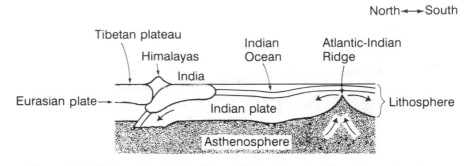

Figure C. Collision of the Indian Plate with the Eurasian Plate, illustrating the uplifting of the Himalayan Mountains.

Plate) will move south of the present latitude of San Diego.

Continental plates. The discussion so far has been mainly concerned with the movements of oceanic plates. They are the ones that are created along rift valleys and destroyed along subduction zones. Continental plates behave differently for two reasons: one, the magma that wells up along rift valleys is basalt and, therefore, only forms oceanic plates; and two, continental plates are too light to be subducted. They stay permanently on the Earth's surface, which is the reason why some parts of them (such as the shields) are billions of years old.

However, continental plates do not remain unaffected by plate movement. They may collide with and override an oceanic plate. In this case, the edge of the continental plate is raised and coastal mountain ranges, dotted with volcanoes, are formed, as already noted (see figure B). Examples are the coastal mountain ranges along the west coast of North America and the Andes of South America. A continental plate may also collide with another continental plate, in which case the edges of both plates along the collision front buckle and are uplifted (see figure C). Examples are the Alps, which formed by the collision between the African and European Plates, and the Himalayas, which formed by the collision between the Indian and Asian Plates. Continental plates may also break up into separate plates through the formation of rift valleys across them. This happens when the forces pulling on

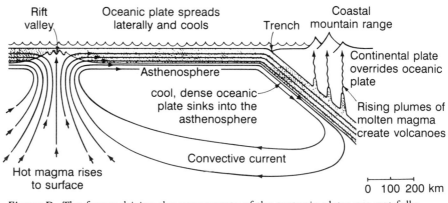

Figure D. The forces driving the movements of the tectonic plates are not fully understood. But it is believed that convective currents in the underlying mantle and the sinking of cooled-off, dense oceanic plates along subduction zones are chiefly responsible.

BOX 9.2 continued

a continental plate are in opposite directions and stretch it beyond the breaking point. An example is the formation of the Atlantic Ocean 200 million years ago, which resulted from the splitting of the original supercontinent into the American Plate and the Eurasian/African Plates. Another example is the present-day rift valley that extends from the Indian Ocean into the Red Sea and southward through eastern Africa, along such large, elongated lakes as Lake Turkana, Lake Tanganyika, and Lake Nyasa. As a consequence of the lateral pulling along this rift valley, the Arabian Plate has already split from Africa, and the northern part of East Africa is in the process of doing so.

To summarize, oceanic plates are created along rift valleys, spread laterally, and descend into the mantle along subduction zones. Continental plates partake in the lateral movements, but today are neither created nor destroyed. They may collide with other plates (continental and oceanic), fuse with each other, or break into two or more plates.

The forces driving these movements are not fully understood. But the convective motions within the Earth surely contribute significantly, by dragging the overlying lithosphere plates along (see figure D). Additional forces arise from the sinking of oceanic plates along subduction zones. They sink because as they spread away from the rift valleys, they cool, become denser, and, hence, less buoyant.

These plate movements have been going on for billions of years, and the accompanying sequence of maps (figure E) illustrates how they have reshaped the Earth's continental distributions over the past 550 million years. At the beginning of this period, the continents were relatively small in size, large in number, and widely dispersed. They then began to converge slowly and, approximately 225 million years ago, fused into a single supercontinent, *Pangaea*. By about 180 million years ago, this supercontinent began to break up again, first into a northern and southern block—*Laurasia* and *Gondwanaland*—and subsequently into the continents that we witness today. These slow, but relentless changes of our planet's surface features have had profound effects on the evolution of life, some of which are discussed in the text.

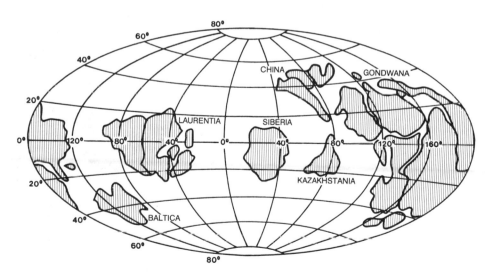

Figure E. Continental Drift. It is now commonly accepted that the Earth's surface features are slowly, but relentlessly changing due to the movements of the tectonic plates. The nine "snapshots" shown here illustrate in an approximate fashion the drifting of the continents from about 550 million years ago to 50 million years into the future. *Map 1.* The continents are →

BOX 9.2 *continued*

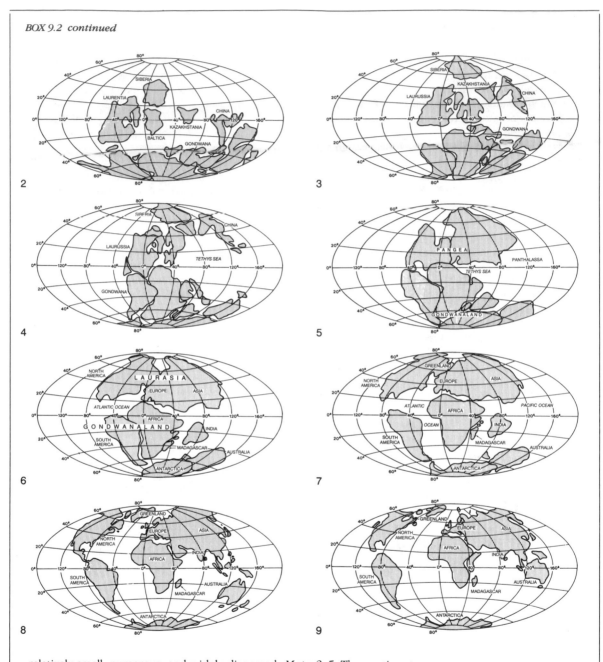

relatively small, numerous, and widely dispersed. *Maps 2–5.* The continents slowly converge into a single supercontinent, Pangaea, surrounded by an enormous ocean, Panthalassa (meaning "all sea"). *Maps 6–8.* Pangaea breaks up into a northern and a southern block—Laurasia and Gondwanaland—and subsequently into the continents existing on Earth today. *Map 9* shows world geography as it may look in 50 million years if present-day plate movements continue.

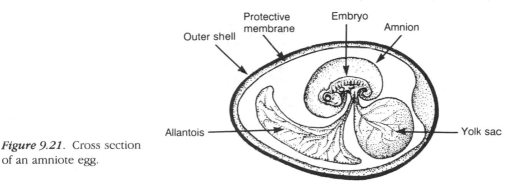

Figure 9.21. Cross section of an amniote egg.

The eggs of reptiles are *amniote eggs* (see figure 9.21). Those of the early reptiles were large and covered by tough protective membranes or shells (see figure 9.22), which prevented desiccation, in contrast to the small, soft eggs laid in water by the amphibians. Furthermore, the eggs contained yolk, the nourishment for the developing embryo. They also contained two sacs: one of them, the *amnion* (from which the eggs get their name), was filled with a cushioning liquid and enclosed the embryo; the other, the *allantois*, was the receptacle for the embryo's waste products. Finally, the membranes enclosing the eggs were porous, which allowed oxygen to diffuse in and carbon dioxide to go out. Thus, the amniote egg eliminated the need for the larval stage in reproduction. It provided instead a protected, self-contained private pool in which the offspring developed. This liberated the animals possessed of such an egg from having to return regularly to water for reproduction, as was the case with the amphibians, and opened up to them vast new territories for colonization.

In addition to changes in reproduction, some reptiles also evolved longer and more slender limbs, whose orientation slowly changed from the horizontal to a more vertical direction. This made running much more efficient, in comparison to the labored waddle of the amphibians. The sense of sight, smell, and taste

Figure 9.22. Broken dinosaur egg discovered in the Gobi Desert, Mongolia.

Eusthenopteron, a lobe-finned fish
from the Devonian

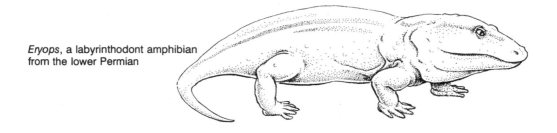

Ichthyostega, a vertebrate at the
transition stage from fish to amphibian, from the Devonian

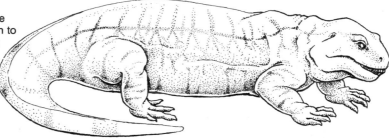

Eryops, a labyrinthodont amphibian
from the lower Permian

Diadectes, a seymourian, at the
transition stage from amphibian to
reptile, from the lower Permian

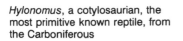

Hylonomus, a cotylosaurian, the
most primitive known reptile, from
the Carboniferous

Figure 9.23. The fossil record allows us to recognize a number of distinct stages in the evolution from the fishes to the amphibians and the reptiles, as illustrated here by five examples.

improved and became better suited for the requirements of air, in contrast to those imposed by water. Finally, there were a great many changes in the organ-system structures, particularly in the urinary, respiratory, and nervous systems, that contributed still further to the reptiles' outstanding adaptation to life on land.

Interestingly, fossils found in early Permian sediments north of the town of Seymour, Texas, appear to stem from animals that bridged the gap between the two classes of vertebrates. The animals (which have been named *Seymouria*) possessed skull bones and teeth similar to those of the labyrinthodont amphibians, but their backbones and vertebrae resembled those of the ancient reptiles. Unfortunately, no fossil eggs of *Seymouria* have been found so far, so we do not know whether these animals reproduced by laying and fertilizing their eggs in water or by laying amniote eggs on land.

Despite their particular features, the seymourians are not considered part of the trunk lineage that led from the amphibians to the reptiles. They were probably just one of several groups of amphibians that independently evolved reptilian characteristics and eventually became extinct, similar to the many other specialized and now extinct groups that are part of the general pattern of evolutionary experimentation.

The oldest known vertebrate that probably lay on the trunk lineage to the reptiles was a small animal about 30 cm long, with the name *Hylonomus* (see figure 9.23). Its fossils have been found in Carboniferous swamp deposits at Joggins, Nova Scotia, indicating that this animal preceded *Seymouria* by perhaps 10 million years or more and confirming that the evolution of reptilian features took place in a number of different amphibian lineages and geographic localities. *Hylonomus* possessed a solidly built skull, with laterally placed eyes. The teeth, which rimmed elongated jaws, were sharp and well suited for catching prey. The trunk, vertebrae, and limbs were strong and similar to those of later reptiles, except that the limbs were still rather sprawling. Finally, the tail was long and tapered. No fossil egg of *Hylonomus* has been found so far. However, the animal's anatomical features leave no doubt that it was indeed a reptile, albeit a very primitive one.*

Hylonomus belonged to the order Cotylosauria, whose members are regarded by many paleontologists as the ancestral stock from which all of the more advanced reptiles, as well as the birds and mammals, evolved. Other experts disagree. They think that the various reptilian lineages arose separately from different amphibian ancestors. Whichever was the case, the reptiles' unique adaptations — reproduction by the amniote egg, strong backbone and limbs, sense organs adapted to life in air, and organ-systems particularly suited for life on dry land — led to an explosive evolutionary diversification, as they

---

*The oldest known fossil specimen that most paleontologists interpret as being the remains of a reptilian amniote egg dates from early Permian times and was found in the Texas Redbeds. Recent examination of the egg's structure and chemistry by Karl F. Hirsch of the University of Colorado Museum suggests that it was soft-shelled. Presumably, all of the eggs of the earliest reptiles were soft-shelled and, therefore, did not fossilize well. That probably is the reason why they are rare. In contrast, the eggs of reptiles of later times had leathery shells and those of the birds had hard, calcareous shells similar to today's chicken eggs. These eggs fossilized more readily and are quite abundant in the fossil record.

Subclass	Distinguishing Characteristic	Fossil Representatives	Modern Representatives	Skulls of Ancient Representatives
Anapsida	No temporal opening in the skull behind the eye	cotylosaurs mesosaurs	turtles	Cotylosaur
Synapsida	A single lateral temporal opening	pelycosaurs therapsids (mammal-like reptiles)	(mammals)	opening Pelycosaur
Euryapsida	A single superior temporal opening	protorosaurs ichthyosaurs placodonts sauropterygians (incl. plesiosaurs)	none	opening Plesiosaur
Diapsida	Two temporal openings	lepidosaurs thecodonts dinosaurs flying reptiles	lizards snakes crocodiles and alligators tuataras (birds)	openings Thecodont

began to colonize our planet's land areas. This diversification began early in the Permian, accelerated rapidly late in this period, and lasted throughout the Mesozoic era. It made the reptiles the dominant land vertebrates for approximately 180 million years, a time span known as the "Age of Reptiles."

The number of reptile species that arose during the Mesozoic was very large and classifying them is a difficult task. Briefly stated, there are four subclasses of reptiles based on the presence or absence of temporal (from "temple" or "side of the head") openings in the skull behind the eye, as illustrated in figure 9.24.

Subclass Anapsida. The most ancient of these subclasses was probably Anapsida, which included the cotylosaurs mentioned above and the *mesosaurs* (figure 9.25). The mesosaurs lived during late Carboniferous and early Permian times. Fossil specimens indicate that these reptiles were quite small with long, slender jaws and long, tapered tails. They were probably well adapted to swimming and catching fish, and are likely to have inhabited freshwater lakes and ponds. The only modern survivors of the anapsids are the turtles

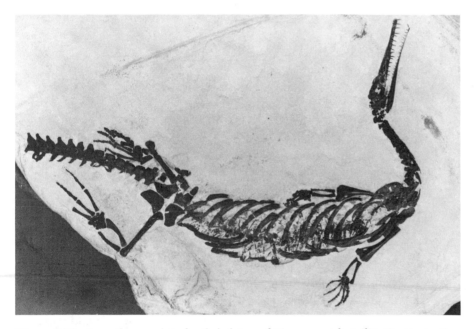

Figure 9.25. A nearly complete fossil skeleton of *Mesosaurus brasiliensis*, an aquatic reptile from the Permian, Irati Formation, Brazil.

Subclass Synapsida. It may be surprising, but one of the earliest groups of reptiles was the one that eventually led to the mammals. They were the *pelycosaurs*, of the subclass Synapsida (see figure 9.26). They first appeared in the late Carboniferous and closely resembled the cotylosaurs, from whom they probably evolved. Many of them grew to 2 to 3 meters in length. And some of

← *Figure 9.24.* The four subclasses of reptiles and their distinguishing characteristics. (Examples in parentheses have evolved into separate classes.)

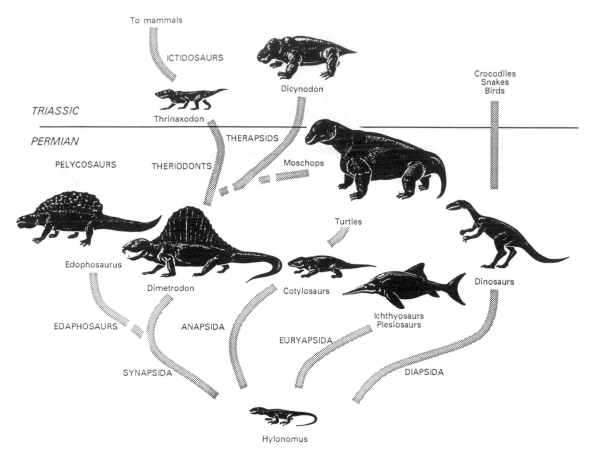

Figure 9.26. Evolution of the reptiles, with details given for the synapsids, or mammal-like reptiles.

these larger ones, such as *Dimetrodon* and *Edaphosaurus*, distinguished themselves by dorsal "sails," which may have been temperature regulators. If this interpretation is correct, the animals probably stood in the sun and used the sails as heat collectors when it was cold, and they sought out shade and used the sails as heat radiators when it was hot.

From the pelycosaurs evolved the therapsids of Permian and Triassic times, which are thought to have been the direct ancestors of the mammals. The evidence comes from the evolutionary changes that took place in their skulls, jaws, teeth (which became differentiated into incisors, canines, and molars and clearly served a mammal-like, carnivorous mode of life), and general skeletal features. Slowly these structures became more and more mammal-like and, by late Triassic and Jurassic times they had reached forms that make it a matter of mere definition whether these animals should be regarded as reptiles or mammals. Their further evolution will be discussed in the following section.

Subclass Euryapsida. Interestingly, some groups of reptiles returned to life in water. Among them were the *ichthyosaurs* ("fish lizards") and *plesiosaurs*, of the subclass Euryapsida. The ichthyosaurs readapted to an aquatic mode of life much as today's mammalian dolphins and porpoises have done (figure 9.27). Their streamlined bodies, with limbs resembling fins, a dorsal fin, and a fishlike tail, suggest that these animals rarely, if ever, came onto land. And fossilized skeletons with babies inside indicate that they reproduced by giving birth to live young in the sea. The plesiosaurs were probably somewhat less water-bound than the ichthyosaurs and some of them may have come ashore occasionally, perhaps to lay their eggs. But their flipperlike limbs would have made walking very clumsy. None of these ancient aquatic reptiles has survived to modern times.

Subclass Diapsida. The most spectacular reptiles of the Mesozoic were the *dinosaurs*, or "terrible lizards," of the subclass Diapsida (see figure 9.28). They were so named because some of them grew to enormous sizes, such as *Apatosaurus* (also known as *Brontosaurus* ["thunder lizard"], see box 9.3), *Brachiosaurus*, and *Diplodocus*. They reached up to 25 meters in length and weighed as much as 35 metric tons. These giants were all herbivores, as is indicated by their blunt teeth. Many other dinosaurs, that were only slightly smaller, were herbivores, too. Examples are the heavily armored *Stegosaurus* ("roofed lizard"), the three-horned *Triceratops, Scolosaurus* with its cover of spiked bony plates, and *Iguanodon*, which was a biped and had large, horny thumbs. Some of them probably roamed in large herds and fed on the lush vegetation of the swamps and forests, which by then included mosses, ferns, conifers, and, in the later Cretaceous, flowering plants. Other dinosaurs were hunters that ate herbivores and each other. Examples were *Tyrannosaurus* and *Allosaurus*, both of which were bipedal, with massive hind legs and tiny clawed front legs used for grabbing. Still other dinosaurs were much smaller and

Figure 9.27. The mesozoic reptiles were a diverse group, much like today's mammals. And like them, some mesozoic reptiles returned to life in water, such as *Ichthyosaurus*.

(a) **Herbivore dinosaurs**

Apatosaurus
(25 meters, 35 metric tons)

Figure 9.28. Examples of dinosaurs that were herbivores (*Apatosaurus* and *Tricera-tops*), and of hunters that ate the herbivores and each other (*Allosaurus* and *Tyran-nosaurus*). Estimated lengths and weights of the animals are given in parentheses.

Triceratops
(9 meters, 10 metric tons)

(b) Carnivore dinosaurs

Allosaurus
(9 meters, 10 metric tons)

Figure 9.28. continued

Tyrannosaurus
(12 meters, 15 metric tons)

probably filled ecological niches that today are occupied by deer, rabbits, and dogs.

In addition to spreading out over the land and reinvading the water, some Mesozoic reptiles took to the air (see figure 9.29). These flying reptiles were close relatives of the dinosaurs and also belonged to the subclass Diapsida. They

---

BOX 9.3.
Apatosaurus

THE search for dinosaur fossils has a colorful and often stormy history. In the United States the search began in the mid-1800s, when fur traders came back with large fossilized bones from the upper Missouri River country, an area that today includes parts of Colorado, Wyoming, and Montana. By the late 1870s, attempts to be first in discovering new dinosaur fossils erupted into the "Dinosaur Wars" between two American paleontologists, Othniel Charles Marsh (1831–1899) of Yale's Peabody Museum of Natural History and Edward Drinker Cope (1840–1897) of Philadelphia. In their haste to outdo each other, both men sometimes mislabeled or misinterpreted their fossil finds. For example, Marsh first found a specimen of *Apatosaurus* in 1877. Two years later, when he found another specimen, he called it *Brontosaurus*, not realizing that both animals were the same.

Perhaps Marsh's greatest error was associating the wrong head with *Apatosaurus*'s body. In his restoration of this large animal, he relied on two skulls found 4 miles and 400 miles from the rest of the skeleton, details he never mentioned in his publications. The skulls were, in fact from a different and slightly smaller dinosaur, *Camarasaurus*. As a result, the exhibits of *Apatosaurus* in museums around the world displayed the wrong head. Though there have been suspicions since about 1915 that something was wrong with *Apatosaurus*'s head, only recently was the mistake fully recognized, mainly through the efforts of John S. McIntosh of Wesleyan University in Middletown, Connecticut, and David S. Berman of the Carnegie Museum of Natural History in Pittsburgh. In 1979, the *Apatosaurus* skeleton exhibited in the Carnegie Museum was the first to obtain the correct head, one that is longer and more slender than the old head. Other museums are intending to follow suit.

---

evolved wings composed of skin that stretched from their forelimbs, including a greatly prolonged fourth finger (corresponding to our "ring" finger), to their hind legs; and their bones became hollow, which reduced their weight. One of the earliest of the flying reptiles was *Rhamphorhynchus* from the Jurassic, which had a long reptilian tail and a beak studded with large teeth. In later forms, such as *Pterodactylus, Pteranodon*, and *Pterosaurus* ("winged lizard"), the tail became reduced to a stub and the teeth were either mere bristles or disappeared entirely. The sizes of the flying reptiles varied greatly, from giant *Pterosaurus* with a wingspan of up to twelve meters, down to animals resembling today's pigeons and sparrows.

None of the flying reptiles possessed feathers, though a moderately closely related group did. Exceptionally well-preserved fossil specimens of this group's prototype, *Archaeopteryx* ("ancient wing"), come from a Jurassic limestone quarry near the Bavarian town of Solnhofen, Germany (see figure 9.30). *Archaeopteryx* was roughly the size of a modern crow, with a long feathered tail, a skull that narrowed into a beak lined with well-developed teeth, strong hind limbs, and enlarged feathered forelimbs. These forelimbs resembled wings, though it is not clear whether they and the associated muscles were sufficiently developed to have allowed the animal to carry out flapping flight. The feathers, which had evolved from scales, also covered the body. They probably served as insulation to retain body heat, suggesting that *Archaeopteryx* was warm-blooded. If true, this was a most significant development. Warm-bloodedness greatly increases metabolism, in comparison to the metabolism of cold-blooded animals, and is of great advantage for flight, with its high demands on energy consumption.

Figure 9.29. Some ancient reptiles took to the air, such as *Rhamphorhynchus* and *Pteranodon*.

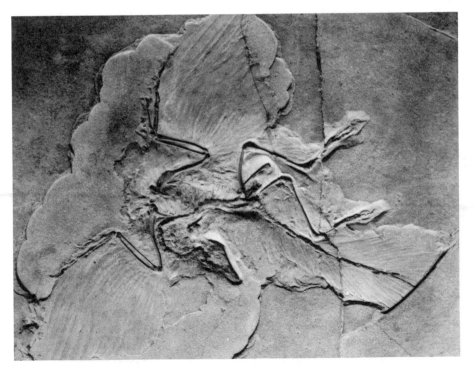

Figure 9.30. *Archaeopteryx*. This fossil skeleton, with the impressions of feathers, is the earliest known record of a bird. It was found in a Late Jurassic limestone quarry near the Bavarian town of Solnhofen, West Germany.

From animals like *Archaeopteryx* evolved all modern birds. They acquired exceptionally complex behavior patterns, such as nesting and singing and, in some species, the ability to make long migrations every year. Further, their physical characteristics allowed them to adapt to a remarkably wide range of habitats from the air to water and land.

By the end of the Mesozoic, roughly 65 million years ago, the dinosaurs and other members of the subclass Diapsida, along with most other reptiles, became extinct (see box 9.4). The only survivors of the diapsids were the lizards, snakes, crocodiles, alligators, tuataras (little-known reptiles from New Zealand), and birds, the last of which had evolved by then into a new class, the *aves*. Of the other reptiles, only the turtles and mammals, which also had become separate classes, survived. Thus, the long reign of the reptiles came to an end. But, as always in evolution, the end of the dominion of one group of organisms ushers in the rise of another. The close of the Mesozoic era was also the beginning of the "Age of the Mammals."

THE CENOZOIC ERA: THE AGE OF MAMMALS

The mammals—bats, shrews, rabbits, mice, dogs, whales, sheep, horses, elephants, and man, to name just a few—have today a wider geographic distribution and are more adaptable than any other class of animals, with the exception of the insects, arachnids, and, possibly, the birds. They are found in all habitats from the equator to the polar regions, from sea level to the snow and glaciers of the highest mountain ranges, under ground and above, in the sea and in the air.

Mammary Glands and Hair

The mammals acquired their remarkable adaptability by evolving numerous anatomical and physiological features that distinguish them in fundamental ways from their reptilian ancestors. Most significant among these features, and the defining characteristic of the class, are the mammals' habit of *nourishing their young with milk, secreted by the females from special mammary glands, and their having a skin covered in varying degrees by hair*. The mammary glands, or breasts in humans, evolved initially from sweat glands. The evolutionary origin of hair is not clear (the idea that hair evolved from the scales of reptiles, which has been proposed, is inconsistent with embryological development), but it probably evolved as insulation against excessive loss of body heat. This is important, for mammals are warm-blooded and their present-day members maintain a nearly constant body temperature of about 26–39° C. In humans, for example, the normal body temperature is 37° C, with fluctuations of about one-half degree in the course of the diurnal cycle.

Metabolic Rate and the Four-Chambered Heart

The evolution of warm-bloodedness greatly increased the mammals' basal metabolic rate (that is, the metabolic rate when the animals are resting and not

BOX 9.4.
The Mass Extinction at the End of the Mesozoic Era

THE dinosaurs were not the only species that became extinct at the end of the Mesozoic era, approximately 65 million years ago. Many other animals and plants, both on land and in the sea, also died out at about that time. It was a mass extinction similar to that at the end of the Paleozoic era, 180 million years earlier.

Over the years, many theories have been offered to explain the mass extinction of 65 million years ago, including a drastic lowering of the dinosaurs' fertility, the explosion of a nearby supernova, an increase in solar UV-radiation reaching the ground, and severe cooling of the world's climate by the flow of cold arctic waters to the south. None of these theories has been convincing enough to be accepted by the scientific community.

Then in 1980, Luis W. Alvarez, his son Walter, Frank Asaro, and Helen V. Michel, all of the Lawrence Berkeley Laboratory in Berkeley, California, proposed that the cause of the mass extinction was a collision of Earth with an asteroid or comet that was perhaps ten kilometers across. The force of the impact was so severe that several 10^{14} tons of dust and steam were thrown into the upper atmosphere, blocking out sunlight and creating nearly total darkness everywhere on the surface of the Earth for about six months. As a result, temperatures plummeted to well below freezing, even at the equator, and photosynthesis came to a halt. All animals that required a warm climate and depended on fresh vegetation for food succumbed. Likewise, carnivores that ate the herbivores died. The dinosaurs belonged to one or the other of these categories and, hence, they all became extinct. Animals that could cope with the cold and whose diet consisted chiefly of nuts, seeds, and insects stood a much better chance of surviving, particularly if they were also small so that they required little food. Most of the mammals belonged to this category and, hence, many of them survived the catastrophe.

Alvarez and his colleagues based their suggestion on the discovery of an unusual layer of clay, 65 million years old, in a limestone cliff near Gubbio, a small town in the Apennine Mountains of Italy. The clay layer contains a 30-fold increase in iridium, compared with the limestone above and below. Since 1980, similar or still higher iridium enrichments in rock deposits 65 million years old have been found at approximately forty sites all over the globe. The point is that iridium is a rather rare element on the surface of the Earth, but is much more common in meteorites. (Iridium is a siderophile element and most of it sank, along with iron, to the core of our planet during its formation, see chapter 5, p. 212). The most consistent explanation is that the source of the iridium in these rock deposits was an asteroid or comet that collided with Earth and blanketed the entire planet with fine, pulverized debris, first in the upper atmosphere and later, after the matter had drifted downward, on the ground and in the ocean. Incidentally, no crater has so far been found that can be unequivocally associated with the proposed impact, even though the probability of finding remnants of such an impact have been estimated as roughly 70 percent.

Despite the lack of definite proof that a large asteroid or comet collided with Earth roughly 65 million years ago, the worldwide iridium concentrations in rock layers of roughly that age are accepted by most scientists as sufficient evidence for such a collision. In fact, based on the known flux of Earth-crossing asteroids (such as the Apollo asteroids), such an event is expected to occur on the average once every 100 million years.

Not all scientists agree that such an impact was responsible for the mass extinction at the end of the Mesozoic. They argue that the dinosaurs and many other species had already begun to decline millions of years earlier and would have become extinct sooner or later in any case, though the cold and darkness following the meteorite impact might have speeded up their demise. Besides, they point to evidence suggesting that the mass extinction took place over tens of thousands of years, rather than suddenly, as would have been the case if an asteroid had been the chief cause. They also point to the fact that some dinosaurs lived north of the Arctic Circle, which implies that these animals must have been well adapted to long periods of darkness and, perhaps, to rather severe seasonal changes in climate. Therefore, they are unlikely to have become extinct on account of conditions caused by an asteroid or comet impact. Many of these scientists believe that gradual and cumulative changes in the Earth's surface conditions at the close of the Mesozoic—such as a drop in the sea level by 50 to 100 meters, an increase in volcanic activity, rapid cycles of atmospheric heating and cooling, and widespread deoxygenation of ocean water—are more likely to have been the main causes of the mass extinction.

digesting food), in comparison with that of their cold-blooded reptilian ancestors.* For instance, today's shrew, a very tiny mammal, has such a high metabolic rate that it consumes its weight in food every three to four hours. Larger mammals—in particular the giant herbivores such as the elephants, hippopotamuses, and bisons—have a much lower metabolic rate, weight for weight. But, except during periods of inactivity such as hibernation, it still is generally higher than that of the reptiles.

Increasing the metabolic rate required increased intakes of oxygen for respiration. This problem was solved by the evolution of a four-chambered heart (figure 9.31), in which oxygenated and deoxygenated blood never mixed, as the blood did (and still does) in the hearts of reptiles. (Birds, incidentally, also evolved a four-chambered heart.) Furthermore, the mammals acquired a diaphragm that separated the heart and lungs from the abdomen and assisted through muscle action in the inhaling and exhaling of air.

Jaw and Tooth Evolution

Additional evolutionary changes among the mammals included the development of a direct hinge between the lower jaw and the skull (see figure 9.32). In the process, the two bones that formed the hinge in reptiles (the articular and quadrate) became transformed into two of the tiny bones in the mammalian ear, the hammer and anvil. Furthermore, mammalian teeth became highly differentiated and shaped for specific tasks. The incisors in the front became sharp (cutting teeth); the canines remained conical and pointed (piercing and tearing teeth, also useful in defense); and the cheek teeth, or molars, in the back became broad and flat (grinding and chewing teeth). The singly hinged jaw and differentiated teeth, combined with improved locomotor skills (see below), made the mammals very efficient eaters and greatly increased the variety of food accessible to them.

Improved Nervous Systems and Locomotion

Increases in metabolism were of little use unless they could be applied to giving the animals greater sustained speed and agility in going after food, defending themselves or their young against predators, escaping from danger, and carrying out the numerous routine activities that are part of daily life. The two most important developments in this regard were improvements in the nervous systems and in bodily movements. The nervous system improved mainly by enlargement of the cerebrum, the upper part of the brain, and extension of its control over the various body parts. Along with improvements of the nervous systems, the senses for picking up information about the environment became

---

*It has been estimated that for every drop of 10° C in the body temperature of cold-blooded animals the metabolic rate is reduced by half. This explains why, for example, snakes, lizards, and frogs are extremely sluggish on cold mornings and become active only after the Sun has warmed them. (Metabolism refers to the sum total of biochemical reactions of an organism, including those of energy generation and of the synthesis of complex molecules.)

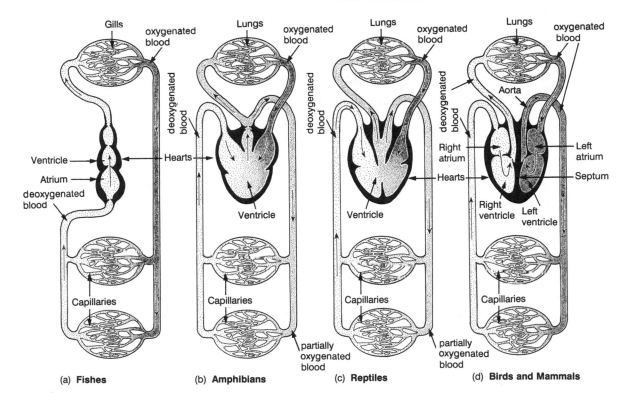

Figure 9.31. Diagrammatic illustration of the circulatory systems and hearts of (*a*) fishes, (*b*) amphibians, (*c*) reptiles, and (*d*) birds and mammals, indicating progressive levels of complexity and efficiency. The heart of the fishes is essentially a pulsating tube that pumps blood through both the gills and the capillaries of the body. Amphibians, reptiles, birds, and mammals have hearts with two parts, one pumping the blood through the lungs and the other pumping it through the capillaries of the body. In amphibians and reptiles the two parts of the heart have a common chamber, the ventricle, and oxygenated and deoxygenated blood mix (meaning that some oxygenated blood is pumped again through the lungs and some deoxygenated blood is pumped again through the capillaries of the body). In birds and mammals, a septum separates the two parts of the heart and prevents mixing of oxygenated and deoxygenated blood. Interestingly, the embryonic development of the mammalian heart begins with a tube and passes through stages analogous to the amphibian and reptilian hearts.

sharper also. For example, the development of three bones in the middle ear increased the animals' hearing, and changes in the nasal structure suggest that the sense of smell became more acute as well.

Bodily movements were improved by numerous changes in the skeletal structure, as illustrated in figure 9.33. Of particular importance was the gradual shifting of the two pairs of limbs more and more beneath the body, with the knees of the hind legs pointing forward and the elbows of the forelegs pointing backward (see figure 9.34). It was a process that had already begun with ancient mammal-like reptiles, but was carried through to completion only in the true mammals. As a result, the animals were raised off the ground, allowing them to

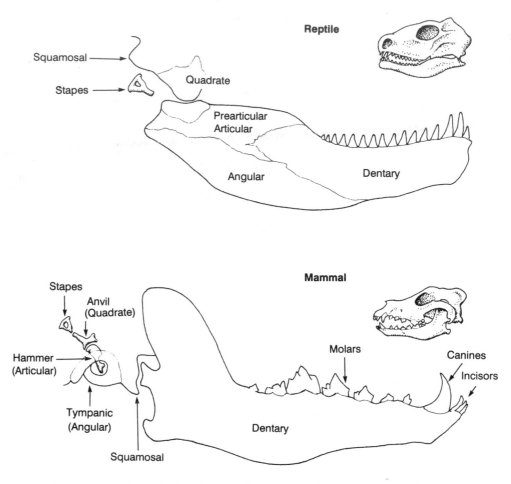

Figure 9.32. The evolution from reptile to mammal was accompanied by significant changes in the lower jaw and ear region. In reptiles, the lower jaw contains six or more bones (top diagram), the ear contains one sound-transmitting bone (the stapes), and the teeth are generally simple cones and rather uniform. In mammals, the lower jaw is a single bone (the mandible), the ear contains three sound-transmitting bones (the hammer, anvil, and stapes), two of which were derived from reptilian jaw bones (the articular and quadrate), and the teeth are highly differentiated.

carry their weight with a minimum of fatigue. This generalized limb structure gradually produced numerous adaptations specialized for particular kinds of locomotions and modes of life: the walking and running of rats and hedgehogs, the sustained high-speed running of horses and dogs, the leaping of rabbits and kangaroos, the burrowing of moles, the swimming of beavers, otters, whales, and dolphins, the flying of bats, the climbing of the primates, and the upright gait of humans.

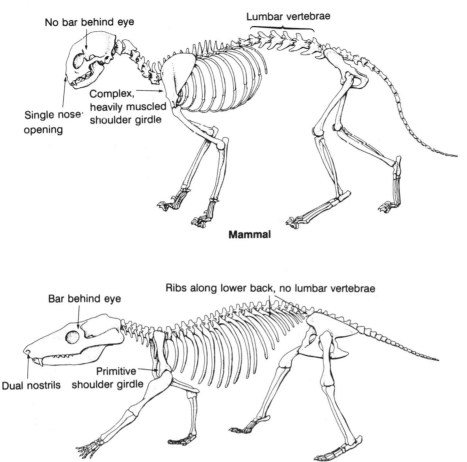

Figure 9.33. The skeletons of a therapsid and an early mammal, showing both similarities as well as important differences between the two. The mammalian skeleton allows greater flexibility, agility, and activity.

Reproduction

Another important characteristic of the mammals was the evolution of a number of new reproductive strategies, which gave rise to new and distinct orders. For example, each of the three surviving orders of mammals—the monotremes, marsupials, and placentals—has its own distinct style of reproduction.

The *monotremes* have the most primitive and reptilelike reproduction method. A female lays eggs. Depending on the species, she places the eggs either in a nest or in a temporary pouch on her underside. The eggs are then incubated by her body heat and partially developed young are hatched. The young remain in the nest or pouch and feed on milk secreted by the mother. Feeding by the mother continues until the offspring are ready for an independent life.

The fossil evidence of the monotremes is very meager, suggesting that they probably never constituted a large order. Their only modern descendants are the platypuses of Australia, which lay their eggs in nests (see figure 9.35), and the echidnas or spiny anteaters of Australia and New Guinea, which lay their eggs in temporary pouches.

The reproduction of the *marsupials* proceeds without the laying of eggs. Instead, the fetuses develop within the mother's womb from eggs and then leave the womb. At that point they are still in a very unfinished state, without hair, blind, and resembling tiny worms. Their hind legs are mere buds, but their forelimbs are more developed and with them they crawl to a pouch that, as in the monotremes, is located on the mother's abdomen. There they fasten themselves onto rather long teats and nurse on milk. After the babies become strong enough to walk, they begin to make their first excursions out into the world, though for some time they continue to return to the pouch for feeding, warmth, and safety.

The place of origin of the marsupials, like that of the other mammalian orders, is not known. However, their greatest early development occurred in North America. They were also present in Gondwanaland (today's South America, Africa, and Australia, see box 9.2), and there is some evidence that they spread into Antarctica. With the exception of the opossums and caenolestids, which today live in South and Central America (though one species, the Virginia opossum, ranges as far north as southern Canada), their only modern descendants are found in Australia. Examples are the kangaroos (see figure 9.36), wombats, bandicoots, marsupial wolf, phalangers, and koalas.

The most elegant process of reproduction is that of the *placental mammals*. The fetus grows in the mother's womb (figure 9.37), protected by a fluid-filled sac, the amnion, comparable to the amnion in reptilian eggs. It remains there

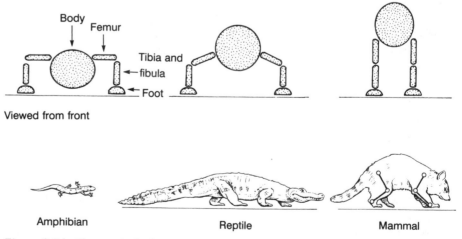

Figure 9.34. Changes in limb posture and locomotion in the evolution from amphibians to reptiles and mammals. Amphibians have a sprawling posture close to the ground and during walking their bodies execute S-shaped waves, when viewed from above. In contrast, mammals are raised high off the ground, with the limbs beneath the body, the knees of the hind legs pointing forward, and the elbows of the forelegs pointing backward, thereby greatly increasing the efficiency of walking and running.

Figure 9.35. A platypus, one of the two surviving members of the monotremes. The platypus is a small, densely furred, aquatic animal with a fleshy bill resembling that of a duck. It lives in southern and eastern Australia and Tasmania.

Figure 9.36. A kangaroo. Kangaroos are herbivorous leaping marsupials, found in Australia, Tasmania, New Guinea, and on the islands east of the Bismark Archipelago. They occupy niches that elsewhere are held by grazing and browsing animals.

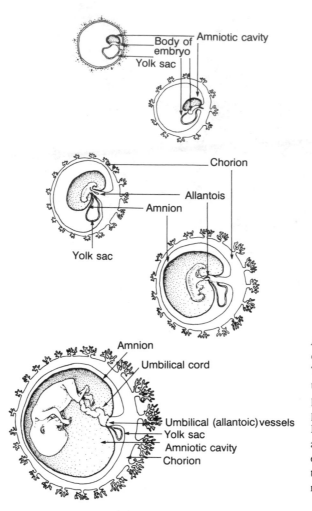

Figure 9.37. Five stages in the development of a human embryo. The *amnion* is the membrane lining the *amniotic cavity*, which is the protective, water-filled bag that protects and cushions the embryo. From the *allontois* develop the arteries and veins that run from the embryo through the *umbilical cord* to the *chorion*, a highly vascular membrane and part of the placenta.

for a longer time and develops to a much higher degree than in the case of the marsupials. For example, in the Asiatic elephant the gestation period is nearly 2 years (the longest of any mammal), in humans it is 9 months, and in the hamster it is 16 days (the shortest) on the average, in comparison to 12 days for the Virginia opossum. While in the womb, the fetuses are nourished, receive oxygen, and have their waste products removed through an intricate vascular organ, the *placenta* (see figure 9.38). In the placenta the mother's and the fetus's blood streams circulate around each other in interlocking networks of capillaries, though the two blood streams do not mix. They merely exchange nutrients, gases (O_2 and CO_2), and wastes. After birth, the young live by drinking milk secreted from the mother's mammary glands and for months or even years they require parental care and protection. Without question, the placental mammals are today the most successful members of their class, having wide geographical distribution and adaptability.

Animals as complex as mammals require long periods of development, compared with reptiles and amphibians. Furthermore, mammalian fetuses are

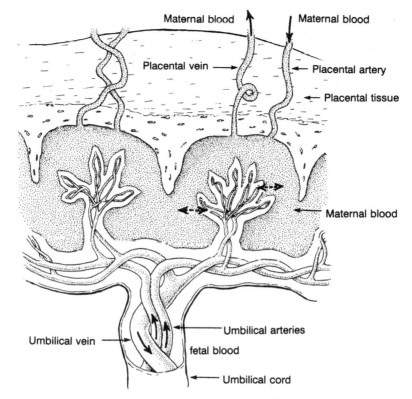

Maternal blood

Maternal blood

Placental vein ⟶

⟵ Placental artery

⟵ Placental tissue

⟵ Maternal blood

Umbilical vein ⟶

Umbilical arteries

fetal blood

⟵ Umbilical cord

Figure 9.38. Detail of circulation of maternal and fetal blood through the placenta. The two blood streams are never in direct contact. Nourishment and waste materials are exchanged by diffusion across capillary membranes (dashed arrows).

warm-blooded, so large amounts of nourishment have to be supplied. In fact, the nourishment needed by the fetuses and newborn is much more than can be stored as yolk in eggs. Consequently, in the course of the evolution of the mammalian characteristics, it became progressively more difficult to procreate by simply laying eggs and letting nature take care of incubation and the hatching of young capable of independent survival. The evolution of new ways of reproduction was essential.

Origin of the Mammals

The many features that characterize today's mammals did not evolve all at once. Progress was made slowly and intermittently, first with some of the features and then with others, depending on the selective demands made at the time by the environment. The process began with the *pelycosaurs*, the most primitive of the mammal-like reptiles, more than 245 million years ago during the Permian period. One group of mammal-like reptiles, which evolved from the pelyco-saurs, were the *therapsids* (see the discussion of the subclass Synapsida above). The teeth in the most advanced therapsids were differentiated into incisors, canines, and molars, indicating that their eating habits were similar to those of

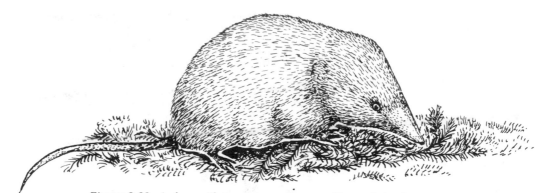

Figure 9.39. A shrew. Shrews are small, mouselike and chiefly nocturnal mammals with long, pointed snouts, small eyes, and velvety furs. They belong to the mammalian order *Insectivora*, which also includes moles and hedgehogs. They have an exceptionally high metabolic rate that forces them to feed nearly incessantly.

some mammals. Furthermore, their vertebral column, ribs, shoulder girdles, pelvis, and limbs were also very mammal-like.

Finally, during late Triassic and early Jurassic times, roughly 220 to 175 million years ago, a group of animals appeared whose skeletal and dental features indicate that they were crossing the threshold separating the mammals from the reptiles. Their lower jaws were essentially mammalian in structure, but the reptilian articular and quadrate bones were still present, though in much reduced form. These animals were the *ictidosaurs*.

From those times forward, fossilized teeth and bones give increasingly more frequent evidence of the existence and evolution of the mammalian lineage. These earliest mammals were quite small and roughly resembled today's shrews (see figure 9.39). They had five toes on each foot and a well-developed tail. Their jaws were relatively long and slender, and their teeth suggest that their diet consisted mainly of insects and seeds. Their sense of smell was highly developed and, very likely, most of them were nocturnal animals that hid in the undergrowth, in the cracks of rocks, and in burrows, trying to outwit the smaller of the carnivorous reptiles (see figure 9.40). They were an inconspicuous group of animals, not highly specialized for any particular way of life that would have required major modifications of the anatomy they inherited from their reptilian ancestors.

For more than 100 million years, the mammals remained small and were completely overshadowed by the dinosaurs and other reptiles, which during this period reached the height of their development. However, it was during this long stretch of time that the mammals' many remarkable features became perfected and a number of distinct subclasses arose, five of which are known from Triassic and Jurassic times: the *docodonts, symmetrodonts, eupantotheres, triconodonts,* and *multituberculates* (terms meaning "beam teeth," "symmetrical teeth," "true all animals," "triple-cone teeth," and "many-lumped teeth"). None of these subclasses have survived to the present. It is not quite clear when the docodonts became extinct. Some paleontologists think they evolved into today's *monotremes*, though this is a matter of controversy. The symmetrodonts and eupantotheres became extinct during the Jurassic period, the triconodonts during the Cretaceous period, and the multituberculates early in the Cenozoic era. Even though the eupantotheres died out, it appears that a side branch of them gave rise, early in the Cretaceous, to two new groups of mammals that have survived to the present — the *marsupials* and *placentals*.

When the dominance of the reptiles ended at the close of the Mesozoic, the marsupials and placental mammals were ready to move into the ecological niches that had become free. This resulted in a phase of very rapid evolutionary diversification and the emergence of a large number of new taxa, each with specialized features depending on the particular modes of life its members adopted (see figure 9.41). For example, hares and rabbits evolved teeth specialized for gnawing and relatively long legs for fast acceleration. Horses, zebras, and deer acquired teeth suitable for eating leaves and grass and long, powerful legs that were good for running at high speeds across open terrain. Cats, dogs, and bears became equipped for preying on other animals. Bats evolved wings and strong chest muscles that allowed them to fly. Whales and dolphins lost their hair, and their bodies became fishlike and well suited for life in the sea. And primates grew grasping hands and feet that allowed them to move about safely in trees.

Figure 9.40. *Morganucodon* ("morning-tooth"), one of the earliest mammals. The entire appearance of this animal suggests a nocturnal, smell-oriented way of life. The animal was about 10 centimeters long.

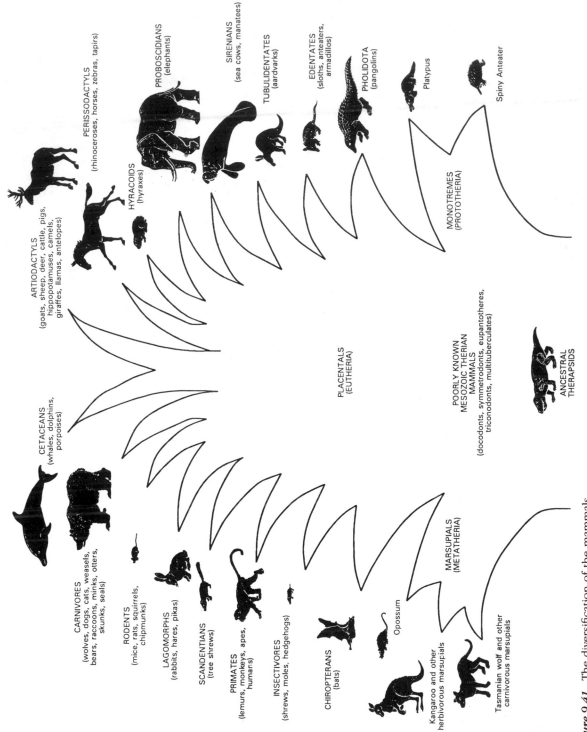

Figure 9.41. The diversification of the mammals.

This chapter followed the evolution of the vertebrates from the ancient jawless fishes that swam in the Cambrian sea to the amphibians, reptiles, birds, and mammals of the present. The success of the vertebrates, like that of all other animal phyla, depended on the fundamental equipment they acquired at their origin, or, put differently, on the range of problems that could be solved with the equipment, which included a notochord, dorsal nerve cord, and gill slits. Over the ages these structures evolved into strong yet flexible internal skeletons, appendages such as fins, limbs, and wings, jaws with teeth, and efficient nervous systems, including finely tuned sensory organs.

The equipment inherited by the vertebrates not only affected their successes but also placed limits on their evolutionary potentials and contributed to their failures. In particular, species in which the application of the equipment was pushed to extremes eventually succumbed because of their specializations, even though many were temporarily highly successful. Examples are the placoderms, which were protected by such large and heavy plates that speed and agility were sacrificed, and the dinosaurs, in many of which the size of their internal skeletons developed to the limit of the possible. Both of these specialists were backed into dead-end evolutionary avenues, from which there was no escape when environmental conditions changed. In contrast, species in which the fundamental equipment was applied in a less extreme manner, were often more successful in the long run. Examples are the cartilaginous and bony fishes, which began to rule the sea after the demise of the placoderms, and the mammals, which began to dominate the land after the extinction of the dinosaurs.

Exercises

1 (a) Name and briefly describe the earliest known fishes.
(b) Approximately how long ago did these fishes live?
(c) Who are their modern survivors?

2 (a) Which fishes were the first with true jaws?
(b) Approximately when did these fishes live?
(c) What impact did the origin of vertebrate jaws have on the subsequent course of evolution?

3 (a) Draw a phylogenetic diagram outlining the evolution from the earliest fishes to the amphibians. Include in your diagram the jawless fishes, placoderms, cartilaginous fishes, bony fishes, ray-finned fishes, fleshy-finned fishes, lungfishes, and lobe-finned fishes.
(b) Also include such specific examples as *Jamoytius, Dinichthys, Cladoselache*, salmon, trout, stur-

geon, *Neoceratodus, Latimeria, Eusthenopteron,* and *Ichthyostega*.
(c) Indicate the branch that led to the reptiles.

4 Why was the discovery of *Latimeria* in 1938 hailed as a major scientific success and reported in newspapers around the world? (See, for example, *The New York Times*, April 2, 1939, III, 4:2, and April 9, 1939, III, 4:3.)

5 Briefly describe the earliest land plants. Which structural characteristics had to evolve before the land plants could grow to great heights and successfully spread across the continents?

6 Land-based forests were necessary before animals could spread far beyond their ancestral water habitat. However, in the case of the vertebrates, the forests were probably not the chief cause of this spreading. What, according to the theory presented in the text, was the chief cause?

7 Both *Eusthenopteron* and *Ichthyostega* lie close to the phylogenetic dividing line between the fishes and amphibians, but the former is usually classified as a fish, the latter as an amphibian. Describe the reasons for this difference in classification.

8 Life on land presented new obstacles for the amphibians and their descendants that the fishes never had to face. Describe the most significant of these obstacles and how they were overcome in the course of evolution.

9 Discuss the pros and cons of the theory that an asteroid impact was responsible for the end of the Age of Reptiles roughly 65 million years ago.

10 The origin of the mammals was accompanied by the evolution of a number of significant anatomical and physiological characteristics that eventually allowed them to become the dominant land animals on Earth. Describe some of those characteristics and their competitive significance.

11 (*a*) Describe the size, appearance, and way of life of the earliest mammals.
(*b*) Approximately how long ago did they live?

12 (*a*) Name the three surviving orders of mammals, give examples of each, and indicate their geographic distributions.
(*b*) Briefly describe the reproductive processes characterizing each of the three orders.

13 Distinguish between the following geological terms:
(*a*) Crust, mantle, and core.
(*b*) Asthenosphere and lithosphere.
(*c*) Subduction zone, suboceanic ridge, and transform fault.
(*d*) Basalt and granite.
(*e*) Pangaea, Panthalassa, Gondwanaland, Laurasia, Tethys Sea.

14 The Earth is a dynamic and, at times, violent planet, as earthquakes, volcanism, and the slow but steady movements of the continents amply demonstrate. Summarize the essential features of the modern theory of plate tectonics, which explains this dynamism and violence.

Suggestions for Further Reading

The reader should also consult the Suggestions for Further Reading, Chapter 8.

Bakker, R. T. 1975. "Dinosaur Renaissance." *Scientific American*, April, 58–78. The author argues that dinosaurs were not obsolete reptiles but a novel group of warm-blooded animals and that birds are their descendants.

Buffetaut, E. 1979. "The Evolution of the Crocodillians." *Scientific American*, October, 130–44.

Clarke, E. 1981. "Sharks, Magnificent and Misunderstood." *National Geographic*, August, 138–87. The author describes the excitement of studying and photographing sharks as well as the behavior and remarkable sensing abilities of these cartilaginous fishes, which began to evolve more than 300 million years ago.

Colbert, E. H. 1983. *Dinosaurs: An Illustrated History*. Maplewood, N.J.: Hammond Inc., 1983.

———. 1980. *Evolution of the Vertebrates: A History of the Backboned Animals Through Time*. 3d ed. New York: John Wiley & Sons. An excellent, well-illustrated text, for advanced students.

Dietz, R. S., and **J. C. Holden.** 1970. "The Breakup of Pangaea." *Scientific American*, October, 30–41.

"The Dynamic Earth." 1983. *Scientific American*, September. Eight articles on the structure, dynamic changes, ocean, atmosphere, and biosphere of our planet.

Dziewonski, A. M., and **D. L. Anderson.** 1984. "Seismic Tomography of the Earth's Interior." *American Scientist*, September–October, 483–94. By analyzing many earthquake waves, the authors derive three-dimensional maps of the Earth's mantle and gain some understanding of the convective flow that propels the crustal plates. (Also see "Seismic Tomography" by the same authors, *Scientific American*, October 1984, 60–68.)

Glut, D. F. 1982. *The New Dinosaur Dictionary*. Secaucus, N.J.: Citadel Press. An alphabetical listing of all presently known dinosaur genera, with brief summaries of important data and many superb illustrations.

Graham, L. E. 1985. "The Origin of the Life Cycle of Land Plants." *American Scientist*, March–April, 178–86. The author suggests that a simple modification in the life cycle of an extinct alga was the likely origin of the first land plants.

Hopson, J. L. 1981. "A Queer Mammal of Ducklike Bill and Reptilian Walk." *Smithsonian*, January, 62-69. Australia's platypus.

Horner, J. R. 1984. "The Nesting Behavior of Dinosaurs." *Scientific American*, April, 130–37. The discovery of large numbers of dinosaur eggs and skeletons of young dinosaurs in 80-million-year-old sediments in Montana has led to a novel interpretation of the social relations of these extinct reptiles.

Jackson, D. D. 1985. "It Is About Time the Shrew Stood Up to Be Recognized." *Smithsonian*, October, 147–53. The behavior and physiology of the shrew, a little-studied animal until recently, even though it probably resembles the first placental mammals to evolve during the age of the reptiles.

Johansen, K. 1968. "Air-Breathing Fishes." *Scientific American*, October, 102–11. The physiology and adaptations of these remarkable fishes, from Devonian times to the present.

Kemp, T. S. 1982. *Mammal-like Reptiles and the Origin of Mammals*. London: Academic Press. A comprehensive treatment of the origin of the mammals, for the advanced student.

Langston, W., Jr. 1981. "Pterosaurs." *Scientific American*, February, 122-36. Neither dinosaurs nor birds, these creatures — some with wingspans of 12 meters — were flying reptiles that endured for 135 million years.

Maxwell, J. C. 1985. "What is the Lithosphere?" *Physics Today*, September, 32–40. An excellent review of our current understanding and of some of the unresolved problems of the theory of plate tectonics.

McLaughlin, J. C. 1980. *Synapsida: A New Look into the Origins of Mammals*. New York: The Viking Press. A well-written and beautifully illustrated book.

Mossman, D. J., and W. A. S. Sarjeant. 1983. "The Footprints of Extinct Animals." *Scientific American*, January, 74–85. Vertebrate animals have left tracks in sediments ever since they first appeared on dry land nearly 400 million years ago. Indeed most of the known extinct species are known only from their footprints.

McFarland, W. N., et al. 1979. *Vertebrate Life*. New York: MacMillan. An excellent text that offers a broad and detailed view of vertebrate biology. In particular, it describes vertebrate evolution and shows how vertebrates function in their environment by integrating such traditionally separate specialties as physiology and behavior, and ecology and morphology.

Niklas, K. J. 1986. "Computer-Simulated Plant Evolution." *Scientific American*, March, 78–86. How the major trends in plant evolution can be re-created with a desktop computer.

Nilsson, L. 1984. "Prehistoric Flowers Bloom Again." *Smithsonian*, November, 100–107. Fossilized pieces of stems, leaves, and whole flowers, about 80 million years old, have been found in a quarry in southern Sweden.

Ostrom, J. H. 1979. "Bird Flight: How Did It Begin?" *American Scientist*, January–February, 46–56.

Press, F., and R. Siever. 1982. *Earth*. 3d ed. San Francisco: W. H. Freeman and Co. A college-level geology text.

Russell, D. A. 1982. "The Mass Extinctions of the Late Mesozoic." *Scientific American*, January, 58–65. The mass extinction of 65 million years ago.

Additional References

National Geographic Atlas of the World, 5th ed., 1981. Washington, D.C., 30–31.

Retallack, G. J., and C. R. Feakes. 1987. "Trace Fossil Evidence for Late Ordovician Animals on Land." *Science* 235 (2 January): 61–63.

Figure 10.1. Version of a cave painting of galloping horses, a giant bull, and a black stag with antlers on a wall in the Hall of the Bulls of the *Grotte de Lascaux* near Montignac, Dordogne, France. This cave is one of the most outstanding sites of prehistoric art and served for thousands of years (ca. 17,000–12,000 B.C.) as a center of hunting and magical rites.

Origin and Evolution of the Primates

THE FINAL TOPIC of Part Two of this book addresses the origin and evolution of the primates and, in particular, the origin and evolution of our own species *Homo sapiens*, or "wise man." Before starting, let us recognize that this book has taken a rather *anthropocentric*—or human-centered—point of view from the first. For example, our Milky Way galaxy was described in more detail than any of the other galaxies in the Universe. Two chapters were devoted to orbital layout, structure, and origin of our planetary system. Chapters 8 and 9 discussed the origin and evolution of animal phyla, with special emphasis on the vertebrates, to the near exclusion of the plants, fungi, protists, and bacteria. The present chapter continues this self-centered point of view by focusing on our own order and species.

The anthropocentric point of view is not looked upon favorably by most twentieth-century scientists, mainly because science is supposedly concerned with the establishment of generally valid laws, verifiable by observation and experiment. Science is not meant to be biased or to accord special significance to any object, part of space, or period in time. And that includes man and our current historical period.

Throughout the history of science, however, assumptions of man's centrality in nature have been difficult to overcome. As human beings with feelings and emotions we are more interested in our own species, our past, and our particular place in the Universe than in others. No doubt, if we were intelligent dolphins or crustaceans, with the ability to write and do science, we would focus on the origin and evolution of the order *Cetacea* or the class *Crustacea*. If we lived on a planet circling another star in another galaxy, we would focus on that planet and that galaxy. We would do this not just because of partiality, but also because our own species, the planet we happen to live on, and our Galaxy would be closer and more available for intensive study than others.

Thus, the emphasis in the present text on animal life, our planetary system, and our galaxy should not be interpreted as meaning that they have any central significance in nature. They are merely of particular interest to us. Likewise, the primates have no central significance compared to other organisms, except that we happen to belong to this order. Therefore, primates are of particular interest to us, and that is why this book concludes by discussing their origin and evolution.

THE PRESENTLY LIVING PRIMATES

Numerous definitions of the primates can be found. Many of them use highly technical terms, making them rather formidable for the nonexpert. A brief, yet informative definition is given in *Webster's Third New International Dictionary*:

> Primates are members of an order of eutherian mammals including man, apes, monkeys, [tarsiers, lorises, aye-ayes,] lemurs, and living and extinct related forms that are all thought to be derived from generalized arboreal ancestors descended in turn from shrewlike precursors during the Paleocene and that are in general characterized by [an acutely developed sense] of binocular vision, specialization of the appendages for grasping, and enlargement and differentiation of the brain.

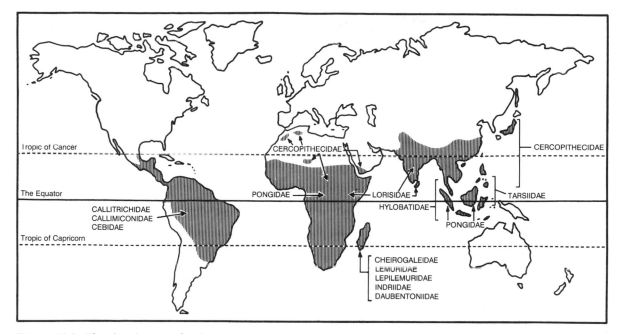

Figure 10.2. The distribution of today's nonhuman primates.

This definition consists of three parts. It lists the order and major groups of primates, it makes a statement about their ancestry, and it enumerates their three most distinguishing characteristics. The first of these three parts will be discussed in the present section, the distinguishing characteristics in the next, and the ancestry in the remainder of this chapter.

Traditionally, the primates have been divided into two suborders, the *prosimians* (meaning "before apes") and the *anthropoids* (meaning "humanlike"). The prosimians included the tree shrews, lemurs, aye-ayes, lorises, and tarsiers. The anthropoids included the New and Old World monkeys and the hominoids (which comprise the apes and humans). Recent improvements in our understanding of the morphology and evolutionary history of the primates is changing this classification scheme somewhat. In particular, tree shrews are no longer classified as primates, but are accorded the status of a separate order, Scandentia (from the Latin, meaning "climbers"). Many zoologists now group the tarsiers with the anthropoids (for reasons discussed below), and the resulting enlarged suborder is called *Haplorhini* (meaning "one-nosed"). The suborder of the remaining prosimians—the lemurs, aye-ayes, and lorises—is called *Strepsirhini* (meaning "twisted nose"). This new classification scheme, summarized in table 10.1, is the one followed in this book.

The geographical distribution of the nonhuman primates is illustrated in figure 10.2. Most of them live in a broad belt centered on the equator, extending approximately north to the Tropic of Cancer and south to the Tropic of Capricorn. The conditions in this belt are typically tropical rain forest and savannah.

Table 10.1. The Presently Living Primates

Kingdom	— Animalia
Phylum	— Chordata
Subphylum	— Vertebrata
Class	— Mammalia
Subclass	— Eutheria (placental mammals)
Order	— Primata

Suborders	Strepsirhini (Prosimians)					
Superfamilies	Lemuroidea				Daubentonioidea	Lorisoidea
Families (no. of genera)	Cheirogaleidae (3)	Lemuridae (3)	Lepilemuridae (1)	Indriidae (3)	Daubentoniidae (1)	Lorisidae (5)
Common names of genera (No. of species)	Dwarf lemurs (4) Mouse lemurs (3)	Lemurs (9)	Lepilemurs (7)	Avahis (1) Indris (1) Sifakas (2)	Aye-ayes (1)	Lorises (2) Angwantibos (1) Galagos (6) Pottos (1)
Geographic distributions	Madagascar and nearby islands					Eastern Asia (Lorises) Africa (2nd group)

Source: Michael Kavanagh, *A Complete Guide to Monkeys, Apes and Other Primates* (New York: Viking Press, 1984), 216–221.

Strepsirhines

Of the two suborders of primates, the strepsirhines are, on the average, the smaller kinds of animals and they retain more of the ancient mammalian characteristics (see figure 10.3). Some of them are the size of a mouse and

Callimiconidae (Gk)	— beautiful marmosets	lepilemur	— pleasing lemur
Callitrichidae (Gk)	— beautiful haired	loris, loeris (F)	— simpleton
cerco-	— tail	orangutan (M)	— man of the forest
Cercopithecidae (Gk)	— long-tailed monkeys	*pithekos* (Gk)	— monkey, ape
cheir-, chir- (Gk)	— hand	prosimii (L)	— before apes
chimpanzee	— derived from a Kongo dialect	*rhin-, rhis-* (Gk)	— nose
Daubenton, Louis J. M.	— French naturalist (1716–1800)	siamang	— derived from the Malay
Gorillai (Gk)	— an African tribe of hairy women	*simia* (L)	— ape
haplo- (Gk)	— one	*simus* (L)	— snub-nosed
homo (L)	— human being	*strepsi-* (Gk)	— to turn, twist
Hylobatidae (Gk)	— walkers in the woods	*tarsos* (Gk)	— flat of the foot, ankle
lemures (L)	— nocturnal spirits	*talapoin* (F)	— Buddhist monk

Meanings and roots of some of the terms of this table
(F = French, Gk = Greek, L = Latin, M = Malay)

	Haplorhini (Simians)							
Tarsioidea	Ceboidea (New World monkeys)			Cercopithecoidea (Old World monkeys)	Hominoidea			
Tarsiidae (1)	Callitrichidae (4)	Callimiconidae (1)	Cebidae (11)	Cercopithecidae (13)	Hylobatidae (1) (Lesser apes)	Pongidae (3) (Great apes)	Hominidae (1) (Humans)	
Tarsiers (3)	Marmosets (8) Tamarins (12)	Goeldi's monkeys (1)	Capuchins (4) Howler monkeys (6) Night monkeys (1) Sakis (6) Spider monkeys (4) Squirrel monkeys (1) Titis (3) Uakaris (2) Woolly monkeys (2) Woolly spider monkeys (1)	Doucs (1) Langurs (16) Macaques (16) Proboscis monkeys (1) Simakobu (1) Snub-nosed monkeys (2) Baboons (7) Colobus monkeys (6) Geladas (1) Guenons (19) Macaques (1) Mangabeys (4) Patas monkeys (1) Talapoins (1)	Gibbons (8) Siamangs (1)	Orangutans (1) Chimpanzees (2) Gorillas (1)	*Homo sapiens* (1)	
Borneo, Philippines and other islands of the East Indies	South America			Asia: India, China, Japan, Southeast Asia (1st group) Africa (2nd group)	Eastern Asia, from Assam in West to Borneo in East	Asia (Orangutan) Africa (Chimp., Gorilla)	Worldwide	

weigh as little as 40 to 100 grams (for example, the mouse lemurs) while the largest, the indris, may weigh 10 kilograms or slightly more. Most of them weigh between a few 100 grams and 1 or 2 kilograms.

Aye-Aye

Figure 10.3. Representatives of the strepsirhines.

Lepilemur

Sifakas

The ancient mammalian characteristics that distinguish the strepsirhines are relatively long snouts and long tails (except the indris which have rather stumpy tails), a well-developed sense of smell, and special scent glands, with which they mark their home range and which play important roles in their communications. Many of them are nocturnal.

The habitats of the strepsirhines are mostly the tropical rain forests of eastern Asia, Africa, and Madagascar. Several species of lemurs live in deciduous forests or in semidesert environments of Madagascar, and two species of galagos live in woodland savannah regions of equatorial Africa. Most strepsirhines occupy rather specialized ecological niches, which has led to the evolution of a great variety of specialized adaptations and forms.

Haplorhines

As already noted, the haplorhines include the *tarsiers, New World monkeys, Old World monkeys*, and *hominoids* (see figure 10.4). These animals distinguish themselves from the strepsirhines mainly by having larger and more complex brains, more forward pointing orbits (eye sockets), noses that are dry, and mammary glands that are always located on the chest. They see in color, which greatly increases the acuity of their vision and the amount of information they can pick up about their environments. Their snouts or faces vary greatly in protrusion, being most marked in baboons and least in humans. Tails are present in some but absent in others. There exist great differences in size and weight between some of the species. Furthermore, except for the tarsiers and night monkeys, the haplorhines are active during the day.

The *tarsiers* are nocturnal hunters of the rain forests of Borneo, the Philippines, and some of the other East Indian islands, living off insects, birds' eggs, and other small prey. They certainly are the most unusual members of the haplorhines. They are very small, with an average body weight of only about 130 grams. They have rather short bodies, round heads with large, saucerlike eyes, long fingers and toes for clinging to branches and trees, and powerful legs for leaping. In overall appearance they resemble some of the smaller lemurs, which is why they used to be classified as prosimians.

However, close examination reveals that the tarsiers' resemblance to the lemurs is only a very superficial one. For example, the tarsiers' noses are dry; and their retinas (the light-sensitive areas of their eyes) contain a fovea, which is a small, roundish depression where vision—in particular, color vision—is most acute. Both of these characteristics are adaptations to a diurnal existence, and they are found in all of the monkeys and hominoids, but not in the strepsirhines. It seems the tarsiers were at one time diurnal animals, but in the course of evolution adapted to a nocturnal way of life. That is probably why they resemble so closely some of the lemurs, many of which are active at night. Yet another feature that the tarsiers share with the monkeys and hominoids, but not with the strepsirhines, is the structure of the placenta. The chorionic membrane of the embryo (see figures 9.37 and 9.38) is much more intimately associated with the uterus, allowing a freer exchange of nutrients and wastes between the fetal and maternal blood than in the case of the strepsirhines. Thus, there exist

TARSIOIDS

NEW WORLD MONKEYS

Tarsier

Tamarin

Woolly
Monkey

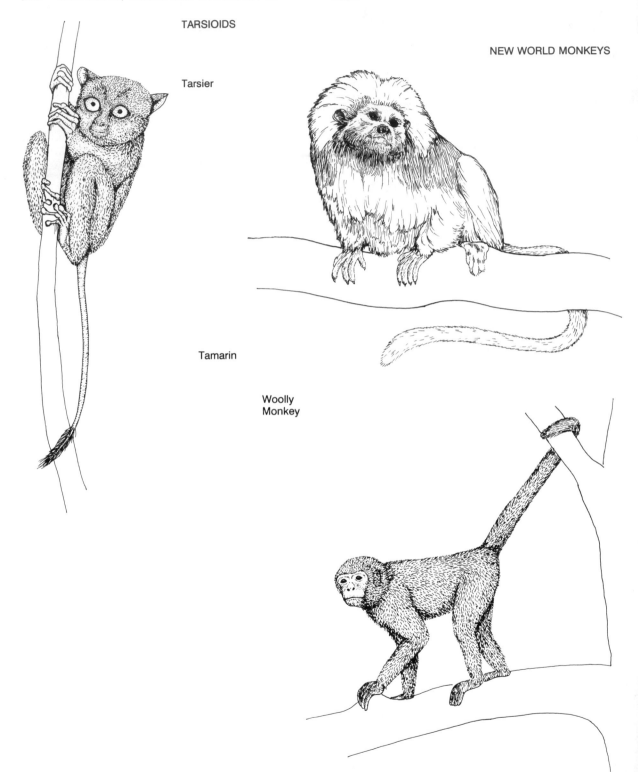

Figure 10.4. Representatives of the haplorhines.

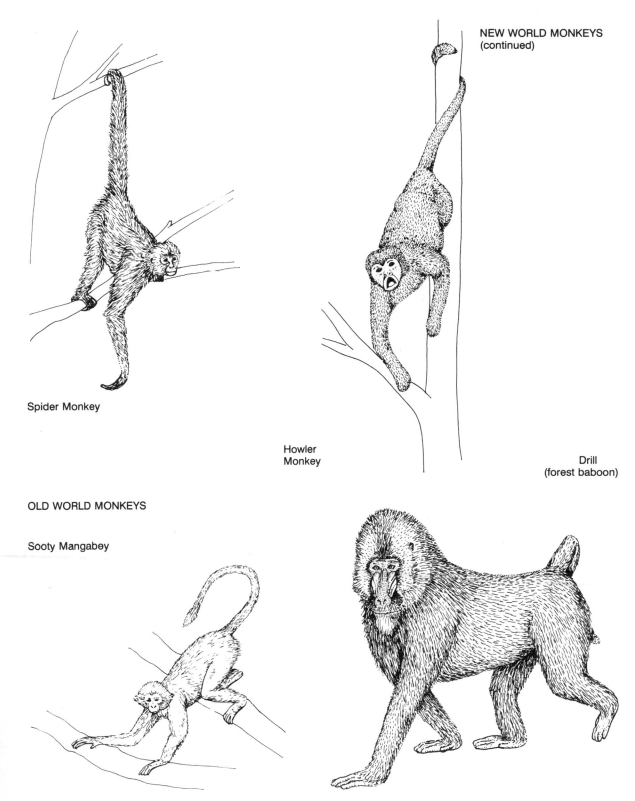

Spider Monkey

NEW WORLD MONKEYS
(continued)

Howler
Monkey

Drill
(forest baboon)

OLD WORLD MONKEYS

Sooty Mangabey

HOMINOIDS

Chimpanzee

Figure 10.4. continued   Gorilla

Siamang

Orangutan

compelling reasons for placing the tarsiers into a common suborder with the monkeys and hominoids.

The *New World monkeys* live exclusively in the tropical forests of South and Central America, reaching as far north as southern Mexico. Their most obvious characteristics are their flat noses and widely spaced nostrils. They have rounded heads and relatively small jaws. They all are arboreal and some, such as the spider, howler, and woolly monkeys, have prehensile tails, which they can curl around branches and use as a fifth limb in climbing and hanging. They are on the average larger than the strepsirhines, with weights in the range from approximately 100 grams (pygmy marmosets) to 15 kilograms (woolly spider monkeys).

Today's *Old World monkeys* are native to Africa and Asia, though fossil evidence indicates that in the past they also lived in Europe. Their habitats range from low lying tropical rain forests (most species) to woodland savannahs (baboons, vervet monkeys — a species of guenons — and patas monkeys), subdesert environments (baboons), and mountainous areas (geladas, guenons, langurs, and macaques). A single species of macaques, the Barbary "apes" of North Africa, live on Gibraltar. Most species are arboreal (see figure 10.5), but some live at least part of the time on the ground (baboons, geladas, macaques, patas monkeys), and others inhabit temples and villages (langurs and rhesus monkeys, a species of macaque).

The Old World monkeys have a wide range of physical characteristics, some of which will be discussed in the following section. Unlike their New World relatives, they have nostrils that are more narrowly spaced and downward pointed, none have prehensile tails, and they all have conspicuous sitting pads on their hindquarters. Their weights range from approximately 1 kilogram (guenons) to 20 kilograms (geladas).

The *hominoids* constitute the superfamily that includes, besides the *lesser apes* (*gibbons* and *siamangs*) and *great apes* (*orangutans, gorillas*, and *chimpanzees*), our family — *Hominidae* (note the distinction between *homin-oi-d* and *homin-i-d*). They are tailless. Their teeth and their external and internal anatomies are very similar to ours as are the stages of their embryological developments.

The *gibbons* are the smallest of the hominoids, with a range of body weight from about 4 to 8 kilograms and little difference between males and females. They are slender and monkeylike and have enormously elongated arms with hooklike fingers. They are highly arboreal and among the most agile and acrobatic climbers and leapers of the primates (see figure 10.6). On the ground they usually walk bipedally, with arms held up high and to the sides for balance. Their natural habitats are the rain forests of Southeast Asia, from Burma, Thailand, and southern China to Sumatra, Java, and Borneo. The siamangs are close relatives of the gibbons, though they are somewhat larger and more heavily built.

The *orangutans* (from the Malay, meaning "man of the forest") are moderate-sized hominoids, with adult males weighing between about 50 and 100 kilograms and females half of that. They have relatively short legs and long, muscular arms that extend to the ankles when the animals stand erect. They are also arboreal, but they do not leap. They usually swing slowly and deliberately

Figure 10.5. Savannah baboons in a stand of trees, seeking refuge from a lion. It is not difficult to imagine that the early hominids, the australopithecines of some 4 to 3 million years ago, relied similarly on climbing trees to escape pursuing predators (see discussion on p. 525).

from branch to branch, though they also walk quadrupedally on branches or bipedally with arms holding on above. Males come to the ground quite often, where they walk rather clumsily, usually on all fours and using their fists for support. They spend the nights sleeping in nests, which they build anew every day. Today these rather shy animals are found in the wild only in the rain and swamp forests of Sumatra and Borneo, but during the Pleistocene they ranged north into China. Only a few thousand of them survive.

The *gorillas* are the largest of the apes. Males weigh approximately 160 kilograms, though some reach weights of 200 kilograms or more. Females are smaller and typically weigh between about 75 and 110 kilograms. These animals have massive bones, broad shoulders, long arms, strong jaws with huge teeth (particularly the males), and heavy brow ridges overhanging the eyes. Adult males have prominent crests on the tops of the skulls, giving the false impression of greatly expanded brain volumes. Occasionally these giant animals climb

Figure 10.6. Gibbons are the master acrobats among the primates. This photograph shows one of them brachiating, or arm swinging, from branch to branch. Hanging by its forelimbs, the gibbon releases one hold, swings the arm swiftly forward, grasps the next hold, and then repeats the motion with the other arm. In this way, gibbons (as well as their close relatives, the siamangs) can travel at enormous speeds through the forest canopy, especially on a downward incline when they sometimes hurtle themselves for up to 15 meters through the air between supports. (On broad branches or on the ground, gibbons and sia-mangs are also capable of bipedal walking, as illustrated in figure 10.18.)

into trees to forage for certain food items, such as particular fruits and flowers, and females and young have been observed to sleep in nests they build in the trees. However, they are primarily terrestrial and spend much of the day wandering about in the undergrowth of the jungle. They walk on all fours, supporting themselves on their knuckles. Their hands are prehensile, with opposable thumbs, and they use them extensively in manipulating food and building nests. But there is no evidence of either tool making or using.

When alarmed, male gorillas beat their chests, roaring menacingly and pretending to attack. But they rarely do. Nevertheless, these displays, combined with the animals' huge size and powerful appearance, have given rise to a reputation for fierceness and savagery. In reality, gorillas are peaceable creatures, unless unduly disturbed.

The *chimpanzees* are our closest primate relatives. They are considerably smaller and more agile than the gorillas, weighing typically between 40 and 50 kilograms, with females only slightly lighter than the males. They are highly arboreal animals, and sleep and feed in trees. On the ground they walk on all fours, using their knuckles for support, much like the gorillas. Occasionally they stand erect on their hindlegs, probably to improve their visual range. But they run or walk bipedally only rarely. Their hands are highly dexterous and they use them skillfully for grooming, nest building, and tool handling. They are exceptionally intelligent and inquisitive animals, and they have a wide range of facial expressions, hand gestures, body postures, and vocalizations.

Both gorillas and chimpanzees are native to partially overlapping regions of equatorial Africa, from the lowlands of southeastern Nigeria south to the mouth of the Congo and from the upper Congo (the river Lualaba) eastward to the mountain ranges near Lakes Edward, Kivu, and Tanganyika. In addition, one species of chimpanzees lives in the forested regions of western Africa, from Sierra Leone and Guinea to the Niger.

The *hominids* are by far the most intelligent of the primates. They distinguish themselves mainly by habitually walking upright and bipedally. Furthermore, they have large, highly fissured brains, communicate by language, and possess rich and complex cultures. Their thumbs are fully opposable, making the hands more dexterous than those of any of the other primates. The face is nonprotrusive, with relatively weak jaws and small teeth. Except for the top of the head, pubic area, arm pits, and the faces of males, they are relatively hairless. Their heights and weights have a fairly broad range, from about 1.2 meters (pygmy race) to 2.0 meters (tallest African race) and from 40 to more than 100 kilograms. Females are slightly smaller than males. The only presently living hominids are we, *Homo sapiens*. Our distribution is worldwide, with a total current population of about 5 billion and a yearly increase of 1.65 percent, which corresponds to an increase of 1 million people every 4 to 5 days.*

THE CHARACTERISTICS OF THE PRIMATES

The primates evolved from the same insectivorelike ancestral stock as did the other mammals. However, whereas the members of most mammalian orders acquired highly specialized characteristics—such as hoofs for running across grassy plains, teeth specialized for particular diets, wings for flying, or flippers for swimming—the primates have done so to a much lesser degree. For example, they still have four mobile limbs, each ending in five digits. They can freely rotate and tilt their heads. Their tooth pattern allows them to eat virtually everything. They have excellent visual, audio, and tactile senses; and they still can smell quite well, though not as well as many other mammals. Furthermore, they are curious and intelligent, and they respond adaptively to a wide range of stimuli.

The one specialization that does distinguish the primates is their adaptation to arboreal ways of life (followed in some species by a readaptation to life on the ground). This specialization has led to the development of numerous physical and behavioral characteristics, many of which vary significantly from one group of primates to the next. Let us examine some of these characteristics and their varying degrees of expression.

*Data reported by the World Population Institute, Washington, D.C. (*The New York Times*, 7 July 1986, I, 7:4).

Lengthening of the Fore- and Hindlimbs and Shortening of the Trunk

One of the most striking features of many primates is their short trunk and relatively long arms and legs. These developments are adaptations to locomotion either in the trees or on the ground (see figures 10.7 and 10.8). For example, long legs are an adaptation to the arboreal locomotion of *vertical clinging and leaping* practiced by the members of the indriid family, the galagos, lepilemurs, and tarsiers (see figure 10.9). Long legs many also be an adaptation to bipedal walking on the ground, as in the case of humans. Long arms, like those of the gibbons and orangutans, are adaptations to the arboreal locomotion of *arm swinging* (also known as *brachiation*), in which the animals hang suspended from branches and swing from one hold to another by the arms. Long arms are clearly an advantage here, for they increase the length of the arm stride and, therefore, the speed and efficiency of travel. Long arms as well as long legs, such as those of baboons, are adaptations to *quadrupedal walking* on the ground.

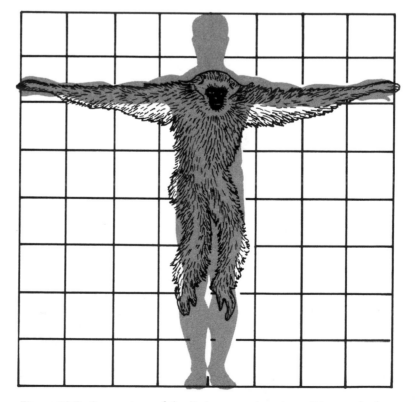

Figure 10.7. Comparison of the limb proportions in a gibbon and a human.

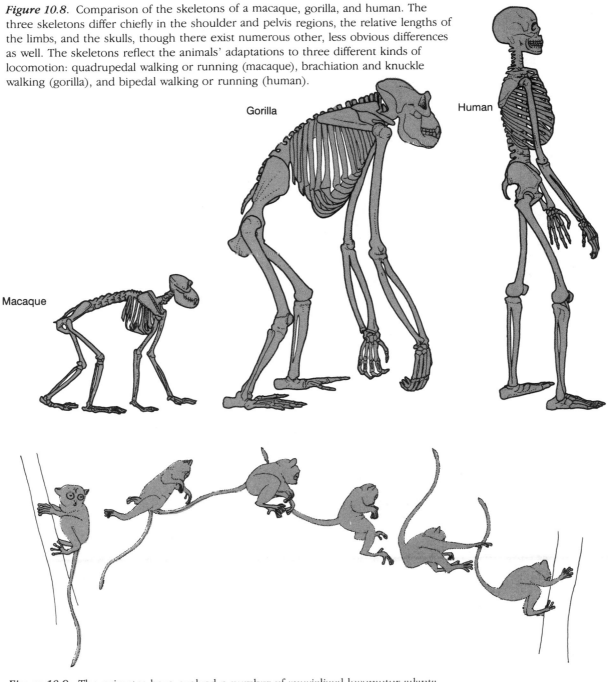

Figure 10.8. Comparison of the skeletons of a macaque, gorilla, and human. The three skeletons differ chiefly in the shoulder and pelvis regions, the relative lengths of the limbs, and the skulls, though there exist numerous other, less obvious differences as well. The skeletons reflect the animals' adaptations to three different kinds of locomotion: quadrupedal walking or running (macaque), brachiation and knuckle walking (gorilla), and bipedal walking or running (human).

Gorilla

Human

Macaque

Figure 10.9. The primates have evolved a number of specialized locomotor adaptations. One of them is *vertical clinging and leaping,* as illustrated here by drawings of frames from a motion picture taken of a tarsier. The animal's typical resting position is clinging vertically to tree trunks, while its favored locomotion through the forest canopy consists of gigantic leaps from tree trunk to tree trunk. The animal executes these leaps by kicking itself off tree trunks with its long, powerful hind limbs, turning in midair, and landing on the next tree trunk with feet first.

Increased Prehensile Skills of Hands and Feet

The primitive insectivorelike ancestors from which the primates evolved had digits (fingers and toes) that ended in claws. Among today's primates, only the aye-ayes, needle-nailed galagos, marmosets, and tamarins have claws, at least on some of their digits. In all others, the claws have been replaced by flat nails. The advantage of nails is that they allow the digits to end in rather broad surfaces with a much improved sense of touch, which is important when climbing and swinging from branch to branch. Digits ending in nails are also useful in handling small food items, such as flowers, seeds, nuts, birds' eggs, and insects.

The disadvantage of nails is that they cannot be used in hooklike fashion in clinging. To compensate, the digits of the primates grew longer and the thumbs and big toes became opposable in varying degrees to the other digits, making the hands and feet prehensile. In humans the big toes are no longer opposable, but fully aligned with the other toes. This is an adaptation to bipedal walking, in which opposable toes would get in the way.

In the case of the strepsirhines, tarsiers, and New World monkeys, the grabbing and holding with the hands is restricted to a curling of the fingers, producing the *power grip* (see figure 10.10). This grip is strong, but not delicate and precise. The Old World monkeys and hominoids also possess the power grip. In addition, they have a *precision grip*, which is achieved by moving the thumbs toward the fingers. This grip makes the hands highly dexterous and allows the animals to pick up and handle small objects with precision. It is particularly well developed in chimpanzees and humans and a key factor in these animals' ability to make and use tools.

Progression Toward Upright Posture and Bipedalism

All primates are able to raise their trunk into a vertical position. In fact, most of them commonly assume such vertical postures as sitting, clinging, or hanging. And a few, such as the apes and humans, are able to stand and walk with the trunk held erect. The evolution of erect posture and bipedalism was accompanied by a number of anatomical changes, such as the shortening of the vertebral

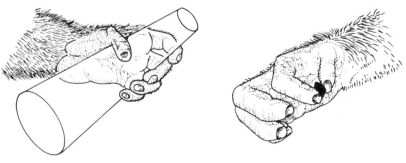

Macaque Power Grip   Chimpanzee Precision Grip

Figure 10.10. Comparison of the *power grip* and *precision grip* of primates. All primates can execute the power grip, which is strong. Old World monkeys and the hominoids can also execute the precision grip, which is delicate and precise.

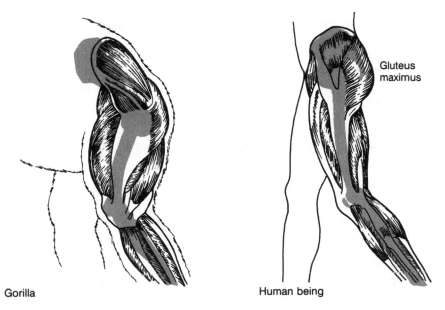

Gorilla   Human being

Figure 10.11. Comparison of the pelvis and leg bones, as well as the associated muscles, in a human and a gorilla. The differences shown are due to the fact that gorillas are quadrupeds, while humans are bipeds.

column, changes in the structure of the pelvis and the lower limbs, and a forward movement of the foramen magnum (see below).

In humans, the changes in the structure of the pelvis and the lower limbs were accompanied by the evolution of strong buttock muscles, the *glutei maximi* (plural of *gluteus maximus*), which along with other smaller muscles in the hip and thigh region are responsible for maintaining erect posture (see figure 10.11). Furthermore, humans evolved the ability to lock their knees so when they stand erect they do not have to use their leg muscles to hold themselves up. In contrast, gorillas, for whom standing erect is much more an effort than for humans, the buttock muscles have remained relatively underdeveloped and they cannot lock their knees.

The foramen magnum is the opening on the underside of the skull through which the spinal cord passes and makes connection with the brain stem (see figure 10.12). In the primates, the foramen magnum is located forward of where

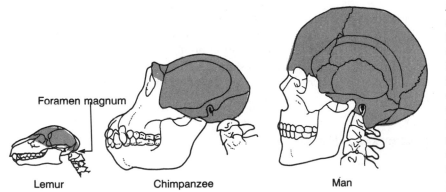

Lemur   Chimpanzee   Man

Figure 10.12. Comparison of the skull characteristics of a lemur, chimpanzee, and human. The diagram illustrates the differences in the protrusion of the jaws, dentition, location of the foramen magnum, and the size of the brain (shaded areas) of these three primates.

it is in other mammals (see figure 9.33). But in most primates this forward location is only partial, which means that when they assume an upright posture the neck muscles need to be continuously flexed to offset the weight of the face and jaws. Only in humans is the foramen magnum located directly under the skull, so that during upright posture the vertebral column balances the skull almost perfectly.

Improved Stereoscopic and Color Vision

Life in trees makes much greater demands on the sense of sight than does life on the ground. For example, safe jumping and leaping depend on the ability to judge distances accurately and to discern fine detail among the background of foliage and branches. Consequently, primates have evolved far better vision, in particular stereo and color vision, than any other mammals.

Stereoscopic vision, the ability to see in three dimensions, results from a pair of eyes that point forward and have an overlapping field of view. The small differences in direction in which objects at close and intermediate ranges (up to about 500 meters for humans) are seen by the two eyes form the basis of distance judgment and the sense of three dimensions.

Color vision results from the ability of the eyes to distinguish between different wavelengths of light. This is achieved by the presence of two or more types of photoreceptors (the *cones*) at the focusing area (the retina) of the eyes, each one sensitive to different wavelength regions. Both stereo and color vision are more highly developed in humans and the other haplorhines (for instance, their retinas have a fovea) than in the strepsirhines.

Reduction of the Sense of Smell

Whereas the tactile sensitivity of the hands and good vision are important to an arboreal existence, the sense of smell is less so. Consequently, as the manual dexterity and eye sight of the primates improved, the sense of smell diminished. This development occurred to a greater degree in the Old World monkeys and hominoids than in the rest of the primates. For example, most of today's strepsirhines are still rather smell-oriented animals, and a number of them, such as the lemurs and lorises, mark their territories either with the use of special scent glands or urine. So do the New World monkeys, though they use only urine.

With the reduction in the sense of smell, the haplorhines lost the moist, hairless skin at the tip of the snout that is present in the strepsirhines and other smell-oriented mammals. This development was accompanied by the acquisition of a mobile upper lip, which significantly broadened the repertoire of facial expressions and vocal communications, so typical, for instance, of apes and humans.

Changes in Dentition

The primates have retained the general dental features of their primitive mammalian ancestors. They still possess incisors, canines, and molars, though the number of teeth has been reduced from 44 to 32 or 36. For example, adult humans usually have 2 incisors, 1 canine, 2 premolars or bicuspids, and 3 molars in each quadrant of their dentition, making a total of 32 teeth.

As a result of their generalized dentition, primates are able to eat a great variety of foods: fruits, flowers, leaves, grass, roots, tubers, bark, pith, seeds, nuts, birds, birds' eggs, lizards, spiders, insects, frogs, and crustaceans. In fact, many primates are omnivorous when necessary, though in the wild they restrict their diet, depending on what is readily available in their habitat.

Increased Volume and Complexity of the Brain

The brains of all primates, including humans, consist of the same basic structures—cerebrum, corpus callosum, thalamus, and so forth (see figure 10.13 and table 10.2). The only differences from the lemurs to humans are in size and organization of the brain. For example, humans have by far the largest and most complex brain, with an average brain volume of about 1400 cm³. In comparison, the great apes have average brain volumes about one-third that size (300 to 600 cm³). Those of the gibbons are approximately 100 cm³. The monkeys and strepsirhines have still smaller brains, though they are larger than those of other mammals of comparable size.

Brain size alone does not determine the quality of the brain. The relative sizes of the various sensory and association areas, the constitution of the individual brain cells, the number and kinds of interconnections between individual brain cells, and the general layout and organization of the brain are other factors. For example, from the lemurs to the hominoids, the sensory centers responsible for vision and tactile acuity tend to become enlarged, while those responsible for smell tend to become reduced. The cerebral cortex, which controls memory,

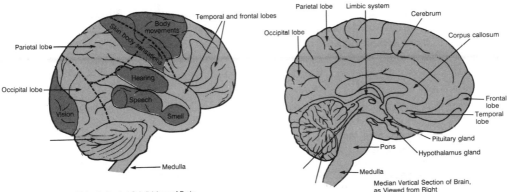

Figure 10.13. The structure of the human brain. The human brain may be thought to consist of three major geographic areas: the *forebrain, midbrain*, and *hindbrain*. Each area is composed of several parts, the most important of which are shown in these two diagrams. They are briefly described in table 10.2.

Table 10.2. Structure of the Human Brain

Structure	Location and Description	Function
Forebrain Cerebrum	Uppermost part of brain, greatly enlarged in humans compared to other primates; divided into left and right hemispheres	
Cortex (outer layer of cerebrum)	Thin, deeply folded layer with large surface area; contains about 14 billion nerve cells; divided into several lobes (frontal, temporal, parietal, occipital); most recent evolutionary development	Center of conscious thought, memory, speech, personality, and judgment
White matter (inner cerebrum)	Nerve tracts connecting parts of cortex to each other and to rest of brain and spinal cord	Center of sensation, including sight, hearing, taste
Corpus callosum	Band of about 300 million separate neuronal lines	Connection between the cerebral hemispheres
Thalamus	Groups of cell bodies arranged in shape of a football	Important integrating center for most sensory input on its way to cortex (vision, hearing, touch, taste, but not smell); regulation of sleep and wakefulness
Limbic system	Internal region of forebrain, divided into several parts	Involved in emotion, motivation, and reinforcement
Hypothalamus	Below the thalamus and linked to pituitary gland	Control of basic drives (eating, drinking, sleeping, sexual behavior); regulation of temperature, blood pressure, heart rate
Midbrain Mesencephalon	Greatly reduced in humans and other mammals compared to other vertebrates	Connection between forebrain and hindbrain
Hindbrain Cerebellum ("little brain")	Baseball-sized lump of gray-white tissue lying on both sides of brain stem	Unconscious coordination of muscle movements and maintenance of bodily equilibrium
Pons	Situated at front and upper end of medulla	Relay station between cerebral cortex and medulla
Medulla oblongata (brain stem)	Lowest part of brain; connects brain through the foramen magnum (opening in lower part of skull) with the spinal cord	Control of respiration, heart rate, and gastrointestinal function

conscious thought, vocalization, judgment, and personality, also increases in size. In fact, the greatly enlarged and highly folded cerebral cortex is the most distinguishing characteristic of the human brain. In the other primates, the cerebral cortex is smaller, less folded, and much simpler in organization.

Table 10.3. Durations of Stages of Life of Five Representative Primates

	Fetal stage (weeks)	Infantile stage (years)	Juvenile stage (years)	Adult stage (years)	Life span (years)
Lemurs (strepsirhines)	18	0.75	1.75	11	14
Macaques (Old World monkeys)	24	1.5	6	20	28
Gibbons (lesser apes)	30	2	6.5	20	29
Chimpanzees (great apes)	32	3	7	30	40
Humans	38	6	14	50	70

Source: J. R. Napier and P. H. Napier, *A Handbook of Living Primates*. (New York: Academic Press, 1967), 40.

Note: The durations are averages. The actual durations vary somewhat from one individual to another.

Lengthening of the Duration of All Stages of Life

From the lemurs to the monkeys, apes, and man, there occurs a steady lengthening of the duration of all stages of life (see table 10.3). For example, the human fetal stage is approximately twice that of the lemur; our infantile and juvenile stages exceed the lemurs' by about a factor of eight; our adult stage exceeds the lemurs' by a factor of 4.5; and our entire life span is roughly five times as long as the lemurs'.

Interestingly, the smallest differences between us and the strepsirhines, with regard to duration of stages of life, occurs in the fetal stage. If this were otherwise—say, if human pregnancies lasted eight times as long as the lemurs' (33 months instead of 9)—babies would be much bigger and females would need enormously large pelvises to allow the babies to make it through the birth canal. Such large pelvises would make upright bipedal walking very difficult, if not impossible, and, hence, have not evolved. Thus, human babies and, to a lesser degree, those of the other hominoids are born in a quite unfinished state and require years of nurturing before they reach independence. The years of nurturing are, however, not just periods of physical growth. A great deal of learning and conditioning takes place then, which are crucial for the individuals' healthy growth and integration into their society.

Along with extended infantile and juvenile phases, the onset of puberty is postponed as well. This characteristic probably evolved because it permits raising the young without the intragroup aggression and fighting that sexual competition inevitably brings about.

Reduction in the Number of Offspring

The long and rather intensive adult–child interaction typical of primates would be difficult to sustain if these animals had as many offspring as, for example, their close relatives the rodents and lagomorphs (litters of six to ten every few months are not uncommon for them). Therefore, along with the evolution of their other characteristics, the primates' reproductive pattern changed toward a greatly reduced number of offspring. Today, most primates give birth to a single baby at a time, though multiple births do occur occasionally as we know from our human experiences.

In conclusion, remember that the characteristics that distinguish the primates are primarily adaptations to arboreal ways of life, and that none of these characteristics evolved independently. They evolved together and form an integrated adaptive pattern. Despite enormous evolutionary changes experienced by the primates, these animals have retained many of the ancient mammalian characteristics. This lack of specialization is the primates' most distinguishing trait, compared with the other mammalian orders. It is their mechanism in the competition for survival.

PRIMATE EVOLUTION, PART ONE: THE ORIGIN AND DIVERSIFICATION OF THE PRIMATES

This section examines the evolution of the primates from their earliest known beginnings, nearly 70 million years ago, to the arrival of the hominoids. However, even further back in early Cretaceous times, roughly 120 million years ago, major geologic and climatic changes were taking place that affected the origin and diversification of the placental mammals.

During the early Cretaceous, the breakup of Pangaea, the all-inclusive supercontinent of the Triassic, was already well under way (see box 9.2). The northern landmass (Laurasia) had separated from the southern landmass (Gondwana) and opened up the North Atlantic and Indian Oceans. North America was still joined to the Eurasian land mass, South America and Africa were connected as were Antarctica and Australia, and the Indian subcontinent was separate and drifting northeastward toward Asia. The climate was hot and humid over much of the globe.

The placental mammals arose in this setting. Like all of the early mammals, those ancient placentals were small, nocturnal, and smell-oriented animals. They lived in the undergrowth, hiding and competing the best they could with the dinosaurs and other reptiles, which still dominated the land and continued to do so for about another 55 million years. During that time the first of the placental orders arose, one of which was the primates.

We cannot be certain which environmental factors were primarily responsible for the origin of the primates. Quite probably, the appearance during the Cretaceous of flowering plants, along with a great variety of fruits, nuts, nectars, and flowers as well as new forms of insects that fed upon the new flora, was one of the reasons. The flowering plants, and especially the flowering shrubs and trees, created brand new habitats, and, as observed in previous chapters, new habitats do not remain unoccupied for long.

It seems likely that primitive and rather generalized members of a stock of insectivores, perhaps resembling today's shrews, were among the first placental mammals to invade the new arboreal habitats successfully. In time, these animals evolved prehensile hands and feet, excellent eye sight, improved eye–hand coordination, and other adaptations that allowed them to climb and to make the trees and undergrowth of tropical forests their habitat. In particular, these physical characteristics gave the animals access to the terminal branches of the flowering shrubs and trees, where most of the fruits, nuts, nectars, and flowers were concentrated and where many of the insects visited. For instance, grasping hands and feet would have been very "advantageous to animals that habitually forage[d] in terminal branches, since they permit[ted] these animals to suspend themselves by their hind limbs (and tail, if prehensile) while using the forelimbs to reach and manipulate food items," as Matt Cartmill (1972), Professor of Anatomy of Duke University, put it. Animals such as these are thought to have been the ancestors of the primate lineage.

The main competition that the early primates faced in the terminal branch habitat probably came from the members of another placental order, the bats, which during the early part of the Cenozoic were rapidly evolving and diversifying. Both the bats and primates were nocturnal, and they both fed on the same fruits and flowers. Little is known at present of how this competition affected the evolution of the primates. However, it seems reasonable to assume that it contributed quite significantly to selection pressures and thereby speeded up the primates' adaptations to the new arboreal environment, just as the presence of the primates probably speeded up the evolution of bats (and, perhaps, even brought about the evolution of bat flight). Of the other mammals that also acquired arboreal abilities, such as certain members of rodents (such as squirrels) and carnivores (such as weasels and some of the cats), none specialized in terminal branch feeding. They exploited different niches in the forest canopy and acquired different adaptive characteristics.

Strepsirhine Evolution

The assumption that the earliest primates evolved from primitive and rather unspecialized insectivores is supported by fossil evidence from nearly 70-million-year-old deposits in North America and Europe. It consists of teeth, jaws, and a few postcranial fragments that have been interpreted as being the remains of very primitive primates. The sizes of these animals ranged from approximately those of field mice to those of large domestic cats. They possessed long snouts, relatively large smell apparatuses, eyes pointed to the sides, small brains, and clawed digits similar to those of squirrels. Their teeth suggest that they ate mainly plant material and insects.

A number of families of these ancient primates have been identified, and together they constitute the superfamily *Plesiadapoidea*, of the suborder Strepsirhini. These animals flourished during late Cretaceous and Paleocene times, when the world climate was still hot and humid, even in the northern latitudes, and well suited for primate life. However, by the early Eocene, some 55 to 50 million years ago, the plesiadapines had all became extinct (see figure 10.14).

As the earliest known primates were dying out, new groups arose that were better adapted to the arboreal way of life. Most widely represented among them were the *adapines* and *anaptomorphines*. They lived in North America, Europe, and Asia during Paleocene and Eocene times and were surprisingly diversified, with about sixty genera identified to date. They, too, were very small and, like their predecessors, they were strepsirhines. But they looked more like today's lemurs and tarsiers than primitive insectivores. They were mostly frugivorous, had large, forwardly pointed eyes, and their brains were noticeably enlarged compared with other mammals of similar size, with emphasis on vision and some reduction in smell. They had a divergent toe and thumb, and flattened nails rather than claws. Their long, slender legs suggest that they were skilled climbers. These animals are the earliest known primates of "modern aspect," with characteristics that allowed them to move much more skillfully through the terminal branches of shrubs and trees and to forage there.

Haplorhine Evolution

During the Oligocene, the epoch following the Eocene, major changes took placed in the geology and climates of the Earth, with far-reaching effects on plant and animal life. North America separated from Europe. South America drifted further west. Antarctica and Australia divided. Africa moved closer toward Europe and western Asia, and the Indian subcontinent collided with central Asia, creating major uplifts from the Alps in the west to the Taurus Mountains in Turkey, the Zagros Mountains in Iran, and the Himalayas in the east. These geologic changes were accompanied by a steady cooling trend, which produced drier and more seasonal climates in North America, Europe, and Northern Asia. The tropical forests that had covered these regions became subtropical or temperate, interspersed by grasslands (savannah or prairie). Rodents became abundant as did browsing and grazing animals, including horses, zebras, deer, and camellike forms. Furthermore, the strepsirhines, which had flourished in the northern latitudes, declined and retreated southward.

In a broad belt north and south of the equator and including parts of Africa, Asia, and South America, the weather remained tropical, and in these areas the primates continued to thrive and evolve. The fossil record suggests that a major adaptive shift was then taking place among the primates from small-bodied, nocturnal strepsirhines to large-bodied, diurnal creatures, resembling monkeys and apes. At the same time, their diet was becoming more diverse than that of their ancestors. It included leaves and a large variety of pulpy fruits, pods, and seeds. According to Robert W. Sussman (personal communication 1985) of Washington University and Peter H. Raven, Director of the Missouri Botanical Garden, St. Louis, the primates reaped significant benefits from this adaptive shift. It allowed them to move into new ecological niches and, thereby, reduce the resource overlap with the bats, without seriously intruding on the niches already occupied by other arboreal mammals and by the birds.

The origin of Old World monkeys and hominoids. The fossil evidence of Oligocene primates is rather scanty and widely dispersed. However, there exists one famous site in which such fossils are rather abundant. It lies in the

Fayum province, some 100 kilometers south of Cairo, Egypt, which today is mainly desert, but which during the Oligocene consisted of dense tropical forests, swamps, and occasional savannah. The Fayum primate fossils date from approximately 30 to 25 million years ago and represent quite a diverse group of genera: *Apidium, Propliopithecus, Oligopithecus, Parapithecus, Aeolopithecus*, and *Aegyptopithecus*. The phylogenetic relationship of these primates to later forms is not entirely clear, but it appears that *Parapithecus* probably lay on, or close to, the lineage that led to the Old World monkeys, while most of the other Fayum primates were early members of the hominoid lineage.

Perhaps the most interesting of the Fayum primates is *Aegyptopithecus*, of which a complete skeleton, lower jaw, and teeth have been found. *Aegyptopithecus* was approximately the size of a modern gibbon. It had large, forwardly directed eyes and an expanded cranium. Its teeth suggest that it may have been an ancestor of the dryopithecines, which were great apes of the Miocene (see below).

The origin of the New World monkeys. Before continuing the discussion of the evolution of the Old World primates, let us briefly turn to the New World monkeys. On the whole, their fossil record is rather scanty also. It begins with a few isolated jaw and teeth fragments (of the genus *Branisella*) from about 35 million years ago, found in Bolivia, and continues with additional finds (of the genera *Homunculus, Cebupithecia*, and others) from the Miocene. Little can be said about the kinship of *Branisella* to modern primates. But it seems likely that *Homunculus* and *Cebupithecia* were early ancestors of the New World monkey family *Cebidae*.

An important question with regard to the New World primates pertains to their origin. They could not have been present in South America when this continent separated from Africa approximately 90 million years ago. This would be inconsistent with primate phylogeny as derived from protein comparisons (see below). Quite possibly, the New World monkeys migrated to South America from North America via the archipelago that rose between these two continents during the early part of the Tertiary, perhaps by accidentally rafting from island to island on fallen logs. This hypothesis is supported by evidence showing that some early interchange of faunas occurred between the two Americas. However, no monkey forms have yet been found in North America and, unless they are, the suggestion that this continent was the place of origin of the New World monkeys must be regarded as conjectural.

Divergence of the hominoids. The early part of the Miocene, between approximately 25 and 15 million years ago, was marked by a highly significant event: a land bridge was established between Africa and Asia, which allowed the free migration of animals, including primates, between the two continents. The linkup was the result of the northward movement of the African continent. These geologic events were accompanied by the continued cooling of the world's climate, though during the Miocene it did not become as cool as it is today. The grasslands expanded further. Most of the archaic forms of mammals, including the strepsirhines of the northern latitudes, died out and the survivors acquired essentially modern forms. This was also the time when, according to the fossil record from Africa, Southern Europe, and Asia, the hominoids first began to diverge into their modern lineages.

One of the earliest Miocene hominoids was *Dendropithecus*. This tailed ape, which resembled today's gibbons, lived approximately 23 million years ago in Africa. Along with other gibbonlike forms of later times, such as *Pliopithecus* of about 15 million years ago from Europe, *Dendropithecus* may well represent the beginning of the lineage that led to the modern gibbons and siamangs of Southeast Asia.

From about mid-Miocene times until the close of the era, the fossil record of hominoids is particularly extensive, with well over a thousand finds reported to date. These remains consist mostly of teeth and jaws, and therefore their phylogenetic relations are not fully clear. Very approximately, the primates from which these fossils come may be divided into two broad groups: the *dryopithecines* and the *ramapithecines*. The dryopithecines are the older of the two groups. Their fossils span the time from approximately 20 to 10 million years ago and come mainly from East Africa and Southern Europe, though there are a few finds from India. The fossil record of the ramapithecines starts approximately 15 million years ago and extends well into the late Miocene. Most specimens of this group have been found in the Siwalik Range south of the Himalayas and other Asian sites, but a few come from East Africa.

The dryopithecines are believed to be distantly related to some of the Fayum primates, such as *Aegyptopithecus*. They were rather generalized apes, without any specialization for either extreme brachiation or terrestrial life. Their most distinguishing features were strongly protruding incisors, probably used for opening hard-shelled fruit, and large projecting canines. They were noticeably bigger than the earlier primates, with sizes in the range from those of today's baboons to those of gorillas. The best known dryopithecine ape was *Proconsul africanus*. It was relatively small in size, lived in trees, and ate fruits. Its elbows, shoulder joints, and feet resembled those of modern chimpanzees, its wrists were like a monkey's, and its lumbar vertebrae were similar to a gibbon's. *Proconsul* was clearly a most interesting creature and, like all of the dryopithecines, is difficult to link phylogenetically with any of the later species of great apes.

The ramapithecines, too, confront us with a number of uncertainties. For example, we do not know of their origin, though it has been suggested that they evolved from one of the earlier species of dryopithecines. Nor is there agreement concerning their descendants. Until the early 1970s, the ramapithecines were known only from a few jaw and tooth fragments. The jaws were rather robust and the molars were large and thickly enameled, features that are similar to those of the Pleistocene hominids of the genus *Australopithecus*. It was, therefore, commonly assumed that the ramapithecines were either themselves hominids or direct ancestors of the hominids.

Numerous recent discoveries of ramapithecine fossils by David Pilbeam, an anthropologist from Harvard University, in collaboration with the Geological Survey of Pakistan, have cast doubt on this interpretation. According to Pilbeam (1984, 93) and one of his colleagues, Steven C. Ward at Kent State University, the ramapithecines had skulls and limbs that were quite unlike those of the australopithecines, but closely resembled those of today's orangutans, the sole

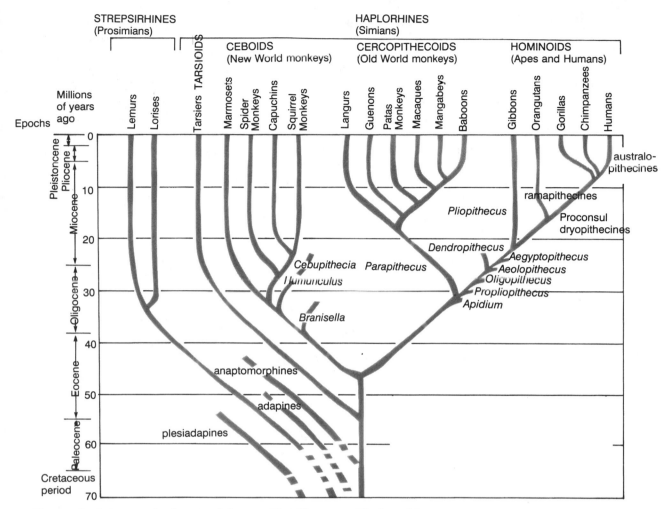

Figure 10.14. Primate family tree of the past 70 million years. The branching sequence is based on protein comparison data of living species. The specific dating of the branches is based on the ages of fossils and other geochronological data.

survivors of the Asian great ape lineage. They suggest that the ramapithecines may have been early members of that lineage.

Furthermore, Pilbeam and his colleagues, along with most other primatologists, believe that the Asian great apes share a common ancestor with the African great apes, which include today's gorillas and chimpanzees. From the age of the earliest ramapithecine fossils, Pilbeam concludes that the African and Asian great ape lineages diverged approximately 16 million years ago and that migration across the Afro–Asian land bridge contributed significantly to this divergence. It led first to geographic and later to the phylogenetic separation that we witness today.

Figure 10.14 summarizes our current understanding of primate evolution. The tree is not based entirely on the fossil record, however. Only the specific dating of the branches was taken from that record. The phylogenetic linkages among the branches were established by comparing the molecular structures of proteins taken from different presently living primates. Presumably the closer the protein structures are to each other, the more closely related are the primates from which they were taken. The larger the difference, the more distant is the phylogenetic relationship. The protein comparison technique was used because the primate fossil record alone is too incomplete to allow deduction of the full pattern of the family tree. But through the pooling of techniques from two distinct disciplines — molecular biology and physical anthropology — a reasonably cohesive understanding of the diversification of primates has been achieved. Note that this understanding is still far from complete. New fossil finds as well as the development of new techniques of interpreting the fossils and extant primates will surely lead to modifications in the picture presented here.

PRIMATE EVOLUTION, PART TWO: THE EARLY HOMINIDS

The Laetoli Footprints

The oldest undisputed evidence of hominids was found in 1976 by Mary D. Leakey[*] and an international team of scientists while searching for fossils in the Laetoli Beds of northern Tanzania (see figure 10.15). For two field seasons, the search had turned up very little of importance. Then, to the team's immense surprise, they found something far more striking and personal than fossils. They found footprints that were left in a layer of gray ash by people who walked there some 3.8 to 3.6 million years ago. One of the sets of footprints is described by Mary Leakey in these words:[†]

> It's a trail left by three people who walked across a flat expanse of volcanic ash [more than] three and a half million years ago. They are far and away the earliest human prints known anywhere in the world, and most probably the most important find that has been made by my team during the whole of my career.
> . . . there's a smallish person on the left . . . the rather oddly shaped tracks on . . . [the right are] . . . a large print with a smaller one superimposed inside it. . . . It's quite a small foot. It's not bigger than the length of my hand. And this was a group of people walking together. One assumes that they were perhaps holding hands. They're so evenly spaced, the tracks, and they're keeping step — always left foot for left foot, and right foot for right foot — that it may, for all we know, have been a family party.

Besides hominid footprints, the Laetoli Beds harbor prints from a host of other animals — hares, baboons, elephants, antelopes, rhinoceroses, pigs, hye-

[*]Mary Leakey is the widow of Louis S. B. Leakey (1903–1972), who carried out archaeological and anthropological research in East Africa from 1924 until his death. Aided by his wife and sons (Richard and Philip), his discoveries proved that man is far older than previously believed and that early human evolution was probably centered in Africa.

[†]"The Making of Mankind," Program 2, p. 2. 1983. A BBC-TV production in association with Time-Life Films. PTV Publications, P. O. Box 701, Kent, Ohio 44240.

Figure 10.15. Mary Leakey measuring 3.8 to 3.6 million year old footprints of *A. afarensis* in the Laetoli Beds of northern Tanzania. The prints clearly show the raised arch, rounded heel, pronounced ball, and forward-pointing big toe characteristic of the genus *Homo*. There can be no doubt that the individuals who left the prints walked upright and bipedally.

nas, ostriches, and others—attesting to the fact that the Laetoli environment of mid-Pliocene times was open savannah, much as today. Many of the footprints are surprisingly sharp and clear, as if they were from yesterday. Apparently, they were made in a fresh layer of ash ejected from Sadiman, a nearby volcano. Rains wetted the ash, and when the people and animals later crossed it, they left behind crisp imprints. Subsequent drying, hardening, and burial under more ash preserved them. Potassium–argon dating of the ash layers just above and just below the layers containing the footprints yields ages of 3.6 and 3.8 million years, respectively, indicating that the prints were made during the 200,000 year interval spanning these two limiting ages (see table 5.4).

The question raised by the Laetoli discovery is, "Who were the people that left the footprints?" Even a cursory examination shows that the prints are very similar to those of modern man, and careful analysis confirms it. The human raised arch, rounded heel, pronounced ball, and forward-pointing big toe are clearly in evidence. Thus, the consensus of anthropologists who have studied the footprints is that they could only have been made by bipedal individuals striding upright across the plain, individuals who either were our direct ancestors or very closely related to them.

Australopithecus afarensis

This assessment of the Laetoli people is confirmed by hominid fossils found at Hadar, an archeological site some 2000 km north of Laetoli in the Afar triangle

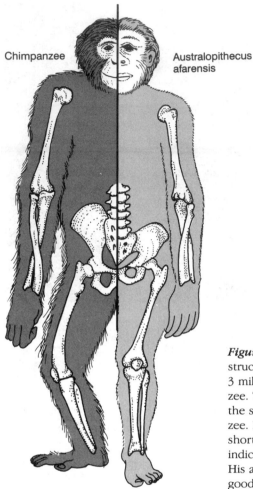

Chimpanzee

Australopithecus
afarensis

Figure 10.16. Comparison of a reconstruction of *Australopithecus afarensis* of 3 million years ago with today's chimpanzee. The australopithecine was roughly of the same size and build as the chimpanzee. However, his arms were slightly shorter, and his hip, knee, and ankle joints indicate that he was able to walk bipedally. His anatomy also suggests that he still had good climbing abilities.

of Ethiopia and dated from approximately 3.7 to nearly 3.0 million years ago. At this site, the remains of dozens of individuals—men, women, and children—were discovered during the 1970s by the International Afar Research Expedition, including one group of five to seven people at a single location. Comparison of the teeth and jaws of the Laetoli and Hadar people shows that they were very similar and probably belonged to the same species. (A few tooth and jaw fragments were found of the Laetoli people, but few other fossils.) Donald C. Johanson, one of the leaders of the Afar expedition, named this species *Australopithecus afarensis*, meaning "southern ape of Afar."

The Hadar hominids were quite small compared to us, with heights from about 1.0 to 1.2 meters and weights from 25 to 45 kilograms (see figures 10.16 and 10.17). Males were larger than females. Relative to the legs, their arms were longer than ours, but not as long as those of apes, the arms were attached to the shoulder blades at a slightly higher point than in us, and their hands were large and strong. Their teeth were thickly enameled and quite primitive. The molars were very large compared to the incisors and canines and they were strongly worn, pointing to a diet containing mainly roots, pods, tubers, and other tough, fibrous edibles. Their facial and skull features, including brain size, were similar to those of chimpanzees. However, their hip, knee, and ankle joints clearly indicate that they walked upright on two legs, though perhaps not quite as erect and with as much ease as we do.

The complex of anatomical characteristics just described—in particular the long arms, the attachment of the arms high on the shoulder blades, and the powerful hands—suggest that even though the early australopithecines walked bipedally, they were probably still good climbers. In fact, they may well have been climbers more than they were habitual bipeds. Anthropologists who have studied the fossil remains of A. afarensis agree that even though these people walked upright over open savannah, perhaps to forage and to get to the next clump of trees, they probably slept and sought refuge in trees. This is what savannah primates do today, such as the savannah baboons (see figure 10.5), vervet monkeys, and patas monkeys, except that they walk quadrupedally.

No stones or bones altered for use as tools have been found with the remains of A. afarensis, indicating that these people were still very primitive. Of course, this does not rule out that they might have used stones, bones, or wood casually in food gathering and the other activities of their daily lives, perhaps in ways like chimpanzees crack open nuts with rocks or extract termites from nests with trimmed sticks. Such casual use of tools would have left no permanent record.

Origin of Hominid Bipedalism

Before continuing with tracing the hominid evolution, let's ask two questions:

1. Why do anthropologists assign the Laetoli and Hadar people to the hominid family, when their brains were no larger than a modern chimpanzee's?
2. How did the hominid lineage arise?

Only the first of these two questions has a definite answer: *The Laetoli and Hadar people were hominids because they habitually walked upright and bipedally.* The fact that their brains and intelligence were modest by our standards and that their faces and other characteristics were apelike has nothing to do with their hominid status. For the same reason, chimpanzees, gorillas, and others of today's primates that are capable of bipedal walking but do so only occasionally and with considerable effort are not hominids.

No one knows for certain the answer to the second question because no fossils or other firm evidence older than the Laetoli footprints and relevant to

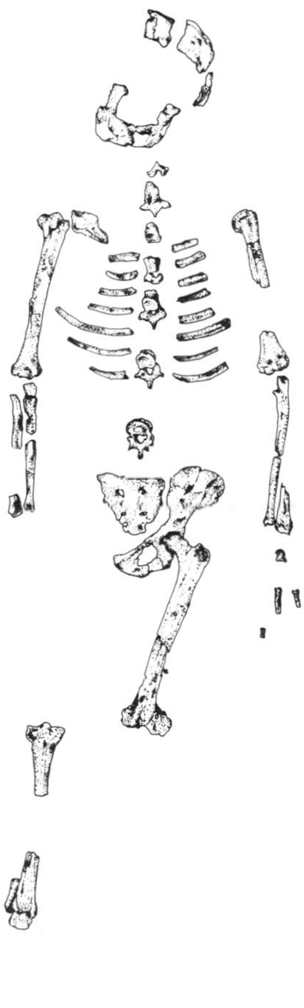

Figure 10.17. The now famous, roughly 3.5 million year old skeletal remains of "Luci," a female member of A. *afarensis*, from the Hadar region of Ethiopia. Luci was found in 1974 by the International Afar Research Expedition, whose members named her after the then popular Beatles song "Lucy in the Sky with Diamonds." The skeleton is nearly 40% complete. It suggests that Luci was an adult, because she had cut her wisdom teeth, and that like her contemporaries she was rather small—a little more than one meter tall. Interestingly, she may have been suffering from arthritis, as indicated by the lipped edges of several of her vertebrae.

hominid origins have yet been found. We can only make educated guesses. To do so, it is best, given the answer to question one, to rephrase question two: *How did the bipedalism of the hominids arise?*

In the past, this question was answered by assuming that hominid bipedalism arose as an adaptation to life in the savannah. It was argued that bipedalism would have freed the hands for tool and weapons use, made our ancestors better hunters, and offered them an increased range of view. However, none of these answers stands up to close scrutiny. For example, there is no evidence that the early hominids used tools and weapons intensively. Without tools, the early hominids would have been very poor hunters on the savannah and could not have competed successfully with such fast animals as cheetahs and lions. More often than not they probably would have been prey rather than hunter. Upright posture would have increased their range of view, but it would hardly have offered enough of a selective advantage to have led to the origin of bipedalism. Besides, the ability of raising oneself into an upright posture does not necessarily imply the ability of upright, bipedal walking or evolution toward this ability.

According to yet another theory, now rejected, put forth in the early 1980s, hominid bipedalism arose because it freed our ancestors' hands, allowing males to carry food items (consisting mainly of bulky vegetable matter) and to provision the females and their offspring. The females could then raise more than one young at a time, which increased their reproductive success. As part of this development, the females lost their estrus, became sexually continuously receptive, and male–female bonds (monogamy) became established.

This theory has been heavily criticized because there exists almost no evidence among today's primates or human hunting and gathering societies in its support and much evidence against it. (Note that today's hunting and gathering societies are probably our closest contemporary analogue to the societies of the ancient australopithecines.) For example, as Robert Sussman (personal communication 1985) has pointed out, in today's human hunting and gathering societies children are typically spaced four to six years apart, which makes their raising under the conditions these people encounter possible. Males in these societies do not provision females and their offspring; all members share in the task of finding food. The same is true of nonhuman primates. Furthermore, it would be very difficult, if not impossible, for the anatomy of quadrupedal terrestrial primates to change so as to allow habitual bipedalism. Finally, more than 80 percent of contemporary human societies are polygamous, and so are nearly all primate species. Monogamy in humans is commonly found only in western industrialized societies.

How then did hominid bipedalism arise? The theory most widely accepted today is that *hominid bipedalism arose as a preadaptation among our arboreal apelike ancestors, because of the way they climbed and moved about in trees*. The evidence for this theory comes from observations of similar preadaptations among today's primates that specialize in the arboreal locomotions of *vertical clinging and leaping* and of *brachiation*.

Examples of vertical clingers and leapers are the galagos, lepilemurs, tarsiers, and members of the indriid family. The typical resting position of these animals is clinging vertically to tree trunks, while their favored locomotion through the

Figure 10.18. The gibbons and siamangs are *brachiators:* their typical locomotion through the forest canopy is by hanging from branches with their arms and alternately reaching from hold to hold, thereby propelling themselves forward. This form of locomotion has *preadapted* these animals for bipedal walking, which they do either on broad branches (as shown here in the case of a siamang) or on the ground.

forest canopy consists of gigantic leaps from tree trunk to tree trunk. They execute these leaps by kicking themselves off the tree trunks with their long, powerful hind limbs, turning in midair, and landing on the next tree trunk with feet first. When these animals come down to the ground, they move about bipedally either by a series of kangaroolike hops or by slow walking. The anatomical requirements of these animals' arboreal locomotion has *preadapted* them for bipedalism on the ground.

Of the brachiators among today's primates, the most skillful ones are the gibbons and siamangs (see figure 10.6). These small apes of southeastern Asia have extremely long and slender forelimbs, with which they suspend themselves from branches and move through the forest canopy by swinging from hold to hold. When the gibbons come down onto the ground or when they travel along broad branches, they run or walk bipedally with arms stretched out overhead and to the sides (see figure 10.18). As in the case of vertical clingers and leapers, their mode of arboreal locomotion, too, has *preadapted* them to bipedalism.

Yet other examples of brachiators with preadaptations for bipedalism are today's great apes—the chimpanzees, gorillas, and orangutans. However, they have evolved such enormous upper torsos that they cannot sustain bipedal standing or walking for very long and must come down on their forelimbs for support.

These examples from today's primates suggest that the earliest members of the hominid lineage, too, were preadapted for bipedal walking on the ground. However, the arboreal locomotion that was primarily responsible for this pre-adaptation was probably neither vertical clinging and leaping nor brachiating. According to Jack H. Prost (1980) of the University of Illinois, it probably was quadrupedal vertical climbing. Prost bases his theory on the fact that quadru-pedal vertical climbing requires anatomical traits in the pelvis and lower limbs very similar to those required in upright bipedalism. He thus postulates that *the ancestors of the hominid lineage were probably apes that were skilled in quadrupedal vertical climbing, though they may have also been good bra-chiators. When these animals came down on the ground, they walked about upright and bipedally because this was their most efficient mode of locomo-tion on the ground.*

Of today's primates, none have the particular adaptation of quadrupedal vertical climbing that, according to Prost, characterized the ancestors of the hominid lineage, but the chimpanzees come quite close. They are capable of bipedal walking, though due to their massive upper torsos they cannot sustain bipedalism for very long.

The theory of the origin of hominid bipedalism just described has found wide acceptance during the past few years. As Robert Sussman (personal communi-cation 1985) puts it, "The theory is simple, it is consistent with the factual data we have of early hominids and of today's primates, and it does not require postulating any complex or highly unlikely evolutionary scenario."

When Did the Hominids Originate?

The next question to consider is, "When and under what circumstances did the hominid lineage originate?" Protein comparison data suggest that chimpanzees and gorillas are our closest primate relatives (for instance, chimpanzee and human DNA differ by no more than 2 percent) and that they and we go back to a common ancestor of roughly 5 to 8 million years ago. By then the weather in Africa, where the hominids most likely arose, had become drier and more seasonal and the original tropical forest was changing locally to a mosaic of woodland and savannah (figure 10.19).

The change from tropical forest to woodland and savannah would have been most conducive for the origin of the hominids according to the theory just presented. It would have compelled our arboreal, apelike ancestors to come down from the trees onto the ground, at least occasionally and temporarily. Because they were presumably preadapted for bipedal walking, they would have used this mode of locomotion while on the ground. At first, of course, their

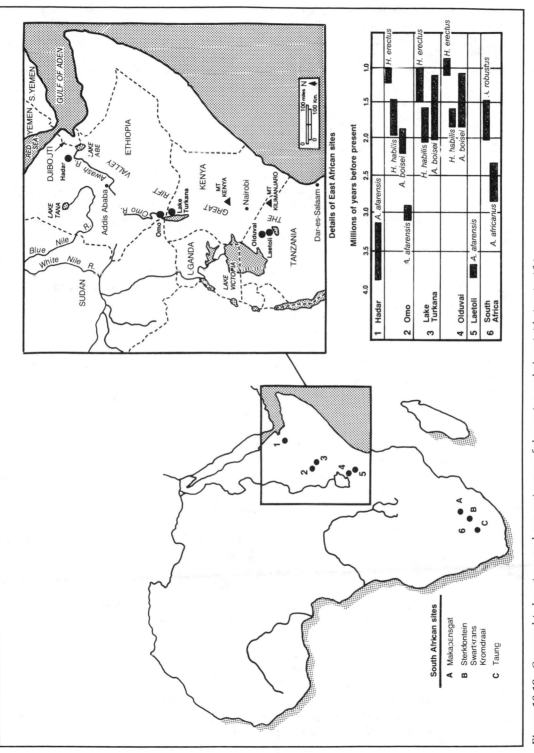

Figure 10.19. Geographic locations and age estimates of the major early hominid sites in Africa.

bipedal walking was probably quite awkward and inefficient by modern human standards, but in the course of time, perhaps over a period of a million years or so, their anatomy changed due to natural selection so that they were able to walk much more naturally and with more ease. The footprints and fossils of *A. afarensis* found at Laetoli and Hadar, dated between nearly 4 and 3 million years ago, are at present the earliest evidence of this evolutionary adaptation.

Australopithecine Evolution

Some time between 3 and 2 million years ago the early australopithecines evolved into two and, possibly, more phylogenetic lineages (see figure 10.20). The members of one of these lineages retained their basic ancestral features — among them small brains, primitive dentition, and long arms — and, hence, are also classified as australopithecines. They are *A. africanus*, *A. robustus*, and, somewhat tentatively (see below), *A. boisei*. They lived in the woodlands and savannah of southern and eastern Africa. *A. africanus* was of small build and so similar to *A. afarensis* that his placement into a new taxon has been questioned (see figure 10.21). *A. robustus* was much bigger and more robust than *A. afarensis*. His facial skeleton was unusually massive, with powerful jaws and huge molars. These features were even more pronounced in *A. boisei*. But the brains of both of these robust species were only slightly larger than those of *A. afarensis*. What we see here, as we go from *A. afarensis* to *A. robustus* and *A. boisei*, appears to be a progressively more specialized adaptation to a tough, fibrous vegetarian diet that required much repetitive chewing. At the same time, there apparently occurred little or no increase in intelligence.

A second, though more tentative lineage that is suspected to have evolved from the early australopithecines is known from chiefly one fossil skull. The skull — labelled KNM-WT 17000 — was discovered in 1985 by Alan Walker of Johns Hopkins University and several colleagues in 2.5 million-year-old sediments west of Lake Turkana (formerly Lake Rudolf), Kenya. The skull has a massively built cranium and a very small brain case. The teeth are large and highly specialized for a tough, fibrous plant diet, similar to the teeth of *A. robustus* and *A. boisei*. Furthermore, the skull's huge crests suggest that they anchored exceptionally powerful chewing muscles. The overall impression of skull WT 17000 is that it comes from a very robust australopithecine that possessed both ancestral and evolved traits. Its great age and robustness suggest that it may represent a lineage distinct from that of *A. africanus*, which is of comparable age but much less robust. Alan Walker and his colleagues proposed that the skull may be of a lineage whose later members included *A. boisei*. However, at present the evidence is too meager to be certain about any of the proposed phylogenetic relationships among the australopithecines. Therefore, the family tree shown in figure 10.20 should be regarded as provisional, and updates or corrections are to be expected as new fossils of early hominids are discovered. Donald Johanson expressed this current state of uncertainty as follows (1986, 84): "What this specimen [i.e., skull WT 17000] does is to precipitate changes in the family tree. But that's what always happens in this field when you get new things you didn't expect to find."

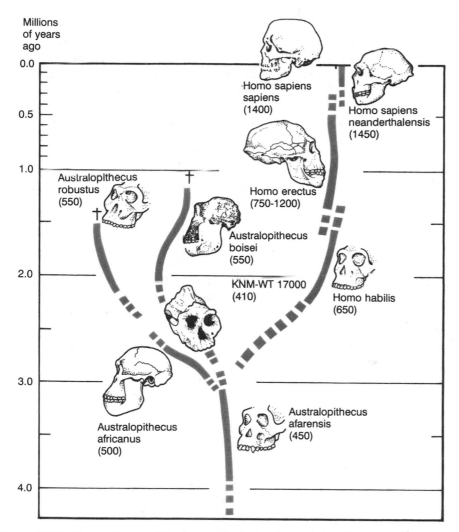

Figure 10.20. The evolution of the hominids during the past 4 million years. The dashes signify uncertainties in our understanding of the phylogenetic relationships. Estimates of average brain sizes in cubic centimeters are given in parentheses.

None of the australopithecines were successful in the long run, and by about 1 million years ago they all had disappeared. It seems that their extreme specialization for a highly specialized diet, combined with their limited intelligence, kept them from adapting successfully to the changing conditions of their environment.

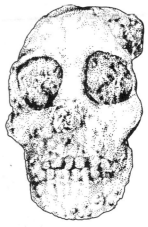

Figure 10.21. The first australopithecine remains were discovered in 1924 by Raymond Dart of the University of Witwatersrand, at Taung, South Africa. The best known of these early discoveries is the skull shown here. It is of a child of about six years of age and of the species *A. africanus.* It contains all of the milk teeth, the first permanent molars, and a nearly complete cast of its brain.

THE GENUS HOMO

The evolution of the hominids during the past 2 million years — the *Pleistocene* and *Recent Epochs* — needs to be viewed in the context of the general cooling trend that characterized the climate of those times and, in particular, of the repeated ice ages of the past 1 million years. During each ice age temperatures fell well below normal and vast regions from the poles to the mid-latitudes were covered by ice. In the Northern Hemisphere, the ice reached as far south as England, central Europe and Siberia, and the northern regions of the United States (see figure 10.22). In the Southern Hemisphere, glaciation was less extensive, mainly because large parts of this hemisphere were covered by oceans. In Africa, glaciers were absent except on the highest peaks, but the vegetation belts, including the animals that lived in them, were lowered by as much as 1000 meters. The ice ages lasted typically for a few ten thousand years and were followed temporarily by warmer interglacial periods, during which the ice retreated toward the poles and high-mountain chains.

We do not know precisely to what degree the climatic cooling trend affected the evolution of the hominids during the Pleistocene. It probably was no coincidence that yet another lineage of hominids, which is thought to have also evolved from *A. afarensis* and which was more intelligent than any of the australopithecines, developed such cold-weather survival techniques as building substantial shelters, wearing clothing, hunting small and large game, and using fire. This lineage is the genus *Homo* and its first member was *Homo habilis*. It and its descendants survived and even prospered during the Pleistocene, whereas the australopithecines did not.

The Species Homo habilis

The earliest signs of *H. habilis* are tooth and skeletal fragments that were discovered in 2.2 to 1.5 million-year-old layers at Lake Turkana, Omo, and Olduvai Gorge in eastern Africa (see figure 10.23). They are the remains of people who superficially looked like *A. africanus*, with heights of about 1.3 meters and weights between 30 and 50 kilograms. Their arms were slightly shorter and their legs longer, indicating that they were losing their apelike appearance and evolving in a direction closer to man. Their jaws and teeth, too, were becoming more like ours. Particularly the molars were much smaller than those of their australopithecine ancestors and contemporaries. It was a dentition suitable for a broadly based diet that included meat as well as fruits and vegetables. Most importantly, these people had significantly enlarged brains, with an average volume of about 650 cm^3.

The name *Homo habilis* was first proposed in 1964 by Louis S. B. Leakey, Philip V. Tobias of South Africa, and John R. Napier of England, the discoverers of the first fossils of this species. They chose the specific name *habilis* (Latin for "able, handy, skillful") in order to emphasize that in intelligence and manual skills *H. habilis* had advanced considerably beyond the australopithecines. This advance is most clearly demonstrated by the presence of concentrations of animal bones and altered stones, often brought from some distance away, with the remains of these people. The stones are obviously tools, which were crudely

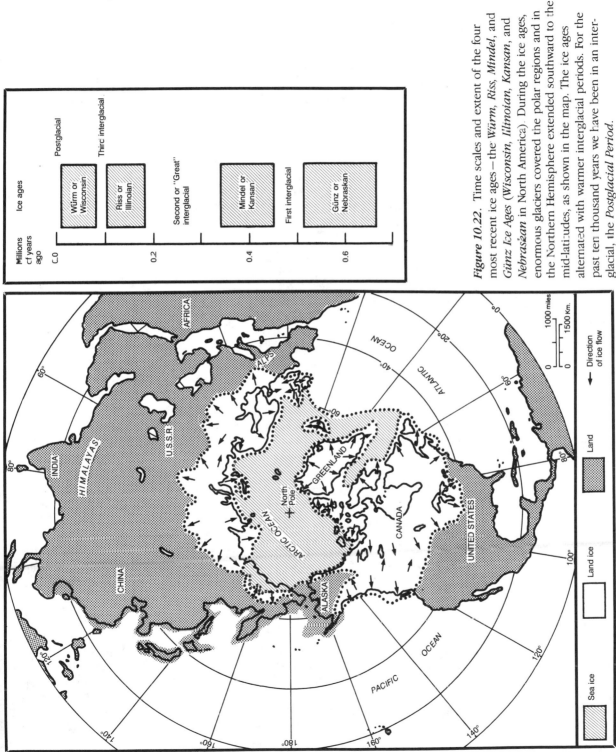

Figure 10.22. Time scales and extent of the four most recent ice ages — the *Würm, Riss, Mindel,* and *Günz Ice Ages* (*Wisconsin, Illinoian, Kansan,* and *Nebraskan* in North America). During the ice ages, enormous glaciers covered the polar regions and in the Northern Hemisphere extended southward to the mid-latitudes, as shown in the map. The ice ages alternated with warmer interglacial periods. For the past ten thousand years we have been in an interglacial, the *Postglacial Period.*

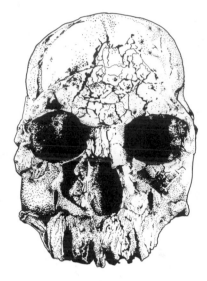

Figure 10.23. Homo habilis skull. This skull, which was unearthed in 1972 at Lake Turkana in northern Kenya, is more than 2 million years old and dates back to near the beginning of the *Homo* lineage.

flaked on one side and probably served as cutting and pounding instruments in the preparation of meat and vegetable foods. These tools belong to the *Oldowan tradition* (a name derived from "Olduvai Gorge," where tools of this type were first found) and mark the beginnings of the *Paleolithic* — the first phase of the *Stone Age* — which lasted until about 35,000 years ago (see figure 10.24). In addition to making tools, *H. habilis* built primitive shelters, as indicated, for example, by an artificial circle of loosely piled stones found at one excavation site. It seems that *H. habilis* was indeed "handy" and not only used tools, but also knew how to protect himself from the elements.

Most of the early stone tools used by *H. habilis* have been found in the savannah regions of eastern Africa. However, in 1986 Noel T. Boaz of the Virginia Museum of Natural History in Martinsville, John W. K. Harris of the University of Wisconsin in Milwaukee, and Alison S. Brooks of George Washington University in Washington, D. C., reported the discovery of nearly 300 quartz tools along with remains of animal bones and teeth farther south in the rain forests of eastern Zaire. Like the tools from eastern Africa, they are simple cobbles, flakes, and cores. They are roughly 2.5 to 2.0 million years old and most likely have also been made by *H. habilis*, though no human remains have yet been found with them. Anthropologists are particularly excited about these new discoveries because they suggest that *H. habilis* inhabited not only the dry savannahs of eastern Africa, but also the wet rain forest of central Africa.

The discoveries of these stone tools require us to ask the following questions: Was *H. habilis* a hunter-gatherer who actively hunted game and transported it back to his home base, to be shared with family and tribe? Or were these people scavengers who got meat the same way vultures and hyenas do, by eating whatever remained of the kills made by lions and other predators? Was meat, in fact, a major component of their diet or just an occasional supplement to a basically vegetarian diet? At present we don't know the answers. But judging from the life styles of most of today's hunting and gathering societies, animal protein probably was only a small part of the *H. habilis* people's diet.

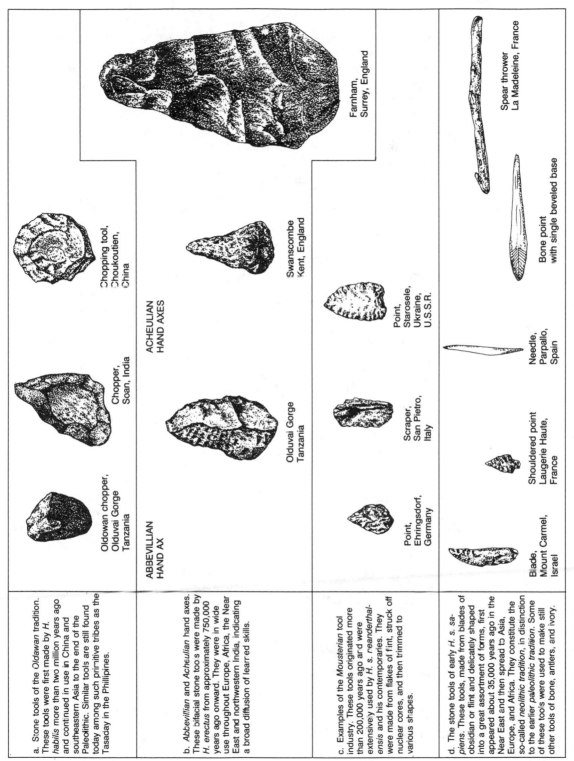

a. Stone tools of the *Oldawan* tradition. These tools were first made by *H. habilis* more than two million years ago and continued in use in China and southeastern Asia to the end of the Paleolithic. Similar tools are still found today among such primitive tribes as the Tasaday in the Phillipines.

Oldowan chopper, Olduvai Gorge Tanzania

Chopper, Soan, India

Chopping tool, Choukoutien, China

b. *Abbevillian* and *Acheullian* hand axes. These bifacial stone tool's were made by *H. erectus* from approximately 750,000 years ago onward. They were in wide use throughout Europe, Africa, the Near East and northwestern India, indicating a broad diffusion of learned skills.

ABBEVILLIAN HAND AX

ACHEULIAN HAND AXES

Olduvai Gorge Tanzania

Swanscombe Kent, England

Farnham, Surrey, England

c. Examples of the *Mousterian* tool industry. These tools originated more than 200,000 years ago and were extensively used by *H. s. reanderthalensis* and his contemporaries. They were made from flakes of flint, struck off nuclear cores, and then trimmed to various shapes.

Point, Ehringsdorf, Germany

Scraper, San Pietro, Italy

Point, Starosele, Ukraine, U.S.S.R.

d. The stone tools of early *H. s. sapiens.* These tools, made from blades of obsidian or flint and delicately shaped into a great assortment of forms, first appeared about 35,000 years ago in the Near East and then spread to Asia, Europe, and Africa. They constitute the so-called *neolithic tradition,* in distinction to the earlier *paleolithic tradition.* Some of these tools were used to make still other tools of bone, antlers, and ivory.

Blade, Mount Carmel, Israel

Shouldered point, Laugerie Haute, France

Needle, Parpallo, Spain

Bone point with single beveled base

Spear thrower La Madeleine, France

Figure 10.24. Tools of the Stone Age.

The Species Homo erectus

Though we do not know exactly how the *H. habilis* people lived, it seems safe to assume that their evolutionary success was due largely to their superior intelligence, compared with the australopithecines', and the adoption of meat for part of their sustenance. Support for this assumption comes from the remains of a second member of the homo lineage, *Homo erectus*, who evolved from *H. habilis* and who had both a larger brain and ate meat regularly. Apparently, these were characteristics that offered adaptive advantages and were selected for during the early evolution of the *Homo* genus.

Homo erectus populations appeared in eastern Africa around 1.6 million years ago. By 1 million years ago they were present in China and southeastern Asia. And somewhat later they had spread to the eastern parts of the Mediterranean, northern Africa, and many parts of Europe. In many of these areas they survived until roughly 300,000 years ago.

In size and build, these people resembled the smaller members of modern man, with heights of about 1.4 to 1.5 meters and weights between 40 and 55 kilograms. Their anatomy was no longer characterized by adaptations to arboreal living, as was the case with the earlier australopithecines. These people were as fully adapted to habitual upright and bipedal walking as we are. However, their skulls, jaws, and teeth were not yet modern. The skull was thickly boned, with heavy brow ridges and a receding forehead (see figure 10.25). The front teeth were as large as those of the earlier hominids, but the jaws and molars were smaller, though not as small as ours. Their brain volumes ranged from about 750 to more than 1200 cm^3, which brought them quite close to ours. This large spread was in part due to evolution, with the later members of *H. erectus* tending to possess larger brains, but it was also due to such factors as sexual dimorphism and differences from one population to another, factors that produced variations in other characteristics as well.

In their daily lives, too, the *H. erectus* people were becoming more like us. For example, when their teeth are examined under the microscope, they show enormous wear, with deep scratches and chunks of enamel broken off. Alan Walker of the Johns Hopkins Medical School, who investigated the teeth of *H. erectus*, compares their appearance under the microscope to a "cement slab

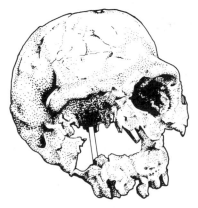

Figure 10.25. *Homo erectus* skull from China, ca. 700,000 years old. The prominent brow ridges of this skull are a characteristic of the *H. erectus* species, which survived for more than a million years, until about 300,000 years ago, in Africa, Europe, and Asia.

that's been crunched with a sledge hammer."* Only a diet that contained plenty of meat, bone, and grit could have done this kind of damage. Furthermore, these people made larger and more symmetrical and versatile tools than their *H. habilis* ancestors, and from about 750,000 years ago onward they made hand axes (the *Abbevillian* and *Acheulian stone tool traditions*, see figure 10.24). Many of the tools were used for hunting game, skinning and butchering it, and breaking open bones to get at the marrow inside.

At the archaeological site of Zhoukoudian near Beijing (Peking), China, which dates from approximately 800,000 to 500,000 years ago, a number of hearths as well as charred bones were found. Thus, at least by that time, some *H. erectus* populations had discovered the use of fire. Hearths and charred bones of deer, elephants, boars, and other wild animals are also present at various European sites. At Terra Amata, an archaeological site about 300,000 years old near Nice, France, post holes and oval arrangements of stone blocks were discovered that look like the remnants of shelters made of poles and covered with branches or hides. It seems that by a few hundred thousand years ago, the *H. erectus* people were establishing villagelike home bases, from which they carried out their hunting and gathering activities.

The Species Homo sapiens

In such activities as tool making, hunting game, using fire, and establishing home bases, the *H. erectus* people left reliable signs that they already possessed quite a rich and complex culture. It was a culture that probably had its origin with the *H. habilis* people or, possibly, even the australopithecines.

The existence of culture suggests that the *H. erectus* and *habilis* people communicated through language.† In fact, language was probably the key to their successes as hunters, inventors, and survivors under the rigors of repeated ice ages. It allowed them to warn each other of impending dangers, pass on information, discuss issues of common concern, teach the younger generation, and share feelings and emotions. No doubt, the language (or languages) was primitive, with a considerably smaller vocabulary and simpler syntax than ours. But it was language nevertheless and, probably, the single most important factor in the appearance of our species—*Homo sapiens*.

The major taxonomic characteristic defining *H. sapiens* is the large brain, which in contemporary members of the species has an average volume of 1400 cm^3. This makes it approximately 200 to 600 cm^3 larger than the brain of *H. erectus*. In comparison, the other anatomic differences between *H. erectus* and *H. sapiens*—such as size of jaws and teeth, shape of skull, and robustness of body frame—are relatively minor.

*"The Making of Mankind," Program 2, p. 5. 1983. A BBC-TV production in association with Time-Life Films. PTV Publications, P. O. Box 701, Kent, Ohio 44240.

†*Webster's Third New International Dictionary* defines culture as "the total pattern of human behavior and its products embodied in thought, speech, action, and artifacts and dependent upon man's capacity for learning and transmitting knowledge to succeeding generations through the use of tools, language, and systems of abstract thought."

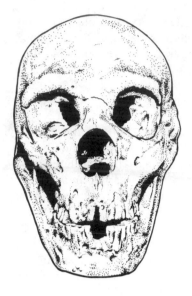

Figure 10.26. Neanderthal skull. The Neanderthals were early members of our species, *Homo sapiens*.

The earliest fossil evidence of *H. sapiens* is rather sparse. It comes from sites near Steinheim, Germany, and Swanscombe, England, both dated between 200,000 and 150,000 years in age; a site at Ngandong near the river Solo, Java, with a very uncertain age of somewhat less than 250,000 years; and a few other sites in Europe. Only in strata of about 100,000 years ago and younger does the fossil record of *H. sapiens* become widespread. The most abundant of these younger fossils are from the Neanderthal people, so named after the Neander Valley in Western Germany, where the 1856 workmen found the first specimen. The Neanderthals lived in Europe, the Near East, and in the southern parts of Russia until about 35,000 years ago. We do not know exactly how or where they originated, but their anatomic features strongly suggest that they had evolved from *H. erectus*.

The Neanderthals were roughly the same size we are, but they were stockier and much stronger, as indicated by the muscle attachments on their bones. They still had prominent brow ridges, a somewhat sloping forehead, and a slightly larger and more protruding dentition than we have, though it was smaller than that of *H. erectus* (see figure 10.26). Their lower jaws had a distinct bony chin like ours, which was absent in the earlier hominids. Surprisingly, their brain capacity was on the average slightly larger than ours, a feature probably related to their greater body weight and not to any superiority in intellectual or behavioral capacities (though this interpretation may be just a prejudice held by us).

Early in the twentieth century, after numerous additional fossil specimens of Neanderthals had been discovered, Marcellin Boule (1861–1942), the leading French anthropologist of the time, interpreted these people as brutish, subhuman creatures, closely related to the apes and barely capable of walking erect. This view gained wide popularity among laymen and experts alike. Today we know that Boule was wrong. True, the Neanderthals had facial and cranial features reminiscent of *H. erectus*, but their brains were as large or larger than ours and they had the same posturing abilities, manual dexterity, and range and quality of movement as we do. It has been suggested that they were people who, if dressed in modern clothes, would probably pass unnoticed in any New York City subway. Today they are classified as a subspecies of *H. sapiens* and referred to scientifically as *Homo sapiens neanderthalensis*.

During the last ice age (the Würm glacial), which began about 75,000 years ago and lasted until nearly 10,000 years ago, the Neanderthals stayed in the northern latitudes and met the challenges of increasing cold. They became more competent hunters than their ancestors. They refined the tool-making skills they had inherited and developed the *Mousterian tool industry* (from *Le Moustier*, a cave in southwestern France). This tool industry was based on the technique of striking flakes off a core of flint and then trimming the flakes to various shapes—projectile points, knives, scrapers, and hand axes (see figure 10.24). They protected themselves against the cold by wearing clothing and living in tents made of animal skins, though they also took shelter in caves. They built hearths and used fire.

Most remarkably, the Neanderthals buried their dead and placed ceremonial offerings at the grave sites. For example, at Teshik-Tash, in central Asia, the remains of a Neanderthal boy were found buried in a shallow pit, with pairs of goat horns surrounding it, and the ashes from a fire nearby. In another grave at the Shanidar Cave at the foothills of the Zagros Mountains of Iraq, dense clusters of pollen were discovered with the skeletal remains of an adult. Analysis of the pollen reveals that the body had been placed on a bed of pine branches and then covered with flowers.

The ceremonial burial of the dead by people who lived 35,000 and more years ago may strike us as unusual, for we tend to think of ceremonies as being unique to modern man. They are part of our traditions and symbolic expressions of our most deeply felt beliefs and emotions. Through them we find meaning in our existence, confirm the unity of our society, and attempt to establish communication between the sacred (the supernatural or divine) and the profane (space and time, cause and effect). We can only speculate that ceremonies had similar meanings for the Neanderthals. If true, then these people's inclusion in *H. sapiens* is all the more justified.

Modern Humans: Homo sapiens sapiens

Approximately 45,000 years ago, modern humans—*Homo sapiens sapiens*—made their appearance in the fossil record and within 10,000 years had completely replaced the Neanderthals and their contemporaries. The most ancient fossils of *H. s. sapiens* come from two caves in Israel, Qafzeh and Skhul, and from another cave, Cro-Magnon, near the village of Les Eyzies in southwestern France. Slightly later fossils of *H. s. sapiens*, dated between 35,000 and 30,000 years ago, have been found at various sites throughout central Europe, as, for example, at Velika Pecina in Yugoslavia and Brno in Czechoslovakia. New fossils of these people are still appearing, especially in Africa and Asia.

These early members of our subspecies are now commonly referred to as *Cro-Magnon Man*, after the cave near Les Eyzies where they were first discov-

Figure 10.27. Reconstruction of four hominids.

ered. According to the fossil record, they still had the Neanderthals' stocky build but otherwise they were very similar to us (see figure 10.27). Their teeth, jaws, brow ridges, and faces were smaller than those of the Neanderthals, and their skulls were short (as measured from the face to the back), high, narrow, and rounded in the back (see figure 10.28). However, these are merely descriptions of the average characteristics of the Cro-Magnon people. If we could observe them alive, we probably would find as great a racial diversity among them as we do among ourselves.

With the arrival of *H. s. sapiens*, the pace of cultural evolution, which had been relatively slow and conservative for the previous 2 million years, began to quicken. For example, within roughly ten thousand years, these people advanced the art of tool making to levels similar to those of the Eskimos and

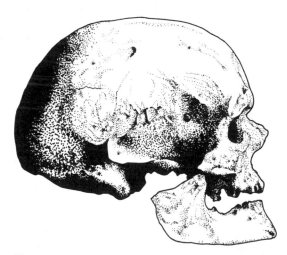

Figure 10.28. Skull of *Homo sapiens sapiens* (Cro-Magnon Man), from the Dordogne region of France, ca. 40,000 years old. This excellently preserved skull shows that the facial features and head proportions of the earliest members of our subspecies were very similar to ours.

Native Americans before the coming of the white man. From thin, symmetrical blades, which they struck off cores of obsidian and flint, they made exquisitely shaped spear points, knife blades, borers, and scrapers (see figure 10.24). They then used some of these tools to work such materials as bone and antlers, fashioning them into still other tools—harpoons for hunting, hooked levers for throwing spears, and needles and awls for sewing. It was the "new" or second phase of man's stone age technology—the *Neolithic*.

Many of the early members of *H. s. sapiens* were also masters of the hunt, particularly those living in the colder climates. For instance, in Europe, where the Würm Ice Age had created conditions favorable for a great variety of grazing and browsing animals, *H. s. sapiens* hunted such large game as the musk ox, woolly rhinoceros, mammoth, bison, bear, wild boar, reindeer, and horse. We know how significant these animals and the hunting of them was to these people not only from the many animals bones found along with human fossils, but also from elaborate paintings of animals and hunting scenes in dozens of caves in France, Spain, Africa, and Asia (see figure 10.1). Many of these paintings rival our own fine art in vitality and design. However, they probably were painted not just for art's sake, but as part of mystical rituals meant to invoke success in the hunt.

Besides painting, these early people expressed their artistic talents by sculpturing and carving. Low-relief carvings have been found on the walls of numerous rock shelters and caves (see figure 10.29), and statuettes of humans

Figure 10.29. *Venus of Laussel*. This 40 centimeter high figure is carved on the wall of a limestone overhang (the "rock shelter of Laussel") in the Dordogne region of France and is approximately 20,000 years old. The carving has been described as "the most vigorously sculptured representation of the human body in the whole of primeval art."

and animals carved out of rock, bone, and ivory come from several sites in Europe and western Siberia. Some of the statuettes are surprisingly realistic, while others show greatly exaggerated bellies, breasts, and buttocks and apparently were symbols of life and fertility. Artifacts such as these have been found in graves along with bodies that had been painted with red ochre, suggesting the existence of myth and a belief in an afterlife.

Clearly, the cultural accomplishments of these people were remarkably high and varied. To be so successful, they must have reached comparably high levels in the ways they organized themselves into societies. We may assume that families were joined into tribes and tribes into still larger groupings, all speaking a common language.

TRANSITION TO AGRICULTURE, CITIES, AND MODERN CULTURE

It was through achievements such as those just described that our ancestors laid the foundation for the emergence of our modern cultures and societies. Up to about ten thousand years ago, all human populations had subsisted by hunting and gathering. They had lived off the land, eating the fruits, grains, and vegetables that were in season and following the herds in their yearly migrations. But now, as the last ice age was coming to a close and the climate grew milder, some groups of *H. s. sapiens* were beginning to cultivate plants and domesticate animals — barley and wheat, and goats, sheep, cattle, and pigs. They were turning to a farming way of life and settling down on a piece of land they called their own. No longer were they forced to follow the migrations of animals or the growth of grains and fruit. They were now raising their own food. They gained independence from the nomadic life.

The earliest human settlements signifying the change to an agricultural way of life were built in the Fertile Crescent in the Middle East, which arches from the Jordan Valley and Syria through eastern Turkey to Iraq and Iran. Here we find stone ruins of walls, houses, and towers of nearly ten thousand years in age. An example is Jericho, which had a more than 5000-year-old history before the biblical writers made reference to it (see figure 10.30). In its original form, this city extended over an area of roughly 200 meters on each side and was occupied by as many as 2000 people.

Along with the construction of these early settlements and the development of the agricultural technology necessary to support the inhabitants, progress was made in other areas as well. Hierarchical social structures evolved, headed by kings and priests. Myths and religions arose, intended to strengthen the power of the rulers and to unify the society. Crafts, such as weaving and pottery, were introduced. Commerce developed between neighboring settlements. Weapons and defenses were constructed to ward off invaders. Rules were enacted for dividing up the harvest and securing everyone's participation in the tasks facing the society as a whole. In short, the traditions necessary for regulating life in a city society became established.

In the course of the next few thousand years, many of the earlier cities declined, new ones arose, and some merged with others into larger, more

Figure 10.30. Remains of a house in Jericho, ca. 7000 B.C. The ancient ruins of Jericho include the remains of solid city walls and towers as well as houses, attesting that by about nine thousand years ago some of our ancestors had made the transition from a hunting-gathering to a sedentary, agricultural way of life.

powerful associations. From about 5000 years ago onward, the first great civilizations appeared: the Egyptian Civilization at the mouth of the Nile (3100–1090 B.C.), the Sumerian and Babylonian Civilizations of Mesopotamia (3100–1200 B.C.), the Indus Valley Civilization in India-Pakistan (2500–1700 B.C.), and the Huang Ho Valley Civilization in China (1500–1000 B.C.). Yet another great civilization developed in the highlands and coastal areas of western and northwestern South America. This Andean civilization flourished from about 3500 B.C. to the 16th century A.D., when it was destroyed by the Spanish invaders.

It was during the times of these early civilizations that the first cities of large scale were constructed (figure 10.31); sophisticated political, social, and religious systems arose; formal statutes of law were laid down; modern technology (e.g., metallurgy) got its start; and sea trading originated. They were also the times when writing was introduced, at first probably merely as a means for keeping inventory of possessions and tabulating commercial transactions. Before long, writing was also used to record stories, myths, laws, and religious doctrines and to pass on information in general (see figures 10.32 and 10.33).

Together, these developments marked the final break with the hunting–gathering way of life, which stressed above everything else the need to adapt to the environment. Now humans were learning to construct their own environment. Along with this learning came a new attitude, based on the belief that man is the rightful master over land and life on Earth. Nowhere is this belief more clearly expressed than in the first book of the Bible (next page).

Figure 10.31. The ruins of Uruk (Evech), ca. 3000 B.C., one of the great cities of the Sumerian civilization. The city was built on the then marshy, reed-covered delta of the Tigris and Euphrates Rivers and enclosed by defensive walls of about 10 kilometers in length. According to legend, the walls were built by Gilgamesh, the hero of the oldest known epic.

> . . . and God said unto them, "Be fruitful and multiply, and replenish the earth, and subdue it: and have dominion over the fish of the sea, and over the fowl of the air, and over every living thing that moveth upon the earth." (Genesis 1:28)

We followed this dictum to the letter. We multiplied. We colonized all continents, including the Arctic and Antarctic regions. We explored the oceans' depths and are sending men and women into space. And with our science, atomic power, rapid transportation, high-speed computers, and instantaneous worldwide communication, we tend to feel (at least those of the middle and upper classes in industrialized countries) that we are indeed the supreme masters of the Earth.

This supremacy, however, is much less secure than we often think. We are still subject to nature's ways. Droughts, hurricanes, severe cold, and earthquakes visit us regularly. Besides, we are grossly abusing our newly found powers, with

Figure 10.32. Clay tablet from Nippur, the religious and cultural center of the Sumerian civilization. The tablet describes the creation of Heaven and Earth (Kramer 1981, 81):

> After Heaven had been moved away from
>   Earth,
> After Earth had been separated from Heaven,
> After the name of man had been fixed,
> After [the heaven-god] An carried off the
>   Heaven,
> After [the air-god] Enlil carried off the Earth

unforeseeable consequences for the future. We are polluting the land, water, and air on a global scale. Many regions are severely overpopulated (see figure 10.34). We are destroying the tropical rain forests, thereby altering the world's climate and causing the extinction of countless species of plants and animals. We still harbor racial and cultural prejudices. And in our attempt to protect our nations and ideologies, we are endangering our very existence by aiming thousands of missiles tipped with nuclear warheads at each other.

Figure 10.33. Clay tablet inscribed with the oldest known law codes, issued by the Sumerian king Ur-Nammu (reigned 2112–2095 B.C.). The text is fragmentary, but two of the laws read as follows (Kramer 1981, 55):

> If a man to a man with a weapon his bones
>   of . . . severed,
>   1 silver mina he shall pay.
> If a man to a man with a geshpu-instrument
>   the nose has cut off,
>   2/3 of a silver mina he shall pay.

Thus, in the Ur-Nammu law code the primitive concept of "eye for eye" and "tooth for tooth" had already given way to the more humane punishment by monetary fines.

Where will we go from here? Clearly, as Thomas Wolfe has said, we "can't go home again." We cannot return to the simple, nomadic life of the hunter–gatherer. Yet we cannot go on exploiting the land either and, decade after decade, keep increasing our nuclear arsenals. That surely means courting the dangers of a holocaust. If there is to be hope for the future, we must accept our racial, cultural, and political diversities (see figure 10.35). Furthermore, we must learn to combine the two seemingly contradictory attitudes — modern man's reliance on science and technology and the nomad's maxim of adaptation to the environment — and achieve a new balance in our view of ourselves and our place on Earth. We must make judicious use of our scientific and technological knowledge, and we must learn to live in harmony with our environment and with each other. We must join new information with old, proven wisdom. This will be the most difficult task ever faced by humanity. It will require a profound understanding of our global needs, long-term vision, worldwide planning, sacrifice, and self-discipline on an enormous scale. As the German poet Johann

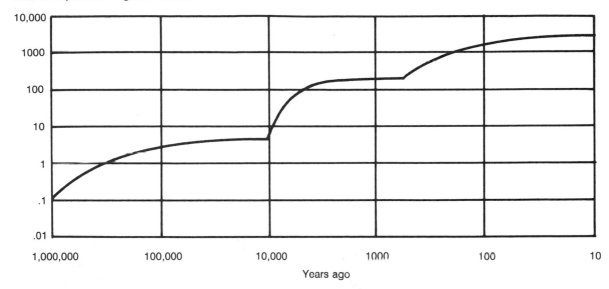

Figure 10.34. Estimates of the growth of the human population during the past 1 million years. The three rather rapid rises were each due to a technological breakthrough that permitted the support of a larger number of people than before. The first rise (ca. 10^6 years ago) was due to the invention of the stone-tool industry. The second (ca. 10^4 years ago) was due to the change from hunting-gathering to the establishment of agricultural settlements. The third (ca. 400 to 300 years ago) was due to the advent of the modern industrial revolution. The current world population is 5 billion and grows at an annual rate of approximately 1.65%, which means that if the growth rate remains the same the population will double roughly every forty years. (Note that the scales of the graph are not linear. Neighboring divisions differ by factors of ten.)

Wolfgang von Goethe (1749–1832) so wisely wrote more than 150 years ago, only he deserves freedom and life who every day conquers them anew:

> A marsh extends along the mountain-chain
> That poisons what so far I've been achieving;
> Were I that noisome pool to drain,
> 'Twould be the highest, last achieving.
> Thus space to many millions I will give
> Where, though not safe, yet free and active they may live.
> Green fertile fields where straightway from their birth
> Both men and beast live happy on the newest earth,
> Settled forthwith along the mighty hill
> Raised by a daring, busy people's will.
>
> . . .
>
> Yes, to this thought I hold unswerving,
> To wisdom's final fruit, profoundly true:
> Of freedom and of life he only is deserving
> Who every day must conquer them anew.
> Thus here, by danger girt, the active day

Of childhood, manhood, age will pass away.
Aye! such a throng I fain would see,
Stand on free soil among a people free.
Then might I say, that moment seeing:
"Ah, linger on, thou art so fair!"
The traces of my earthly being
Can perish not in æons — they are there!

(Goethe, *Faust*, Second Part, Act V.)

Figure 10.35. Ethnic diversity among humans. All living members of the genus *Homo* belong to the same species — the species *Homo sapiens sapiens*. We are all able to interbreed and to have offspring. However, like all species of organisms, ours is characterized by a range of genetic variation and, consequently, we do not all look alike. For example, as a result of adaptation to different climates and habitats, a great number of different human races evolved. Still, in the basic anatomical and behavioral characteristics — among them upright and bipedal gait, large and complex brain, communication through language, and in our feelings and emotions — we are alike and constitute the *family of man*.

Exercises

1 "All human cultures are anthropocentric. In fact, cultural conditioning is so strong that generally we are not even aware of our anthropocentric biases." In a brief essay defend or refute these statements. Support your arguments by examples from religion, literature, art, science, contemporary entertainment, and so forth.

2 Distinguish between:
 (*a*) Strepsirhines and haplorhines.
 (*b*) Tarsiers and lemurs.
 (*c*) Old World and New World monkeys.
 (*d*) Lesser and great apes.
 (*e*) Hominoids, hominids, and *Homo*.

3 Several specialized kinds of locomotions have evolved among primates, as adaptations to arboreal environments or to terrestrial environments.
 (*a*) Name the specialized primate locomotions discussed in the text and give examples of primates using them.
 (*b*) Describe the anatomical characteristics that form the physical basis of these locomotions.

4 Some primates possess the *precision grip*, in addition to the *power grip*.
 (*a*) Describe what is meant by these two terms and illustrate them with a few examples from your personal experiences.
 (*b*) Give examples of primates that possess the power grip and that possess the precision grip.
 (*c*) In which primates is the precision grip most highly developed?

5 (*a*) Select any three of the primate characteristics discussed in the text and describe them in your own words.
 (*b*) Discuss which selection pressures you think were responsible for their evolution.
 (*c*) Have these characteristics also evolved in other animals? If so, give examples and discuss which selection pressures might have been responsible for their evolution.

6 The fossil record indicates that the earliest primates were nocturnal and rather small, with sizes roughly in the range from those of present-day field mice to large domestic cats. Then, during the Oligocene, a major adaptive shift took place among the primates toward larger size and diurnal ways of life. Discuss what environmental changes probably brought about this adaptive shift and what benefits the primates reaped from it.

7 In discussing primate evolution, paleontologists and anthropologists use such terms as:
 terminal branch feeding
 primates of "modern aspect"
 Fayum primates
 dryopithecines and ramapithecines
 Afar triangle
 Oldowan tradition
 Neanderthal
 Würm glacial
 Cro-Magnon Man
 Lascaux Cave
 Fertile Crescent
Briefly explain these terms.

8 Mary Leakey (1979) considers her discovery of the Laetoli footprints as " . . . probably the most important find that has been made by [her] team during the whole of [her] career." Discuss whose footprints Dr. Leakey refers to in this statement, their age, and why she values their discovery so much.

9 (*a*) Name and briefly discuss the defining characteristic of the hominids.
 (*b*) How, according to the theory discussed in the text, did this characteristic arise? (Refer to both the immediate ancestors of the hominids as well as to environmental factors.)
 (*c*) Briefly describe some of the alternative theories that have been proposed in the past to explain the origin of the hominids and state why these theories have been discarded.

10 We frequently hear the statements, "Humans did not evolve from monkeys," or "Humans did not evolve from chimpanzees." Both statements are correct, but for reasons other than those the people who make such statements usually have in mind. Explain.

11 It has been suggested that language appeared very late and rather suddenly in the evolution of the genus *Homo*, and probably only after the brain of the members of this genus had reached its present volume. This hypothesis contradicts the theory presented in this text, according to which the cultural achievements of *Homo habilis* and *Homo erectus* are signs of a much earlier origin of language and of its gradual evolution toward increased vocabulary and more complex syntax. Defend one or the other of these two theories.

12 Select the half dozen or so events in the evolution from the early vertebrates (the jawless fishes) to *Homo sapiens* that you regard as most significant.

Briefly describe these events and discuss why you regard them as so significant. (For my answer to this exercise, turn to the *Epilogue*.)

Suggestions for Further Reading

Adams, R. M. 1960. "The Origins of Cities." *Scientific American*, September, 153–68.

Bogucki, P., and R. Grygiel. 1983. "Early Farmers of the North European Plain." *Scientific American*, April, 104–12. Excavations in Poland reveal the remains of people who farmed, herded, and hunted in the area some 7000 years ago.

Braidwood, R. J. 1960. "The Agricultural Revolution." *Scientific American*, September, 130–48. How the domestication of plants and animals, some 10,000 years ago in the Middle East, freed man from hunting and gathering.

British Museum (Natural History). 1980. *Man's Place in Evolution*. London: Cambridge University Press. A small book that traces our ancestry in a simple, but effective, style.

Begley, S., J. Carey, and R. Sawhill. 1983. "How the Brain Works." *Newsweek*, February 7, 40–47. A serious discussion of our current understanding of the human brain, including brain structure and function, left- and right-brain hemispheres, neurons, mind, thinking, emotions, and the effects of food and drugs on the brain.

Bryant, V. M., Jr., and G. Williams-Dean. 1975. "The Coprolites of Man." *Scientific American*, January, 100–109. About the study of fossil human feces that range in age from a few 100 to 300,000 years and yield information on prehistoric diet, environment, and behavior.

Campbell, J. 1983. *Historical Atlas of World Mythology. Vol. 1. The Way of the Animal Powers*. New York: Alfred Van der Marck Editions, distributed by Harper and Row. A large-format, beautifully illustrated book about the origin of our species, and about the artifacts, mythologies, and hunting and gathering ways of humans past and present.

Covey, C. 1984. "The Earth's Orbit and the Ice Ages." *Scientific American*, February, 58–66. This article identifies and suggests a cause for nine major cycles of ice ages during the past 570 million years: periodic variations in the geometry of the Earth's orbit. The focus is on the most recent cycle, the Pleistocene ice ages, which began a million years ago.

Eaton, G. G. 1976. "The Social Order of Japanese Macaques." *Scientific American*, October, 97–106.

Fairservis, W. A., Jr. 1983. "The Script of the Indus Valley Civilization." *Scientific American*, March, 58–66.

Festinger, L. 1983. *The Human Legacy*. New York: Columbia University Press. Some of the insights and knowledge about human origins, evolution, and social behavior gathered over the author's 40-year research career in psychology, physiology, and animal behavior. The chapters on the transition from hunting–gathering to sedentary living, agriculture, and the coming of technology, trade, religion, and war are particularly worth reading.

Friberg, J. 1984. "Numbers and Measures in the Earliest Written Records." *Scientific American*, February, 110–18. Sumerian and Elamite scribes of some 5000 years ago had well-developed systems of numbers and measures, including precursors of the decimal system.

Ghiglieri, M. P. 1985. "The Social Ecology of Chimpanzees." *Scientific American*, June, 102–13. A study of these wild apes carried out without the use of food as a lure, shows that their social structure is shared only with humans.

Gladkih, M. I., N. L. Kornietz, and O. Soffer. 1984. "Mammoth-Bone Dwellings on the Russian Plain." *Scientific American*, November, 164–75. These dwellings built 15,000 years ago by hunting–gathering bands, suggest that a profound social change was taking place on the steppe at the end of the great Ice Age.

Hammond, N. 1982. "The Exploration of the Maya World." *American Scientist*, September–October, 482–95.

Hay, R. L., and M. D. Leakey. 1982. "The Fossil Footprints of Laetoli." *Scientific American*, February, 50–57.

Jolly, A. 1985. "The Evolution of Primate Behavior." *American Scientist*, May–June, 230–39.

Kalb, J. E., et al. 1984. "Early Hominid Habitation in Ethiopia." *American Scientist*, March–April, 168–78. Recent discoveries of fossils and artifacts in the Middle Awash Valley should yield one of the longest and most complete records of hominid occupation yet described.

Kavanagh, M. 1984. *A Complete Guide to Monkeys, Apes and Other Primates*. New York: The Viking Press. A comprehensive and very readable reference book, with photographs and descriptions of each living primate genus.

Kramer, S. N. 1981. *History Begins at Sumer*. Philadelphia: The University of Pennsylvania Press. An exceptionally interesting account of the spiritual and cultural achievements of one of the earliest and most creative civilizations.

Leakey, M. 1983. "Tanzania's Stone Age Art." *National Geographic*, July, 84–99. An article about prehistoric drawings of uncertain age found on cliffs in Tanzania. The stylized human figures and remarkably realistic animals attest to the great skills and artistry of an unknown people.

———. 1979. "Footprints in the Ashes of Time." *National Geographic*, April, 446–457. The discovery of 3.6-million-year-old Laetoli footprints.

Leakey, R., and A. Walker. 1985. "*Homo Erectus* Unearthed." *National Geographic*, November, 624–29. Kenya's backcountry yields a 1.6-million-year-old fossilized boy—the best preserved, most complete skeleton of an early human yet found.

Leroi-Gourhan, A. 1982. "The Archaeology of Lascaux Cave." *Scientific American*, June, 104–12.

Lewin, R. 1984. *Human Evolution: An Illustrated Introduction*. New York: W. H. Freeman and Co. A concise, up-to-date, and clearly written introduction to human origins. In addition to physical anthropology, the book addresses topics from geology, evolutionary biology, origins of language, prehistoric art, the agricultural revolution, and human culture.

———. 1982. "How Did Humans Evolve Big Brains?" *Science* 216: 840–41. Description of the work of the British primatologist Robert Martin who argues that big brains are energy expensive and, therefore, their evolution can occur only under certain favorable ecological circumstances.

Napier, J. 1980. *Hands*. New York: Pantheon Books. An interesting book about the structure, function, and evolution of the hands of humans and other primates, and about the social and cultural aspects of the hand.

Nauta, W. J. H., and M. Feirtag. 1979. "The Organization of the Brain." *Scientific American*, September, 88–111. About the organization of the network of billions of neurons in the brain and spinal cords of mammals, including humans.

Pfeiffer, J. E. 1984. "A Once-in-a-Lifetime Skull Session." *Smithsonian*, August, 50–57. Paleoanthropologists from around the world examine and discuss 40 priceless fossils of our human ancestors that were part of a 1984 exhibit, "Ancestors: Four Million Years of Humanity," at the American Museum of Natural History in New York City.

———. 1982. *The Creative Explosion: An Inquiry into the Origins of Art and Religion*. New York: Harper & Row, Publishers.

Pilbeam, D. 1984. "The Descent of Hominoids and Hominids." *Scientific American*, March, 84–96.

Potts, R. 1984. "Home Bases and Early Hominids." *American Scientist*, July–August, 338–47. The fossil record of Olduvai Gorge suggests that the concentrations of bones and stone tools do not represent fully formed campsites but an antecedent to them.

Rensberger, B. 1986. "Scientists See Signs of Mass Extinction as Rain Forests Are Cleared." *The Washington Post*, September 29, A, 1:6. Scientists are beginning to realize that the current wholesale destruction of tropical rain forests in the Amazon basin, Indonesia, and central Africa, which are the world's most biologically diverse habitats, may cause a mass extinction of species as great or greater as any in the past. This loss of species is likely to have unforeseeable consequences for our future.

Rukang, W. and L. Shenglong. 1983. "Peking Man." *Scientific American*, June, 86–94. About the fossil finds of *Homo erectus* in a cave near Zhoukoudian, China, which has yielded a wealth of evidence on the biological, technological, and perhaps even social evolution of early humans.

Sandelowsky, B. H. 1983. "Archaeology in Namibia." *American Scientist*, November–December, 606–15. Fossils, stone tools, and rock art testify to the continuous hominid and human occupation of this corner of southwestern Africa for the last 2 million years.

Smith, P. E. L. 1976. "Stone-Age Man on the Nile." *Scientific American*, August, 30–38. Hunters and gatherers lived along the Nile thousands of years before the first pharaohs. The author describes how their adaptation to their environment affected the later development of agriculture and high civilization.

Trinkaus, E., and **W. W. Howells.** 1979. "The Neanderthals." *Scientific American*, December, 118–133.

Weaver, K. F. 1985. "The Search for Early Man." *National Geographic*, November, 560–623. An extensive, well-illustrated article about the latest interpretations of mankind's fossil record.

Wendorf, F., A. E. Close, and **R. Schild.** 1985. "Prehistoric Settlements in the Nubian Desert." *American Scientist*, March–April, 132–41. A region that is now virtually uninhabitable contains a record of human adaptation to arid environments that may be 500,000 years old.

White, R. 1986. *Dark Caves, Bright Visions: Life in Ice Age Europe*. New York and London: The American Museum of Natural History and W.W. Norton and Co. An exceptionally beautifully illustrated book of prehistoric art, mostly from southwestern France and dating back up to 35,000 years. The book is based on a 1986–1987 exhibition at The American Museum of Natural History, New York City.

Zvelebil, M. 1986. "Postglacial Foraging in the Forests of Europe." *Scientific American*, May, 104–15. The author suggests that hunting–gathering was not just a prelude to agriculture, as is often thought. It continued as a way of life for many populations long after the development of farming and in some areas was as productive as early farming.

Additional References

Cartmill, M. 1972. "Arboreal Adaptations and the Origin of the Order Primates." In R. Tuttle, ed., *The Functional and Evolutionary Biology of Primates*, 97–122. Chicago: Aldine-Atherton.

Johanson, D. C. 1986. Quoted in "Human Roots (and Branches)." *Scientific American, Science and the Citizen* Section, October, 80–84.

Prost, H. 1980. "Origin of Bipedalism." *American Journal of Physical Anthropology* 52: 175–89.

We are not alone in the universe, and do not bear alone the whole burden of life and what comes of it. Life is a cosmic event — so far as we know the most complex state of organization that matter has achieved in our cosmos. It has come many times, in many places — places closed off from us by impenetrable distances, probably never to be crossed even with a signal. As men we can attempt to understand it, and even somewhat to control and guide its local manifestations. On this planet that is our home, we have every reason to wish it well. Yet should we fail, all is not lost. Our kind will try again elsewhere.

George Wald (1954)
(*Professor Emeritus*, Harvard University)

Epilogue

"ARE WE ALONE in the Universe, or has life evolved elsewhere also?" Attempting to answer this question will be the final topic of this book. In particular, I shall attempt to estimate how frequently intelligent beings, capable of communicating across interstellar distances, might have evolved on other planets circling other stars. If they have, perhaps we can make contact with them and, thereby, join a larger, cosmic fraternity.

The topic of extraterrestrial life is not a new one. It was already being discussed during the first century B. C. by Titus Lucretius Carus, the Latin poet and philosopher, who had no doubt at all that life, including intelligent life, exists elsewhere in the Universe. Lucretius (see Goldsmith 1980, 4) justified his belief by a simple, yet very persuasive logical argument in his famous poem *De rerum natura* ("On the Nature of the Universe"):

> For since infinite space stretches out on all sides, and atoms of numberless number and incalculable quantity fly about in all directions quickened by eternal movement, it can in no way be considered likely that this is the only heaven and earth created, and all those other atoms there beyond are doing nothing. For this world was created by Nature after atoms had collided spontaneously and at random in a thousand ways, driven together blindly, uselessly, without any results, when at last suddenly the particular ones combined which could become the perpetual starting points of things we know—earth, sea, sky, and the various kinds of living things. Therefore, we must acknowledge that such combinations of other atoms happen elsewhere in the universe to make worlds such as this one, . . . with races of different men and different animals . . .

Today many scientists believe that life, including intelligent life with advanced technology (namely the ability to communicate across interstellar distances), is not unique to Earth. This belief is based on logical arguments that are in principle no different from that of Lucretius. The main difference is that today we express our reasoning in more quantitative terms, mainly because we know much more about the physical structure and content of the Universe as well as about life and its evolution.

However, we must be aware that quantitative estimates of the frequency of extraterrestrial life may be no more right than Lucretius's qualitative statements. They may, in fact, be very misleading because they might delude us into thinking that we know more than we actually do. The reason is that at present Earth is our only example of life's origin and evolution. We know of no other case either on other planetary bodies in our Solar System or elsewhere in the Universe. Therefore, in estimating how widespread life is in the Universe, we must extrapolate from our terrestrial biosphere to other places. Unfortunately, we have no way of knowing how right or wrong such extrapolations are because we have no observational basis to go by. The best we can do is to decide which developments on Earth were essential for the origin and evolution of life, and then to assign low probabilities to those that on our planet required the simultaneous occurrence of rather specialized conditions (for example, geologic and climatic conditions) and proportionately higher probabilities to those that were less constrained. This strategy may give us a reasonably accurate estimate, though there is no guarantee. We simply must accept that, given the limits of our current scientific understanding, this uncertainty cannot be avoided.

Given the present dearth of observational and experimental data concerning extraterrestrial life, we should not expect a consensus among scientists with regard to the probable frequency of such life. In fact, the published numerical estimates range over many orders of magnitude. The numbers I present here do not reflect the judgment of the scientific community as a whole; they merely represent my personal opinion and, possibly, say more about my beliefs and prejudices than about the actual evolution of extraterrestrial life. Because of these uncertainties, readers are as much entitled to their opinions on this subject as I am to mine or the "experts" are to theirs, and I invite them to substitute their own numerical estimates for mine and to draw their own conclusions from the final result.

BASIC ASSUMPTIONS

Before we begin calculating the probable frequency of extraterrestrial intelligent life with advanced technology, we must agree on a number of assumptions·

1. The fundamental unit of life is always the cell, except during the earliest stages of life's origin.

2. Life is never static, but always evolves, and does so as explained by Darwin and Wallace. This requires molecular information carriers and the passage of the information from one generation of organisms to the next, including the continual introduction of changes in the information through mutations and recombinations. The changes in information create diversity, upon which natural selection can act. Very likely, these processes require catalysts to control the rates of chemical reactions. We also assume that only organic chemistry offers such information carriers and catalysts, because we think only carbon bonded to hydrogen, nitrogen, oxygen, phosphorus, sulfur, and numerous other elements provides molecules of sufficient variety, stability, and chemical reactivity to do the job.

3. The chemical reactions of life require not only the ready availability of hydrogen, carbon, nitrogen, oxygen, and so forth, but also liquid water. This is certainly the case on Earth. The condition of liquid water places severe limitations on the environment in which life can arise. It requires a planet circling a star. Furthermore, it requires that the planet has roughly the same mass and composition as Earth, with a solid surface, volcanism, outgassing, and temperature-pressure conditions at which water condenses but does not freeze, at least not everywhere. The limitation on temperature, in turn, requires a stable planetary orbit of low eccentricity and in the right distance range from the central star.

4. The central star must be stable and radiate at a nearly constant rate for at least 3 to 4 billion years (that is, it has to remain on the main sequence for at least that long). Otherwise there probably would be insufficient time for intelligent beings to evolve. We further assume that the star radiates in a wavelength range that peaks neither in the ultraviolet nor in the infrared. Ultraviolet radiation would be too damaging to life, especially land-based life; and infrared radiation might not be sufficiently energetic to induce some of the

chemical reactions crucial during the earliest stages in the evolution of life, before efficient light-gathering (i.e., photosynthetic) molecules have evolved.

5. Intelligent beings capable of developing advanced technology are land-based and have passed through a tree-living phase in their earlier evolution. Only such a phase is likely to produce the large brain and grasping hands, including the ability to execute a precision grip, that appear to be essential for the evolution of such beings. Note that intelligence alone is not sufficient for the development of advanced technology that includes the ability to communicate across interstellar distances. Finely tuned manipulative skills are required as well, to handle the nuts and bolts, to build the transmitters and receivers of electromagnetic signals, and, in general, to assemble the hardware required for such a technology. Of the many millions of species of animals that have arisen on our planet, only among the primates have manipulative appendages evolved, namely our human hands, that are up to the task. For example, such appendages as the trunks of elephants, the eating instruments of insects and other arthropods, or the beaks of birds are not so capable, despite their many unusual and often surprisingly versatile uses. Nor, in my opinion, do these or other nonprimate appendages have much of a potential to evolve to a stage at which they could be used in the development of advanced technology.

6. Finally, let us agree to consider only our Galaxy in estimating the probability of finding extraterrestrial intelligent life. This will simplify the calculations. We assume that the result can be readily extended to other galaxies, provided they contain, besides hydrogen, sufficient amounts of carbon, nitrogen, oxygen, and other heavy elements.

Most scientists probably would agree that these assumptions are quite reasonable, though some might argue that we have set our limits too stringently. Maybe life does not necessarily have to be cellular or be based on organic chemistry, or maybe it can arise in places other than earthlike planets, such as in the upper atmospheric layers of jovian planets or in interstellar clouds. Maybe the manipulative skills required for the development of advanced technology might arise along evolutionary routes quite different from the one we assumed. However, at present we have no scientific basis to believe in such scenarios. Hence, I prefer to take the conservative approach and use the assumptions listed.

METHOD OF ESTIMATING THE PROBABILITIES

Next, let us discuss a method of estimating how frequently we expect that life with advanced technology has arisen on other planets circling other stars in our Galaxy. This method was first suggested in 1965 by Frank D. Drake of Cornell University, one of the pioneers in the search for extraterrestrial intelligence. The method involves two steps. In the first step, the values of several probabilities are estimated:

1. The probability that any given star in our Galaxy has planets suitable for supporting life.

2. The probability that on such planets life arises.

3. The probability that such life evolves into complex multicelled forms.

4. The probability that such evolution leads to intelligent beings with advanced technology.

In the second step, the four probabilities are multiplied together, which gives an estimate of the probability per star in our Galaxy that intelligent beings with advanced technology might arise. Multiplying this result by the number of stars in the Galaxy yields the number of times that technologically advanced civilizations are likely to have arisen in our Galaxy.

CALCULATING THE PROBABILITIES

We are now ready to begin the actual calculations.

Calculation 1: The Probability That Any Given Star in Our Galaxy Has Planets Suitable for Supporting Life

This probability is itself the product of three numbers, namely the fraction of stars in our Galaxy capable of sustaining life in their vicinity, the probability that such stars are circled by planets, and the probability that one or several of such planets are suitable for the origin and evolution of life.

For numerous reasons, not all stars are capable of sustaining life in their vicinity. First, none of the oldest stars in the Galaxy, such as those in globular star clusters, are good candidates. They are made of mostly primordial material (that is, hydrogen and helium) and contain very little carbon, nitrogen, oxygen, silicon, metals, or other elements needed for the making of planets with a solid surface and for the origin and evolution of life. Second, only about fifteen percent of all stars are single. The rest are members of binary or higher order systems, which reduces the probability of finding stable planetary orbits at suitable distances from those stars. Third, all stars more massive than about $1.2\,M_\odot$ or less massive than about $0.8\,M_\odot$ should be excluded. The more massive stars emit too much UV radiation, which would probably be harmful to life. Besides, such stars become red giants in fewer than 3 to 4 billion years, which allows insufficient time for intelligent life to evolve. Stars less massive than about $0.8\,M_\odot$ emit mainly red and IR radiation, which might be insufficiently energetic to induce some of the chemical reactions essential for the origin of life. Furthermore, such low-mass stars may have no continuously habitable zones in their vicinity. This means that planets circling them may not be able to maintain for any great length of time the mild climatic conditions that have characterized our planet. They probably either freeze over or heat up excessively due to runaway greenhouse effects (a fate that may yet befall Earth). Not all scientists agree with this assessment. Nevertheless, even if habitable zones surround low-mass stars, those zones would be quite close to the stars. Planets orbiting there would experience enormous tides, which would slow their rotation rates until they were synchronous with the orbital rates (that is, the durations of the planets' days and years would become equal, as has happened in the case of the Moon with respect to Earth). Such planets would then forever point the same hemispheres toward the central stars. These hemispheres would heat up and probably turn into permanent deserts, while the opposite hemispheres would be in darkness and freeze. If we take these three restrictions into account, it turns out that only *about two percent of all stars in our Galaxy are capable of sustaining life in their vicinity*.

Next let us estimate the probability that a given star is circled by planets. We have little firm observational data to go by in this estimate, for in the case of only two stars other than the Sun, namely Beta Pictoris and Van Biesbroeck 8 (see footnote on p. 248), do we at present have evidence that they might be circled by planets. And even that evidence is not fully established. This lack of firm observational evidence is almost certainly due to the great technical difficulties involved in detecting distant planets, rather than to their nonexistence. Judging from the preponderance of binary and higher order star systems, the fact that even in our Solar System there exist several "planetary" systems (for example, those of Jupiter and Saturn), and the discovery of rocky debris circling Vega, Fomalhaut, and Beta Pictoris, many astronomers think that perhaps all stars between 0.8 and 1.2 $M_\odot$, which are potentially capable of sustaining life in their vicinity, are surrounded by planets. Other astronomers are a bit more cautious and would reduce this estimate somewhat. I, too, shall be cautious and assume that *two-thirds of all stars between 0.8 and 1.2* $M_\odot$ *are circled by planets*.

But not all planets are suitable for the origin of life. They need to satisfy the conditions described in the assumptions above. For example, massive planets like the Sun's Jovian gas giants or very small planets like Mercury and Mars will not do. Probably their masses have to lie within plus or minus a few ten percent of the mass of Earth, and their surface layers must contain water and other molecules rich in hydrogen, nitrogen, oxygen, silicon, phosphorus, sulfur, and various other kinds of elements. Only such combinations of mass and composition would give rise to a solid crust, volcanism, outgassing, oceans, and a suitable atmosphere. Furthermore, planets capable of supporting life must lie within the central star's habitable zone, which in the case of the Solar System extends at most from somewhere between 0.8 and 0.9 to between 1.2 and 1.3 astronomical units from the Sun. For example, Venus, which orbits at 0.72 AU from the Sun, developed a runaway greenhouse effect; Mars, which orbits at 1.52 AU, is frozen. Very likely, the planet also must rotate at a moderate rate, similar to Earth. If it rotates much more slowly, one hemisphere would heat up excessively; if it rotates much more rapidly, hurricanelike storms would blow almost incessantly in the lower atmosphere. Thus, the restrictions on a planet to be suitable for life are rather severe and, consequently, the probability that they are satisfied is quite small. How small is difficult to judge, given our current limited knowledge of planetary system formation and evolution. However, it appears that *the probability of a given star being circled by a planet suitable for supporting life cannot be assumed to be greater than about one-tenth*.

Multiplying together these three numbers (2/100, 2/3, and 1/10) gives 1/750. This means that according to my estimate *on the average one star out of every 750 in our Galaxy is surrounded by a planet suitable for supporting life. And because there are approximately 150 billion stars in our Galaxy, there may exist 200 million such planets*.

Calculation 2: The Probability That on Suitable Planets Life Actually Arises

We might be tempted to assume that if a planet is suitable for supporting life, life will inevitably arise. This may well be true. Nevertheless, our understanding of the processes involved in the origin of life is so limited that we cannot be certain of this conclusion. We simply don't know under what range of conditions, or sequences of conditions, life can come into existence. For example, can life originate on a planet that is entirely covered by water (perhaps to depths of several kilometers), without any islands or continents; or are shorelines needed? What if there is much less water than on Earth, collected in a few salty little pools, and the rest of the planet is arid land? (Either of these conditions might arise just as likely as those on Earth, which has two-thirds of its surface covered by water, see p. 229.) Does the planet have to be circled by a massive satellite to cause the daily rising and falling of large (but not too large) tides? Does the origin of life depend on special atmospheric conditions or on particular seasonal variations in its climate? Are, perhaps, unusual clay crystals needed to act as catalysts during prebiotic chemical synthesis? We don't know the answer to any of these questions. Besides, there are probably many other questions of this kind that we don't even know how to ask.

Thus, we cannot simply take it for granted that life will inevitably arise on all planets that are suitable for supporting life once it has arisen. Put differently, we cannot assume that time will always be "the hero of the plot," as George Wald suggested in his discussion of terrestrial prebiotic evolution (Wald 1954, 48). On some planets, time alone may not be able to bring about the origin of life.

Given these uncertainties, but also recognizing that life did arise on at least one planet, namely Earth, I shall assume that the probability of this happening on other similar planets is *one-fifth*. Some biologists would assign a much lower value to this probability, while others might set it closer to one. At present, we have to accept this uncertainty. Multiplying this probability by the previously derived number of suitable planets in our Galaxy yields *40 million. This is the number of planets in our Galaxy on which I expect that life has actually arisen.*

Calculation 3: The Probability That Life, Once It Has Arisen on a Planet, Evolves into Complex Multicelled Forms

Judging from the events as they occurred on Earth, it seems very probable that once life has arisen on a planet competition for the limited resources available will sooner or later lead to the development of photosynthesis and respiration (both anaerobic and aerobic). Of course, aerobic respiration would evolve only after molecular oxygen had begun to accumulate in the planet's atmosphere and oceans, due to its release by aerobic photosynthesis. With the accumulation of molecular oxygen, it seems probable that eukaryotic cells as well as mitotic cell division would evolve. If this happened, meiosis and sexual reproduction could evolve (or some other, comparable mechanism that would shuffle the genetic information of two, or possibly more, parents during reproduction). It is then very likely that complex multicelled life will arise.

As simple as it sounds, this sequence of events is not easily accomplished. It takes time. On Earth it took more than 3 billion years, from roughly 4 billion to somewhat less than 1 billion years ago (see summary on p. 343). Would it also be just a matter of time on other planets, or might there arise insurmountable obstacles on some of them? We don't really know. But it seems unlikely that such obstacles would arise often. As George Gaylord Simpson (1967, 242), the noted American paleontologist and evolutionist, put it, there exists " . . . a tendency for life to expand, to fill in all the available spaces in the livable environments, including those created by the process of that expansion itself." (Simpson meant here life as a whole, not just particular groups or species of organisms.) Expressed differently, natural selection will probably always force life to evolve in directions that allow it to make use of the widest possible resource base, including the resources created by the process of evolution itself. That would probably be an evolution toward eukaryotic cells, as well as toward sexual reproduction and complex multicelled forms. I shall assume that this kind of evolution takes place approximately *three-fourths* of the time.

We must still take into account the fact that only a fraction of the planets in our Galaxy on which life may have arisen are roughly 3 to 4 billion years old (or older), which we assumed is the time required for the evolution of complex multicelled and, possibly, intelligent life. Very approximately this fraction is *two-thirds*. Multiplying together this number, the number three-fourths, and my earlier estimate of the number of planets in our Galaxy on which life may have arisen gives *20 million. This is the number of times I expect that complex multicelled life has evolved in our Galaxy*.

Calculation 4: The Probability That Complex Multicelled Life Evolves into Intelligent Beings with Advanced Technology

Traditionally, most physical scientists who have concerned themselves with the issue of extraterrestrial life have tended to oversimplify the issue of the probability that complex multicelled life would give rise to intelligent beings.[*] Their reasoning has essentially been as follows: Life always evolves from the simple toward the complex; intelligent beings are the most complex of all; therefore, the probability is high that intelligent beings, including ones capable of developing advanced technology, will evolve in time.

This argument is flawed on two accounts. First, life does not always evolve toward greater complexity. It evolves in directions that allow it to exploit the widest possible resource base, as noted above. If increased complexity in certain species helps to accomplish this, then evolution will progress in this direction;

---

[*]Since Frank Drake published his now famous equation (the probability formula we are using), the main users of it have been astronomers and other physical scientists. Relatively few biologists, anthropologists, and social scientists have, even though their expertise is very pertinent to the calculation of the probability that life will originate and evolve on other planets circling other stars.

if it does not, then different evolutionary routes will be followed. For instance, quite often complexity has decreased in the course of evolution, as in the case of certain cave fishes whose eyes have become nonfunctional, primates whose sense of smell has been reduced, cetaceans whose limbs have evolved into tails and flippers (which are anatomically simpler than walking legs), and certain mollusks who have lost their ancestral shells. Furthermore, the complexity of bacteria does not appear to have increased measurably during the past billion or so years, even though these organisms have been evolving and adapting to their changing environments.

Second, implicit in the above argument is the assumption that intelligence is generally an advantage in the struggle for survival and, hence, is selected for in the course of evolution. This is not always the case. For instance, whales and dolphins are the most intelligent animals in the sea, but by number they constitute only a minuscule fraction of all large marine animals. Elephants, too, are very intelligent. Yet today only two species survive of a once much larger order. In contrast, worms, arthropods, and fishes, which are not characterized by great intelligence, have been enormously successful, if we accept number and longevity of phylogenetic lineages as measures of success.

It appears that high intelligence is an advantage in the competition for survival mainly if it is combined with suitable physical traits allowing the animals to make effective use of it. For example, in humans the combination of intelligence together with the precision grip of our hands, our bipedalism (which frees the hands for uses other than locomotion), and the wide range of our vocalization abilities (which is crucial for language development) has made us successful. In fact, no other terrestrial animal, past or present, comes even close to matching us or our *Homo* ancestors (such as *H. habilis, H. erectus,* and the Neanderthals) in intelligence and its effective application.

Thus, it seems not justified to assume *a priori* that complex multicelled life will necessarily, or very likely, give rise to intelligent beings capable of developing advanced technology. If we use our terrestrial experiences as a guide (see Assumption 5 above), only evolution on land, including a tree-living phase, is likely to accomplish this.

In order to estimate the probability that complex multicelled life will evolve into intelligent beings with advanced technology, following approximately the route we assumed, let us begin by listing those developments that were central to this evolution on Earth.

1. Emergence of the chordate phylum during the early diversification of complex multicelled life, with members characterized by an internal skeleton and specialized for a swimming mode of life.

2. Evolution of several distinct lineages among the early chordates, including one—the vertebrates—that allowed some of its members successfully to invade the forest environments that were spreading across the land from the middle of the Paleozoic Era onward.

3. Evolution among the terrestrial vertebrates of a wide range of adaptations to life in their new habitat, including a stronger internal skeleton suitable for quadrupedal locomotion on land, sense organs adapted to life in the air, changes in jaw structure and dentition, warm-bloodedness, a four-chambered

heart, enlargement of the cerebrum, and new reproductive strategies (for example, mammary glands, placenta).

4. Evolution of terrestrial flowering trees and the successful invasion by several groups of vertebrates of the arboreal habitats thus created, among them animals that evolved into the primate order.

5. Evolution among the primates of prehensile hands and feet (including the precision grip in some of them), excellent eye–hand coordination, a wide range of specialized arboreal locomotions (including quadrupedal walking on branches, brachiation, vertical clinging and leaping, and vertical quadrupedal climbing), increased intelligence, and reproductive strategies characterized by the intensive caring for a small number of offspring.

6. Periodic readaptation of some species of primates to terrestrial living, including one group, the australopithecines, that was preadapted for upright, bipedal walking when, some 8 to 5 million years ago, it came out of the trees and onto the ground.

7. Reliance of one group of australopithecines on their intelligence for survival and their evolution into the *Homo* lineage. This evolution was characterized by an extraordinarily rapid increase in brain capacity (particularly of the cerebral cortex) and intelligence, and by the development of tool use and language.

The probability that these or roughly similar developments take place on a planet on which complex multicelled life has arisen is very difficult to estimate. To come anywhere close to estimating it realistically, we must recognize that each of the key developments just described depends on the simultaneous occurrence of sometimes rather stringent geologic, climatic, and biologic conditions. For example, life can invade land only if the planet has islands or continents and is not entirely covered by water. Extensive diversification of land animals probably occurs only if the landmasses have sizes as large or nearly as large as our continents (only then is the competition among the species sufficiently intense) and are not just groups of small islands. Flowering trees suitable for arboreal animal life will probably evolve only if the climate is warm and wet. The return of arboreal animals to the ground as upright, bipedal walkers will probably take place only if the forests change to open savannah *and* simultaneously some of the animals have become preadapted to bipedalism (a condition that is not found among any of the nearly 200 species of today's nonhuman primates, at least not to the degree that it existed in the australopithecines). Finally, the evolution of bipedal land animals into humanoids (beings resembling those of the genus *Homo* on Earth) requires suitable ecological conditions in which reliance on intelligence proves to be advantageous and, hence, is selected for (which did not happen in the cases of *A. robustus* and *A. boisei*). Still other factors that might affect the course of evolution are such random and largely unpredictable events as episodes of severe earthquake activities and volcanism, asteroid or comet impacts, fluctuations in climate, mass extinctions, and the chance evolution of organisms capable of impeding or, perhaps, even terminating the sequence of developments outlined above (such as the evolution of bacteria or viruses that could cause fatal infections among animal species, or of predators that keep bipedal arboreal animals from moving onto the ground).

Only developments *1, 3,* and *5* in the above list appear to have a good chance of occurring on a planet on which complex multicelled life has arisen, though even in their cases we cannot be entirely certain. Thus I shall assign the value one-half to their probabilities. Developments *2, 4,* and *7* seem to have somewhat smaller chances of taking place, mainly because they depend more strongly on suitable geologic, climatic, and biologic conditions. I shall assign the value one-fifth to each of them. Development *6* has the smallest probability of all to be repeated on another planet because it depends most strongly on the simultaneous occurrence of rather specific geologic, climatic, and biologic conditions. Therefore, I shall assign to it a probability of 1/100.

Multiplying these seven numbers together gives 1/100,000 which is my estimate of the probability that complex multicelled life evolves into humanoids possessing language and tool-making abilities. This is the most uncertain of the numbers derived. Some might think it is too small and, perhaps, should be closer to 1/10,000 or even larger. But most paleontologists and primatologists would probably agree that at best it is an upper limit and that a value ten or a hundred times smaller is more realistic. For lack of any observational evidence to guide us, let us work with the value I have derived, acknowledging, however, the great range of uncertainty inherent in it.

To complete this calculation, we still need to multiply this number by the probability that humanoids will, in fact, develop advanced technology. At present we have little understanding of the range of conditions under which this kind of evolution will occur. Surely, geologic, climatic, biologic, as well as historic conditions affect it. Perhaps, the best we can do is to assume that conditions that made possible the origin of humanoids would probably also favor the subsequent evolution toward civilizations with advanced technology. Accepting this assumption as likely, I shall assign the value one-half to the probability that such an evolution takes place. Multiplying this number by 1/100,000 gives *1/200,000,* which is *my estimate of the probability that complex multicelled life evolves into beings with advanced technology*. Finally, multiplying this probability by 20 million, which is my earlier estimate of the number of times complex multicelled life has evolved in our Galaxy, yields the number *100. This is how many times, according to my estimate, civilizations with advanced technology are likely to have evolved in our Galaxy*.

What does my final result mean? It is the product of probability estimates obtained from data belonging to the fields of astronomy, geology, biochemistry, microbiology, paleontology, primatology, anthropology, and history. All of the estimates are based on extrapolations from our terrestrial environment and experiences to other places and contain progressively greater uncertainties as we come closer to *Homo*-like beings. Clearly, little trust can be placed in the final result, as I warned at the outset. Certainly answers of 10,000 or 1 are just as defensible as mine. Some physical scientists even think that an answer in the millions is not out of the question, but biologists tend to be more conservative in this regard. Most of those who have concerned themselves with this topic believe that an answer of 100, like mine, constitutes at best an upper limit of the number of times intelligent life with advanced technology is likely to have evolved in our or other similar galaxies, and that very probably a value smaller by several orders of magnitude lies closer to the truth.

The biologists' view is generally based not on multiplying together a string of probabilities, but on the realization that evolution takes place in the context of ecosystems, all of whose countless components — both biological and physical — are evolving and continually affecting each other. To a large extent this evolution is a statistical process and, hence, at least in detail is neither predictable nor repeatable (see p. 259). Should life start over again on Earth, with identical conditions, the probability that any particular species alive today would evolve again is essentially zero, and that surely includes *Homo sapiens*. This is why George Gaylord Simpson (1964, 267) wrote, "This essential nonrepeatability of evolution on earth obviously has a decisive bearing on the chances that it has been repeated or closely paralleled on any other planet. The assumption, so freely made by astronomers, physicists, and some biochemists, that once life gets started anywhere, humanoids will eventually and inevitably appear is plainly false."

However, neither Simpson nor other responsible biologists deny categorically that intelligent beings, including ones capable of developing advanced technology, could not have arisen elsewhere. Such beings might be quite different from us and have evolved along quite different routes than we assumed. This sentiment of caution has been clearly expressed by Francisco J. Ayala and Theodosius Dobzhansky (Gould 1983, 65): "Granting that the possibility of obtaining a man-like creature is vanishingly small even given an astronomical number of attempts . . . there is still some small possibility that another intelligent species has arisen, one that is capable of achieving a technological civilization."

Keeping in mind the range of opinions that exist on this subject and the inherent uncertainties, and also remaining true to my own rather conservative tendency on such matters, I would interpret the numbers derived above as follows:

> Life is not uncommon in the Universe. It probably has arisen tens of millions of times — perhaps even hundreds of millions of times — in our Galaxy alone. In many cases, life probably managed to evolve into complex multicelled forms. In some cases, evolution probably continued and produced ecosystems comparable to ours in complexity and biological diversity. But probably only in the rarest of instances did evolution lead to intelligent beings with culture and advanced technology. It may have happened a hundred times in our Galaxy, though this is likely too optimistic an estimate. If we also keep in mind that, like all biological species, beings with intelligence probably endure for only a limited stretch of time (particularly if they have the technology to destroy themselves), it appears conceivable that we are the only civilization with advanced technology presently existing in our Galaxy. However, just as likely, given the billions of galaxies that populate the observable part of the Universe, we are not entirely alone. Surely our kind has evolved, and survived, elsewhere also.

If this interpretation is correct, then the probability that we will make radio contact with another civilization, separated from us by interstellar or perhaps intergalactic distances, is rather slim indeed. Still, if we have learned anything at all from the study of science, it is to expect the unexpected. Maybe by pure chance there presently exist intelligent beings on another planet circling a nearby star, sending out messages just in case someone is listening. And because we have the ability to listen, without undue expense compared to many of our other technological undertakings, it surely is worth the effort. In fact, radio astronomers in the United States and abroad have been doing this since 1960, when Frank Drake first pointed the 26-meter radio telescope of the U. S. National Radio Observatory in Green Bank, West Virginia, at two nearby sunlike stars, Epsilon Eridani and Tau Ceti, and listened for signals bearing the signature of another civilization. But, to date none of these attempts has been successful.

Should we some time in the future detect such signals, it may well turn out to be "the most cataclysmic event in our entire intellectual history," as Stephen Jay Gould (1983, 65) so aptly put it. We would then know for certain that we are not alone. This might give us hope for our own future, for chances are good that any extraterrestrial civilization with whom we establish communication is technologically more advanced than we. And if they were able to survive the dangers that seem to accompany the early years of technological development, so can we. We would also then know that "we do not bear alone the whole burden of life and what comes of it" (Wald 1954, 53). Perhaps this thought might allow us to take ourselves less seriously and to be less self-centered. It might even motivate us to put more of our talents and intelligence into solving some of the social, ecological, and political problems we have brought upon ourselves.

But even if we should fail to make contact, benefits would still be reaped from a search for extraterrestrial civilizations. Like all large scientific enterprises, such a search would probably be of international scope and, thereby, contribute to bringing nations of different political ideologies closer together. Furthermore, in its broadest sense such a search is a very interdisciplinary endeavor and would promote the lowering of some of the barriers that have arisen among physical, biological, and social scientists. Finally, like all other great scientific undertakings of the past, a search for extraterrestrial civilizations is bound to yield unanticipated discoveries and deepen our understanding of the Universe at large as well as of ourselves and our terrestrial environment. This kind of understanding can only help to enrich our lives now and in the future.

Suggestions for Further Reading

Ball, J. A. 1980. "Extraterrestrial Intelligence: Where Is Everybody?" *American Scientist*, November–December, 656–63.

Barrow, J. D., and F. J. Tipler. 1985. *The Anthropic Cosmological Principle*. Fair Lawn, N.J.: Oxford University Press. The anthropic cosmological principle, based in Protagoras's assertion that "man is the measure of all things," has been interpreted more recently to mean that life of any sort is impossible unless the basic laws of nature are exactly as they are. Barrow and Tipler discuss four versions of the anthropic cosmological principle. This gets them deeply into the underlying philosophical principles of quantum mechanics and cosmology, the question of the existence of a Creator, and the problem of extraterrestrial life, whose existence they deny. The book is controversial, but worth reading for anyone with patience and a philosophical bent. (Also see "The Anthropic Principle" by G. Gale, referenced in chapter 1).

Billingham, J., ed. 1981. *Life in the Universe*. Washington, D.C.: Scientific and Technical Information Branch, NASA. A collection of reports presented at a 1979 conference held at the NASA Ames Research Center (Moffett Field, California) on the broad question of the evolution of life on Earth and possibly elsewhere, and on the possibility of detecting the existence of extraterrestrial life.

Bracewell, R. N. 1975. *The Galactic Club: Intelligent Life in Outer Space*. San Francisco: W. H. Freeman and Co. An excitingly written little book about the possible existence of intelligent civilizations beyond the Solar System and our joining this "galactic club."

Cameron, A. G. W., ed. 1963. *Interstellar Communication: A Collection of Reprints and Original Contributions*. New York: W. A. Benjamin. Though published more than 20 years ago, many of the 32 articles in this collection—all concerned directly or indirectly with the question, "How many technologically advanced societies exist in our galaxy and how can we communicate with them?"—are still timely and worth reading.

Feinberg, G., and R. Shapiro. 1980. *Life Beyond Earth: The Intelligent Earthling's Guide to Life in the Universe*. New York: William Morrow and Co. The nature of life, its possible forms, and its likely distribution in the Universe.

Goldsmith, D., comp. 1980. *The Quest for Extraterrestrial Life: A Book of Readings*. Mill Valley, Calif.: University Science Books. A collection of 58 articles by authors both modern and from the past about the possibility of the existence of extraterrestrial life.

Goldsmith, D., and T. Owen. 1980. *The Search for Life in the Universe*. Menlo Park, Calif.: Benjamin/Cummings Publishing Co.

Gould, S. J. 1983. "The Wisdom of Casey Stengel." *Discover*, March 1983, 62–65. A critique of probability estimates of the existence of extraterrestrial life, focusing on the repeatability of "humanoids" elsewhere and on the much more general issue of the repeatability of "intelligence."

Milne, D., et al., eds. 1985. *The Evolution of Complex and Higher Organisms*. Washington, D.C.: Scientific and Technical Information Branch, NASA. A report prepared by the participants of workshops held in 1981 and 1982 at the NASA Ames Research Center (Moffett Field, California) on the likelihood that extraterrestrial events played a role in the development of complex life on Earth and on whether knowledge of such events could contribute to a better understanding of the nature and distribution of complex extraterrestrial life.

Morrison, P., J. Billingham, and J. Wolfe, eds. 1977. *The Search for Extraterrestrial Intelligence* (*SETI*). Washington, D.C.: Scientific and Technical Information Branch, NASA. A report from a series of workshops sponsored by the NASA Ames Research Center (Moffett Field, California). For advanced students.

O'Neill, G. K. 1976. *The High Frontier: Human Colonies in Space*. New York: William Morrow and Co.

Papagiannis, M. D., ed. 1980. *Strategies for the Search for Life in the Universe*. Dordrecht, Holland: D. Reidel Publishing Co. A collection of reports presented during the International Astronomical Union General Assembly (Montreal, Canada, 1979). For the advanced student.

———. ed. 1985. *The Search for Extraterrestrial Life: Recent Developments*. Dordrecht, Holland: D. Reidel Publishing Co. A collection of reports presented at the 112th symposium of the International Astronomical Union (Boston University, 1984). Topics include the history of the search for extraterrestrial life, current search efforts, extraterrestrial organic matter, universal aspects of biological evolution, and space colonization.

Sagan, C., and F. **Drake**. 1975. "The Search For Extraterrestrial Intelligence." *Scientific American*, May, 80–89.

Shapley, H. 1964. *Of Stars and Men: The Human Response to an Expanding Universe*, revised edition. Boston: Beacon Press. A beautiful little book, in an easy, popular style, by one of the great astronomers of the twentieth century. Though many of the facts presented are now dated, the book still lives up to its intended purpose: "To present some . . . ideas, new and old, bearing on the position of mankind in the universe of physics and sensation; [and to write] a tentative obituary . . . of anthropocentrism in our description of the universe."

Shklovskii, I. S., and C. **Sagan**. 1966. *Intelligent Life in the Universe*. San Francisco: Holden-Day. A classic on the topic of life on Earth and the possibility of its existence beyond, by two eminent scientists, one Russian and the other American. Though much of the information in this book has been superseded by new discoveries, the spirit of adventure in which it was written has not.

Additional References

Lucretius (1st century B.C.). 1980. *De rerum natura*, trans. Mary-Kay Gamel Orlandi. In D. Goldsmith, ed., *The Quest for Extraterrestrial Life: A Book of Readings*, 4. Mill Valley, Calif.: University Science Books.

Simpson, G. G. 1967. *The Meaning of Evolution: A Study of the History of Life and of Its Significance for Man*, rev. ed. New Haven and London: Yale University Press.

———. 1964. *This View of Life: The World of an Evolutionist*. New York: Harcourt, Brace & World.

Wald, G. 1954. "The Origin of Life." *Scientific American*, August, 44–53.

Author Index*

*Numbers in italics refer to figures and tables; numbers followed by *n* refer to footnotes.

Subject Index*

*Numbers in italics refer to figures and tables; numbers followed by *e* refer to exercises; numbers followed by *n* refer to footnotes; numbers followed by *r* (*author's name*) refer to Suggestions for Further Reading.